Recent Titles in This Series

W9-AFD-398

(Continued in the back of this publication)

Multivariable Operator Theory

CONTEMPORARY MATHEMATICS

185

Multivariable Operator Theory

A Joint Summer Research Conference on
Multivariable Operator Theory
July 10–18, 1993
University of Washington, Seattle

Raúl E. Curto
Ronald G. Douglas
Joel D. Pincus
Norberto Salinas
Editors

American Mathematical Society
Providence, Rhode Island

The Joint Summer Research Conference on Multivariable Operator Theory was held at the University of Washington, Seattle, Washington, from July 10–18, 1993, with support from the Division of Mathematical Sciences of the National Science Foundation, Grant No. DMS-9221892.

1991 *Mathematics Subject Classification*. Primary 47–XX, 32–XX, 46–XX; Secondary 30–XX, 43–XX, 19–XX.

Library of Congress Cataloging-in-Publication Data

Multivariable operator theory : a joint summer research conference on multivariable operator theory, July 10–18, 1993. University of Washington, Seattle / Raúl E. Curto ... [et al.], editors.

p. cm. — (Contemporary mathematics, ISSN 0271-4132; v. 185)

Includes bibliographical references.

ISBN 0-8218-0298-4

1. Operator theory—Congresses. 2. Functions of several complex variables—Congresses. I. Curto, Raúl E., 1954– . II. Series: Contemporary mathematics (American Mathematical Society); v. 185.

QA329.M85 1995

515′.94–dc20

95-2345
CIP

Contents

Preface

This volume consists of a collection of papers presented at the 1993 Summer Research Conference on Multivariable Operator Theory, held at the University of Washington in Seattle, under the auspices of the American Mathematical Society.

The articles represent contributions to a variety of areas and topics, which may be viewed as forming an emerging new subject, involving the study of geometric rather than topological invariants associated with the general theme of operator theory in several variables.

Beginning with J. L. Taylor's discovery in 1970 of the right notion of joint spectrum and analytic functional calculus for commuting families of Banach space operators, a number of significant developments have taken place. For instance, a Bochner-Martinelli formula has been generalized to commuting n-tuples in arbitrary C^*-algebras; various forms of the multivariable index theorem have been proved using non-commutative differential geometry and algebraic geometry; and systems of Toeplitz and Hankel operators on Reinhardt domains, bounded symmetric domains, and domains of finite type have been substantially understood from the spectral and algebraic viewpoints, including the discovery of concrete Toeplitz operators with irrational index.

These developments have been applied successfully to various types of quantizations, and functional spaces on Cartan domains and on pseudoconvex domains with smooth boundary have been thoroughly studied. A generalization of the Berger-Shaw formula to several variables has been proved; and connections with the local multiplicative Lefshetz numbers, analytic torsion, and curvature relations of canonically associated hermitian vector bundles have been established. Moreover, a sophisticated machinery of functional homological algebra suitable for the study of multivariable phenomena has been developed, and a rich theory for invariant pseudodifferential operators on domains with transverse symmetry has been produced.

Much of multivariable operator theory involves the interaction between the subspace geometry of defect spaces and algebraic K-theory. For one example, a multivariable index theorem corresponding to commuting pairs of elements with finite defect $A, B \in \text{End}(H)$, where H is a vector space over an arbitrary field F, has emerged in connection with the Quillen algebraic K-theory. The joint torsion $\tau(A, B; \mathcal{H})$ is a multiplicative Euler characteristic defined solely in terms of the homology of the Koszul complex $K_*(A, B, H)$. It has been found to be the obstruction in the homotopy lifting problem from the classifying space $BGL(\text{End}(H))^+ \to BGL(\text{End}(H)/\mathcal{F}))^+$, where $\mathcal{F}$ is the finite rank ideal; for it has been shown that $\tau(A, B; \mathcal{H}) = \partial\{A + \mathcal{F}, B + \mathcal{F}\}$, where $\{A + \mathcal{F}, B + \mathcal{F}\}$ is the Steinberg element in $BGL(\text{End}(H)/\mathcal{F}))^+$.

On a different direction, Calabi-like rigidity phenomena for analytically invariant subspaces of the Hardy and Bergman spaces have been discovered, and sheaf models for subnormal n-tuples have been formulated, which have led to a substantial understanding of their spectral properties. Results from polynomial convexity have been used to solve intriguing problems on joint quasitriangularity, and the polynomially hyponormal conjecture for single operators has been settled using ideas from joint hyponormality.

As probably expected during the early stages of a new subject, the recent years have seen the rise of many new approaches (all quite different) to multivariable operator theory, certainly connected, but with relationships not well understood. The subject has developed in several directions and with the aid of many and varied techniques, and a good number of the advances have been made through cross pollination among different areas of mathematics.

The principal goal of the conference was to provide a forum for the discussion of the actual connections among the various approaches, which one hopes will allow researchers to combine their efforts in finding an understanding of the above mentioned relationships and new directions for future research. These proceedings represent the products of those discussions.

The Editors
December 1994

Contemporary Mathematics
Volume **185**, 1995

EXPLICIT FORMULAE FOR
TAYLOR'S FUNCTIONAL CALCULUS

D. W. ALBRECHT

ABSTRACT. In this paper integral formulae, based on Taylor's functional calculus for several operators, are found. Special cases of these formulae include those of Vasilescu and Janas, and an integral formula for commuting operators with real spectra.

§1 Introduction. Let $a = (a_1, \ldots, a_n)$ be a commuting tuple of bounded linear operators on a complex Banach space X, let Ω be an open subset containing the Taylor's joint spectrum of a, $Sp(a, X)$, and let f be a holomorphic function defined on Ω. Then Taylor [T] defined $f(a)$ as follows:

$$f(a)x = \frac{1}{(2\pi i)^n} \int_\Omega (R_{\alpha(z)} f(z)x)dz,$$

for each $x \in X$.

However, since $R_{\alpha(z)}$ is a homomorphism of cohomology, this means, in contrast with the Dunford-Schwartz calculus, that the functional calculus of Taylor is rather inexplicit.

In this paper we will show how explicit formulae for Taylor's functional calculus can be obtained, under certain circumstances which include the following:

(1) X is a Hilbert space;
(2) a has real spectrum, i.e. $Sp(a, X) \subset \mathbb{R}^n$;
(3) Ω is a bounded convex domain given by a smooth defining function.

§2 Notation. Let $\sigma = (s_1, \ldots, s_n)$ and $d\bar{z} = (d\bar{z}_1, \ldots, d\bar{z}_n)$ denote tuples of indeterminates, and $\Lambda[\sigma \cup d\bar{z}]$ denote the exterior algebra, over $\mathbb{C}$, generated by these tuples. Let $\Lambda^p[\sigma \cup d\bar{z}]$ denote the subspace of p-forms in $\Lambda[\sigma \cup d\bar{z}]$. For $1 \leq i \leq n$, let S_i be operators on $\Lambda[\sigma \cup d\bar{z}]$ defined by $S_i(\xi) = s_i \wedge \xi$, for every $\xi \in \Lambda[\sigma \cup d\bar{z}]$. We shall call these the creation operators for σ on $\Lambda[\sigma \cup d\bar{z}]$. Let $d\bar{z}_1, \ldots, d\bar{z}_n$ also denote the creation operators for $d\bar{z}$ on $\Lambda[\sigma \cup d\bar{z}]$, and for each $1 \leq i \leq n$, let S_i^* denote the operators on $\Lambda[\sigma \cup d\bar{z}]$ which satisfy the following anti-commutation relations:

$$S_i^* S_j^* + S_j^* S_i^* = 0$$

$$S_i S_j^* + S_j^* S_i = \begin{cases} 1 & \text{if } i = j \\ 0 & \text{if } i \neq j \end{cases},$$

1991 *Mathematics Subject Classification.* Primary 47A60; Secondary 47A56.
The detailed version of this paper has been submitted for publication elsewhere

for every $1 \leq i, j \leq n$.

We consider $a_1, \ldots, a_n$, and $\frac{\partial}{\partial \bar{z}_1}, \ldots, \frac{\partial}{\partial \bar{z}_n}$ to be operators on $C^\infty(\Omega, X)$, and set

$$\bar{\partial}_z = \frac{\partial}{\partial \bar{z}_1} d\bar{z}_1 + \cdots + \frac{\partial}{\partial \bar{z}_n} d\bar{z}_n$$

and

$$\alpha(z) = (z_1 - a_1)S_1 + \cdots + (z_n - a_n)S_n.$$

It follows from Taylor [T] and Vasilescu [V1] that $(C^\infty(\Omega, X), C_0^\infty(\Omega, X), \alpha, \bar{\partial}_z)$ is a Cauchy–Weil system.

Let

$$R_\alpha(z) = (-1)^n \pi^* (i^*)^{-1} s^*,$$

where s^* is the homomorphism induced by the morphism $s(\xi) = \xi \wedge s_1 \wedge \cdots \wedge s_n$, for $\xi \in C^\infty(\Omega, X) \otimes \Lambda[d\bar{z}]$, i^* is the homomorphism induced by the inclusion mapping i in the exact sequence

$$0 \longrightarrow C_0^\infty(\Omega, X) \stackrel{i}{\longrightarrow} C^\infty(\Omega, X) \stackrel{r}{\longrightarrow} C^\infty(\Omega, X)/C_0^\infty(\Omega, X) \longrightarrow 0,$$

and π^* is the homomorphism induced by the special transformation π, where π^0 is the identity on $C_0^\infty(\Omega, X)$, $\pi(s_i) = 0$ for $1 \leq i \leq n$, and $\pi(d\bar{z}_j) = d\bar{z}_j$ for $1 \leq j \leq n$.

§3 **An Explicit Formula.** Let $\Lambda[C_0^\infty]$, $\Lambda[C^\infty]$ and $\Lambda[C^\infty/C_0^\infty]$ denote respectively $C_0^\infty(\Omega, X) \otimes \Lambda[\sigma \cup d\bar{z}]$, $C^\infty(\Omega, X) \otimes \Lambda[\sigma \cup d\bar{z}]$ and $C^\infty(\Omega, X)/C_0^\infty(\Omega, X) \otimes \Lambda[\sigma \cup d\bar{z}]$, and consider the following commuting diagram:

We can show that

$$(i^*)^{-1}[f] = [f - (\alpha + \bar{\partial}_z)g],$$

where

$$(\alpha + \bar\partial_z)r(g) = r(f),$$
$$(\alpha + \bar\partial_z)r(f) = 0,$$

and [ ] denotes the cohomology class.

Thus, we see that the question of an explicit expression $(i^*)^{-1}$, and therefore an explicit formula for $f(a)$ is closely tied to finding an explicit expression for a solution of the above equations. This in turn leads to the following result, which was suggested by results in [V2] and is a special case of a result in [A].

For each $z \in \Omega \backslash Sp(a, X)$, let $(b_1(z), \ldots, b_n(z))$ be a commuting tuple of bounded linear operators on X. Moreover, for $z \in \Omega \setminus Sp(a, X)$, let

(1) $\beta(z) = b_1(z)S_1^* + \ldots b_n(z)S_n^*$,
(2) $z \mapsto \beta(z)$ be an infinitely-differentiable mapping, and
(3) $\alpha(z) + \beta(z)$ be invertible on $\Lambda[\sigma, X]$.

Lemma 1. *Suppose*

$$f = f_0 + \cdots + f_n,$$

$f_p \in \Lambda[d\bar z, C^\infty] \otimes \Lambda^p[\sigma]$, *and*

$$(\alpha + \bar\partial_z)r(f) = 0.$$

Then

$$r(g) = \sum_{k=0}^{n-1} \sum_{m=0}^{n-k-1} (-1)^m (\alpha + \beta)^{-1} (\bar\partial_z(\alpha+\beta)^{-1})^m r(f_{k+m+1})$$

is a solution of $(\alpha + \bar\partial_z)r(g) = r(f)$.

We can now obtain explicit formulae for Taylor's functional calculus.

Theorem 1. *Let D be an open subset of Ω, with compact closure, piecewise C^1 boundary, and containing $Sp(a, X)$. Then*

$$(1)\qquad f(a)x = \frac{1}{(2\pi i)^n} \int_{\partial D} (\alpha + \beta)^{-1} (\bar\partial_z(\alpha+\beta)^{-1})^{n-1} f(z) x\, s_1 \wedge \cdots \wedge s_n dz,$$

for every $x \in X$.

Proof. Let F_1, F_2 be compact subsets of Ω, such that $Sp(a, X)$ is contained in the interior of F_1, F_1 is contained in the interior of F_2, and F_2 is contained in D. Suppose now that $\phi \in C^\infty(\Omega)$ is such that $\phi = 0$ on F_1 and $\phi = 1$ on $\Omega \setminus F_2$. Let

$$g(z) = \phi(z)(\alpha(z) + \beta(z))^{-1}(\bar\partial_z(\alpha(z) + \beta(z))^{-1})^{n-1} f(z)s_1 \wedge \ldots s_n.$$

Then by Lemma 1, we can show the following:

$$
\begin{aligned}
f(a)x &= \frac{1}{(2\pi i)^n} \int_\Omega R_{\alpha(z)} f(z) x\, dz \\
&= \frac{1}{(2\pi i)^n} \int_\Omega (-1)^n \pi^* (i^*)^{-1} s^* f(z) x\, dz \\
&= \frac{1}{(2\pi i)^n} \int_\Omega \bar{\partial}_z g(z) x\, dz \\
&= \frac{1}{(2\pi i)^n} \int_D d(g(z)x)\, dz \\
&= \frac{1}{(2\pi i)^n} \int_{\partial D} g(z) x\, dz,
\end{aligned}
$$

for every $x \in X$.

Example 1. Let X be a Hilbert space, and

$$
\beta(z) = (z_1 - a_1)^* S_1^* + \cdots + (z_n - a_n)^* S_n^*,
$$

for $z \in \mathbb{C}^n$. Then we can show that β satisfies the conditions above, and that the explicit formula (1) reduces to Vasilescu's Martinelli type formula in [V2].

Example 2. Let a have real spectrum, i.e. $Sp(a, X) \subset \mathbb{R}^n$, and

$$
\beta(z) = M(z,a)^{-1}(\bar{z}_1 - a_1) S_1^* + \cdots + M(z,a)^{-1}(\bar{z}_n - a_n) S_n^*,
$$

where

$$
M(z,a) = \sum_{i=1}^n (z_i - a_i)(\bar{z}_i - a_i).
$$

Then we can show that β satisfies the conditions above, and that the formula (1) reduces to the following expression:

$$
f(a) = \frac{1}{(2\pi i)^n} \int_{\partial D} f(z) M(z,a)^{-n} W(z,a)\, dz,
$$

where

$$
W(z,a) = (n-1)! \sum_{p=1}^n (-1)^{p-1} (\bar{z}_p - a_p) \wedge \bigwedge_{q \neq p} d\bar{z}_q.
$$

Example 3. Let $\Omega \subset \mathbb{C}^n$ be a bounded convex domain given by a defining function ρ, i.e. $\Omega = \{z \in \mathbb{C}^n : \rho(z) < 0\}$, where ρ is infinitely differentiable on an open neighbourhood of $\bar{\Omega}$ and

$$
\left(\frac{\partial \rho}{\partial z_1}, \ldots, \frac{\partial \rho}{\partial z_n}\right) \neq 0 \quad \text{on } \partial\Omega = \{z \in \mathbb{C}^n : \rho(z) = 0\}.
$$

We will now show that we can use the formula (1) to obtain the integral formula of Janas [J].

Let f be continuous on $\bar{\Omega}$ and holomorphic on Ω. Then we wish to show the following:

$$f(a) = \frac{1}{(2\pi i)^n} \int_{\partial\Omega} f(z)M(z,a)^{-n}W(z)dz,$$

where

$$M(z,a) = \sum_{i=1}^{n} \frac{\partial\rho}{\partial z_i}(z)(z_i - a_i),$$

$$W(z) = (n-1)! \sum_{p=1}^{n} (-1)^{p-1} \frac{\partial\rho}{\partial z_p}(z) \wedge \bigwedge_{q \neq p} \frac{\partial^2\rho}{\partial\bar{z}_q\partial z_p}(z)d\bar{z}_q.$$

By Taylor's spectral mapping theorem [T], we have that $M(z,a)$ is invertible in a neighbourhood, N, of $\partial\Omega$. Let $\{D_k\}_{k=1}^{\infty}$ be a suitable exhaustion of $\bar{\Omega}$ by open subsets of Ω which contain $Sp(a,X)$, have compact closure and piecewise C^1 boundaries contained in N, and

$$\beta(z) = M(z,a)^{-1}\frac{\partial\rho}{\partial z_1}(z)S_1^* + \cdots + M(z,a)^{-1}\frac{\partial\rho}{\partial z_n}(z)S_n^*,$$

for $z \in N$. Then we can show that formula (1), for each k, reduces to the following:

$$f(a) = \frac{1}{(2\pi i)^n} \int_{\partial D_k} f(z)M(z,a)^{-n}W(z)dz.$$

Hence, passing to the limit $k \to \infty$, we obtain the result.

REFERENCES

[A]   D.W.Albrecht, *Integral formulae for special cases of Taylor's functional calculus*, Studia Mathematica **105** (1993), 51–68.

[J]   J.Janas, *On integral formulas and their applications in functional calculus*, J. Math. Anal. Appl. **114** (1986), 328–339.

[T]   J.L.Taylor, *Analytic-functional calculus for several commuting operators*, Acta Math. **125** (1970), 1–38.

[V1]   F.-H. Vasilescu, *On a class of vector-valued functions*, (Romanian) St. cere. mat. **28** (1976), 121–127.

[V2]   F.-H. Vasilescu, *A Martinelli type formula for the analytic functional calculus*, Rev. Roum. Math. Pures Appl. **23** (1978), 1587–1605.

MONASH UNIVERSITY, CLAYTON, AUSTRALIA, 3168
E-mail address: **dwa@euler.maths.monash.edu.au**

Contemporary Mathematics
Volume **185**, 1995

A Survey of Invariant Hilbert Spaces of Analytic
Functions on Bounded Symmetric Domains

Jonathan Arazy[1]

We survey here some basic facts and recent developments concerning invariant Hilbert spaces of analytic functions on bounded symmetric domains. The survey is written for the general audience of functional analysts, and it contains function and Jordan theoretical background material.

The invariant Hilbert spaces can be used to establish duality theory of invariant Banach spaces of analytic functions. They are also very interesting in studying questions concerning quantization on the bounded symmetric domains, as well as the study of the irreducible unitary representations of the automorphism groups of these domains.

The survey is organized in the following way. In the introductory Section 1 we use the Bergman kernel of a Cartan domain D to define certain Hilbert spaces of analytic functions on D which are invariant under an appropriate weighted actions of $\mathrm{Aut}(D)$, the group of biholomorphic automorphisms of D. Some preliminary information on these spaces is obtained without difficulty; the detailed study of these spaces require some preparation. In Section 2 we give some Jordan theoretic background (which connects the bounded symmetric domains to the JB*-triples). This is used in Section 3 to study the structure of the analytic polynomials on D. In particular, we describe the irreducible orbits of the polynomials under the action of the maximal compact subgroup K of G $(= \mathrm{Aut}(D)_0)$. The decomposition of an analytic function on D according to these orbits refines the homogeneous expansion and plays a central role in the theory; it is called the Peter–Weyl decomposition. Section 4 is devoted to the survey of the work [FK1] in which a complete description of the invariant Hilbert spaces is obtained in terms of the Peter–Weyl expansions of the functions. Of particular importance is the occurrence of

1991 Mathematics Subject Classification. Primary : 46E20, 32M15.
Secondary : 43A85, 46H70, 32H10.
[1] Research supported in part by a grant from NSF No. DMS 8702371.

a non-trivial composition series of invariant subspaces in the case when the rank of D is greater than 1. Section 5 is devoted to the survey of the joint work of the author with S.D. Fisher [AF3] in which we prove the uniqueness of the invariant Hilbert spaces. Sections 6 and 7 contain various integral formulas for the invariant inner product in the important special cases of bounded symmetric domains of tube type and the unit ball of $\mathbb{C}^n$, and uses material from [A1] and [A2] (which is based on the unpublished manuscript [P3]).

In section 8 we survey the theory of a class of invariant differential operators on symmetric convex cones introduced and studied by Z. Yan in [Y2] and [Y3], and its application to the theory of invariant Hilbert spaces of analytic functions on the associated tube-type Cartan domains. This material is taken from [Y2], Y3] and [AU].

Finally, in section 9 we review briefly some recent works in various areas in analysis on symmetric domains in which the theory of invariant Hilbert spaces of analytic functions is used.

The author benefited from many discussions with his colleagues S. D. Fisher, J. Peetre, and H. Upmeier. He is also grateful to J. Faraut and A. Koranyi for sending him the preliminary version of their book [FK2], and to Z. Yan and A. Koranyi for sending him their papers ([Y1], [Y2] and [Ko3], respectively) and for their comments on the first draft of this survey.

Acknowledgment: The first version of this survey as written while the author visited the University of Kansas at Lawrence and the Mittag Leffler Institute (during the academic year 1989-90 and the Fall of 1990, respectively), and was published as a report No. 19, 1990/91 of the Mittag Leffler Institute. The author thanks both institutions for their hospitality and support.

§ 1. Introduction

A bounded domain D in $\mathbb{C}^d$ is *symmetric if every point in* D *is an isolated fixed point of a biholomorphic automorphism φ of* D *of period two* (i.e. $\varphi \circ \varphi$ = identity). This is known to be equivalent to the following property: *every two points in* D *can be interchanged by a biholomorphic automorphism of* D *of period two.*

D *is reducible* if it is biholomorphically isomorphic to a product of two non-trivial domains. Otherwise D is *irreducible*.

The irreducible bounded symmetric domains were completely classified up to a biholomorphic isomorphism by E. Cartan [C], see also [Ha]. The list *of Cartan domains* is the following.

<u>Type I$_{n,m}$</u> (n $\leq$ m): n $\times$ m complex matrices z with $\|z\| < 1$;
<u>Type II$_n$</u> (5 $\leq$ n): n $\times$ n anti-symmetric complex matrices z with $\|z\| < 1$;

<u>Type III$_n$</u> $(2 \leq n)$: $n \times n$ symmetric complex matrices with $\|z\| < 1$;

<u>Type IV$_n$</u> $(5 \leq n)$ (the Lie ball): all $z \in \mathbb{C}^n$ so that

(1.1)
$$\left(\left(\sum_{j=1}^{n} |z_j|^2 \right) - \left| \sum_{j=1}^{n} z_j^2 \right|^2 \right)^{1/2} < 1 - \sum_{j=1}^{n} |z_j|^2 \; ;$$

<u>Type V</u> 1×2 matrices z over the 8-dimensional Cayley algebra, with $\|z\| < 1$.

<u>Type VI</u> 3×3 Hermitian matrices z with entries in the 8-dimensional Cayley algebra, with $\|z\| < 1$.

The norm used in the definitions of the Cartan domains is the *spectral norm*, i.e., $\|z\| = s_1(z)$, the largest *singular number* of z (see §2 below for the notion of singular numbers). The Cartan domains of types I–IV are called *classical*, while the domains of types V and VI (of dimensions 16 and 27 respectively) are *exceptional*.

An important property of the Cartan domains is that they contain 0 and are *circular* and *convex* (thus they are in the *Harish–Chandra form*). Therefore a Cartan domain is the open unit ball of a certain complex Banach space, called a *Cartan factor*. A Cartan factor carries a unique triple product $Z \times Z \times Z \to Z$ which induces the structure of a *JB*-*triple*. Important subclasses of the JB*-triples are the C*-algebras, the JB*-algebras and Hilbert spaces. In the matrix factors of types I–III the triple product is given by $\{abc\} = (ab*c + cb*a)/2$. In the Cartan factor of type IV, *the spin factor*, the triple product is given by $\{abc\} = (a,b)c + (c,b)a - (a,\bar{c})\bar{b}$.

The classical Cartan factors admit realizations as subtriple of $B(H)$, H a Hilbert space. This is not true for the exceptional factors. While the theory we present is unified and considers all Cartan factors simultaneously, our examples will be taken from the classical factors which are more interesting from our point of view.

The Jordan theoretic approach to the bounded symmetric domains is presented in [L2], see also [U2] and [L1]. For the connection between bounded symmetric domains and JB*-triples in infinite dimensional spaces, see [Ka] and [U1]. The recent book [FK2] provides a natural approach to analysis on symmetric cones and Cartan domains of tube type, by means of Jordan theoretical tools. This approach is different from that of [L2], and will not be discussed here. [FK2] contains also basic material in Harmonic analysis on asymmetric cones and Cartan domains of tube type.

Let D be a Cartan domain in $\mathbb{C}^d$, and let Z be the corresponding Cartan factor, i.e., $Z = \mathbb{C}^d$ with D as the open unit ball. Let Aut(D) denote the group of all biholomorphic automorphisms of D, equipped with the topology of uniform convergence on compact subsets of D. Let G =

$(\mathrm{Aut}(D))_0$ be the connected component of the identity in $\mathrm{Aut}(D)$. Due to the symmetry of D, G acts transitively on D. Both $\mathrm{Aut}(D)$ and G are semi-simple real Lie groups. Let $K = G \cap GL(\mathbb{C}^d)$ be the subgroup of linear automorphisms in G. By *Cartan's linearity theorem*, $K = \{\varphi \in G; \varphi(0) = 0\}$; K is known to be a *maximal compact subgroup* of G. All other maximal compact subgroups of G are conjugate to K; thus they are the isotropy subgroups of points in D. The evaluation map $G \to D$ defined by $\varphi \longmapsto \varphi(0)$ realizes D as the quotient G/K. It also carries the Haar measure $d\varphi$ of G onto the unique (up to multiplicative constant) G-invariant measure μ on D.

The automorphism groups of the classical Cartan domains admit realizations as classical groups of matrices, see [Hu], [Kn], and [Py]. We will use this concrete realization only via formula (1.3) below, therefore we state these results without details. In a type $\mathrm{I}_{n,m}$-domain, $G = SU(n,m)$ and $K = S(U(n) \times U(m))$. The action of G on D is by Potapov-Möbius transformations: If $A = \begin{pmatrix} a & b \\ c & d \end{pmatrix} \in SU(n,m)$ (where $a \in M_{n,n}(\mathbb{C})$, $b \in M_{n,m}(\mathbb{C})$, $c \in M_{m,n}(\mathbb{C})$, and $d \in M_{m,m}(\mathbb{C})$ and $A^* \begin{pmatrix} I_n & 0 \\ 0 & -I_m \end{pmatrix} A = \begin{pmatrix} I_n & 0 \\ 0 & -I_m \end{pmatrix}$), then for $z \in D$

$$(1.2) \qquad A \cdot z = (az + b)(cz + d)^{-1}$$

For a type II_n domain, $G = SO^*(2n)$, and for type III_n, $G = Sp(n,\mathbb{R})$. In both cases $K = \left\{ \begin{pmatrix} a & 0 \\ 0 & \bar{a} \end{pmatrix}; a \in U(n) \right\}$ and the action of G on D is via (1.2). The situation is a little more complicated in type IV_n, see [Hu].

For any $a \in D$ there is a unique biholomorphic symmetry $\varphi_a \in G$ (i.e., $\varphi_a \circ \varphi_a =$ identity) which interchanges 0 and a, for which the geodesic midpoint b between 0 and a is an isolated fixed point. Clearly, $\varphi_a'(b) = -I$. The uniqueness of φ_a follows from *Cartan's uniqueness theorem* which says that if $\varphi, \psi \in G$ satisfy $\varphi(b) = \psi(b)$ and $\varphi'(b) = \psi'(b)$ for some $b \in D$, then $\varphi = \psi$. In case D is the unit disk in $\mathbb{C}$, $\varphi_a(z) = (a - z)/(1 - \bar{a}z)$. In case of the matrix domains (types I – III)

$$(1.3) \qquad \varphi_a(a) = (1 - aa^*)^{-1/2}(a-z)(I-a^*z)^{-1}(I-a^*a)^{1/2};$$

see [IS]. (It can be shown that (1.3) can be written in the form (1.2).) It is easy to verify the desired properties in a domain of type I. By expanding $\varphi_a(z)$ in a Taylor series in z one can express $\varphi_a(z)$ via triple products only. This shows that φ_a has the desired properties also in the domains of types II and III. Moreover, in the matrix domains a direct computation using (1.3) shows that

$$(1.4) \qquad \varphi_a'(z)w = -(1 - aa^*)^{1/2}(I - za^*)^{-1} w(I - a^*z)^{-1}(I - a^*a)^{1/2}.$$

We denote by dV the Lebesgue measure on $\mathbb{C}^d$, normalized so that $V(D) = 1$. Let $L^2(D) = L^2(D, dV)$ and let $L_a^2(D)$ denote the closed subspace of analytic functions in $L^2(D)$. It is elementary to show that point evaluations are continuous linear functionals on $L_a^2(D)$. Thus for every $w \in D$ there is a unique element $K_w = K(\cdot, w) \in L_a^2(D)$ so that

$$(1.5) \qquad f(w) = (f, K_w), \quad f \in L_a^2(D).$$

The family of functions $K_w(z) = K(z, w)$, $w \in D$, is called the *Bergman kernel of* D. $L_a^2(D)$ is called the *Bergman space*. Thus $L_a^2(D)$ is a reproducing kernel Hilbert space with reproducing kernel K. The orthogonal projection $P: L^2(D) \to L_a^2(D)$ is given by $(Pf)(w) = (f, K_w)$, $w \in D$.

The Bergman kernel is closely related to the determinant function (we shall discuss this in greater detail in Section 3 below) as the following examples show:

Type $I_{n,m}$: $\quad K(z,w) = \det(I_n - zw^*)^{-(n+m)}$

Type II_n: $\quad K(z,w) = \det(I_n - zw^*)^{-(n-1)}$

Type III_n: $\quad K(z,w) = \det(I_n - zw^*)^{-(n+1)}$

Type IV_n: $\quad K(z,w) = \left(1 + \sum_{j=1}^{n} z_j^2 \sum_{j=1}^{n} \overline{w}_j^2 - 2 \sum_{j=1}^{n} z_j \overline{w}_j \right)^{-n}$

A fundamental property of the Bergman kernel is its transformation rule under the action of automorphisms; namely,

$$(1.6) \qquad J\varphi(z) \, K(\varphi(z), \varphi(w)) \, \overline{J\varphi(w)} = K(z,w)$$

for all $\varphi \in \text{Aut}(D)$ and all $z, w \in D$. Here $J\varphi(z) = \det(\varphi'(z))$ is the *complex Jacobian* of φ at z. The proof of this formula uses the fact that the real *Jacobian* of φ is related to the complex Jacobian via $J_{\mathbb{R}}\varphi(z) = |J\varphi(z)|^2$. Hence, the map $U(\varphi)f = (f \circ \varphi) \cdot J\varphi$ is an isometry of $L^2(D)$ which leaves $L_a^2(D)$ invariant. Now if $f \in L_a^2(D)$, $\varphi \in \text{Aut}(D)$ and $w \in D$ then

$$\begin{aligned}
f(w) &= J(\varphi^{-1})(\varphi(w))f(\varphi^{-1}(\varphi(w))) \cdot J\varphi(w) \\
&= (U(\varphi^{-1})f, K_{\varphi(w)}\overline{J\varphi(w)}) \\
&= (f, U(\varphi)(K_{\varphi(w)})\overline{J\varphi(w)})
\end{aligned}$$

Thus $U(\varphi)(K_{\varphi(w)})\overline{J\varphi(w)} = K_w$ and (1.6) is proved.

In particular, the Bergman kernel is invariant under linear automorphisms:

$$(1.7) \qquad K(k(z), k(w)) = K(z,w), \quad k \in K.$$

From (1.6) and the fact that $K_0 = 1$ it follows that

$$(1.8) \qquad J\varphi_w(0) = (-1)^d K(w,w)^{-1/2}; \quad w \in D.$$

Therefore, using (1.6) again, one obtains

$$(1.9) \qquad J\varphi_w(z) = (-1)^d K(z,w) \, K(w,w)^{-1/2}.$$

(The argument to compute the sign in (1.8) is due to A. Koranyi [Ko3]). Notice that (1.6) and (1.9) imply that $K(z,w) \neq 0$ and $J\varphi(z) \neq 0$ for all $z,w \in D$ and $\varphi \in \text{Aut}(D)$.

The map U from G into the group of unitary operators on $L^2(D)$ given by $U(\varphi)f = (f \circ \varphi)J\varphi$ ($\varphi \in G$, $f \in L^2(D)$) is a strongly continuous antihomomorphism. Thus $U(\varphi \circ \psi) = U(\psi)U(\varphi)$ for all $\varphi, \psi \in G$, and $U(\text{id}) = I$, where "id" is the identity function on D. For every $\lambda \in \mathbb{C}$ we define a map $U^{(\lambda)}$: $\varphi \mapsto U^{(\lambda)}(\varphi)$ from G to operators on analytic functions on D, by

$$(1.10) \qquad U^{(\lambda)}(\varphi)f := (f \circ \varphi) \cdot (J\varphi)^{\lambda/g} .$$

Here g is a certain positive integer, called the *genus* of D, which will be discussed later. Since $J\varphi(z) \neq 0$, the principal branch of $(J\varphi(z))^{\lambda/g}$ is well defined and analytic in z for fixed φ. If λ/g is an integer then the map $\varphi \mapsto U^{(\lambda)}_{(\varphi)}$ is even anti-multiplicative. However, if λ/g is not an integer this is not true, nor is $\varphi \mapsto J\varphi(z)^{\lambda/g}$ continuous. To overcome this, let $\tilde{G}$ be *the universal covering group* of G and let $\pi: \tilde{G} \to G$ be the covering map. Since $\tilde{G}$ is simply connected there is a lifting of $(J\varphi(z))^{\lambda/g}$ to $\tilde{G} \times D$, denoted by $J(\tilde{\varphi},z)^{\lambda/g}$ (the principal branch of the power function), which is analytic in $z \in D$ and continuous (in fact, real analytic) in $\tilde{\varphi} \in \tilde{G}$, and satisfies

$$J(\tilde{\varphi} \circ \tilde{\psi},z)^{\lambda/g} = J(\tilde{\varphi},z)^{\lambda/g} \cdot J(\tilde{\psi},z)^{\lambda/g}$$

for all $\tilde{\varphi}, \tilde{\psi} \in \tilde{G}$ and $z \in D$. This permits one to define an action $\tilde{U}^{(\lambda)}$ of $\tilde{G}$ on functions on D via

$$(1.11) \qquad \tilde{U}^{(\lambda)}(\tilde{\varphi})f := (f \circ \pi(\tilde{\varphi}))J(\tilde{\varphi},\cdot)^{\lambda/g}$$

which satisfies

$$(1.12) \qquad \tilde{U}^{(\lambda)}(\tilde{\varphi} \circ \tilde{\psi}) = \tilde{U}^{(\lambda)}(\tilde{\psi})\tilde{U}^{(\lambda)}(\tilde{\varphi}) , \qquad \tilde{U}^{(\lambda)}(\text{id}) = I$$

for all $\tilde{\varphi}, \tilde{\psi} \in \tilde{G}$. From this it follows that

$$(1.13) \qquad U^{(\lambda)}(\varphi\psi) = c(\varphi,\psi,\lambda)U^{(\lambda)}(\psi)U^{(\lambda)}(\varphi)$$

where $c(\varphi,\psi,\lambda)$ is a unimodular scalar, which transforms as a cocycle, i.e. $U^{(\lambda)}$ is a *projective anti-homomorphism* of G.

Since $D \times D$ is simply connected and $K(z,w) \neq 0$, there is an unambiguous definition of $K(z,w)^{\lambda/g}$ (principal branch of the power function) which is analytic in z and conjugate analytic in w. Moreover, by (1.6) and (1.9)

$$(1.14) \qquad J\varphi(z)^{\lambda/g}K(\varphi(z),\varphi(w))^{\lambda/g}\overline{J\varphi(w)}^{\lambda/g} = K(z,w)^{\lambda/g}$$

for all $\varphi \in G$ and $z, w \in D$.

Let $\mathscr{H}^{(0)}_\lambda = \text{span } \{K_w^{\lambda/g}; w \in D\}$ and define a sesqui-linear form $(\cdot,\cdot)_\lambda$ on

$\mathcal{H}_\lambda^{(0)}$ by

(1.15) $$(K_w^{\lambda/g}, K_z^{\lambda/g})_\lambda = K(z,w)^{\lambda/g}.$$

Certainly, $(\cdot,\cdot)_\lambda$ is $U^{(\lambda)}$-*invariant*, i.e.

(1.16) $$(U^{(\lambda)}(\varphi)f, U^{(\lambda)}(\varphi)h)_\lambda = (f,h)_\lambda.$$

for all $f,h \in \mathcal{H}_\lambda^{(0)}$ and $\varphi \in G$. This is an easy consequence of (1.14) and (1.15).

Definition: The *Wallach set* (W(D) of D is the set of all $\lambda \in \mathbb{C}$ so that $K(z,w)^{\lambda/g}$ is a positive definite function of $z, w \in D$.

Clearly, for $\lambda \in W(D)$, $\mathcal{H}_\lambda^{(0)}$ is a pre-Hilbert space. We denote by $\mathcal{H}_\lambda$ the associated Hilbert space, i.e. the completion of $(\mathcal{H}_\lambda^{(0)}(\cdot,\cdot)_\lambda)/\{f \in \mathcal{H}_\lambda^{(0)}; (f,f)_\lambda = 0\}$. As will be shown in Section 4 below, $(f,f)_\lambda \neq 0$ for $f \neq 0$ in $\mathcal{H}_\lambda^{(0)}$. It is obvious that $\mathcal{H}_\lambda$ is a Hilbert space of analytic functions on D and that $K(z,w)^{\lambda/g}$ is its reproducing kernel. Clearly, $U^{(\lambda)}$ extends to a projective, unitary anti-representation of G on $\mathcal{H}_\lambda$, namely (1.13) is satisfied, rather than the anti-multiplicativity. If λ/g is an integer, then $U^{(\lambda)}$ is unitary antirepresentation of G on $\mathcal{H}_\lambda$. For general $\lambda \in W(D)$, $\tilde{U}^{(\lambda)}$ extends to a unitary anti-representation of $\tilde{G}$ on $\mathcal{H}_\lambda$. These facts are proved by using the continuity of the Bergman kernel $K(z,w)$.

The determination of $W(D)$ will be presented in Section 4. We would like to make here few preliminary remarks concerning its structure. It is well known that the Schur–Hadamard product of two positive matrices is again positive definite. Every reproducing kernel is positive definite (since $\sum_{j,\ell} a_j \bar{a}_\ell K(z_j, z_\ell) = \left\| \sum_\ell \bar{a}_\ell K(\cdot, z_\ell) \right\|^2 \geq 0$). Thus $K(z,w)$, $K(z,w)^2$, $K(z,w)^3$, $\cdots$ are all positive definite. Thus $\{ng\}_{n=1}^\infty \subseteq W(D)$. In fact, much more can be proved without much effort. The standard Cauchy–Schwarz inequality $|K(z,w)|^2 \leq K(z,z)K(w,w)$ with $w = 0$ yields $1 \leq K(z,z)$. Hence the integral

(1.17) $$c(\lambda)^{-1} := \int_D K(z,z)^{1-\lambda/g} dV(z).$$

is finite for all $\lambda \geq g$ (in fact, for all $\lambda > g - 1$, see [FK1]). Let $d\mu_\lambda(z) = c(\lambda) K(z,z)^{1-\lambda/g} dV(z)$ and consider the *weighted Bergman spaces* $L_a^2(\mu_\lambda)$, consisting of all analytic functions in $L^2(\mu_\lambda) := L^2(D,\mu_\lambda)$. As in the case of the Bergman space ($\lambda = g$), $U^{(\lambda)}$ is an isometric action of G on $L^2(\mu_\lambda)$ which maps $L_a^2(\mu_\lambda)$ onto itself. $L_a^2(\mu_\lambda)$ is a reproducing kernel Hilbert space, with reproducing kernel $K(z,w)^{\lambda/g}$. This follow easily from $f(0) = (f,1)_{L^2(\mu_\lambda)}$ for $f \in L_a^2(\mu_\lambda)$ and by the isometric $U^{(\lambda)}$-action of G. Thus, we also find that $(g-1, \infty) \subset W(D)$.

Another well known example of an invariant Hilbert space of analytic functions on D is the *Hardy space* $H^2(S)$. Here S is the *Shilov boundary* of D (i.e. if f is continuous on $\overline{D}$ and analytic in D, then $|f|$ assumes its maximum on S). $H^2(S)$ is the closure of the analytic polynomials in the space $L^2(S,\sigma)$, where σ is the unique, K invariant probability measure on S. By [Ko1], or [FK1, Lemma 3.3 and Corollary 3.5] (which will be explained in Section 4) it follows that $H^2(S) = \mathscr{H}_{d/r}$ where d = $\dim_{\mathbb{C}}(D)$ and r is the *rank* of D (to be defined below). Thus $d/r \in$ W(D) as well. If D is the open unit ball in $\mathbb{C}^d$, then r = 1, S = ∂D, the topological boundary and $H^2(S)$ is the traditional Hardy space.

§ 2. Jordan theoretic background

In order to proceed with our exposition we need some more Jordan theoretic tools. The material presented here is studied in detail in [U2] and [L2]. By definition, a *JB*-triple* is a complex Banach space Z, equipped with a triple product $Z \times Z \times Z \longrightarrow Z$, written as $(x, y, z) \mapsto \{xyz\}$, so that:

(1) $\{xyz\}$ is bilinear and symmetric in x, z and conjugate-linear in y;

(2)   for all x, y, u, v, w $\in$ Z the *Jordan Triple Identity* holds:

(2.1) $\{\{uvw\}yx\} + \{\{uvx\}yw\} - \{uv\{wyx\}\} = \{w\{vuy\}x\}$

(3)   For every $z \in Z$ the operator $D(z)y := \{zzy\}$ is hermitian (i.e. $\{\exp(itD(z))\}_{t\in\mathbb{R}}$ is a one-parameter group of isometries of Z), and has a positive spectrum;

(4)   For every $z \in Z$, $\|D(z)\| = \|z\|^2$ (the "C*-axiom").

We remark that the meaning of the Jordan triple identity is that the operators $iD(z)$, $z \in D$, are *triple derivations*, i.e. for all u, v, w $\in$ Z

(2.2) $iD(z)(\{u,v,w\})=\{iD(z)u, v, w\}+\{u, iD(z)v, w\}+\{u,v,iD(z)w\}$

The most important example of JB*-triples are the *JC*-triples*, i.e. the norm closed subspaces Z of B(H) (= the bounded operators on a Hilbert space H) which are closed under the cubic map $z \mapsto zz^*z$. By polarization it follows that such Z are closed under the *triple product of B(H)*

(2.3) $\{xyz\} = (xy*z + zy*x)/2$.

Thus C*-subalgebras, spaces of the form Z = eAf where A is a C*-algebra and e,f are projections in A, the Cartan factors (in finite or infinite dimensions) – are all JC*-triples.

An element $v \in Z$ is a *tripotent* of $\{vvv\} = v$. In the Cartan factors of types I–III the tripotents are simply the partial isometries. Two tripotents v, u are *orthogonal* if $\{vuz\} = 0$ for all $z \in Z$. It follows easily from the Jordan Triple Identity that orthogonality is a symmetric relation. If v is a tripotent then by the Jordan Triple Identity again, the operator D(v) satisfies

(2.4) $$D(v)(2D(v) - I)(D(v) - I) = 0.$$

It follows that the spectrum of $D(v)$ is contained in $\{0, \frac{1}{2}, 1\}$, and that Z admits a direct sum decomposition

(2.5) $$Z = Z_2(v) \oplus Z_1(v) \oplus Z_0(v) ,$$

called the *Peirce decomposition* associated with v, where $Z_j(v)$ is the eigenspace of $D(v)$ corresponding to the eigenvalue $j/2$, $j=0, 1, 2$. The corresponding spectral projections $P_j(v)$, $j = 0, 1, 2$. are the *Peirce projections* of v.

If $\{v_1, \cdots , v_n\}$ are pairwise orthogonal tripotents, then there is a direct sum decomposition, called the *joint Peirce decomposition*, associated with $\{v_j\}_{j=1}^n$,

(2.6) $$Z = \sum_{0 \le i \le j \le n} \oplus Z_{i,j}$$

where

$$Z_{j,j} := Z_2(v_j) \ (j = 1, 2, \cdots , n), \quad Z_{i,j} := Z_1(v_i) \cap Z_1(v_j) \ (1 \le i < j \le n),$$

$$Z_{0,j} := Z_1(v_j) \cap \bigcap_{\substack{1 \le i \le n \\ i \ne j}} Z_0(v_i) \quad \text{and} \quad Z_{0,0} := \bigcap_{j=1}^n Z_0(v_j).$$

The *joint Peirce projections* $\{P_{i,j}\}_{0 \le i \le j \le n}$ are defined by $P_{i,j}|_{Z_{i,j}} = I_{Z_{i,j}}$ and $P_{i,j}|_{Z_{k,\ell}} = 0$ for all $(k,\ell) \ne (i,j)$ with $0 \le k \le \ell \le n$. The $Z_{i,j}$ are subtriples of Z.

A closed subspace W of a JB*-triple Z is an *ideal* if $\{Z\ Z\ W\} \subseteq W$ and $\{Z, W, Z\} \subseteq W$. A JB*-triple is a *factor* if its only ideals are the trivial ones: $\{0\}$ and Z. The category of JB*-triples is equivalent to that of bounded symmetric domains (see [Ka]). On one hand the open unit ball of a JB*-triple is a bounded symmetric domain. On the other hand – every bounded symmetric domain in a complex Banach space is biholomorphically equivalent to the open unit ball of a unique JB*-triple. Moreover, the open unit ball of a JB*-factor is an irreducible bounded symmetric domain, and two JB*-triples are isometric if and only if their open unit balls are biholomorphically equivalent. The *Cartan factors* are the JB*-factors whose open unit balls are the Cartan domains (which were described in section 1 above).

From now on we assume that Z is the Cartan factor whose open unit ball is the Cartan domain D.

A tripotent $v \in Z$ is *minimal* if $Z_2(v) = \mathbb{C}v$. Since Z is finite dimensional it contains minimal tripotents. A maximal family of orthogonal, minimal tripotents is called a *frame*. It is known that if $\{e_j\}_{j=1}^r$ and $\{v_j\}_{j=1}^n$ are two frames, then $n = r$ and there is a $k \in K$ so that $ke_j = v_j$ for $1 \le j \le r$. The *rank* of Z (= the rank of D) is the number of elements in a frame, and is denoted by r. The rank can be characterized as the maximal r so that ℓ_r^∞ (the r-dimensional ℓ^∞-space) is isometric to a subspace of Z.

Fix a frame $\{e_j\}_{j=1}^r$ and let $e = \sum_{j=1}^r e_j$. Clearly, e is a maximal tripotent. Let $Z = \sum_{0 \le i \le j \le r} \oplus Z_{i,j}$ be the joint Peirce decomposition asso-ciated with $\{e_j\}_{j=1}^r$. Notice that $Z_{0,0} = Z_0(e) = \{0\}$. We define a parameter $a = a(D)$ by $a = 0$ if $r = 1$ and

(2.7) $a := \dim_{\mathbb{C}}(Z_{i,j})$ for some $1 \le i < j \le r$ if $1 < r$.

The parameter a is well defined and independent of (i,j) and of the frame $\{e_j\}_{j=1}^r$, because of transitivity of the action of K on the frames.

We define also a parameter $b = b(D)$ by

(2.8) $b = \dim(Z_{0,j})$

where $1 \le j \le r$. Again, b is a well defined integral invariant of D. The *genus* of D is

(2.9) $g = g(D) = (r-1)a + b + 2$.

The *type* of D is the triple

(2.10) (r, a, b),

and is known to be a complete biholomorphic invariant of D.

In the matrix factors there is a canonical and natural choice of frames. In the factors $I_{n,m}$ one considers the standard matrix units $\{e_{i,j}; 1 \le i \le n, 1 \le j \le m\}$ and defines $e_j = e_{j,j}, 1 \le j \le n = r$. Then $Z_{j,j} = \mathbb{C}e_j$, $Z_{i,j} = \mathbb{C}e_{i,j} + \mathbb{C}e_{j,i}$ $(1 \le i < j \le n)$, and $Z_{0,j} = $ span $\{e_{j,i}; n < i \le m\}$. In the factor of type III_n one takes also $e_j = e_{j,j}, 1 \le j \le n = r$ and get $Z_{j,j} = \mathbb{C}e_j$ and $Z_{i,j} = \mathbb{C}(e_{i,j} + e_{j,i}), 1 \le i < j \le n = r$. In case of the factor type II_n one devines $v_{i,j} = e_{i,j} = e_{j,i}$, and chooses $e_j = v_{2j-1,2j}$ $(1 \le j \le [\frac{n}{2}])$, and in case n is odd – $Z_{0,j} = \mathbb{C}v_{2j-1,n} + \mathbb{C}v_{2j,n}$ $(1 \le j \le [\frac{n}{2}])$.

The Cartan factor of the type IVn, called the *Spin factor*, is $\mathbb{C}^n$ with the triple product
$$\{x,y,z\} = (x,y)z + (z,y)x - (x,\bar{z})\bar{y}$$
where $\bar{x} = (\bar{x}_1, \cdots, \bar{x}_n)$. In this case the rank is $r = 2$. The minimal tripotents (of rank 1) are the elements v for which $\|v\| = 1/\sqrt{2}$ and $(v,\bar{v}) = \sum_{j=1}^n v_j^2 = 0$. The maximal triponents (of rank 2) have the form v $= e^{i\theta}x$, where $x \in \mathbb{R}^n$, $\|x\| = 1$ and $\theta \in \mathbb{R}$. The canonical frame is

$$e_1 = \frac{1}{2}(1,i,0,\cdots,0), \qquad e_2 = \frac{1}{2}(1,-i,0,\cdots,0).$$

The Peirce spaces are

$$Z_{j,j} = \mathbb{C}e_j \ (j=1,2) \quad \text{and} \quad Z_{1,2} = \{(0,0,\alpha_3,\alpha_4,\cdots,\alpha_n); \ \alpha_j \in \mathbb{C}\} \equiv \mathbb{C}^{n-1}.$$

The *parameters of the Cartan domains* are shown in the next table.

type parameter	$I_{n,m}$ $(n \leq m)$	II_n $(5 \leq n)$	III_n $(2 \leq n)$	IV_n $(5 \leq n)$	V	VI
d (dimension)	nm	$\frac{(n-1)n}{2}$	$\frac{n(n+1)}{2}$	n	16	27
r (rank)	n	$\left[\frac{n}{2}\right]$	n	2	2	3
a	2, if $2 \leq n$ 0, if $1 = n$	4	1	n−2	6	8
b	m−n	0, if n even 2, if n odd	0	0	4	0
g (genus)	m+n	2n−2	n+1	n	12	18

Notice that the unit ball of $\mathbb{C}^m$ is the special case of the domain of type $I_{1,m}$. It is the only Cartan domain with rank one.

If b = 0 then $Z_{0,j} = 0$ for $1 \leq j \leq r$ and e becomes a *unitary tripotent*, i.e. $Z = Z_2(e)$. In this case D is a *tube-type domain* and Z is a *JB*-algebra* with (binary) Jordan product $x \circ y = \{xey\}$, involution $x^* = \{exe\}$ and unit e. The tube-type domains correspond under the *Cayley transform* with respect to e, $\gamma(z) = i(e+z) \circ (e-z)^{-1}$, to a *Siegel domains of type I*, called also tube domains, while the general Cartan domains correspond via the general Cayley transform to *Siegel domains of type II*; see [U2] and [FK2] for details.

Every element $z \in Z$ admits a *singular value decomposition* $z = \sum_{j=1}^{\ell} s_j v_j$ where $\{v_j\}_{j=1}^{\ell}$ are pairwise tripotents and $s_1 > s_2 > \cdots > s > 0$ are the distinct, non-zero, *singular numbers* of z. Let rank $(v_j) = \text{rank}(Z_2(v_j)) = r_j$. Refining $\{v_j\}_{j=1}^{\ell}$ to obtain a frame, we get by the transitivity of K on the frames an element $k \in K$ so that

(2.11)
$$k\left(\sum_{j=1} s_j \sum_{i \in \sigma_j} e_j\right) = z$$

where $\{\sigma_j\}$ are disjoint subsets of $\{1, 2, \cdots r\}$ and $|\sigma_j| = r_j$. It follows that D is the K-orbit of $\left\{t = \sum_{j=1}^{r} t_j e_j \ ; \ 0 \leq t_j < 1\right\}$. This latter set is canonically identified with the cube $I^r = [0,1)^r$, so we can write

(2.12) $D = KI^r$.

There is a complete description of the *boundary* ∂D *of* D. For every tripotent v let $D(v) = D \cap Z_0(v)$ and $B(v) = v + D(V)$. $B(v)$ is the *boundary component* of D corresponding to v. Let

(2.13) $S_j = \bigcup_{rank(v)=j} B(v), \quad 1 \leq j \leq r$.

Each S_j is real, connected manifold and the boundary of D is the disjoint union.

(2.14) $\partial D = \bigcup_{j=1}^{r} S_j$.

$S := S_r$ is the *Shilov boundary* and it is the manifold of maximal tripotents of Z. For example, in the context of the factor of type $I_{n,m}$ $(1 \leq n \leq m)$ S is the set of maximal partial isometries, i.e. the co-isometries. If n = m, then S = U(n) is the set of all n × n unitary matrices. If n = 1, then D is the unit ball of $\mathbb{C}^m$ and S = ∂D. This is the only case where the Shilov boundary coincides with the full topological boundary of D.

It is a well known fact that every $\varphi \in G$ extends to an analytic (in fact, rational) function in a certain neighborhood of $\overline{D}$, and that it is a homeomorphism of $\overline{D}$, and thus of ∂D. Moreover, for each tripotent v, $\varphi(B(v))$ is also a boundary component of D. Thus there is a unique tripotent $\tilde{\varphi}(v)$ with rank(v) = rank($\tilde{\varphi}(v)$) so that

(2.15) $\varphi(B(v)) = B(\tilde{\varphi}(v))$

Letting $S_0 = D$, we see that for $0 \leq j \leq r$ $\varphi(S_j) = S_j$ for every $\varphi \in G$ and $Gz = \{\varphi(z); \ \varphi \in G\} = S_j$ for every $z \in S_j$. That is, the $\{S_j\}_{j=0}^{r}$ are the irreducible G-orbits in $\overline{D}$.

In general, if v is a tripotent then $\varphi(v)$ need not be a tripotent and in particular need not coincide with $\tilde{\varphi}(v)$. But if $k \in K$, then $k(v) = \tilde{k}(v)$ and

(2.16) $k(B(v)) = B(k(v))$.

Let T_j denote the set of all tripotents of rank j, $1 \leq j \leq r$. Then T_j is an irreducible K-orbit, it is a real manifold and carries a unique K-invariant probability measure σ_j, defined by

(2.17)
$$\int_{T_j} f \, d\sigma_j = \int_K f(kv) dk \, ,$$

where $v \in T_j$ is an arbitrary base–point. Notice that $T_r = S_r = S$. We shall denote σ_r by σ.

The set of tripotents is *ordered* in the following way

u $<$ v   if there is a tripotent w orthogonal to u so that v = u + w.

Given an increasing sequence of tripotents $0 < v_1 < v_2 < \cdots v_\ell$ one considers the associated *parabolic subgroup* of G

(2.18) $P(v_1, \cdots, v_\ell) = \{\varphi \in G; \, \bar{\varphi}(v_j) = v_j \text{ for } j = 1, \, \cdots \ell\} \, .$

That is, $P(v_1, \cdots, v_\ell)$ is the subgroup of all $\varphi \in G$ so that $\varphi(B(v_j)) = B(v_j)$ for $j = 1, \cdots, \ell$. If $0 < u_1 < u_2 < \cdots < u_m$ is another increasing sequence of tripotents which refines $v_1 < \cdots < v$, then clearly $P(u_1, \cdots, u_m) \subseteq P(v_1 \cdots v)$. Thus a *minimal parabolic group* is $P(v_1, \cdots, v_r)$ where $\{v_j\}_{j=1}^r$ is a frame. Since K acts transitively on the frames it is clear that *every two minimal parabolic subgroups are conjugate by an element of K.* For this reason we denote "the" minimal parabolic subgroup associated with the fixed frame $\{e_j\}_{j=1}^r$ simply by P. Notice that if D is of rank one (i.e. the open unit ball of $\mathbb{C}^d$), then P is simply the isotropy group of a point on the boundary S of D.

The minimal parabolic subgroup can be described also in terms of the *Iwasawa decomposition* G = KAN. Here A is a maximal abelian subgroup, N is a maximal nilpotent subgroup and K is a maximal compact subgroup. It is known that A,N and K are unique up to a conjugation and that AN is a maximal solvable subgroup. Let M be the centralizer of A in K. Then

(2.19)   P = MAN,

and P is the normalizer of AN in G. Clearly $G = G^{-1} = ANK$. Since K(0) = 0,

$$D = G(0) = AN(0) = P(0) \, .$$

Hence, *both AN and P act transitively on D.*

A *left invariant mean* of a group Γ is a state m on $\ell^\infty(\Gamma)$. That is, m is a positive, normalized linear functional on $\ell^\infty(\Gamma))$ which is left *invariant* under Γ: $m(x_\gamma) = m(x)$ for all $x \in \ell^\infty(\Gamma)$ and $\gamma \in \Gamma$, where $x_\gamma(\beta) = x(\gamma\beta)$. The notion of a right invariant mean is defined in the same way. Γ is *amenable* if there is a right or a left invariant mean of Γ. See [G] and [Pi] for the theory of amenable groups. It is known that abelian groups and compact groups are amenable. Also, a subgroup of amenable group is amenable and a semidirect product of amenable groups is amenable. It follows that a solvable group is amenable. On the other hand, a free group of two or more generators is known to be non-amenable. In our case, $G = (Aut(D))_0$ contains an isomorphic copy of the free group of two generators, thus G *is not amenable*.

The discussion above ensures that *AN is amenable*. As a compact extension of a solvable group, P is also amenable. In fact, both K and P are maximal amenable subgroups of G. Further, K is the unique (up to a conjugation) compact maximal amenable subgroup; see [M] where a complete description of the 2^r maximal amenable subgroups of G is presented.

§3. The structure of polynomials

Let D be a Cartan domain and let Z be the associated Cartan factor. Let $\mathcal{P}(= \mathcal{P}(D) = \mathcal{P}(Z))$ be the space of all analytic polynomials on Z and let $\mathcal{P}_n$ denote the subspace of $\mathcal{P}$ consisting of all homogeneous polynomials of degree n,   n = 0, 1, 2, $\cdots$. Clearly $\mathcal{P} = \sum_{n=0}^{\infty} \oplus \; \mathcal{P}_n$.

To describe the finer structure of $\mathcal{P}$ we need the notion of the determinant polynomials. Let Z be a JB* algebra with unit e. To each $z \in Z$ one associates an operator P(z) on Z via P(z)x = {zx*z}. The map $z \mapsto P(z)$ is called the *quadratic representation*, and it satisfies P(P(z)x) = P(z)P(x)P(z). $z \in Z$ is said to be *invertible* if P(z) is invertible, and the inverse of z is defined by $z^{-1} := P(z)^{-1}z$. Clearly $z \circ z^{-1} = e$. It is known [U2, Chapter 4] that the inverse map in Z is a rational map, written in exact (i.e. reduced) form as $z^{-1} = n(z)/N(z)$, where N(z) is the *determinant* (or the Koecher *norm*) of z, and n(z) plays the role of the *adjoint* of z. Among the Cartan factors the JB*-algebras are those of types $I_{n,n}$, II_n (n even), III_n, IV_n, and VI. In types $I_{n,n}$ and III_n, N(z) = det(z), the ordinary determinant. In type II_n (n even) N(z) is the Pfaffian polynomial, and it satisfies $N(z)^2 = det(z)$. In type VI_n, N(z) = $(z,\bar{z}) = \sum_{j=1}^{n} z_j^2$. In the exceptional algebra of type VI, N(z) is the Freudenthal determinant.

Fix a frame $\{e_j\}_{j=1}^r$ of orthogonal minimal tripotents in a Cartan factor Z. For j=1,2, $\cdots$,r consider $A_j := Z_2(e_1+e_2 \cdots +e_j) = P_2(e_1+ \cdots +e_j)Z$. It is a JB*-algebra with unit $e_1 + \cdots + e_j$. We denote by N_j the determinant polynomial of A_j and we extend it to all of Z by defining $N_j(z) = N_j(P_2(e_1+...+e_j)z)$. The case j=r is of particular importance and we write $e=e+ \cdots +e_r$, $N=N_r$. N is called the *highest determinant*, or, simply, the *determinant*. Notice that if $t = \sum_{j=1}^{r} t_j e_j$ then $N_j(t) = \prod_{i=1}^{j} t_i$.

A *signature* is a sequence $\underline{m} = (m_1, m_2, \cdots, m_r)$ of integers so that

$m_1 \geq m_2 \geq \cdots \geq m_r \geq 0$. The *conical polynomial* associated with $\underline{m}$ is

(3.1) $$N_{\underline{m}} = N_1^{m_1-m_2} \cdot N_2^{m_2-m_3} \cdot \cdots \cdot N_r^{m_r} .$$

Notice that if $t = \sum_{j=1}^{r} t_j e_j$ then $N_{\underline{m}}(t) = \prod_{j=1}^{r} t_j^{m_j}$. The linear span of the K-orbit of $N_{\underline{m}}$ is

(3.2) $$P_{\underline{m}} = \text{span } \{N_{\underline{m}} \circ k; \, k \in K\}$$

Note that $\deg(N_{\underline{m}}) = |\underline{m}| = m_1 + m_2 + \cdots + m_r$. Thus $P_{\underline{m}} \subseteq \mathcal{P}_{|\underline{m}|}$.

A fundamental result of Schmid [S] (see also [U3]) is that *the K-irreducible subspaces of the Hardy space $H^2(S)$ are precisely the $P_{\underline{m}}$, where $\underline{m}$ ranges over all signatures, and they are mutually K-inequivalent.* The arguments in [S] and [U3] extend to general K-invariant Hilbert space $\mathcal{H}$ of analytic functions on D. The decomposition

(3.3) $$\mathcal{H} = \sum_{\underline{m}} \oplus P_{\underline{m}}$$

is called in [U3] then *Peter–Weyl decomposition*. In case $r = 1$, where D is the open unit ball of $\mathbb{C}^d$, $P_m = \mathcal{P}_m$ and the Peter–Weyl decomposition reduced to the homogeneous decomposition.

Let

(3.4) $$L = \{k \in L; \, k(e) = e\} .$$

Clearly, L is a closed subgroup of K, hence compact. For example, if D is of type $I_{n,m}$ with $n \leq m$, then each $k \in K$ has the form $k(z) = uzv*$, where $u \in U(n)$, $v \in U(m)$. Thus $k \in L$ if and only if $v = \begin{pmatrix} u & 0 \\ 0 & w \end{pmatrix}$ with $u \in U(n)$ and $w \in U(m-n)$.

Since K acts transitively on the Shilov boundary S, the map $K \to S$ defined by $k \mapsto k(e)$ is surjective and identifies S with K/L. Every continuous function f on $\bar{D}$ which is analytic in D is uniquely determined by its values on S. Conversely, if f is a continuous function on S it can be extended to a continuous function on $\bar{D}$ and analytic on D via the *Szegö integral*

(3.5) $$f(z) = \int_S f(u) \, S(z,u) \, d\sigma(u), \quad z \in D,$$

where $S(z,u) = h(z,u)^{-d/r}$ is the *Szegö kernel* namely, the reproducing kernel of $H^2(S)$, and σ is the unique K-invariant probability measure on S. Thus, such an f can be identified with a right L-invariant continuous function on K. f is called *spherical* if $f(\ell(z)) = f(z)$ for all $z \in D$ and $\ell \in L$. Via the above identification, a spherical function is a continuous function on K which is both left and right L-invariant.

For each signature $\underline{m}$ let

$$(3.6) \qquad \varphi_m(z) = \int_L N_m(\ell(z)) \, d\ell, \qquad z \in D,$$

where $d\ell$ is the Haar measure of L. Obviously, φ_m is spherical. It is known that φ_m is the only spherical polynomial in P_m which satisfies $\varphi_m(e) = 1$. Thus the spherical polynomials in P_m are the multiples of φ_m and $\{\varphi_m\}$ forms a basis for the space of spherical polynomials.

The group K is identified with the connected component of the identity in the group of linear isometries of Z . It is a fundamental fact in JB*-triples that the linear isometries are precisely the *triple automorphisms*. Thus, every $k \in K$ satisfies

$$(3.7) \qquad k\{x, y, z\} = \{kx, ky, kz\}$$

for all x, y, z $\in$ Z. In particular, every $\ell \in L$ leaves the spaces $Z_j(e)$ invariant (j = 0, 1, 2), and it is an automorphism of the JB*-algebra, $Z_2(e)$ whose unit is e. Thus, ℓ leaves the highest determinant polynomial $N = N_r$ invariant: $N(\ell(z)) = N(z)$, $z \in D$. It follows that N is spherical, and

$$(3.8) \qquad \varphi_{(m,m,\ldots,m)} = N^m, \qquad m = 0, 1, 2, \cdots$$

Let Z be a JB*-algebra. Using the uniqueness of the representation of the inverse as a reduced quotient $z^{-1} = n(z)/N(z)$, one proves that for all $k \in K$ and $z \in D$

$$(3.9) \qquad N(k(z)) = \chi(k)N(z)$$

where $\chi: K \to \mathbb{T} = \{\lambda \in \mathbb{C}; |\lambda| = 1\}$ is a homomorphism and thus a one-dimensional, unitary character. It follows that in this case

$$(3.10) \qquad P_{(m,m,\ldots,m)} = \mathbb{C}N^m ; m = 0, 1, 2, \cdots .$$

As will be seen below the highest determinant polynomial N plays a very special role in the structure of the polynomials on the JB*-algebra Z. Let

$$(3.11) \qquad K_1 = \{k \in K ; \chi(k) = 1\}$$

K_1 is a closed normal subgroup of K so that $K/K_1 = \mathbb{T}$.

For every polynomial q(z) on Z, let $\partial_q = q\left(\dfrac{\partial}{\partial z}\right)$ by the differential operator obtained by replacing the variables (say z_j) in q(z) by the corresponding differential operators (namely, $\dfrac{\partial}{\partial z}$). Precisely, fix a K-invariant inner product $(\cdot, \cdot)$ on Z, normalized so that $\|v\|^2 = (v,v) = 1$ for all minimal tripotents. Let $z \mapsto \bar{z}$ be a conjugation on Z, and set $e_w(z) = e^{(z,w)}$, $z, w \in Z$. Then ∂_q is completely determined by $\partial_q e_w = q(\bar{w}) e_z$, $w \in Z$. In particular, ∂_q is independent of the choice of the coordinates (i.e. orthonormal basis) in Z. A polynomial p is *invariant* if $p \circ k = p$ for all $k \in K_1$; p is *harmonic* if $\partial_q(p) = 0$ for all invariant polynomial q with $q(0) = 0$. Denote by $I(Z)$ the space of the invariant polynomials and by $H(Z)$ the space of harmonic polynomials.

Lemma 3.1 [U3]: *(i) $I(Z) = \mathbb{C}[N]$, the polynomials in N over $\mathbb{C}$;*
(ii) $J(Z) := \{p \in \mathcal{P}; \; p|_{N^{-1}(\{0\})} = 0\} = N\mathcal{P}$.

Theorem 3.2 [U3]: *(i) $\mathcal{P} = J(Z) \oplus H(Z) = N\mathcal{P} \oplus H(Z)$ (direct sum as linear spaces);*
(ii) The mapping $(p,q) \mapsto p \cdot q$ induces an isomorphism of algebras over $\mathbb{C}$

$$\mathcal{P} = I(Z) \otimes H(Z) \; ;$$

(iii) Each $p \in \mathcal{P}$ has a unique representation

$$p = \sum_{j=0}^{\infty} p_j N^j \quad \text{with} \quad p_j \in H(Z) \, .$$

Even in the general case (i.e. when D is not of tube type, equivalently Z is not a JB*–algebra) the highest determinant N is closely related to the Bergman kernel and thus plays a fundamental role. It is known (see [FK1, section 3]) that there exists a unique real polynomial h on Z which is K–invariant and its restriction to span $\{e_j\}_{j=1}^{r}$ is given by

$$(3.12) \qquad h\left(\sum_{j=1}^{r} t_j e_j\right) = N\left(e - \sum_{j=1}^{r} |t_j|^2 e_j\right) = \prod_{j=1}^{r}(1 - |t_j|^2) \, .$$

h can be extended to a polynomial $h(z,w)$ which is analytic in z and conjugate analytic in w and satisfies $h(z,z) = h(z)$, see [FK1]. In the Cartan factors of type $I_{n,m}$ and III_n .

$$(3.13) \qquad\qquad h(z,w) = \det(I_n - zw*)$$

In the Cartan factor of type $II_n, h(z,w)^2 = \det(I_n - zw*)$. In general, the connection between h and the Bergman kernel is

$$(3.14) \qquad\qquad K(z,w) = h(z,w)^{-g}, \qquad z,w \in D \, ,$$

where g is the genus of D.

The determinant polynomial of a JB*–algebra transforms nicely under composition with elements of G.

Proposition 3.3 [Y1]: *For every $a, z \in D$*

$$(3.15) \qquad\qquad N(\varphi_a(z)) = \frac{N(a - z)}{h(z,a)} \, .$$

Since every $\varphi \in G$ has a unique representation $\varphi = k \circ \varphi_a$ with $k \in K$ and $a \in D$, (3.9) and (3.15) provide the general transformation rule for $N \circ \varphi$. The proof for the matrix Cartan factors (types $I_{n,n}$, II_n (even), and III_n) is a straightforward application of (1.3) and the explicit forms of

$N(z)$ and $h(z,a)$, and it was established in [A1]. The general case treated in [Y1] requires more effort, but it is enough to establish (3.15) for a fixed $a \in D$ and $z \in S$, the Shilov boundary, because both sides of (3.15) are analytic in $z \in D$ and continuous in $\bar{D}$. Since $h(a,e) = N(e-a)$ (see [FK1]), the K-invariance of both N and h and (3.15) imply

$$(3.16) \qquad N(-\varphi_a(u)) = \frac{N(u-a)}{h(u,a)} = N(u)\frac{h(a,u)}{h(u,a)}; \qquad a \in D, u \in S.$$

Let ε be the unique ellipsoid of maximal volume contained in D and let $|\cdot|$ be the Euclidean norm on $\mathbb{C}^d$ whose open unit ball is ε. Then each $k \in K$ preserves ϵ and thus becomes a unitary transformation on $(\mathbb{C}^d, |\cdot|)$. The points of contact of $\partial \varepsilon$ and ∂D are precisely the tripotents of rank one: $\partial \varepsilon \cap \partial \Omega = T_1$. In the case of the matrix Cartan factors, $|\cdot|$ is simply the Hilbert–Schmidt norm. In general, $\left|k\sum_{j=1}^{r}t_je_j\right| = \left(\sum_{j=1}^{r}|t_j|^2\right)^{1/2}$ for all $(t_j) \in \mathbb{C}^r$ and $k \in K$.

It is obvious that orthogonality in the triple sense implies orthogonality with respect to the Euclidean structure induced by ε. Thus, the joint Pierce decomposition $Z = \sum_{0 \le i \le j \le r} \oplus Z_{i,j}$ is orthogonal in the Euclidean sense. Let $dm(z)$ be the unnormalized Lebesgue measure, i.e. $m\left\{\sum_{j=1}^{r} a_je_j; |a_j| < 1\right\} = \pi^d$, and $m(\varepsilon) = \pi^d/d!$.

The *Fischer–Fock inner product* on $\mathcal{P}$ is defined by

$$(3.17) \qquad (p,q)_F = \frac{1}{\pi^d}\int_{\mathbb{C}^d} p(z)\overline{q(z)}e^{-|z|^2}\,dm(z).$$

If $z^\alpha = \prod_{j=1}^{d} z_j^{\alpha_j}$ (with respect to any choice of coordinates) then

$$(3.18) \qquad (z^\alpha, z^\beta)_F = \delta_{\alpha,\beta}\alpha!$$

where $\alpha! = \prod_{j=1}^{d} \alpha_j!$.

For every $p \in \mathcal{P}$ let $\partial_p = p(\partial)$ be the corresponding differential operator with constant coefficients. If $p(z) = \sum_\alpha a_\alpha z^\alpha$, then $\partial_p = \sum_\alpha a_\alpha D^\alpha$, where $D^\alpha = \prod_{j=1}^{d}\left(\frac{\partial}{\partial z_j}\right)^{\alpha_j}$. If $p \in \mathcal{P}$ with $p(z) = \sum_\alpha a_\alpha z^\alpha$, we define $p* \in P$ by $p*(z) = \sum_\alpha \bar{a}_\alpha z^\alpha$. If Z is a JB*-algebra, then $p*(z) = \overline{p(z*)}$ where $z* = \{e,z,e\}$. Similar formula in the general case is based on the theory of *Jordan pairs* (see [L1]).

The Fischer–Fock inner product admits a differential form:

(3.19) $(p,q)_F = \partial_p(q*)(0)$; $p,q \in SP$.

To prove this just check that $\partial_{z^\alpha}(z^\beta)(0) = \delta_{a,\beta} \, \alpha \, !$.

It is easy to verify that the inner product $(\cdot,\cdot)_F$ *is unitarily invariant*

(3.20) $(p \circ u, q \circ u)_F = (p,q)_F$

for all $p,q \in \mathcal{P}$ and a unitary map u of $(\mathbb{C}^d, E)$. In particular, $(\cdot,\cdot)_F$ *is K–invariant*.

Since point evaluations are continuous linear functionals with respect to $(\cdot,\cdot)_F$, each signature space P_m has a reproducing kernel, denoted by $K^m(z,w)$. By the K–invariance of $(\cdot,\cdot)_F$ we get for all $z,w \in D$ and $k \in K$

(3.21) $K^m(k(z), k(w)) = K^m(z,w)$.

In particular, $K^m(\cdot,e)$ is L–invariant and thus $K^m(\cdot,e) = c_m \varphi_m$. The determination of c_m will be described in the next section.

We conclude this section with a result based on the K–irreducibility of the P_m's and on Schur's lemma in representation theory. Recall that a semi inner product is a sequi–linear hermitian form which is non–negative.

Proposition 3.4: *Let $(\cdot,\cdot)$ be a K–invariant semi inner product on $\mathcal{P}$. Then:*
(i) $\{P_m\}$ are pairwise $(\cdot,\cdot)$-orthogonal.
(ii) For every signature $\underline{m}$ there is a constant $a_m \geq 0$ so that

$$(p,q) = a_m(p,q)_F \quad \text{for all} \quad p,q \in P_m .$$

(iii) $(\cdot,\cdot)$ is an inner product if and only if $a_m > 0$ for all $\underline{m}$.
(iv) If, moreover, evaluations at points of D are continuous linear
    functionals on $(\mathcal{P}, (\cdot,\cdot))$ then the completion $\mathcal{H}$ of $(\mathcal{P}, (\cdot,\cdot))$ has
    a reproducing kernel $L(z,w)$ with the property that the projection
    of $L(\cdot,w)$ on P_m is $a_m^{-1}K^m(\cdot,w)$. Thus

$$L(z,w) = \sum_{\underline{m}} a_m^{-1} \, K^m(z,w)$$

where the convergence is both pointwise, uniformly on compact sub-
sets of $D \times D$ and in norm.

§4. The determination of the Wallach set and the study of the composition series

Fix a Cartan domain D in $\mathbb{C}^d$ of type (r, a, b) and genus g. For a signature $\underline{m}$ and $\lambda \in \mathbb{C}$, let

$$(4.1) \qquad (\lambda)_m = \prod_{j=1}^{r} \prod_{\ell=0}^{m_j-1} \left(\lambda + \ell - (j-1)\frac{a}{2}\right) = \prod_{j=1}^{r} \left(\lambda - (j-1)\frac{a}{2}\right)_{m_j}.$$

The following basic result was proved by several authors, see [Hu] and [0] for special cases, and [FK1] and [La] for the general case. Our presentation follows that of [FK1].

Theorem 4.1: *For every* $\lambda \in \mathbb{C}$ *and* $z, w \in D$

$$(4.2) \qquad h(z,w)^{-\lambda} = K(z,w)^{\lambda/g} = \sum_{m} (\lambda)_m K^m(z,w),$$

where the summation ranges over all signatures. The series converges absolutely and uniformly on compact subsets of $D \times \overline{D}$.

The proof of this result in [FK1] requires the following two lemmas. Here φ_m is the spherical function associated with the signature m and $d_m = \dim(P_m)$.

Lemma 4.2 [FK1, Theorem 3.4]: *For each signature* m.

$$(4.3) \qquad \|\varphi_m\|_F^2 = \frac{(d/r)_m}{d_m}$$

In particular,

$$K^m(z,e) = \frac{\varphi_m(z)}{\|\varphi_m\|_F^2} = \frac{d_m}{(d/r)_m} \varphi_m(z).$$

Lemma 4.3 (FK1, Theorem 3.6]: *For* $\lambda > g-1$, *the integral*

$$(4.4) \qquad c(\lambda)^{-1} := \int_D K(z,z)^{1-\lambda/g} dV(z)$$

is finite. Define $d\mu_\lambda(z) = c(\lambda)K z,z)^{1-\lambda/g} dV(z)$ *and let* $(\cdot,\cdot)_\lambda$ *and* $\|\cdot\|_\lambda$ *denote the inner product and the norm in* $L^2(\mu_\lambda)$. *Then for any signature* m,

$$(4.5) \qquad \|\varphi_m\|_\lambda^2 = \frac{1}{d_m} \frac{(d/r)_m}{(\lambda)_m}.$$

These important facts are proved by *integration in polar coordinates*, a technique developed in [Hu] for the classical Cartan domains and in [He, Chapter I.5] and [FK2, Chapter VI] in greater generality.

In the context of Z, the formula of integration in polar coordinates is
$$\int_Z f(z) dm(z) =$$
$$c \int_0^\infty \cdots \int_0^\infty \int_K f\left(k(\sum_{j=1}^{r} t_j e_j)\right) dk 2^r \prod_{j=1}^{r} t_j^{2b+1} \prod_{1 \le j < \ell \le r} \left|t_j^2 - t_\ell^2\right|^a dt_1 \cdots dt_r,$$

where c is a positive constant whose exact value is not needed below. For instance, the integral (4.4) becomes

$$\int_0^1 \cdots \int_0^1 \prod_{j=1}^r (1-s_j)^{\lambda-g} \prod_{j=1}^r s_j^b \prod_{1 \leq j < \ell \leq r} |s_j - s_\ell|^a \, ds_1 \cdots ds_r ,$$

and this integral is finite if and only if $\lambda > g - 1$.

There is a similar formula of integration in polar coordinates over the positive cone Ω of the JB*-algebra $Z_2(e)$. The actual computations in Lemmas 4.2 and 4.3 involve the *Gindikin gamma function* of Ω, defined in [Gi1] for $\underline{s} = (s_1, \cdots, s_r) \in \mathbb{C}^r$ by

$$\Gamma_\Omega(\underline{s}) = \int_\Omega e^{-(x,e)} N_{\underline{s}-d/r+b}(x) \, dx.$$

It is known [Gi1] that

$$\int_\Omega e^{-(x,y)} N_{\underline{s}-\frac{d}{r}+b}(x) dx = \Gamma_\Omega(\underline{s}) N_{\underline{s}}(x^{-1}), \quad x \in \Omega$$

and

$$\Gamma_\Omega(\underline{s}) = (2\pi)^{\frac{r(r-1)a}{4}} \prod_{j=1}^r \Gamma\left(s_j - (j-1)\frac{a}{2}\right) .$$

Thus

$$(\lambda)_{\underline{m}} = \frac{\Gamma_\Omega(\lambda+\underline{m})}{\Gamma_\Omega(\lambda)} .$$

For $\lambda > g - 1$ (4.2) follows from Proposition 3.3, Lemma 4.2, and Lemma 4.3. The proof of the convergence of the series in (4.2) for general λ is straightforward. Thus the series is analytic in z and conjugate-analytic in w. $K(z,w)^{\lambda/g}$ is also analytic in λ since $K(z,w) \neq 0$. Analytic continuation completes the proof of Theorem 4.1.

Corollary 4.4: ([W], [RV], [FK], [B], [Sh]) *The Wallach set of D is*

$$(4.6) \qquad W(D) = \left\{\frac{ja}{2}\right\}_{j=0}^{r-1} \cup \left((r-1)\frac{a}{2}, \infty\right)$$

Proof. By definition, a complex number λ belongs to $W(D)$ if and only if $K(z,w)^{\lambda/g}$ is a positive definite function of z,w. For each such λ, let $\mathcal{H}_\lambda$ be the associated Hilbert space of analytic functions on D constructed in section 1 from span $\left\{K_z^{\lambda/g}; z \in D\right\}$, with $\left(K_w^{\lambda/g}, K_z^{\lambda/g}\right)_\lambda = K(z,w)^{\lambda/g}$. It is clear by K-invariance of $K(z,w)^{\lambda/g}$ and by Theorem 4.1 that $\mathcal{H}_\lambda = \sum_{\underline{m}} \oplus P_{\underline{m}}$ is an orthogonal decomposition and that the reproducing kernel of $(P_{\underline{m}}, (\cdot \cdot)_\lambda)$ is $(\lambda)_{\underline{m}} K^{\underline{m}}(z,w)$. Thus $K(z,w)^{\lambda/g}$ is positive definite if and only if $(\lambda)_{\underline{m}} \geq 0$ for all $\underline{m}$. By (4.1), this holds if and only if

$$\lambda \in \left\{ j\frac{a}{2} \right\}_{j=0}^{r-1} \cup \left((r-1)\frac{a}{2}, \infty \right).$$ □

Definition 4.5: *The discrete part of the Wallach set* W(D) *is*

(4.7) $W_d(D) = \left\{ j\,\dfrac{a}{2} \right\}_{j=0}^{r-1}.$

The *continuous part of the Wallach set* W(D) *is*

(4.8) $W_c(D) = \left((4-1)\dfrac{a}{2}, \infty \right)$

   Berezin [B] found the Wallach set W(D) for the classical Cartan domain by different methods. His theory of quantization over bounded symmetric domains uses the one–parameter family of Hilbert spaces $\{\mathcal{H}_\lambda; \lambda \in W(D)\}$, where the parameter λ plays the role of c/h, h being Planck's constant. It is a crucial fact in this theory that one can let h approach zero through an interval $(0, h_0)$.

Corollary 4.6: (i) *If* $\lambda \in W_c(D)$, *then* $\mathcal{P}$ *is dense in* $\mathcal{H}_\lambda$ *and* $\mathcal{H}_\lambda = \sum_{\mathbf{m}} \oplus P_{\mathbf{m}}.$

(ii)   *If* $\lambda \in W_{d(D)}$, *say* $\lambda = j\dfrac{a}{2}$ *for some* $0 \leq j \leq r-1$, *then*

(4.9) $K(z,w)^{\lambda/g} = \displaystyle\sum_{\substack{m_1 \geq m_2 \geq \cdots \geq m_j \geq 0 \\ m_{j+1} = \cdots = m_r = 0}} (\lambda)_{\mathbf{m}} K^{\mathbf{m}}(z,w)$

and

(4.10) $\mathcal{H} = \displaystyle\sum_{\substack{m_1 \geq m_2 \geq \cdots \geq m_j \geq 0 \\ m_{j+1} = \cdots = m_r = 0}} \oplus P_{\mathbf{m}}$

(iii)   *For any* $\lambda \in W(D)$ *and* $f, g \in \mathcal{H}_\lambda$ *with orthogonal decompositions*

$f = \displaystyle\sum_{\mathbf{m}} f_{\mathbf{m}}$, $g = \displaystyle\sum_{\mathbf{m}} g_{\mathbf{m}}$ *where* $(f_{\mathbf{m}}, g_{\mathbf{m}} \in P_{\mathbf{m}})$,

(4.11) $(f,g)_\lambda = \displaystyle\sum_{\mathbf{m}} \dfrac{(f_{\mathbf{m}}, g_{\mathbf{m}})_F}{(\lambda)_{\mathbf{m}}} \,;$

(iv) $\|f\|_\lambda = 0$ *if and only if* $f = 0.$

Remark: If D is of tube type, then $\mathcal{H}_{(r-1)\frac{a}{2}}$ is the completion of the space H(D) of harmonic polynomials (see Section 3); this follows from (4.10). As we mentioned in Section 1, $\mathcal{H}_{d/r}$ is the *Hardy space* $H^2(S)$, the closure of $\mathcal{P}$ in $L^2(S,\sigma)$. The *Szegö–kernel* $S_z(\xi) = S(\xi,z)$ $(z \in D,\ \xi \in S)$ is its reproducing kernel:

$$f(z) = (f, S_z)_{L^2(S)} = \int_S f(\xi) S(z,\xi) d\sigma(\xi), \qquad f \in H^2(S), \quad z \in D.$$

It follows from [FK1, Lemma 3.3] that for all $f, g \in H^2(S)$

$$(4.12) \qquad (f,g)_{L^2(S)} = \sum_{\underline{m}} \frac{(f_{\underline{m}}, g_{\underline{m}})_F}{\left(\frac{d}{r}\right)_{\underline{m}}}$$

and that

$$(4.13) \qquad S(\xi, z) = K(\xi, z)^{\frac{d}{rg}} = h(\xi, z)^{-\frac{d}{r}}; \quad z \in D, \ \xi \in S.$$

Let us denote by $\mathbb{P}(D)$ the set of all $\lambda \in \mathbb{C}$ for which $(\lambda)_{\underline{m}} = 0$ for at least one signature $\underline{m}$. It follows from the definition (4.1) of $(\lambda)_{\underline{m}}$ that

$$(4.14) \qquad \mathbb{P}(D) = \bigcup_{j=0}^{r-1} \left(j\frac{a}{2} - \mathbb{N}\right)$$

where $\mathbb{N} = \{0, 1, 2, \cdots\}$, and that $\mathbb{P}(D)$ coincides with the set of poles of the function $\Gamma_\Omega(\lambda)$.

We use (4.11) to define a sesqui-linear form $(\cdot, \cdot)_\lambda$ on $\mathcal{P}$ for every $\lambda \in \mathbb{C} \backslash \mathbb{P}(D)$. For fixed $f, g \in \mathcal{P}$ the map $\lambda \mapsto (f,g)_\lambda$ is a meromorphic function whose poles are contained in $\mathbb{P}(D)$. For a signature $\underline{m}$ and $\lambda \in \mathbb{P}(D)$ let $q_{\underline{m}}(\lambda)$ be the order of λ as a zero of the map $\xi \mapsto (\xi)_{\underline{m}}$, and let $q(\lambda) = \sup_{\underline{m}} q_{\underline{m}}(\lambda)$. Notice that by (4.14) $1 \le q(\lambda) \le r$. For $\lambda \notin \mathbb{P}(D)$, we define $q(\lambda) = 0$.

Recall that for $\lambda \in \mathbb{C}$ and $\varphi \in G = (\text{Aut}(D))_0$, $U^{(\lambda)}(\varphi)$ is the operator defined by $U^{(\lambda)}(\varphi)f = (f \circ \varphi) \cdot (J\varphi)^{\lambda/g}$. We denote

$$\mathcal{P}^{(\lambda)} = \text{span}\{U^{(\lambda)}(\varphi)\mathcal{P} ; \varphi \in G\}$$

Definition 4.7 : Let $\lambda \in \mathbb{C}$ and let $0 \le j \le q(\lambda)$. Set

$$S_j^{(\lambda)} = \{\underline{m}; q_{\underline{m}}(\lambda) \le j\}$$

$$M_j^{(\lambda)} = \left\{f \in \mathcal{P}^{(\lambda)}; f = \sum_{\underline{m} \in S_j^{(\lambda)}} f_{\underline{m}}\right\}$$

Clearly, $M_j^{(\lambda)}$ consists of all those $f \in \mathcal{P}^{(\lambda)}$ so that for every $h \in \mathcal{P}$ the map $\xi \mapsto (f,h)_\xi$ has a pole at $\xi = \lambda$ of order at most j.

Theorem 4.8 [FK1, Theorems 5.3 and 5.4]: *Let $\lambda \in \mathbb{C}$. Then:*

(i) $M_{-1}^{(\lambda)} := \{0\} \subset M_0^{(\lambda)} \subset M_1^{(\lambda)} \subset \cdots \subset M_{q(\lambda)}^{(\lambda)} = \mathcal{P}^{(\lambda)}$

(ii)   *The spaces $\{M_j^{(\lambda)}\}_{j=0}^{q(\lambda)}$ are $U^{(\lambda)}$-invariant and every non-zero, $U^{(\lambda)}$-invariant subspace of $\mathcal{P}^{(\lambda)}$ is one of the $\{M_j^{(\lambda)}\}_{j=0}^{q(\lambda)}$.*

(iii)   The quotients $M_j^{(\lambda)}/M_{j-1}^{(\lambda)}$, $0 \le j \le q(\lambda)$, are irreducible;

(iv) $(f,h)_{\lambda,j} := \lim_{\xi \to \lambda} (f,h)_\xi (\xi - \lambda)^j$ is a $U^{(\lambda)}$-invariant hermitian form

on $M_j^{(\lambda)}/M_{j-1}^{(\lambda)}$, $0 \le j \le q(\lambda)$.

(v) $M_j^{(\lambda)}/M_{j-1}^{(\lambda)}$ is unitarizable (i.e. $(\cdot,\cdot)_{\lambda,j}$ is either positive definite or negative definite) if and only if either $j = 0$ and $\lambda \in W(D)$ or $j = q(\lambda)$ and $(r-1)\frac{a}{2} - \lambda \in \mathbb{N}$. In the latter case the quotient $M_{q(\lambda)}^{(\lambda)}/M_{q(\lambda)-1}^{(\lambda)}$ is equivalent to $M_0^{(g-\lambda)}$.

The proofs in [FK1] use the analytic continuation of the action $u^{(\lambda)}$ of the Lie algebra aut(D) of Aut(D) (or, of $G = \text{Aut}(D)_0$). The elements of aut(D) are *complete holomorphic vector fields*, i.e. generators of one-parameter subgroups of G. If $\{\varphi_t\}_{t \in \mathbb{R}}$ is such a subgroup and $X(z) = \frac{\partial}{\partial t} \varphi_t(z)|_{t=0}$ is its generator then for every $f \in \mathcal{P}^{(\lambda)}$,

$$(u^{(\lambda)}(X)f)(z) = f'(z)X(z) + \frac{\lambda}{g}f(z) \text{ trace } X'(z), \quad z \in D.$$

Moreover, $u^{(\lambda)}(X)$ is a $(\cdot,\cdot)_\lambda$-skew-hermitian (possibly unbounded) operator, i.e. for all $f,h \in \mathcal{P}(\lambda)$,

$$(u^{(\lambda)}(X)f,h)_\lambda + (f,u^{(\lambda)}(X)h)_\lambda = 0.$$

Equivalently, one can continue analytically the $\tilde{U}^{(\lambda)}$-action of $\tilde{G}$ and use the invariance

$$(U^{(\lambda)}(\varphi)f,h)_\lambda = (f,U^{(\bar\lambda)}(\varphi^{-1})h)_\lambda$$

for $\lambda \in \mathbb{C}\backslash P(D)$, $\varphi \in G$ and $f,h \in \mathcal{P}^{(\lambda)}$.

The proofs in [FK1] cover all of Theorem 4.8 except for the assertion that *every* $U^{(\lambda)}$-invariant subspace is one of the spaces in (i). To prove this, let $X \subseteq \mathcal{P}^{(\lambda)}$ be a $U^{(\lambda)}$-invariant subspace. By the K-invariance,

$$X = \sum_{m \in A} \oplus P_m$$

for a certain subset A of signatures. Let

$$A_\ell = A \cap (S^{(\lambda)}\backslash S_{\ell-1}^{(\lambda)}), \qquad \ell = 0,1,2,\dots, q(\lambda),$$

and let $j = \max\{\ell; A_\ell \ne \emptyset\}$. By (iii) of Theorem 4.8 and the $U^{(\lambda)}$-invariance of X, $X/M_{j-1}^{(\lambda)} = M_j^{(\lambda)}/M_{j-1}^{(\lambda)}$. From this it follows that $A_\ell = S_\ell^{(\lambda)} \backslash S_{\ell-1}^{(\lambda)}$, $1 \le \ell \le j$. Hence $X = M_j^{(\lambda)}$.

Definition 4.9: Let $\lambda \in P(D)$. The sequence

$$\{0\} = M_{-1}^{(\lambda)} \subset M_0^{(\lambda)} \subset M_1^{(\lambda)} \subset \dots \subset M_{q(\lambda)}^{(\lambda)} = \mathcal{P}^{(\lambda)}$$

is called the *composition series* of $\mathcal{P}^{(\lambda)}$. If $\lambda \in \mathbb{C}\backslash P(D)$, then $q(\lambda) = 0$ and the composition series is trivial: $\{0\} = M_{-1}^{(\lambda)} \subset M_0^{(\lambda)} = \mathcal{P}^{(\lambda)}$.

We denote by $\mathcal{H}_{\lambda,j}$ the completion of $M_j^{(\lambda)}/M_{j-1}^{(\lambda)}$ with respect to $(\cdot,\cdot)_{\lambda,j}$, where λ and j are as in Theorem 4.8. Note that $\mathcal{H}_\lambda = \mathcal{H}_{\lambda,0}$ for all $\lambda \in W(D)$.

The sets $S_j^{(\lambda)}$ and $M_j^{(\lambda)}$ ($\lambda \in P(D)$, $0 \le j \le q(\lambda)$) admit explicit descriptions. The following proposition and its corollary provide a very detailed information on their structure in the case where $\lambda \in W_d(D)$, which is our main concern. We omit the straightforward proofs as well as the formulation of the analogous results for $\lambda \in P(D) \backslash W(D)$.

Proposition 4.10: Let $\lambda \in W_d(D)$, $\lambda = \dfrac{\ell a}{2}$, $0 \le \ell \le r - 1$.

(i) *Suppose that* a *is even. Then* $q(\lambda) = r - \ell$, *and for* $0 \le j \le r - \ell$

$$M_j^{(\lambda)} = \sum_{m_{\ell+1+j} \le ja/2} \oplus P_m .$$

(ii) *If* D *is the Cartan domain of type* III_n *then* $a = 1$, $q(\lambda) = [(n-\ell-1)/2] + 1$, *and*

$$M_j^{(\lambda)} = \sum_{m_{\ell+1+2j} \le j} \oplus P_m . \qquad 0 \le j \le q(\lambda) .$$

(iii) *If* D *is the Cartan domain of type* IV_n *for* n *odd, then* $a = n-2$, $q(\lambda) = 1$ *for* $\ell = 0, 1$ *and*

$$M_0^{(0)} = \mathbb{C}1, \quad M_1^{(0)} = \mathcal{P}^{(0)}, \quad M_0^{(a/2)} = \sum_{m_2 = 0} \oplus P_m, \quad M_1^{(a/2)} = \mathcal{P}^{(a/2)}$$

Recall that $N = N_r$ is the (highest) determinant polynomial.

Corollary 4.11: *Let* $\lambda \in W_d(D)$, $0 \le j \le q(\lambda)$, *and let* $A_{j,\lambda}$ *denote the set of all* $s \in \mathbb{N}$ *so that* $N^s \in M_j^{(\lambda)} \backslash M_{j-1}^{(\lambda)}$. *Then*

(i) $s \in A_{j,\lambda}$ *if and only if* span $\{U^{(\lambda)}(\varphi)N^s; \varphi \in G\} = M_j^{(\lambda)}$.

(ii) *If* a *is even then* $A_{j,\lambda} = ((j-1)\frac{a}{2} , j\frac{a}{2}] \cap \mathbb{N}$ *for* $0 \le j < q(\lambda)$

and $A_{q(\lambda),\lambda} = (q(\lambda)\frac{a}{2}, \infty) \cap \mathbb{N}$;

(iii) *If* D *is of type* III_n *then* $A_{j,\lambda} = \{j\}$ *for* $0 \le j < q(\lambda)$ *and* $A_{j(\lambda),\lambda} = [q(\lambda), \infty) \cap \mathbb{N}$;

(iv) *If* D *is of type* IV_n *for* n *odd then* $A_{0,\lambda} = \{0\}$ *and* $A_{1,\lambda} = (1, 2, 3, \ldots)$.

Corollary 4.11 exhibits the spaces $M_j^{(\lambda)}$ of the composition series associated with $\lambda \in W_d(D)$ as $U^{(\lambda)}$-orbits of the polynomials N^s for appropriate choices of $s \in \mathbb{N}$. We use this result below to show that the $M_j^{(\lambda)}$ are kernels of inavariant differential operators.

Fix $\lambda \in W_d(D)$, $0 \le j \le q(\lambda)$ and $s \in A_{j,\lambda}$. Define a differential oeprator $D_s^{(\lambda)}$ from analytic functions on D to real analytic functions on G by

(4.15) $$D_s^{(\lambda)}(f)(\varphi) = \partial_{N^s}(U^{(\lambda)}(\varphi)f)(0) = (U^{(\lambda)}(\varphi)f, N^s)_F$$

$D_s^{(\lambda)}$ is $U^{(\lambda)}$-invariant in the sense that

(4.16) $$D_s^{(\lambda)}(U^{(\lambda)}(\psi)f)(\varphi) = c(\varphi,\psi,\lambda)D_s^{(\lambda)}(f)(\psi\varphi)$$

for all $\varphi, \psi \in G$ and analytic function f on D. Here $c(\varphi,\psi,\lambda)$ is the unimodular constant appearing in (1.13); it can be avoided if one defines $D_s^{(\lambda)}$ is the context of the representation $\tilde{U}^{(\lambda)}$ of covering group $\tilde{G}$. It follows that ker $D_s^{(\lambda)}$ is a $U^{(\lambda)}$-invariant subspace. Precisely,

Proposition 4.12: *Let* $\lambda \in W_d(D)$, $0 \le j \le q(\lambda)$ *and* $s \in A_{j,\lambda}$. *Then*

$$\ker D_s^{(\lambda)} = M_{j-1}^{(\lambda)} \ .$$

Indeed, $D_s^{(\lambda)}(f) = 0$ if and only if N^s is orthogonal to span $\{U^{(\lambda)}(\varphi)f;\ \varphi \in G\}$ in the Fischer inner product. Since the latter space is one of the $M_\ell^{(\lambda)}$'s the proposition follows. Notice that in case $j = 0$, Proposition 4.12 reduces to the fact that the operator $D_0^{(\lambda)}(f)(\varphi) = f(\varphi(0))J\varphi(0)^{\lambda/g}$ is one-to-one.

We remark that Proposition 4.12 extends to $\lambda \in \mathbb{P}(D) \backslash W(D)$. This follows from the variant of Proposition 4.10 and Corollary 4.11 for $\lambda \in \mathbb{P}(P) \backslash W(D)$.

§ 5. Uniqueness of the invariant inner products

In this section we show that for $\lambda \in \mathbb{P}(D)$ the canonical $U^{(\lambda)}$-invariant Hilbert spaces $\mathcal{H}_{\lambda,0}$ and $\mathcal{H}_{\lambda,q(\lambda)}$ (in case $(r-1)\frac{a}{2} - \lambda \in \mathbb{N}$) are essentially unique. The uniqueness property has two variants: isometric and isomorphic. In the isometric version the operators $\{U^{(\lambda)}(\varphi);\ \varphi \in G\}$ are assumed to be isometries, while in the isomorphic version one assumes that they are only uniformly bounded. The isometric uniqueness is somewhat stronger than the irreducibility of the spaces $\mathcal{H}_{\lambda,0}$ and $\mathcal{H}_{\lambda,q(\lambda)}$; its proof is basically a simple argument involving cyclic vectors. The isomorphic uniqueness is a stronger property and is harder to prove. In the natural attempt to reduce it to the isometric uniqueness by averaging with respect to an invariant mean on an amenable and transitive subgroup (like P, the minimal parabolic subgroup of G) one has to give up the assumption of the orthogonality of the Peter–Weyl decomposition.

The isomorphic uniqueness in the case of the unit disk (d = 1) is established in [AF1] for $\lambda = 0$ and in [AF2] for general $\lambda \in W(D) = [0,\infty)$. The isometric uniqueness in the context of the unit disk and $\lambda \in -N$ is established in [P4]. The isometric uniqueness in the context of the unit ball and $\lambda = 0$ is proved in both [P1] and [Zhu1]. Isometric uniqueness in the case of the type I Cartan domain is proved in [P2]. The uniqueness results in the context of a general Cartan domain D and a general $\lambda \in W(D)$ are established in [AF3] and presented below.

Theorem 5.1 [AF3, Th. 3]: *Let* $\lambda \in W(D)$. *Let* H $\neq$ {0} *be a Hilbert space of analytic functions on* D *such that :*

(1) *For every* $\varphi \in G$, $U^{(\lambda)}(\varphi)$ *acts on* H *as a bounded operator, and*

(5.1)
$$\gamma := \sup_{\varphi \in G} \|U^{(\lambda)}(\varphi)\|_{B(H)} < \infty ;$$

(2) *For any finite Borel measure* ν *on* K *the operator*

(5.2)
$$(T_\nu f)(z) = \int_K f(kz) d\nu(k)$$

maps H *into itself continuously, and the integral converges in the norm of* H.

       Then H $= \mathcal{H}_{\lambda,0}$ *with equivalent norms. Mroeover, if* $\gamma = 1$ *(i.e.* $U^{(\lambda)}(\varphi)$ *is a unitary operator on* H *for all* $\varphi \in G$) *then the norm of* H *is proportional to that of* $\mathcal{H}_{\lambda,.0}$.

Proof. A variant of (5.2): $f(0) = \int_0^{2\pi} f(e^{i\theta}z)\dfrac{d\theta}{2\pi}$, yields $1 \in$ H and

(5.3)
$$|f(0)|\,\|1\|_H \leq \gamma\|f\|_H , \qquad f \in H .$$

Since H consists of functions (not equivalency classes) and $\|\cdot\|_H$ is a norm, $\|1\|_H > 0$. Thus

(5.4)
$$|f(0)| \leq \gamma\|1\|_H^{-1}\|f\|_H .$$

Using (5.4) with $f \circ \varphi_a$ instead of f we get by (5.1)

(5.5)
$$|f(a)| \leq (\gamma^2\|1\|_H^{-1}|J\varphi_a(0)|^{-\lambda/g})\|f\|_H , \qquad a \in D .$$

thus point evaluations are continuous linear functionals on H.

       Recall (see §2) that the minimal parabolic subgroup P of G is amenable and transitive on D. Let m be an invariant mean on P, and define

(5.6)
$$\langle f,h \rangle_H = \underset{\varphi \in P}{m} \ (U^{(\lambda)}(\varphi)f, U^{(\lambda)}(\varphi)h)_H .$$

Then $\langle \cdot,\cdot \rangle_H$ is an equivalent inner product on H, i.e.

(5.7)
$$\gamma^{-2}\langle f,f \rangle_H \leq (f,f)_H \leq \gamma^2\langle f,f \rangle_H , \quad f \in H,$$

Moreover, $\langle \cdot, \cdot \rangle_H$ is $U^{(\lambda)}$–P–invariant, namely :

(5.8) $$\langle U^{(\lambda)}(\varphi)f, \ U^{(\lambda)}(\varphi)h \rangle_H = \langle f,h \rangle_H$$

for all $f,h \equiv H$ and $\varphi \in P$. Let $L(z,w)$ be the reproducing kernel of $(H, \langle \cdot, \cdot \rangle_H)$. Clearly, (5.8) implies

(5.9) $$L(\varphi(z), \ \varphi(w)) \ J\varphi(z)^{\lambda/g} J\overline{\varphi}(w)^{\lambda/g} = L(z,w),$$

for all $z,w \in D$ and $\varphi \in P$. Define

(5.10) $$A(z,w) = L(z,w) \ / \ K(z,w)^{\lambda/g} ; \ \ z,w \in D.$$

It follows by (5.9) and (1.14) that

(5.11) $$A(\varphi(z), \ \varphi(w)) = A(z,w)$$

for all $z,w \in D$ and $\varphi \in P$. Thus, by the transitivity of P on D,

(5.12) $$c := A(0,0) = A(z,z), \qquad z \in D.$$

Since $A(z,w)$ is analytic in z and conjugate–analytic in w, an easy power series argument yields

(5.13) $$A(z,w) = c \ \ \text{for all } z,w \in D.$$

Thus

(5.14) $$L(z,w) = cK(z,w)^{\lambda/g} \qquad \text{for all } \ z,w \in D.$$

Clearly $c > 0$, since $\langle 1,1 \rangle_H > 0$. Now, $H^{(0)} := \text{span}\{L(\cdot,w); \ w \in D\}$ is dense in H (since $f(w) = \langle f, L(\cdot, w) \rangle_H$, $w \in D$, and so $\{L(\cdot, w); \ w \in D\}^\perp = \{0\}$), and for $f,h \in H^{(0)}$

(5.15) $$\langle f,h \rangle_H = c(f,h)_\lambda .$$

From this it follows that $H = \mathcal{H}_{\lambda,0}$ and that (5.15) holds for all $f,h \in H$. □

Remark: The proof of Theorem 5.1 uses some ideas from [Kob1], [Kob2], and [Ku].

Recall that a *semi–Hilbert space* is a linear space H with a sesqui-linear form $(\cdot, \cdot)$ which satisfy all axioms of an inner product except that $(f,f) = 0$ need not imply $f = 0$, and so that $H/\{f; (f,f) = 0\}$ is complete.

Theorem 5.2: *Let* $\lambda \in \mathbb{P}(D)$ *and let* H *be a semi–Hilbert space of analytic functions on* D *so that* $\|1\|_H = 0$ *but that* $\|f\|_H$ *does not vanish for all* $f \in H$. *Assume that conditions (1) and (2) of Theorem 5.1 hold, with* $\gamma = 1$, *that is* $U^{(\lambda)}(\varphi)$ *is a unitary operator on* H *for every* $\varphi \in G$. *Then* $(r-1)\frac{a}{2} - \lambda \in \mathbb{N}$ *and* $H = \mathcal{H}_{\lambda,q(\lambda)}$ *with proportional seminorms.*

The special case where $\lambda \in W_d(D)$ is [AF3, Th. 2].

Proof: Since K acts on H isometrically by composition, it is clear that there is a subset B of signatures so that $\sum_{\underline{m} \in B} \oplus P_{\underline{m}}$ is dense in H. By Theorem 4.8 (ii) and the fact that H is invariant under the isometric $U^{(\lambda)}$-action of G, we find that there is a j, $0 \leq j \leq q(\lambda)$, so that $B = S_j^{(\lambda)}$. Hence $M_j^{(\lambda)}$ is dense in H. The orthogonality of the spaces $\{P_{\underline{m}}\}_{\underline{m} \in S_j^{(\lambda)}}$ in H implies that for any fixed $\underline{m} \in S_j^{(\lambda)}$

$$(5.16) \qquad (f, h)_H = (f_{\underline{m}}, h)_H$$

for all $f \in M_j^{(\lambda)}$ and $h \in P_{\underline{m}}$, where $f_{\underline{m}}$ is the component of f in $P_{\underline{m}}$. Fix $\underline{m} \in S_j^{(\lambda)} \setminus S_{j-1}^{(\lambda)}$. Then $\|h\|_H > 0$ for all non-zero $h \in P_{\underline{m}}$ (otherwise $\|f\|_H = 0$ for all $f \in H$). Thus $(\cdot, \cdot)_H$ is a K-invariant inner product on $P_{\underline{m}}$. By the K-irreducibility of $P_{\underline{m}}$ we find that there exists a constant $\beta \neq 0$ so that $(f, h)_H = \beta(f, h)_{\lambda, j}$, for all $f, h \in P_{\underline{m}}$. Hence,

$$(5.17) \qquad (f, h)_H = \beta(f, h)_{\lambda, j}, \qquad f \in M_j^{(\lambda)}, h \in P_{\underline{m}}.$$

Let $\varphi, \psi \in G$ and let $0 \neq h \in P_{\underline{m}}$. Since $U^{(\lambda)}$ is an isometric action of G on both H and $M_j^{(\lambda)}$, we obtain by (5.17)

$$(U^{(\lambda)}(\varphi)h, U^{(\lambda)}(\psi)h)_H = (U^{(\lambda)}(\psi)^{-1}U^{(\lambda)}(\varphi)h, h)_H$$
$$= \beta(U^{(\lambda)}(\psi)^{-1}U^{(\lambda)}(\varphi)h, h)_H$$
$$= \beta(U^{(\lambda)}(\varphi)h, U^{(\lambda)}(\psi)h)_{\lambda, j}.$$

Since span $\{U^{(\lambda)}(\varphi)h; \varphi \in G\} = M_j^{(\lambda)}$ (by Theorem 4.8 (ii)), it follows that

$$(5.18) \qquad (f_1, f_2)_H = \beta(f_1, f_2)_{\lambda, j}$$

for all $f_1, f_2 \in M_j^{(\lambda)}$.

By assumption $(\cdot, \cdot)_H$ is non-negative definite. Thus, by Theorem 4.8 (v) either $j = 0$ or $j = q(\lambda)$ and $(r-1)\frac{a}{2} - \lambda \in \mathbb{N}$. The first alternative is impossible since $\|1\|_H = 0$. Thus, $j = q(\lambda)$, $(r-1)\frac{a}{2} - \lambda \in \mathbb{N}$ and $H = \mathcal{H}_{\lambda, q(\lambda)}$ with proportional seminorms. $\qquad \square$

The next result is the isomorphic uniqueness of the invariant inner product in the context of the unit ball D of $\mathbb{C}^n$ and $\lambda = 0$. In this case $q(0) = 1$, and by Theorem 4.8 (iv), $\mathcal{H}_{0,1}$ consists of all analytic functions $f(z) = \sum_{\alpha} f_{\alpha} z^{\alpha}$ on D so that

$$(5.19) \qquad \|f\|_{0,1}^2 := \sum_{\alpha} |\alpha| \frac{\alpha!}{|\alpha|!} |f_{\alpha}|^2$$

is finite. We use the standard multi-index notation, see [R].

Theorem 5.3 [AF3, Th. 5]: *Let H be a semi-Hilbert space of analytic functions on the unit ball D of $\mathbb{C}^n$. Assume that*

(i) $\|1\|_H = 0$ *but* $\|f\|_H$ *is not identically zero;*

(ii) *For every* $\varphi \in G$ *the composition operator* $C_\varphi(f) = U^{(0)}(\varphi)f = f \circ \varphi$ *acts continuously on* H, *and* $\gamma := \sup_{\varphi \in G} \|C_\varphi\|_{B(H)} < \infty$;

(iii) *Condition (2) of Theorem 5.1 holds.*

Then $H = \mathcal{H}_{0,1}$ *with equivalent seminorms. Moreover, if* $\gamma = 1$ *then the seminorms of* H *and* $\mathcal{H}_{0,1}$ *are proportional.*

Proof. The arguments in the proof of Theorem 5.2 show that H contains all the polynomials and that they are dense in H. As explained in Section 2, the minimal parabolic subgroup P is the isotropy group of a point in $S = \partial D$, which we may take to be $e_1 = (1, 0, \ldots, 0)$:

(5.20) $P = \{\varphi \in G; \; \varphi(e_1) = e_1\}$

By averaging with respect to an invariant mean on P as in the proof of Theorem 5.1 we can assume that $(\cdot, \cdot)_H$ is P-invariant:

(5.21) $(f_1 \circ \varphi, f_2 \circ \varphi)_H = (f_1, f_2)_H, \qquad \varphi \equiv P$

for all f_1, $f_2 \in H$. We now use the *infinitesimal version of the P-invariance*. Namely, if $\{\varphi_t\}_{t \in \mathbb{R}}$ is a one parameter subgroup of P with generator $h(z) = \frac{\partial}{\partial t}\varphi_t(z)|_{t=0}$, then

(5.22) $([f_1', h], f_2)_H + (f_1, [f_2', h])_H = 0$

for all f_1, f_2 which are analytic in a neighborhood of $\bar{D}$. Here f' is the gradient of f and $[z,w] = \sum_{j=1}^{n} z_j w_j$ is the real inner product of $\mathbb{C}^n$. Thus $[f', h](z) = \sum_{j=1}^{n}\frac{\partial f(z)}{\partial z_j}h(z)_j$. By [L2, Chapter 2] the generators of the one-parameter subgroups of G have the form

(5.23) $h(z) = Az + a - \langle z, a \rangle z$

where A is a skew-hermitian matrix and $a \in \mathbb{C}^n$. Clearly, h is a generator of a one-parameter subgroup of P if and only if (5.22) holds and $h(e_1) = 0$. Thus (5.23) holds with any $a \in \mathbb{C}^n$ and $A = A_1 + A_2$, where

(5.24) $A_1 = \begin{pmatrix} \bar{a}_1 - a_1 & \bar{a}_2 & \cdots & \bar{a}_n \\ -a_2 & & & \\ -a_n & & \bigcirc & \end{pmatrix}$, $A_2 = \begin{pmatrix} 0 & 0 \\ 0 & W \end{pmatrix}$,

and W is an arbitrary $(n-1) \times (n-1)$ skew-hermitian matrix. The rest of the proof is an application of (5.22) to various monomials f_1, f_2, where h is given by (5.23) and (5.24).   We omit the details of the straightforward computation which yields the desired result $\|f\|_H = \beta\|f\|_{0,1}$, $f \in H$. $\square$

Remark: The celebrated *similarity problem* of Dixmier asks whether every uniformly bounded representation on a Hilbert space is similar to a unitary representation. The uniqueness theorems 5.1 and 5.3 give affirmative answers in this special context.

§6. Integral formulas for the invariant inner products: the case of tube–type domains.

The invariant inner products on $\mathcal{H}_{\lambda,0}$ and $\mathcal{H}_{\lambda,q(\lambda)}$, $\lambda \in W(D)$, are given in Theorem 4.8 in terms of the Peter–Weyl expansions of the functions. It is desirable to have more concrete formulas in terms of integrals and derivatives of the function involved.

This is indeed the case for the weighted Bergman spaces and the Hardy space. If $\lambda > g - 1$, then $q(\lambda) = 0$ and $\mathcal{H}_{\lambda,0} = \mathcal{H}_{\lambda,q(\lambda)} = L^2_a(\mu_\lambda)$, with the inner product

$$(6.1) \qquad (f,h)_\lambda = \int_D f(z)\overline{h(z)}\, d\mu_\lambda(z) = c(\lambda) \int_D f(z)\overline{h(z)}\, K(z,z)^{1-\lambda/g} dV(z).$$

Also, $q(d/r) = 0$ and $\mathcal{H}_{d/r,0} = H^2(S)$ with the inner product

$$(6.2) \qquad (f,h)_{d/r} = \int_S f(\xi)\overline{g(\xi)} d\sigma(\xi)$$

Another well known example is the Dirichlet space B_2 on the unit disk $\mathbb{D}$ in $\mathbb{C}$. B_2 is the unique semi–Hilbert space of analytic functions on $\mathbb{D}$ which is invariant under the action $f \mapsto f \circ \varphi = U^{(0)}(\varphi)f$ of $G = \mathrm{Aut}(\mathbb{D})$. Thus $B_2 = \mathcal{H}_{0,1}$. The semi–norm and the inner product of B_2 admit various expressions:

(i) *Via orthogonal expansions:* $\|f\|^2_{B_2} = \sum_{m \geq 1} m|\hat{f}(m)|^2$

(ii) *Via integration on $\mathbb{D}$:*

$$(6.3) \qquad \|f\|^2_{B_2} = \int_{\mathbb{D}} |f'(z)|^2 dA(z),$$

where $dA(z) = \frac{1}{\pi} dxdy = \frac{1}{\pi} rdrd\theta$. The Dirichlet integral (6.3) is the area of $f(\mathbb{D})$, with multiplicity counted.

(iii) *Via integration on the boundary $\mathbb{T}$ of $\mathbb{D}$:*

$$(6.4) \qquad (f,h)_{B_2} = \int_{\mathbb{T}} e^{i\theta} f'(e^{i\theta})\overline{h(e^{i\theta})}\frac{d\theta}{d\pi} = (Rf,h)_{L^2(\mathbb{T})}$$

where Rf is the radial derivative of f.

(iv) *Via double integration on $\mathbb{D}$:*

(6.5)
$$\|f\|^2_{B_2} = \int_{\mathbb{D} \times \mathbb{D}} \int |f(z) - f(w)|^2 |K(z,w)|^\alpha \, d\mu_\alpha(w) \, d\mu_\alpha(z)$$

where $d\mu_\alpha(z) = (\alpha - 1)(1 - |z|^2)^{\alpha-2} \, dA(z)$ and $K(z,w) = (1 - \overline{zw})^{-2}$ is the Bergman kernel. Thus, $\|f\|_{B_2}$ is equal to the Hilbert–Schmidt norm of the Hankel operator $H_{\bar{f}}(h) = (I - P_\alpha)(\bar{f}h)$ on the weighted Bergman space $L^2_a(\mu_\alpha) = L^2(\mu_\alpha) \cap \{\text{analytic functions}\}$, $P_\alpha : L^2(\mu_\alpha) \to L^2_a(\mu_\alpha)$ being the orthogonal projection (see [AFP] and [AFJP]).

(v)   *Via double integration on* $\mathbb{T}$:

(6.6)
$$\|f\|^2_{B_2} = \int_{\mathbb{T} \times \mathbb{T}} \int \frac{|f(e^{i\theta}) - f(e^{it})|^2}{|e^{i\theta} - e^{it}|^2} \, \frac{dt}{2\pi} \frac{d\theta}{2\pi} \, .$$

Again, this expresses the fact that $\|f\|_{B_2}$ is equal to the Hilbert–Schmidt norm of the Hankel operator $H_{\bar{f}}$ on $H^2(\mathbb{T})$, see [P] for this matter.

These formulas in the case of the unit disk motivate our study in the general case. We discuss below generalizations in two different directions. The first is the generalization to a Cartan domain of tube type, established in [A1], and is presented in this section. The second is the generalization to the case of the unit ball of $\mathbb{C}^n$, $n \geq 2$ (which is the simplest non–tube Cartan domain), established in [P3] and [A2]. It is presented in section 7 below. Finally in section 8 we review the theory of a class of invariant differential operators on convex symmetric cones, which were introduced in [Y2] and [Y3], and apply it to obtain further results on integral formulas of the invariant inner products on the associated Cartan domains of tube type [Y2], [Y3] and [AU]).

The case of a tube type domain

Let D be a Cartan domain of tube type in $\mathbb{C}^d$. Let $(r,a,0)$ be the type of D and let $g = (r-1)a + 2 = \frac{2d}{r}$ be its genus. Recall that the list of the Cartan domains of tube type is $\{I_{n,n}\}_{n \geq 1}$, $\{II_{2n}\}_{n \geq 3}$, $\{III_n\}_{n \geq 2}$, $\{IV_n\}_{n \geq 5}$, and VI. Fix a frame $\{e_j\}^r_{j=1}$ of minimal orthogonal tripotents in the JB*-algebra Z $(= \mathbb{C}^d$ with D as an open unit ball), and let $e = \sum_{j=1}^{r} e_j$ be the unit of Z.

Let $\lambda \in \mathbb{P}(D)$ be so that $(r-1)\frac{a}{2} - \lambda \in \mathbb{N}$, i.e. $M^{(\lambda)}_{q(\lambda)}/M^{(\lambda)}_{q(\lambda)-1}$ is unitarizable. Recall that $\mathcal{H}_{\lambda,q(\lambda)}$ denotes the completion of $M^{(\lambda)}_{q(\lambda)}/M^{(\lambda)}_{q(\lambda)-1}$ with respect to the $U^{(\lambda)}$-invariant inner product

(6.7)
$$(f,h)_{\lambda,q(\lambda)} = \sum_{m \in S^{(\lambda)}_{q(\lambda)}} \frac{(f_{\underline{m}}, h_{\underline{m}})_F}{a_{\underline{m}}(\lambda)}$$

where for $\underline{m} \in S_{q(\lambda)}^{(\lambda)}$

(6.8)
$$a_{\underline{m}}(\lambda) = \lim_{\xi \to \lambda} \frac{(\xi)_{\underline{m}}}{\xi - \lambda^{q(\lambda)}} = \prod_{j=1}^{r} \prod_{l=0}^{m_{j-1}} {}' \left(\lambda + l - (j-1)\frac{a}{2} \right) ,$$

and product Π' ranges over all non-zero terms. Define

(6.9)
$$s = \frac{g}{2} - \lambda = \frac{d}{r} - \lambda .$$

Then s is a positive integer and

(6.10)
$$S_{q(\lambda)}^{(\lambda)} = \{\underline{m}; \, m_r \geq s\}.$$

Let $N = N_r$ be the (highest) determinant polynomial. By Proposition 4.10 and Corollary 4.11, s is the first element in $A_{\lambda, q(\lambda)}$, and in particular, span$\{U^{(\lambda)}(\varphi)N^s; \, \varphi \in G\}$ is dense in $\mathcal{H}_{\lambda, q(\lambda)}$. Recall also that $N(kz) = \chi(k)N(z)$ for all $k \in K$ and $z \in D$, where $\chi(k) \in \mathbb{T}$ for all $k \in K$ (see (3.9)).

Theorem 6.1 [A1, Theorem 12]: *Let* $\lambda \in \mathbb{P}(D)$ *be so that* $s := \frac{d}{r} - \lambda \in \mathbb{N}$. *Then for all* $f, h \in \mathcal{H}_{\lambda, q(\lambda)}$,

(6.11)
$$(f, h)_{\lambda, q(\lambda)} = \gamma_1 (N^s \partial_{N^s}(f), h)_{L^2(S)} ,$$

where γ_1 *is a constant depending on* D *and* λ.

Here S is the Shilov boundary of D and $L^2(S) = L^2(S, \sigma)$ where σ is the unique K-invariant probability measure on S.

Notice that (6.11) generalizes (6.4), where $N^s \partial_{N^s}$ reduces to the radial derivative.

The main ingredient of the proof Theorem 5.1 is the analysis of the operator ∂_N. The following result summarizes the analysis in [A1, § 2].

Lemma 6.2: *Let* f *be an analytic function in* D *with a Peter–Weyl expansion* $f = \sum_{\underline{m}} f_{\underline{m}}$. *Then*

(6.12)
$$\partial_N(f) = \sum_{\underline{m}, \, m_r \geq 1} \frac{(d/r)_{\underline{m}}}{(d/r)_{\underline{m}-1}} \frac{f_{\underline{m}}}{N} .$$

In particular

(6.13)
$$N^s \partial_{N^s}(f) = \sum_{\underline{m}, \, m_r \geq s} \frac{(d/r)_{\underline{m}}}{(d/r)_{\underline{m}-s}} f_{\underline{m}} .$$

Here, if $m_r \geq j$ then $\underline{m} - j := (m_1 - j, \, m_2 - j, \, \ldots, \, m_r - j)$ is also a signature. Also, since $P_{\underline{m}} = P_{\underline{m}-j}N^j$, $f_{\underline{m}}/N^j \in P_{\underline{m}-j}$. Thus the right hand side of (6.12) is the Peter–Weyl expansion of $\partial_N(f)$, and (6.13) follows from (6.12) by induction. Notice also that (6.12) exhibits ∂_N as a weighted shift with weights $(d/r)_{\underline{m}}/(d/r)_{\underline{m}-1}$, and $N^s \partial_{N^s}$ is a positive diagonal operator.

The proof of Lemma 6.2 is based on the fact that the adjoint of ∂_N with respect to the Fischer inner product is M_N, the operator of multiplication by N.

Theorem 6.1 yields the following:

Corollary 6.3: *Let λ be as in Theorem 6.1, and let f be an analytic function in D. Then $f \in \mathcal{H}_{\lambda, q(\lambda)}$ if and only if $(N^s \partial_{N^s})^{1/2} f \in H^2(S)$.*

To prove Theorem 6.1 from Lemma 6.2 observe first that by (4.12)

$$(6.14) \qquad (N^s \partial_{N^s} f, h)_{L^2(S)} = \sum_{m, \, m_r \geq s} \frac{(f_m, h_m)_F}{(d/r)_{m-s}} .$$

Since $\frac{d}{r} = s + \lambda$ we get by a direct computation

$$(6.15) \qquad a_m(\lambda) = \left(\frac{d}{r}\right)_{m-s} / \gamma_1 , \qquad (m_r \geq s)$$

with

$$(6.16) \qquad \gamma_1 = \prod_{j=1}^{r} \prod_{\ell=0}^{s-1}{}' \left(\lambda + 1 - (j-1)\frac{a}{2}\right) ,$$

where the product ranges over all non–zero terms. Hence (6.11) follows from (6.7), (6.10), (6.14), and (6.15).

In order to get the generalization of (6.3) we need the following result.

Theorem 6.4 *(the intertwining formula):* *Let D be a Cartan domain of tube type, with dimension d , rank r and genus $g = \frac{2d}{r}$, and let $\lambda \in \mathbb{P}(D)$. Assume that $s := \frac{d}{r} - \lambda \in \mathbb{N}$, i.e., $M_{q(\lambda)}^{(\lambda)} / M_{q(\lambda)-1}^{(\lambda)}$ is unitarizable. Then*

$$(6.17) \qquad \partial_{N^s} U^{(\lambda)}(\varphi) = U^{(g-\lambda)}(\varphi) \partial_{N^s} , \qquad \varphi \in G.$$

Thus, $\mathcal{H}_{\lambda, q(\lambda)}$ with the $U^{(\lambda)}$-action of G is equivalent to $\mathcal{H}_{g-\lambda}$ with the $U^{(g-\lambda)}$-action of G, where the intertwining operator is ∂_{N^s}.

The proof is based on the following result.

Lemma 6.5: *Let D be as above, let $\alpha > (r-1)\frac{a}{2}$ and $\ell \in \mathbb{N}$. Then for every $f \in \mathcal{H}_\alpha$*

$$(6.18) \qquad (\partial_{N^\ell} f)(z) = \frac{\Gamma_\Omega(\alpha + \ell)}{\Gamma_\Omega(\alpha)} (f, N^\ell h(\cdot, z)^{-(\alpha + \ell)})_\ell$$

Proof. Using the fact that

$$N^\ell(w) K^m(w, z) \overline{N^\ell(z)} = \frac{\left(\frac{d}{r}\right)_{m+\ell}}{\left(\frac{d}{r}\right)_m} K^{m+\ell}(w, z),$$

we get for all $w, z \in D$ (with $N(z) \neq 0$)

$$N^{\ell}(w) h(w,z)^{-(\alpha+\ell)})_{\alpha} = \sum_{n \geq 0} \frac{(\alpha+\ell)_n \left(\frac{d}{r}\right)_{n+\ell}}{\left(\frac{d}{r}\right)_n} K^{n+\ell}(w,z) \overline{N(z)}^{-\ell} .$$

Hence,

$$(f, N^{\ell} h(\cdot,z)^{-(\alpha+\ell)})_{\alpha} = \sum_{m_r \geq \ell} \frac{(\alpha+\ell)_{m-\ell} \left(\frac{d}{r}\right)_m}{(\alpha)_m \left(\frac{d}{r}\right)_{m-\ell}} f_m(z) N(z)^{-\ell}$$

$$= \frac{\Gamma_{\Omega}(\alpha)}{\Gamma_{\Omega}(\alpha+\ell)} \sum_{m_r \geq \ell} \frac{\left(\frac{d}{r}\right)_m}{\left(\frac{d}{r}\right)_{m-\ell}} f_m(z) N(z)^{-\ell}$$

$$= \frac{\Gamma_{\Omega}(\alpha)}{\Gamma_{\Omega}(\alpha+\ell)} (\partial_N^{\ell} f)(z) . \qquad \square$$

Proof of Theorem 6.4. Applying Lemma 6.5 with $\alpha = \frac{d}{r}$, we get

$$\partial_{N^s}(U^{(\lambda)}(\varphi_a)f)(z) = \frac{\Gamma_{\Omega}\left(s+\frac{d}{r}\right)}{\Gamma_{\Omega}\left(\frac{d}{r}\right)} (U^{(\lambda)}(\varphi_a)f, N^s h(\cdot,z)^{-\left(\frac{d}{r}+s\right)})_{\frac{d}{r}}$$

$$= \frac{\Gamma_{\Omega}\left(s+\frac{d}{r}\right)}{\Gamma_{\Omega}\left(\frac{d}{r}\right)} \int_S f(\varphi_a(\xi))(J\varphi_a(\xi))^{\lambda/g} N(\xi)^s h(z,\xi)^{-\left(\frac{d}{r}+s\right)} d\sigma(\xi)$$

The change of variables $\xi = \varphi_a(\eta)$, formulas (1.9), (1.14) and (3.15), and the definition of s yield

$$\partial_{N^s}\left(U^{(\lambda)}(\varphi_a)f\right)(z) = \frac{\Gamma_{\Omega}\left(s+\frac{d}{r}\right)}{\Gamma_{\Omega}\left(\frac{d}{r}\right)} (J\varphi_a(z))^{\frac{g-\lambda}{g}} \int_S f(\eta) \overline{N^{\ell}(\eta)} \, h(\varphi_a(z),\eta)^{-\left(\frac{d}{r}+s\right)} d\sigma(\eta)$$

$$= (J\varphi_a(z))^{\frac{g-\lambda}{g}} (\partial_{N^s} f)(\varphi_a(z))$$

$$= U^{(g-\lambda)}(\varphi_a)(\partial_{N^s} f)(z) .$$

Also, for $k \in K$, we have $\text{Det}(k) = \chi(h)^{\frac{d}{r}} = \chi(h)^{\frac{g}{2}}$. Hence

$$\partial_{N^s}(U^{(\lambda)}(k)f)(z) = \text{Det}(k)^{\lambda/g} \partial_{N^s}(f \circ k)(z)$$

$$= \mathrm{Det}(k)^{\frac{\lambda}{g}} \chi(k)^s \partial_{N^s}(f)(kz)$$

$$= U^{(g-\lambda)}(k)(\partial_{N^s}f)(z) \ .$$

Since $\mathrm{Aut}(D)$ is generated by K and $\{\varphi_a;\ a\in D\}$, and (6.17) holds for these generators, we get (6.17) for every $\varphi\in\mathrm{Aut}(D)$.

Next, consider

$$(6.19) \qquad \tilde{\mathcal{H}} := \{f\in\mathcal{H}_{\lambda,q(\lambda)};\ \partial_{N^s}f \in \mathcal{H}_{g-\lambda}\}$$

with the inner product

$$(f_1,f_2)_{\tilde{\mathcal{H}}} = (\partial_{N^s}f_1,\ \partial_{N^s}f_2)_{\mathcal{H}_{g-\lambda}}.$$

$\tilde{\mathcal{H}}$ is a subspace which contains N^s, and (6.17) implies that $\tilde{\mathcal{H}}$ and $(\cdot,\cdot)_{\tilde{\mathcal{H}}}$ are $U^{(\lambda)}$–invariant. Theorem 5.2 implies that $\tilde{\mathcal{H}} = \mathcal{H}_{\lambda,q(\lambda)}$. Thus ∂_{N^s} maps $\mathcal{H}_{\lambda,q(\lambda)}$ isometrically into $\mathcal{H}_{g-\lambda}$. Again, (6.17) implies that $\partial_{N^s}(\mathcal{H}_{\lambda,q(\lambda)})$ is $U^{(g-\lambda)}$–invariant; hence, Theorem 5.1 implies that it coincides with $\mathcal{H}_{g-\lambda}$. $\qquad\square$

Corollary 6.6. *Let* $\lambda\in\mathbb{P}(D)$ *and assume that* $s:= \frac{d}{r} - \lambda\in\mathbb{N}$. *Then the operator* $D_s^{(\lambda)}$ *defined by (4.15) has the form*

$$(6.20) \qquad D_s^{(\lambda)}(f)(\varphi) = \partial_{N^s}(U^{(\lambda)}(\varphi)f)(0) = (\partial_{N^s}f)(\varphi(0))(J\varphi(0))^{\frac{g-\lambda}{g}} \ .$$

In particular, $D_s^{(\lambda)}(f)(\psi_a) = (\partial_{N^s}f)(a)h(a,a)^{\frac{g-\lambda}{g}}$, *where* $\psi_a(z) = \varphi_a(-z)$ *for* $a,z\in D$.

Next, we turn to the generalization of (6.3). Let $\lambda\in\mathbb{P}(D)$ and assume that $s:= \frac{d}{r} -\lambda\in\mathbb{N}$. Let

$$(6.21) \qquad \mathcal{H}^{(\lambda)} := \left\{f\in\mathcal{H}(D);\ D_s^{(\lambda)}\in L^2(G) = L^2(G,d\varphi)\right\}$$

with the seminorm

$$(6.22) \qquad \|f\|_{\mathcal{H}^{(\lambda)}} := \left\|D_s^{(\lambda)}f\right\|_{L^2(G)} \ ,$$

where the Haar measure $d\varphi$ of G is normalized so that

$$\int_G u(\varphi(0))d\varphi = \int_D u(z)h(z,z)^{-g}dV(z)$$

holds for every $u\in L^1(D,h(z,z)^{-g}dV(z))$.

The following result generalizes (6.3) and [A1, Theorem 14].

Theorem 6.7. *Let* $\lambda\in\mathbb{P}(D)$ *and assume that* $s:= \frac{d}{r} - \lambda\in\mathbb{N}$. *Then*

(6.23) $\|f\|_{\mathcal{H}(\lambda)} = \left(\int_D |\partial_{N^s} f(z)|^2 h(z,z)^{-\lambda} dV(z) \right)^{1/2}$.

Consequently, $\mathcal{H}$ is *non-trivial if and only if* $\lambda < 1$. *In this case* $\mathcal{H}^{(\lambda)} = \mathcal{H}_{\lambda,q(\lambda)}$ *with proportional seminorms*

(6.24) $\|f\|_{\mathcal{H}(\lambda)} = \gamma_2 \|f\|_{\mathcal{H}_{\lambda,q(\lambda)}}$, $f \in \mathcal{H}_{\lambda,q(\lambda)}$

where

(6.25) $\gamma_2^2 = \left(\dfrac{d}{r} \right)_{(s,s,\dots,s)} a_{(s,s,\dots s)}(\lambda) c(g-\lambda)^{-1}$.

Notice that Theorem 6.7 says that if $D \neq D(III_n)$ and $\lambda \in W_d(D)$, then $\mathcal{H}^{(\lambda)}$ is non-trivial only if $\lambda = 0$. For $D(III_n)$ and $\lambda \in W_d(D)$, $\mathcal{H}^{(\lambda)}$ is non-trivial only if $\lambda = 0$ and is odd, or if $\lambda = \frac{1}{2}$ and n is even.

The invariance of $D_s^{(\lambda)}$, i.e. (4.16), implies that $\mathcal{H}^{(\lambda)}$ is invariant under the isometric $U^{(\lambda)}$-action of G. Next, s is the first integer in the set $A_{\lambda,q(\lambda)}$ considered in Corollary 4.11. Thus, span$\{U^{(\lambda)}(\varphi)N^s; \varphi \in G\} = \mathcal{P}^{(\lambda)}$. It follows that for $f \equiv \mathcal{H}^{(\lambda)}$, $\|f\|_{\mathcal{H}(\lambda)} = 0$ if and only if $(f,h)_{\lambda,q(\lambda)} = 0$ for all $h \in \mathcal{P}^{(\lambda)}$, i.e. $f = \sum\limits_{m \in S_{q(\lambda)-1}^{(\lambda)}} f_m$. It is also clear by Theorem 5.2 that if $\mathcal{H}^{(\lambda)}$ is non-trivial then $\mathcal{H}^{(\lambda)} = \mathcal{H}_{\lambda,q(\lambda)}$ with proportional seminorms. Finally, the non-triviality of $\mathcal{H}^{(\lambda)}$ is clearly equivalent to $N^s \in \mathcal{H}^{(\lambda)}$.

Next, (1.9) implies that for $\varphi = \varphi_z k \in \text{Aut}(D)$, $|J\varphi(0)| = |J\varphi_z(0)| = h(z,z)^{g/2}$. Thus (6.20) and (6.22) yield (6.23). Applying (6.23) to $f = N^s$, and noticing that $\partial_{N^s}(N^s)(z) \equiv \left(\dfrac{d}{r} \right)_{(s,\dots,s)}$, we conclude that $N^s \in \mathcal{H}^{(\lambda)}$ if and only if $\lambda < 1$. Finally, the proportionate constant γ_2 in (6.24) is computed by taking $f = N^s$.   □

There is also an analog of (6.5) for a general tube domain D. Let $\lambda \in \mathbb{P}(D)$ and assume that $s := \dfrac{d}{r} - \lambda \in \mathbb{N}$. Let

(6.26) $\alpha > g - 1$,

and, as before, let

(6.27) $d\mu_\alpha(z) = c(\alpha) K(z,z)^{1-\frac{\alpha}{g}} dV(z)$.

where $c(\alpha)$ is a normalization constant (making μ_α a probability measure). Let $Q^{(\lambda)}$ be the orthogonal projection on the highest quotient $M_{q(\lambda)}^{(\lambda)} / M_{q(\lambda)-1}^{(\lambda)}$, that is

(6.28) $Q^{(\lambda)}\left(\sum\limits_m f_m \right) = \sum\limits_{m \in S_{q(\lambda)}^{(\lambda)}} f_m = \sum\limits_{m, \, m_r \geq s} f_m$.

We define $\mathcal{H}^{(\lambda)}(\alpha)$ to be the space of all analytic functions f on D so that the map $\varphi \mapsto \|Q^{(\lambda)}(U^{(\lambda)}(\varphi)f)\|^2_{L^2(\mu_\alpha)}$ is in $L^2(G)$, with the seminorm

$$(6.29) \qquad \|f\|_{\mathcal{H}^{(\lambda)}(\alpha)} = \left(\int_G \|Q^{(\lambda)}(U^{(\lambda)}(\varphi)f)\|^2_{L^2(\mu_\alpha)} d\varphi \right)^{1/2}$$

$$= \left(\int_G \|Q^{(\lambda)}(U^{(\lambda)}(\varphi_z)f)\|^2_{L^2(\mu_\alpha)} d\mu(z) \right)^{1/2} .$$

Theorem 6.8: *Let* $\lambda \in \mathbb{P}(D)$ *be so that* $s := \frac{d}{r} - \lambda \in \mathbb{N}$ *and let* $\alpha > \max\{g-1, g-1+\lambda\}$. *Then* $\mathcal{H}^{(\lambda)}(\alpha)$ *is non-trivial if and only if* $\lambda < 1$. *In this case* $\mathcal{H}^{(\lambda)}(\alpha) = \mathcal{H}_{\lambda,q(\lambda)}$ *with proportional seminorms:*

$$\|f\|_{\mathcal{H}^{(\lambda)}(\alpha)} = \gamma_3 \|f\|_{\lambda,q(\lambda)} , \qquad f \in \mathcal{H}_{\lambda,q(\lambda)} .$$

This result generalizes and improves [A1, Theorem 19].

Let us first motivate the definition of $\mathcal{H}^{(\lambda)}(\alpha)$. By (1.9) and a change of variables

$$(6.30) \quad d\mu_\alpha(\varphi_z(w)) = |J\varphi_z(w)|^{\frac{2\alpha}{g}} d\mu_\alpha(w) = |K(w,z)|^{\frac{2\alpha}{g}} K(z,z)^{-\frac{\alpha}{g}} d\mu_\alpha(w).$$

Thus, (6.29) becomes

$$(6.31) \quad \|f\|^2_{\mathcal{H}^{(\lambda)}(\alpha)} =$$

$$c(\alpha)^{-1} \int_{D \times} \int_D |Q^{(\lambda)}(U^{(\lambda)}(\varphi_z)f)(\varphi_z(w))|^2 |K(w,z)^{\frac{\alpha}{g}}|^2 d\mu_\alpha(w) d\mu_\alpha(z).$$

Formula (6.31) says that $c(\alpha)\|f\|^2_{\mathcal{H}^{(\lambda)}(\alpha)}$ is equal to the Hilbert–Schmidt norm of the following integral operator on $L^2(\mu_\alpha)$:

$$(6.32) \qquad A_f\psi(z) = \int_D \overline{a_f(z,w)} K(z,w)^{\alpha/g} \psi(w) d\mu_\alpha(w)$$

where $\psi \in L^2_a(\mu_\alpha)$ and

$$(6.33) \qquad a_f(z,w) = Q^{(\lambda)}(U^{(\lambda)}(\varphi_z)f)(\varphi_z(w))$$

If $\lambda = 0$ (the most interesting case) and $r = 1$, then $(Q^{(0)}f)(z) = f(z) - f(0)$, hence $a_f(z,w) = f(z) - f(w)$ and $A_f = H_{\bar{f}}$, where $H_{\bar{f}}$ is the (ordinary) Hankel operator with symbol $\bar{f}$:

$$(6.34) \qquad H_{\bar{f}}\psi(z) = \int_D (\overline{f(z) - f(w)}) K(z,w)^{\alpha/2} \psi(w) d\mu_\alpha(w) .$$

If $r > 1$ and the analytic function f is not constant, then $H_{\bar{f}}$ is not compact and so surely not Hilbert–Schmidt; see [BCZ]. In case $r=1$, the map $f \mapsto H_{\bar{f}}$ from analytic functions to operators enables the study of the "size" of the analytic function f (growth, approximation by rational functions, smoothness of the boundary values) by means of the "size" of the Hankel operator $H_{\bar{f}}$ (boundedness, compactness, rate of decay of the singular numbers). In order to obtain an interesting theory one has to replace the projection $f \mapsto f - f(0)$ on $M_{q(0)}^{(0)} / M_0^{(0)}$ by the (much smaller) projection $Q^{(0)}$ onto $M_{q(0)}^{(0)} / M_{q(0)-1}^{(0)}$. This leaves a chance for A_f to be a Hilbert–Schmidt operator, provided of course f is "small" enough. $Q^{(0)}$ is not too small, as the "essential part" of an analytic function belongs to the highest quotient of the composition series.

We sketch the proof of Theorem 6.8. by (1.13), $\mathcal{H}^{(\lambda)}(\alpha)$ is invariant under the isometric $U^{(\lambda)}$–action of G. Also if $f \in \mathcal{H}^{(\lambda)}(\alpha)$ then $\|f\|_{\mathcal{H}^{(\lambda)}(\alpha)}$ $= 0$ if and only if f is supported in $S_{q(\lambda)-1}^{(\lambda)}$ (i.e. $f_{\underline{m}} = 0$ for $\underline{m} \notin S_{q(\lambda)-1}^{(\lambda)}$). Thus, applying Theorem 5.2 once more, we see that if $\mathcal{H}^{(\lambda)}(\alpha)$ is non–trivial then $\mathcal{H}^{(\lambda)}(\alpha) = \mathcal{H}_{\lambda,q(\lambda)}$ with proportional semi-norms. As in the case of Theorem 6.7, $\mathcal{H}^{(\lambda)}(\alpha)$ is non–trivial if and only if it contains N^s, where $s = \frac{d}{r} - \lambda$ as before. All what is left is to estimate the integrals in (6.29) for $f = N^s$.

The first step is to obtain the following integral formula for $Q^{(\lambda)}$

$$(6.35) \qquad Q^{(\lambda)}f(z) = \int_S f(u)S(z,u)N(z)^s\overline{N(u)^s}d\sigma(u), \qquad z \in D.$$

Here $S(z,u) = K(z,u)^{1/2} = h(z,u)^{-d/r}$ is the Szegö kernel, and σ is the K–invariant probability measure on the Shilov boundary S. The proof uses the fact that for every signature $\underline{m}$, $l \in \mathbb{N}$ and $z, w \in D$

$$(6.36) \qquad \left(\frac{d}{r}\right)_{\underline{m}} K^{\underline{m}}(z,w)N(z)^l \,\overline{N(w)^l} = \left(\frac{d}{r}\right)_{\underline{m}+l} K^{\underline{m}+l} \,(z,w),$$

where $\underline{m}+l = (m_1+l, m_2+l, \cdots, m_r+l)$. From this (6.35) follows by choosing $l = s$ and using the formula $Q^{(\lambda)}\left(\sum_{\underline{m}} f_{\underline{m}}\right) = \sum_{m_r \geq s} f_{\underline{m}}$.

Next, we will need the following lemma.

Lemma 6.9 [A1, Lemma 23]: *For every* $a, z \in D$

$$(6.37) \qquad Q^{(\lambda)}(U^{(\lambda)}(-\varphi_a)N^s)(z) = K(z,a)^{1/2}N(z)^s K(a,a)^{\frac{\lambda-g}{2g}}$$

Proof: Using (1.9), (3.16), (6.35), and the fact that $|N(u)| = 1$ for $u \in S$, we get

$$Q^{(\lambda)}(U^{(\lambda)}(-\varphi_a)N^s(z) =$$

$$= \int_S N^s(-\varphi_a(u)) \, J(-\varphi_a)(u)^{\lambda/g}S(z,u)N(z)^s\overline{N(u)^s} \, d\sigma(u)$$

$$= \int_S S(u,a)\, K(u,a)^{\lambda/g}\, K(z,u)^{1/2} d\sigma(u) N(z)^s\, K(a,a)^{-\lambda/2g}$$

$$= \int_S S(u,a)\, h(a,u)^s\, K(z,u)^{1/2}\, d\sigma(u) N(z)^s\, K(a,a)^{-\lambda/2g}$$

$$= h(a,a)^s\, K(z,a)^{1/2}\, N(z)^s\, K(a,a)^{-\lambda/2g}$$

$$= K(z,a)^{1/2}\, N(z)^s\, K(a,a)^{\frac{\lambda-g}{2g}} \qquad\qquad \square$$

Conclusion of the proof of Theorem 6.8:

By Lemma 6.9 and (6.36),

$$\|Q^{(\lambda)}(U^{(\lambda)}(-\varphi_a)N^s)\|^2_{L^2(\mu_\alpha)} = \int_D |K(z,a)|\,|N(z)|^{2s} d\mu_\alpha(z)\, K(a,a)^{\frac{\lambda-g}{g}}$$

$$\approx \int_D |K(z,a)| d\mu_\alpha(z) K(a,a)^{\frac{\lambda-g}{g}}$$

From this we get by integraiton over D with respect to $d\mu(a) = K(a,a)dV(a)$

$$\|N^s\|^2_{\mathscr{H}^{(\lambda)}(\alpha)} \approx \int_{D\times D}\int \left|\sum_{\underline{m}}\left(\frac{d}{r}\right)_{\underline{m}} K^{\underline{m}}(z,a)\right|^2 d\mu_\alpha(z) d\mu_{g-\lambda}(a)$$

$$= \sum_{\underline{m}} \left(\frac{d}{r}\right)^2_{\underline{m}} \int_{D\times D}\int |K^{\underline{m}}(z,a)|^2 d\mu_a(z) d\mu_{g-\lambda}(a)$$

It is plain that $\lambda<1$ (i.e. the finiteness of $\mu_{g-\lambda}$) is a necessary condition for the finiteness of $\|N^s\|^2_{\mathscr{H}^{(\lambda)}(\alpha)}$. In case $\lambda<1$ one computes the integrals and finds

$$\|N^s\|^2_{\mathscr{H}^{(\lambda)}(\alpha)} \approx \sum_{\underline{m}} \frac{\left(\frac{d}{r}\right)^2_{\underline{m}}}{(\alpha+s)_{\underline{m}}}\cdot\frac{1}{\|\varphi_{\underline{m}}\|^2_F} = {}_2F_1\left(\frac{d}{r},\frac{d}{r};\alpha+s,e\right).$$

According to [FK1, Th. 4.1] this series converges if and only if $g-\alpha-s < -(r-1)\frac{a}{2}$, i.e. $g+\lambda-1 < \alpha$. $\qquad\qquad \square$

Remark: One can define $\mathscr{H}^{(\lambda)}(\alpha)$ even for $\alpha>(r-1)\frac{a}{2}$ (and $\lambda\in\mathbb{P}(D)$ for which $s:=\frac{d}{r}-\lambda\in\mathbb{N}$) as the space of all $f\in\mathscr{H}(D)$ for which

$$\|f\|_{\mathscr{H}^{(\lambda)}(\alpha)} = \left(\int_D \|Q^{(\lambda)}(U^{(\lambda)}(\varphi_z)f)\|^2_{\mathscr{H}_\alpha} d\mu(z)\right)^{1/2}$$

is finite. Simlar computations lead to the conclusion that $\mathscr{H}^{(\lambda)}(\alpha)$ is non-trivial if and only if $\lambda<1$ and $\alpha>\max\{(r-1)\frac{a}{2}, g-1+\lambda\}$, and in this case $\mathscr{H}^{(\lambda)}(\alpha) = \mathscr{H}_{\lambda,g(\lambda)}$ with proportional semi-norms.

§7. Integral formulas for the invariant inner products: the case of the unit ball of $\mathbb{C}^n$

In this section $D=B$ is the unit ball of $\mathbb{C}^n$. The parameters of B are $r=1$, $a = 0$, $b = n-1$, $g = n+1$, and $d = n$. The Wallach set is $[0,\infty)$ with $W_d(B)=\{0\}$, $W_c(B)=(0,\infty)$, and $\mathbb{P}(B) = -\mathbb{N}$. The Bergman kernel is $K(z,w) = (1-(z,w))^{-(n+1)}$ and the normalized Lebesgue measure on B is $dV(z) = \frac{n!}{\pi^n}dm(z)$. The Shilov boundary S is the topological boundary ∂B and we denote by $d\sigma$ the probability measure on S which is invariant under $K = U(n)$. The group $G = \text{Aut}(B)_0$ can be realized as $SU(1,n)$, and it acts on B via (1.2) (or (1.3), see also [R, Chapter 2]). For $\lambda \in \mathbb{C}$ the $U^{(\lambda)}$–action of G on functions on B is $U^{(\lambda)}(\varphi)f = (f \circ \varphi)(J\varphi)^{\lambda/(n+1)}$.

For $\lambda > n$, the $U^{(\lambda)}$–invariant inner product is

$$(7.1) \qquad (f,g)_\lambda = \binom{\lambda-1}{n} \int_B f(z)\overline{g(z)}(1-|z|^2)^{\lambda-n-1}dV(z).$$

The homogeneous and the Taylor expansions of an analytic function f on B will be denoted by $f(z)=\sum_{m=0}^{\infty}f_m(z)$ and $f(z)=\sum \hat{f}(\alpha)z^\alpha$ respectively. We shall use the standard notation of [R]; thus $z^\alpha = z_1^{\alpha_1} z_1^{\alpha_2} \cdots z_n^{\alpha_n}$, $|\alpha| = \alpha_1 + \cdots + \alpha_n$, and $\alpha! = \alpha_1!\alpha_2! \cdots \alpha_n!$. For general $\lambda \in \mathbb{C}\backslash(-\mathbb{N})$, the $U^{(\lambda)}$–invariant Hermitian form is

$$(7.2) \qquad (f,g)_\lambda = \sum_{m=0}^{\infty}\frac{(f_m,g_m)_F}{(\lambda)_m} = \sum_{m=0}^{\infty}\frac{1}{(\lambda)_m}\sum_{|\alpha|=m}\alpha!\hat{f}(\alpha)\overline{\hat{g}}(\alpha) ,$$

where $(\cdot,\cdot)_F$ is the Fischer inner product, determined by $(z^\alpha,z^\beta) = \delta_{\alpha,\beta}\alpha!$, and

$$(\lambda)_m := \lambda(\lambda + 1)(\lambda + 2) \cdots (\lambda + m - 1).$$

The points $\lambda = -\ell$, $\ell = 0, 1, 2, \ldots$ are simple poles of the form (7.2). In these cases $q(\lambda) = 1$ and the corresponding composition series consists of the two spaces:

$$M_0^{(-\ell)} \sum_{m\leq\ell} \mathcal{P}_m = \text{polynomials of degree} \leq \ell ,$$

$$M_1^{(-\ell)} = \mathcal{P}^{(-\ell)} = \text{span}\{U^{(-\ell)}(\varphi)f; \varphi \in G, f \text{ polynomial}\}.$$

The quotient $M_1^{(-\ell)}/M_0^{(-\ell)}$ is unitarizable and the corresponding $U^{(-\ell)}$–invariant inner product is defined by

$$(7.3) \qquad (f,g)_{(-\ell),1} = (-1)^{(-\ell)} \lim_{\xi \to -\ell} (\xi + \ell)(f,g)_{\xi} = \frac{1}{\ell!} \sum_{m=\ell+1}^{\infty} \frac{(f_m, g_m)_F}{(m-\ell-1)!}$$

In what follows we present the method of J. Peetre to obtain the analytic continuation of the integrals (7.1) to all $\lambda \in \mathbb{C} \setminus (-\mathbb{N})$ with $\mathrm{Re}\lambda \leq n$, using integration by parts in the radial direction. This provides integral formulas for the $U^{(\lambda)}$-invariant forms (7.2) and (7.3). The expanded version [A2] of the original short note [P3] contains the full details.

To describe the results, recall that the *radial derivative* of a differentiable function f is

$$(7.4) \qquad Rf(z) = \frac{\partial}{\partial t} f(tz)_{|t=1}$$

If f is analytic then

$$(7.5) \qquad Rf(z) = \sum_{j=1}^{n} z_j \frac{\partial f}{\partial z_j}(z) = \sum_{m=0}^{\infty} m f_m(z) \ .$$

For $k = 0, 1, 2, \ldots$ we denote

$$(7.6) \qquad \mathcal{R}_k = \prod_{j=1}^{k} (R + n - j) = (R + n - k)_k \ .$$

Theorem 7.1: *Let f, g be analytic functions in a neighborhood of $\overline{B}$. Then*

$$(7.7) \qquad (f,g)_\lambda = \frac{(\lambda-1)}{n!} \int_B (\mathcal{R}_{n-1}f)(z) \overline{g(z)} \frac{(1-|z|^2)^{\lambda-2}}{|z|^{2(n-1)}} dV(z); \qquad \lambda > 1$$

and

$$(7.8) \qquad (f,g)_\lambda = f(0) \overline{g(0)} + \frac{1}{n!} \int_B (\mathcal{R}_n f)(z) \overline{g(z)} \frac{(1-|z|^2)^{\lambda-1}}{|z|^{2n}} dV(z); \lambda > 0.$$

Moreover, for $\lambda = 1, 2, \ldots, n$,

$$(7.9) \qquad (f,g)_\lambda = \frac{(\lambda-1)!}{(n-1)!} \int_S (\mathcal{R}_{n-\lambda}f)(\xi) \overline{g(\xi)} d\sigma(\xi) \ .$$

Theorem 7.2: *Let $\ell = 0, 1, 2, \ldots$ and let f, g be analytic functions in a neighborhood of $\overline{B}$. Then*

$$(7.10) \qquad (f,g)_{-\ell,1} = \frac{1}{\ell!} \left[(f_{\ell+1} g_{\ell+1})_F + \frac{1}{n!} \int_B (\mathcal{R}_{n+\ell+1}f)(z) \frac{\overline{g(z)} \, dV(z)}{|z|^{2(n+\ell+1)}} \right]$$

and

$$(7.11) \qquad (f,g)_{-\ell,1} = \frac{1}{\ell!(n-1)!} \int_S (\mathcal{R}_{n+\ell}f)(\xi) \overline{g(\xi)} d\sigma(\xi)$$

The case $\ell = 0$ is of special importance because $\mathcal{H}_{0,1}$ is invariant under the simplest group action $f \mapsto f \circ \varphi$, $\varphi \in G = \text{Aut}(B)_0$. Here (7.2) takes the form

(7.12)
$$(f,g)_{0,1} = \sum_{\alpha} |\alpha| \frac{\alpha}{|\alpha|!} \hat{f}(\alpha) \overline{\hat{g}(\alpha)}$$

This space was studied in [P1] and [Z]; it is the generalization of the Dirichlet space in the context of the unit ball of $\mathbb{C}^n$.

Corollary 7.3: *Let* f,g *be analytic in a neighborhood of* $\overline{B}$. *Then*

(7.13)
$$(f,g)_{0,1} = (\nabla f(0), \nabla g(0)) + \frac{1}{n!} \int_B (\mathcal{R}_{n+1}f)(z) \frac{\overline{g(z)}\, dV(z)}{|z|^{2(n+1)}}$$

and

(7.14)
$$(f,g)_{0,1} = \frac{1}{(n-1)!} \int_S (\mathcal{R}_n f)(\xi) \overline{g(\xi)}\, d\sigma(\xi).$$

Notice that (7.13) and (7.14) are the natural generalizations of (6.3) and (6.4) respectively. A formula similar to (7.14) was obtained independently by M. Peloso [Pe], using different techniques.

A basic computational tool is the formula of *integration in polar coordinates* [R, 1.4.3]

(7.15)
$$\int_B f(z)dV(z) = 2n \int_0^1 r^{2n-1}dr \int_S f(r\xi)d\sigma(\xi), \quad f \in L^1(B) .$$

The *radialization* of the function f is the function

(7.16)
$$f^{\#}(z) = \int_{U(n)} f(uz)du = \int_S f(r\xi)d\sigma(\xi) , \quad |z| = r.$$

f is *radial* if $f^{\#} = f$, equivalently: $f(z) = f(uz)$ for all $u \in U(n)$. Let us define

(7.17)
$$\tilde{f}(r) = f^{\#}(z), \text{ where } |z| = r^{1/2} .$$

Thus, $f^{\#}(re_1) = \tilde{f}(r^2)$ and if $f \in L^1(B)$ then

(7.18)
$$\int_B f(z)dv(z) = 2n \int_0^1 r^{2n-1}f^{\#}(re_1)\, dr = n \int_0^1 r^{n-1}\tilde{f}(r)\, dr.$$

It is well known that the radial derivative commutes with the action of $U(n)$, i.e. $R(f \circ u) = (Rf) \circ u$ for any differentiable function f and $u \in U(n)$. From this it is obvious that

(7.19)
$$(Rf)^{\#} = R(f^{\#}) = r \frac{\partial f^{\#}}{\partial r} = 2r^2 \tilde{f}'(r^2)$$

where $r = |z|$.

Using (7.18) and the well-known fact (see [R, 1.4.9], or use (4.12)) that

$$(7.20) \qquad (\xi^\alpha, \xi^\beta)_{L^2(S)} = \delta_{\alpha,\beta} \frac{\alpha!}{(n)_{|\alpha|}} ,$$

one can verify all formulas in Theorems 7.1 and 7.2; indeed by bilinearity and orthogonality of monomials, it is enough to take $f(z) = g(z) = z^\alpha$.

We would like to show little more, namely how one actually arrives at formulas (7.7) – (7.11) using integration by part in the radial direction.

Let f, g be analytic functions in a neighborhood of $\overline{B}$. For $m = 0, 1, 2, \ldots$ let

$$(7.21) \qquad c(m) = (f_m, g_m)_{L^2(S)} = \sum_{|\alpha|=m} \frac{\hat{f}(\alpha)\overline{\hat{g}(\alpha)}\alpha!}{(n)_m}$$

and set $c(m) = 0$ for $m < 0$. Then,

$$(7.22) \qquad (f\overline{g})^\#(z) = \sum_{m=0}^{\infty} c(m)r^{2m}, \quad (\widetilde{f\overline{g}})(r) = \sum_{m=0}^{\infty} c(m)r^m$$

For $s \in \mathbb{C}$ with $\mathrm{Re}(s) > -1$ and an integer ℓ so that $c(m) = 0$ whenever $m < -\ell$, consider the integral

$$(7.23) \qquad J(s, \ell, f, g) = \int_B f(z)\, \overline{g(z)}\, \frac{(1-|z|^2)^s}{|z|^{2(n-\ell-1)}}\, dV(z) .$$

Lemma 7.4: *Let f, g, s and ℓ be as above, then*

$$(7.24) \qquad J(s, \ell, f, g) = \frac{1}{s+1} J(s+1, \ell-1, (R + \ell)f, g) + \frac{nc(-\ell)}{s + 1}$$

The proof is a careful integration by parts, starting from

$$(7.25) \qquad J(s, \ell, f, g) = n \int_0^1 (1 - r)^s r^\ell \widetilde{f\overline{g}}(r)\, dr$$

We sketch the main steps in the proofs of Theorems 7.1 and 7.2. Let f, g be analytic in a neighborhood of $\overline{B}$. It follows by induction and Lemma 7.4 that for every $k = 0, 1, \ldots, n - 1$ and every $\lambda \in \mathbb{C}$ with $\mathrm{Re}\,\lambda > n$,

$$(7.26) \qquad (f, g)_\lambda = \frac{\prod_{j=1}^{n-k}(\lambda-j)}{n!} \int_B (\mathcal{R}_k f)(z)\, \overline{g(z)}\, \frac{(1-|z|^2)}{|z|^{2k}}\, dV(z)$$

$$= \frac{\prod_{j=1}^{n-k}(\lambda - j)}{n!} J(\lambda + k - n - 1, n - k - 1, \mathcal{R}_k f, g).$$

By taking $k = n - 1$ in (7.26) we obtain (7.7). To get (7.9), write (7.26) in the form

$$(7.27) \qquad (f,g)_\lambda = \frac{\prod_{j=1}^{n-k}(\lambda-j)}{(n-1)!}\,(\lambda+k-n)\int_0^1 (\mathcal{R}_k f \cdot \bar{g})(r)\frac{(1-r)^{\lambda+k-n-1}}{r^k}dr$$

As $\lambda \to (n - k)_+$, the probability measure $(\lambda + k - n)(1 - r)^{\lambda+k-n-1}dr$ converges (in the weak topology of measures) to the Dirac measure δ_1 at 1. This yields (7.9).

Next, we use Lemma 7.4 to show that for $\ell = -1, 0, 1, 2, \ldots$ and every $\lambda \in \mathbb{C}$ with $\mathrm{Re}\,\lambda > -\ell$

$$(7.28) \qquad (f,g)_\lambda = \frac{J(\lambda + \ell - 1, -\ell - 1, \mathcal{R}_{n+\ell}\,f,g)}{n!(\lambda)_\ell} + \sum_{j=0}^{\ell}\frac{(n)_j}{(\lambda)_j}c(j),$$

where the $c(j)$'s are defined by (7.21). Formula (7.8) is just the case $\ell = 0$ in (7.28). If $\ell = 0, 1, 2, \ldots$ then (7.28) leads by the previous arguments to (7.11). (7.10) is also an easy consequence of (7.28).

A variant of the method described above leads to the following results.

Theorem 7.5: Let f, g *be analytic functions in a neighborhood of* $\overline{\mathrm{B}}$.

(a)   *For* $\ell = 0, 1, 2, \ldots, n-1$ *and* $\lambda \in \mathbb{C}$ *with* $\mathrm{Re}(\lambda) > \ell$

$$(7.29) \quad (f,g)_\lambda = \frac{\prod_{j=1}^{n-k}(\lambda-j)}{n!}\int_B (\mathcal{R}_{n-\ell-1}(R+\lambda)f)(z)\overline{g(z)}\,\frac{(1-|z|^2)^{\lambda-\ell-1}}{|z|^{2(n-\ell-1)}}\,dV(z).$$

(b)   *For* $\ell = 1,2,3,\ldots$ *and* $\lambda \in \mathbb{C}$ *with* $\mathrm{Re}\,\lambda > -\ell$

$$(7.30) \qquad (f,g)_\lambda = \sum_{m=0}^{\ell-1}\frac{(f_m,g_m)_F}{(\lambda)_m} +$$

$$+ \frac{1}{n!(\lambda)_\ell}\int_B (\mathcal{R}_{n+\ell-1}(R+\lambda)f)(z)\overline{g(z)}\frac{(1-|z|^2)^{\lambda+\ell-1}}{|z|^{2(n+\ell-1)}}dV(z)$$

Formulas (7.9) and (7.11) follow from (7.29) and (7.30) respectively by limiting procedures. Theorem 7.5 leads also to the following.

Corollary 7.6: Let f, g *be analytic functions in a neighborhood of* $\overline{\mathrm{B}}$.

(a)   *For* $\ell = 1, 2, \ldots, n$

$$(7.31) \qquad (f,g)_\ell = \frac{(\ell-1)!}{n!}\int_B (\mathcal{R}_{n-\ell}(R+\ell)f)(z)\cdot\overline{g(z)}\,\frac{dV(z)}{|z|^{2(n-\ell)}}\,.$$

(b)   *For* $\ell = 0, 1, 2, \ldots$

(7.32) $(f,g)_{\ell,1} = \dfrac{1}{n!\ell!} \displaystyle\int_B (\mathcal{R}_{n+\ell}(R-\ell)f)(z) \cdot \overline{g(z)} \; \dfrac{dV(z)}{|z|^{2(n+\ell)}}$.

§8. Formulas for the invariant inner products involving invariant differential operators on the associated symmetric cone

In this section we review the theory of a class of differential operators introduced by Yan in [Y2] and [Y3], and studied also in [AU]. These operators are used to obtain formulas for the invariant inner products $(\cdot,\cdot)_\lambda$ in terms of the inner products $(\cdot,\cdot)_{\lambda+k}$ for suitable k. The usefulness of these results stems from the fact that if k is large enough then $(\cdot,\cdot)_{\lambda+k}$ is computed in terms of integration over D, and thus $(\cdot,\cdot)_\lambda$ admits concrete integral formulas.

In what follows D will be a Cartan domain of tube type, $N(z)$ is the determined polynomial and $(\partial_N f)(z) = N(\frac{\partial}{\partial z})f(z)$.

In [Y2] Yan introduces the operators

(8.1) $D^k(\lambda)f := N^{d/r-\lambda}\partial_{N^k}(N^{k+\lambda-d/r}f)$

where $\lambda \in \mathbb{C}$ and $k \in \mathbb{N}$. These operators are diagonal with respect to the Peter-Weyl decomposition:

(8.2) $D^k(\lambda)f_m = \mu_m^k(\lambda)f_m$, $f_m \in P_m$,

where

(8.3) $\mu_m^k(\lambda) = \dfrac{\Gamma_\Omega(\underline{m}+k+\lambda)}{\Gamma_\Omega(\underline{m}+\lambda)} = \displaystyle\prod_{j=1}^{r}\prod_{\ell=0}^{k-1}(m_j+\lambda+\ell-(j-1)\frac{a}{2}) = \gamma\,\dfrac{(\lambda+k)_m}{(\lambda)_m}$,

where $\gamma = (\lambda)_{(k,\ldots,k)}$. It follows that

(8.4) $D_z^k(\lambda)K(z,w)^{\lambda/g} = \gamma K(z,w)^{(\lambda+k)/g}$,

where the subscript "z" denotes differentiation with respect to z. Formulas (8.2) and (8.4) yield the main results of [Y2] and [Y3], which are summarized in the following theorem.

Theorem 8.1: *Let* D *be a tube-type Cartan domain.*

(i)   *If* $\lambda > (r-1)a/2$ *and* $k \in \mathbb{N}$, *then*

(8.5) $(f,h)_\lambda = \gamma(D^k(\lambda)f,h)_{\lambda+k}$; $f,h \in \mathcal{H}_\lambda$.

(ii)   *If* $\lambda \in W_d(D)$, $k \in \mathbb{N}$, $\lambda + k \geq d/r$ *and* $(r-1)a/2 - \lambda \in \mathbb{N}$, *then*

(8.6) $(f,h)_{\lambda,q(\lambda)} = \beta(D^k(\lambda)f,\,h)_{\lambda+k}$; $f,h \in \mathcal{H}_{\lambda,q(\lambda)}$, *where the constant β depends only on λ and k.*

(iii)   *If $\lambda \in W_d(D)$, s = d/r - $\lambda \in \mathbb{N}$, k $\in \mathbb{N}$, k > d/r - 1, and k $\geq$ 3, then*

$$(8.7) \quad (f,h)_{\lambda,q(\lambda)} = \alpha \int_D (D^k(\lambda)f)(z) \cdot \overline{(D^s(\lambda)h)(z)} \frac{K(z,z)^{1-(k+s+\lambda)/g}}{|N(z)|^{2s}} \, dV(z)$$

for all f,h $\in \mathscr{H}_{\lambda,q(\lambda)}$, where α is a constant depending on λ, k, and s.

If $\lambda + k > g - 1$ then (8.5) and (8.6) provide formulas for the invariant inner products in terms of integration over the domain D. If $\lambda + k = d/r$ then the integration is taken over the Shilov boundary S. In this case $D^k(\lambda) = N^k \partial_{N^k}$ and (8.6) reduces to Theorem 6.1. Also, if $\lambda = \frac{d}{r}$ and $\lambda+k>g-1$, then (8.5) expresses the inner product of the Hardy space $H^2(s)$ in terms of integration on D with respect to a weighted volume measure (and if $\lambda+k=g$, the unweighted volume measure).

The work of Yan is extended in [AU] to all $\lambda \in \mathbb{P}(D)$ and all $U^{(\lambda)}$-invariant Hermitian forms on the irreducible quotients $M_j^{(\lambda)}/M_{j-1}^{(\lambda)}$, $1 \leq j \leq q(\lambda)$. To describe this result, let us define for $\lambda \in \mathbb{C}$ and k,$\nu \in \mathbb{N}$

$$(8.8) \quad D_\nu^k(\lambda) := \frac{1}{\gamma!} \left(\frac{\partial}{\partial\xi}\right)^\nu D^k(\xi)\Big|_{\xi=\lambda} \, .$$

Namely, for $f \in \mathscr{H}(D)$ and $z \in D$

$$(8.9) \quad D_\nu^k(\lambda)f(z) = \frac{1}{\nu!} \left(\frac{\partial}{\partial\xi}\right)^\nu D^k(\xi)f(z)\Big|_{\xi=\lambda} \, .$$

The operator $D_\nu^k(\lambda)$ is diagonal with respect to the Peter–Weyl decomposition

$$(8.10) \quad D_\nu^k(\lambda)\left(\sum_m f_m\right) = \sum_m \mu_{m,\nu}^k(\lambda)f_m \, ,$$

where

$$(8.11) \quad \mu_{m,\nu}^k(\lambda) := \frac{1}{\nu!} \left(\frac{\partial}{\partial\xi}\right)^\nu \mu_m^k(\xi)\Big|_{\xi=\lambda} \, .$$

Theorem 8.2 [AU]:   *Let $\lambda \in \mathbb{P}(D)$, k$\in \mathbb{N}$ and assume that $\lambda+k \geq \frac{d}{r}$. Let q=q($\lambda$), 0$\leq$ j$\leq$ q and ν=q-j. Then for all $f_1, f_2 \in \mathscr{H}_{\lambda,j}$*

$$(8.12) \quad (f,g)_{\lambda,j} = \frac{1}{\beta}(D_\nu^k(\lambda)f,g)_{\lambda+k} \, ,$$

where

$$(8.13) \quad \beta=\beta(\lambda,k) = \frac{1}{q!} \left(\frac{\partial}{\partial\xi}\right)^q (\xi)_{(k,k,\ldots,k)}\Big|_{\xi=\lambda} \, .$$

Again, if $\lambda+k>g-1$ then the $U^{(\lambda)}$-invariant hermitian form $(\cdot,\cdot)_{\lambda,j}$ is expressed in terms of integration on D with respect to $\mu_{\lambda+k}$, and if $\lambda+k=d/r$ the integration takes place on the Shilov boundary S with respect to the measure σ.

The proof of Theorem 8.2 is based on the fact that if $\underline{m} \in S_j^{\{\lambda\}} \setminus S_{j-1}^{\{\lambda\}}$ then the polynomial $\mu_{\underline{m}}^k(\xi)$ (defined by (8.3)) vanishes at $\xi = \lambda$ exactly to order $\nu = q - j$.

The case where $j = 1$ in Theorem 8.2 is the generalization of [A1] and [Y2] to $\lambda \in \mathbb{P}(D)$. Indeed, if $s = \frac{d}{r} - \lambda \in \mathbb{N}$ and we take in Theorem 8.2 $k = s$ and $\nu = 0$, then $D_0^s(\lambda) = N^s \partial_{N^s}$ and (8.12) becomes (6.11).

Another aspect of Yan's operators $D^k(\lambda)$ is their invariance properties. Let us consider D as the open unit ball of the JB*-algebra Z with unit e and involution $z \mapsto z^*$, and let Ω be the symmetric cone consisting of all positive elements in the Euclidean Jordan algebra $X = \mathrm{Re}(Z) = \{z \in Z; z^* = z\}$. Let $G(\Omega)$ be the group of invertible, real-linear transformation of X which map Ω onto itself. It is known that $G(\Omega)$ acts transitively on Ω. In fact, let

$$(8.14) \quad P(x)y := \{x, y, x\}, \quad x, y \in Z.$$

Then $P(x) = 2L(x)^2 - L(x^2)$, where $L(x)y = xy = \{x, e, y\}$, and for $x \in \Omega$ we have $P(x^{1/2}) \in G(\Omega)$ and $P(x^{1/2})e = x$. We extend the members of $G(\Omega)$ to $Z = X + iX$ by linearity. This realizes $G(\Omega)$ as a subgroup of biholomorphic automorphisms of the Siegel domain of type I, $X + i\Omega$.

Let $\mathcal{D}(\Omega) = \mathrm{Diff}(\Omega)^{G(\Omega)}$ be the ring of all $G(\Omega)$-invariant differential operators on Ω. It is known that $\mathcal{D}(\Omega)$ is isomorphic to the polynomial ring $\mathbb{C}[X_1, \cdots, X_r]$. In particular, $\mathcal{D}(\Omega)$ is commutative and is algebraically generated by r algebraically-independent elements, and every set of algebraic generators contains at least r elements. See [FK2, Chapter XIV] for more details.

Theorem 8.3 [Y2], [Ko3]: *For every* $k \in \mathbb{N}$ *and* $\lambda \in \mathbb{C}$, $D^k(\lambda)$ *is a* $G(\Omega)$- *invariant differential operator, i.e., a member of* $\mathcal{D}(\Omega)$. *Moreover, for every set* $\{\lambda_1, \cdots, \lambda_r\}$ *of distinct complex numbers, the operators* $D(\lambda_j) = D^1(\lambda_j)$, $1 \leq j \leq r$, *are algebraically-independent generators of* $\mathcal{D}(\Omega)$.

The second statement implies the first, since

$$(8.15) \quad D^k(\lambda) = D(\lambda)D(\lambda+1) \cdots D(\lambda+k-1).$$

We illustrate the proof of Theorem 8.3 by exhibiting the relationship between the operators $D^k(\lambda)$ and the standard algebraic generators of $\mathcal{D}(\Omega)$.

For every $1 \leq \nu \leq r$ let $1_\nu := (1, 1, \cdots, 1, 0, \cdots 0)$, with ν 1's and $r - \nu$ 0's, and let $\varphi_\nu := \varphi_{1_\nu}$ be the corresponding spherical polynomial. Set

$$(8.16) \quad (\Delta_\nu f)(x) = \varphi_\nu\left(\frac{\partial}{\partial y}\right)(f(\psi(y)))\Big|_{y=e},$$

where $\psi \in G(\Omega)$ satisfies $\psi(e) = x$. Δ_ν is well-defined, i.e. independent of the special choice of $\psi \in G(\Omega)$ satisfying $\psi(e) = x$. This is due to the fact that φ_ν is invariant under the group

(8.17) $L = \mathrm{Aut}(X) = \{\psi \in G(\Omega); \psi(e) = e\}$

of Jordan–automorphisms of X (it coincides with the group defined by (3.4)). In particular, we can choose in (8.16) $\psi = P(x^{1/2})$.

Using Theorem 4.1, one gets

(8.18) $N(e+x) = \sum\limits_{\nu=0}^{r} \binom{r}{\nu} \varphi_\nu(x)$.

From this is follows that for $x \in X$

(8.19) $\varphi_\nu(x) = \binom{r}{\nu}^{-1} \sum\limits_{1 \le i_1 < i_2 < \cdots < i_\nu \le r} \lambda_{i_1} \lambda_{i_2} \cdots \lambda_{i_\nu}$,

where $\lambda_j = \lambda_j(x)$ are the eigenvalues of x. Thus, $\{\varphi_\nu(x)\}_{\nu=1}^{r}$ are the (normalized) elementary symmetric polynomials in the eigenvalues of x. In particular $\varphi_1(x) = \frac{1}{r} \mathrm{trace}\ (x)$ and $\varphi_r(x) = \det(x) = N(x)$. From this it is elementary to get

(8.20) $\Delta_1 = \dfrac{R}{r}$ and $\Delta_r = N\partial_N$,

where $Rf(x) = \frac{\partial}{\partial t} f(tx)\big|_{t=1}$ is the radial derivative. It is an interesting open problem to find similar explicit formulas for the operator Δ_ν for $1 < \nu < r$.

Theorem 8.4 [N], [Y2], [FK, Chapter XIV]: *The operator* $\{\Delta_\nu\}_{\nu=1}^{r}$ *are algebraically–independent generators of* $\mathcal{D}(\Omega)$.

Because of (8.16) and (8.19) the operators $\{\Delta_\nu\}_{\nu=1}^{r}$ are considered as the most canonically algebraically–independent generators of $\mathcal{D}(\Omega)$. The following result is established in [Y2] and [AU] independently. Here $\alpha_j = (j-1)\frac{a}{2}$, $1 \le j \le r$ and $\Delta_0 = I$.

Theorem 8.5: *For every* $\lambda \in \mathbb{C}$, *the operator* $D(\lambda) = D^1(\lambda) = N^{\frac{d}{r}-\lambda} \partial_N N^{\lambda+1-\frac{d}{r}}$ *admits the expansion*

(8.21) $D(\lambda) = \sum\limits_{\nu=0}^{r} \binom{r}{\nu} \prod\limits_{j=\nu+1}^{r} (\lambda - \alpha_j) \Delta_\nu$

(Here, we adopt the convention that the empty product is equal to 1). Using (8.15), it follows that

(8.22) $D^k(\lambda) = \prod\limits_{\ell=0}^{k-1} \sum\limits_{\nu=0}^{r} \binom{r}{\nu} \prod\limits_{j=\nu+1}^{r} (\lambda + \ell - \alpha_j) \Delta_\nu$.

It is known ([Y2] and [Y3]) that $\{\varphi_{\underline{m}}\}_{\underline{m}}$ is a linear basis for the space of L-invariant polynomials on X. It follows that the associated $G(\Omega)$-invariant differential operator

$$(8.23) \qquad (\Delta_{\underline{m}}f)_{(x)} = \varphi_{\underline{m}}\left(\frac{\partial}{\partial y}\right)(f(P(x^{1/2})y))|_{y=e} \ ,$$

where $\underline{m}$ ranges over all signatures, form a linear basis of $\mathcal{D}(\Omega)$. The expansion of the operator $D^k(\lambda)$ in this basis is given by

Theorem 8.6 [AU]: *For every* $\lambda \in \mathbb{C}$ *and* $k \in \mathbb{N}$

$$(8.24) \qquad D^k(\lambda) = \frac{\Gamma_\Omega\left(\frac{d}{r}+k\right)}{\Gamma_\Omega\left(\frac{d}{r}-k-\lambda\right)} \sum_{\underline{m},m_1\leq k} (-1)^{rk+|\underline{m}|} \frac{\Gamma_\Omega\left(\frac{d}{r}-\lambda+\underline{m}^*\right)}{\Gamma_\Omega\left(\frac{d}{r}+k+\underline{m}^*\right)} \frac{\Delta_{\underline{m}}}{\|\varphi_m\|_F^2}$$

$$= (\lambda)_{(k,\cdots,k)} \sum_{\underline{m},m_1\leq k} (-1)^{|\underline{m}|} \frac{(-k)_{\underline{m}}}{(\lambda)_{\underline{m}}} \frac{\Delta_{\underline{m}}}{\|\varphi_m\|_F^2}$$

where $\underline{m}^* := (-m_r, -m_{r-1}, \cdots, -m_1)$ *and* $\|\varphi_m\|_F^2$ *is given in (4.3).*

Sketch of the Proof: It is clearly enough to prove that for every $\alpha \in \mathbb{C}$ with $\mathrm{Re}(\alpha) > (r-1)\frac{a}{2}$ and every $k \in \mathbb{N}$

$$(8.25) \qquad N^{\alpha+k}\partial_N{}^k N^{-\alpha} = c \sum_{\underline{m},m_1\leq k} (-1)^{kr+|\underline{m}|} \frac{(\alpha)_{k+\underline{m}^*}}{\left(\frac{d}{r}\right)_{k+\underline{m}^*}} \frac{\Delta_{\underline{m}}}{\|\varphi_m\|_F^2} \ ,$$

where $c = \left(\frac{d}{r}\right)_{(k,k,\cdots,k)} = \Gamma_\Omega\left(\frac{d}{r}+k\right)/\Gamma_\Omega\left(\frac{d}{r}\right)$. Indeed, (8.24) follows from (8.25) by analytic continuation and the substitution $\alpha = \frac{d}{r}-\lambda-k$. Next, the members of $\mathcal{D}(\Omega)$ are determined by their action on the exponential functions $f_y(x) := e^{(x,y)}$, $x,y \in \Omega$. Since

$$\Delta_{\underline{m}}(f_y)(a) = \varphi_{\underline{m}}\left(\frac{\partial}{\partial x}\right)e^{(P(a^{1/2})x,y)}|_{x=e} = \varphi_{\underline{m}}(P(a^{1/2})y)f_y(a),$$

it will suffice to prove

$$(8.26) \qquad (N^{\alpha+k}\partial_N{}^k N^{-\alpha}f_y)(a) = c \sum_{\underline{m},m_1\leq k} \frac{(\alpha)_{k+\underline{m}^*} \, \varphi_{\underline{m}}(P(a^{1/2})y)}{\left(\frac{d}{r}\right)_{k+\underline{m}^*}\|\varphi_m\|_F^2} f_y(a) \ .$$

with c as in (8.25). Using the well-known "binomial formula" (see [FK2, Chapter XIV])

$$(8.27) \quad N^k(e-x) = \sum_{\underline{m}, m_1 \leq k} (-k)_{\underline{m}} \frac{\varphi_{\underline{m}}(x)}{\|\varphi_{\underline{m}}\|_F^2} = c \sum_{\underline{m}, m_1 \leq k} (-1)^{|\underline{m}|} \left(\frac{d}{r}\right)^{-1}_{k+\underline{m}^*} \frac{\varphi_{\underline{m}}(x)}{\|\varphi_{\underline{m}}\|_F^2},$$

the formula $\quad N^{-\alpha}(x) = \Gamma_\Omega(\alpha)^{-1} \int_\Omega e^{-(x,t)} N^{\alpha - \frac{d}{r}}(t) dt \quad$ and the fact that

$p\left(\frac{\partial}{\partial x}\right) f_y(x)\big|_{x=a} = p(y) f_y(a) \quad$ for every polynomial $\quad$ p, we get for very

$y, a \in \Omega$

$(N^{\alpha+k} \partial_{N^k} N^{-\alpha} f_y)(a) =$

$$= \Gamma_\Omega(\alpha)^{-1} f_y(a) N^{\alpha+k}(a) \int_\Omega e^{-(a,t)} N^k(y-t) N^{\alpha - \frac{d}{r}}(t) dt$$

$$= e\Gamma_\Omega(\alpha)^{-1} f_y(a) N^{\alpha+k}(a) N^{\alpha+k}(y) \sum_{\underline{m}, m_1 \leq k} (-1)^{|\underline{m}|}$$

$$\times \frac{\int_\Omega e^{-(P(y^{1/2})a,x)} \varphi_{\underline{m}}(x) N^{\alpha - \frac{d}{r}}(x) dx}{\left(\frac{d}{r}\right)_{k+\underline{m}^*} \|\varphi_{\underline{m}}\|_F^2}$$

(we used the substitution $\quad t = P(y^{1/2})x \quad$ and the fact that $\quad N(P(y^{1/2})z) = N(y)N(z)$). Next, for every $\underline{s} = (s_1, \cdots, s_r) \in \mathbb{C}^r$ with $Re(s_j) > (j-1)\frac{a}{2}$,

$$(8.28) \quad \int_\Omega e^{-(y,x)} \varphi_{\underline{s} - \frac{d}{r}}(x) dx = \Gamma_\Omega(\underline{s}) \varphi_{\underline{s}^*}(y), \quad y \in \Omega,$$

where $\underline{s}^* = (-s_r, -s_{r-1}, \cdots, -s_1)$. This follows from the analogous formula for $N_{\underline{s}}$ instead of $\varphi_{\underline{s}}$, and the fact that $\varphi_{\underline{s}}(x) = \int_L N_{\underline{s}}(\ell x) d\ell$, see [FK2, Chapter VII]. It follows that

$(N^{\alpha+k} \partial_{N^k} N^{-\alpha} f_y)(a) =$

$$= c\Gamma_\Omega(\alpha)^{-1} f_y(a) \sum_{\underline{m}, m_1 \leq k} (-1)^{|\underline{m}|} \frac{\Gamma_\Omega(\underline{m}+\alpha) \varphi_{k+\underline{m}^*}(P(y^{1/2})a)}{\left(\frac{d}{r}\right)_{k+\underline{m}^*} \|\varphi_{\underline{m}}\|_F^2}$$

$$= c f_y(a) \sum_{\underline{n}, n_1 \leq k} (-1)^{kr+|\underline{n}|} \frac{(\alpha)_{k+\underline{n}^*} \varphi_{\underline{n}}(P(a^{1/2})y)}{\left(\frac{d}{r}\right)_{k+\underline{n}^*} \|\varphi_{\underline{m}}\|_F^2},$$

by the fact that $d_{\underline{m}} = d_{k+\underline{m}^*}$ (which follows by the explicit formula for $d_{\underline{m}}$, see [U3] and [FK2, Chapter XI]), and the fact that $P(y^{1/2})a = \ell P(a^{1/2})y$ for some $\ell \in L$ (see [FK2, Lemma XIV.1.2]). This establishes

(8.23) and completes the proof.   □

Our next goal is to express the operators $\{\Delta_\nu\}_{\nu=1}^r$ in terms of the operators $D(\lambda) = D^1(\lambda)$; this will prove Theorem 8.4. This is achieved by using Theorem 8.5 and the techniques of divided differences.

Let f be an analytic function in a subset Λ of $\mathbb{C}$, taking values in some Banach space. The (first-order) divided differences of f are

$$f^{[1]}(\lambda_0, \lambda_1) = \frac{f(\lambda_0) - f(\lambda_1)}{\lambda_0 - \lambda_1}, \qquad \lambda_0, \lambda_1 \in \Lambda,$$

where if $\lambda_0 = \lambda_1$ we interpret the quotient as $f'(\lambda_0)$. The higher order divided differences are defined inductively by

$$f^{[n]}(\lambda_0, \lambda_1, \cdots, \lambda_n) = \frac{f^{[n-1]}(\lambda_0, \cdots, \lambda_{n-2}, \lambda_{n-1}) - f^{[n-1]}(\lambda_0, \cdots, \lambda_{n-2}, \lambda_n)}{\lambda_{n-1} - \lambda_n}$$

with the same interpretation in case $\lambda_{n-1} = \lambda_n$. Alternatively,

$$f^{[n]}(\lambda_0, \cdots, \lambda_n) = \frac{1}{2\pi i} \int_\Gamma \frac{f(\xi)}{\displaystyle\prod_{j=0}^{n} (\xi - \lambda_j)} \, d\xi ,$$

where Γ is a positively oriented Jordan curve in Λ encircling all the points $\lambda_0, \lambda_1, \cdots, \lambda_n$.

Theorem 8.7 [AU]: *Let* $\lambda_1, \lambda_2, \cdots, \lambda_r \in \mathbb{C}$ *be distinct. Then* $\Delta_\nu \in \operatorname{span}\{D(\lambda_j)\}_{j=1}^r$, $\nu = 1, 2, \cdots, r$. *Consequently,* $\{D(\lambda_j)\}_{j=1}^r$ *are algebraically-independent free generators of* $\mathcal{D}(\Omega)$. *In particular, if* $\lambda_j = \alpha_j = (j-1)\frac{a}{2}$ *for* $1 \le j \le r$, *then for* $1 \le \nu \le r$

$$(8.29) \qquad \Delta_\nu = \frac{D^{[r-\nu]}(\alpha_\nu, \alpha_{\nu+1}, \cdots, \alpha_r)}{\binom{r}{\nu}},$$

where $D^{[r-\nu]}(\alpha_\nu, \cdots, \alpha_r)$ *are the divided differences of order* $r-\nu$ *of* $D(\lambda)$, *evaluated at* $\alpha_\nu, \cdots, \alpha_r$.

The proof is based on Theorem 8.5 and the facts that if we define

$$h_k(x) = \prod_{j=k+1}^{r} (x - \alpha_j), \text{ then } h_k^{[m]}(x_0, x_1, \cdots, x_m) = 0 \text{ for } m > r-k,$$

$h_k^{[r-k]}(x_0, x_1, \cdots, x_{r-k}) = 1$ identically, and $h_k^{[r-\ell]}(\alpha_\ell, \cdots, \alpha_r) = 0$ for $k < \ell$.

§9. Recent developments in related areas of analysis on symmetric domains

In this last section we report very briefly on some recent works in areas of analysis on symmetric domains which are related to the theory presented in this survey. Our list of references is surely not exhaustive, but nevertheless we think that this account may be useful for graduate students as well as researchers who want to enter those fields.

As described in the introduction, the theory of invariant Hilbert spaces of analytic functions is closely related to *quantization* on symmetric and more general domains. The papers [P8], [P9] and [U4] as well as the more recent manuscripts [EP], [OZ1] and [UU1] study various aspects of quantization in the sense of Berezin and Weyl, and use heavily the theory of invariant Hilbert spaces of analytic functions.

The theory of invariant Hilbert spaces of analytic functions on symmetric domains can be considered as part of *representation theory* of their automorphism group and its *harmonic analysis*. Some of the topics which were studied most intensively in recent years are Plancherel formulas [Z1], [PZ2]), Capelli identity ([KS1], [KS2] and [Sa3]), the structure of invariant differential operators ([U2], [U3], [UU], and [AU]), and vector valued reproducing kernels and representations ([P6] and [OZ2]). Other interesting recent works in representation theory and harmonic analysis close to our theory include [Sa1], [Sa2], [Z2], [Z3], [PZ1], [OZ3] and [Zhu4].

Recent works on *Hankel operators* and forms on symmetric domains are [Z2] and [PZ]. The recent manuscript [A3] studies the boundedness and compactness of the generalized Hankel operators (6.32) in terms of the membership of their symbols in generalized Bloch and little Bloch spaces respectively. The forthcoming book [US] contains a detailed account on *Toeplitz operators* and their *index theory* in symmetric and much more general domains. The theory of *pseudo differential analysis on symmetric cones* is presented in the forthcoming monograph [UU2].

Yan's operators (see §8) are given integral formulas in [Zhu2] and [Zhu3]. These operators are used to define and study the *holomorphic Besov spaces* on bounded symmetric domains ([Zhu2], [Zhu3]).

The theory of invariant Hilbert spaces of analytic functions is used also in the recent works [BM1] and [BM2] on *tuples of multiplication operators*.

The theory we have presented in this survey deals with non−compact hermitian symmetric spaces. These spaces can be realized as symmetric Siegel domains over a symmetric convex cone. Many concrete examples show that these classes of domains are too narrow, and require the development of Fourier and harmonic analysis over pseudo hermitian symmetric manifolds and over symmetric Siegel domains modeled over symmetric non-convex cones. The recent papers [DG] and [Gi2] are interesting contributions to these two themes respectively.

References

[A1]   J. Arazy, Realization of the invariant inner products on the highest quotients of the composition series, to appear in Arkiv für Matematik, 30(1992), 1-24.

[A2]   J. Arazy, Integral formulas for the invariant inner products in spaces of analytic functions on the unit ball, in Function Spaces (Proceedings of the conference on function spaces, Edwardsville, April 1989, Editor K. Jarosz), Lecture notes in Pure and Applied Mathematics, Vol. 136(1992), Marcel Dekker, pp. 9-23.

[A3]   J. Arazy, Boundedness and compactness of generalized Hankel operators on bounded symmetric domains, preprint (1994).

[AF1]   J. Arazy and S. D. Fisher, The uniqueness of the Dirichlet space among Möbius-invariant Hilbert spaces, Illinois J. Math. 29(1985), 449-462.

[AF2]   J. Arazy and S.D. Fisher, Weighted actions of the Möbius group and their invariant Hilbert spaces, Math. Publication Series, University of Haifa, Series 2 No. 23, 1989.

[AF3]   J. Arazy and S.D. Fisher, Invariant Hilbert spaces of analytic functions on bounded symmetric domains, in Operator Theory: Advances and Applications, Vol. 48, 1990, Birkhauser Verlag, Basel, 67-91.

[AFJP]   J. Arazy, S.D. Fisher, S. Janson, and J. Peetre, An identity for reproducing kernels in a planar domain and Hilbert Schmidt Hankel Operators, J. Reine Angew. Math. 406(1990), 179-199.

[AFP]   J. Arazy, S.D. Fisher, and J. Peetre, Hankel operators on weighted Bergman spaces, Amer. J. Math. 110(1988), 989-1054.

[AU]   J. Arazy and H. Upmeier, Invariant inner products on holomorphic functions on bounded symmetric domains, preprint.

[B]   F.A. Berezin, Quantization in complex symmetric spaces, Izv. Akad. Nauk. SSSR 9(1975), 341-379.

[BCZ]   C.A. Berger, L.A. Coburn, and I.H. Zhu, Function theory on Cartan domains and the Berezin - Toeplitz symbol calculus, Amer. J. Math. 110(1988), 921-953.

[BM1]   B. Bagchi and G. Misra, Homogeneous operators and systems of Imprimitivity, preprint (1994).

[BM2]   B. Bagchi and G. Misra, Homogeneous tuples of multiplication operators on twisted Bergman spaces, preprint (1994).

[C]   E. Cartan, Sur les domains bornés homogénes de l'espace de n variables complexes, Abh. Math. Sem. Univ. Hamburg 11(1935), 116–162.

[DG]   J.E. D'Atri and S. Gindikin, Siegel domain realization of pseudo–hermitian symmetric manifolds, Geomet. Dedicata 46(1993), 91–125.

[EP]   M. Englis and J. Peetre, On the correspondence principle for the quantized annulus, preprint (1994).

[FK1]   J. Faraut and A. Koranyi, Function spaces and reproducing kernels on bounded symmetric domains, J. Funct. Anal. 88(1990), 64–89.

[FK2]   J. Faraut and A. Koranyi, Analysis on symmetric cones, to appear in Oxford University Press.

[G]   F.P. Greenleaf, Invariant Means on Topological Groups. Van Nostrand, New York, 1969.

[Gi1]   S.G. Gindikin, analysis in homogeneous domains, Uspekhi Mat. Nauk. 19, No. 4(1964), 3–92 [in Russian]; English translation: Russian Math. Surveys 19, No. 4(1964), 1–90.

[Gi2]   S. Gindikin, Fourier transform and Hardy spaces of ∂-cohomology in tube domains, C.R. Acad. Sci. Paris 315(1992), 1139–1143.

[Ha]   Harish–Chandra, Representations of semisimple Lie groups VI, Amer. J. Math. 7(1956b),1 564–628.

[He]   S. Helgason, Groups and Geometrical Analysis, Academic Press, New York, 1984.

[Hu]   L.K. Hua, Harmonic Analysis on Functions of Several Complex Variables in the Classical Domains, Amer. Math. Soc., Providence, 1963.

[IS]   J.M. Isidro and L.L. Stacho, Holomorphic Automorphism Groups in Banach Spaces: An Elementary Introduction. North Holland Math. Studies No. 105, Elsevier, New York, 1985.

[Ka]   W. Kaup, A. Riemann mapping theorem for bounded symmetric domains in complex Banach spaces, Math. Z. 183(1983), 503–529.

[Kn]   A.W. Knapp, Bounded Symmetric Domains and Holomorphic
          Discrete Series; in Symmetric Spaces, short courses presented at
          Washington University, Marcel Dekker, New York, 1972, pp.
          211–246.

[Ko1]   A. Koranyi, The Poisson integral for generalized half planes and
          bounded symmetric domains, Ann. of Math. 82(1965),
          332–350.

[Ko2]   A. Koranyi, Hua–type integrals, hypergeometric functions and
          symmetric polynomials, preprint (1990).

[Ko3]   A. Koranyi, personal communication.

[Kob1]   S. Kobayashi, On automorphism groups of homogeneous complex
          manifolds, Proc. Amer. Math. Soc. 12(1961), 359–361.

[Kob2]   S. Kobayashi, Irreducibility of certain unitary representations,
          J. Math. Soc. Japan 20(1968), 638–642.

[Koe]   M. Koecher, An Elementary Approach to Bounded Symmetric
          Domains. Rice Univ., Houston, 1969.

[KS1]   B. Konstant and S. Sahi, The Capelli Identity, Tube domains,
          and the Generalized Laplace Transform, Adv. Math. 87(1991),
          71–92.

[KS2]   B. Konstant and S. Sahi, Jordan algebras and Capelli Identities,
          Invent. Math. 112(1993), 657–664.

[Ku]   R.A. Kunze, On the irreducibility of certain multiplier
          representations, Bull. Amer. Math. Soc. 68(1962), 93–94.

[L1]   O. Loos, Jordan Pairs, Lecture Notes in Math., Springer-
          Verlag, Berlin–Heidelberg–New York, No. 460, 1975.

[L2]   O. Loos, Bounded Symmetric Domains and Jordan Pairs,
          Department of Mathematics, University of California, Irvine,
          California, 1977.

[La]   M. Lassalle, Noyau de Szegö, K–types et algebres de Jordan,
          C.R. Acad. Sci., Paris 303, Série I (1986), 1–4.

[M]   C.C. Moore, Amenable subgroups of semi–simple groups and
          proximal flows, Israel J. Math. 34(1979), 121–138.

[N]   T. Nomura, Algebraically independent generators of invariant
          differential operators on symmetric cones, J. reine Agnew.
          Math., 400(1989), 122–133.

[O]   B. Ørstead , Composition series for analytic continuation of
          holomorphic discrete series representations of SU(n,n), Trans.
          Amer. Math. Soc. 260(1980), 563–573.

[OZ1]   B. Orsted and G. Zhang, Weyl quantization and tensor products of Fock and Bergman spaces, preprint, Odense University (1993).

[OZ2]   B. Orsted and G. Zhang, Reproducing Kernels and Composition Series for Spaces of Vector-Valued Holomorphic Functions on Tube Domains, to appear in J. Funct. Anal.

[OZ3]   B. Orsted and G. Zhang, Generalized Principal Series Representations and Tube Domains, preprint, Odense University (1994).

[P1]   J. Peetre, Möbius invariant function spaces in several variables, preprint (1982).

[P2]   J. Peetre, Type I, preprint (1988).

[P3]   J. Peetre, Analytic continuation of norms, preprint (1986).

[P4]   J. Peetre, Invariant function spaces connected with holomorphic discrete series, in P.L. Butzer, Anniversary Volume on Approximation and Functional Analysis, International Series of Numerical Mathematics, Vol. 65, Birkhauser Verlag Basel (1984), 119-134.

[P5]   J. Peetre, Moebius invariant function spaces – the case of hyperbolic space, Proc. R. Ir. Acad. Soc. 92A (1992), 243-265.

[P6]   J. Peetre, A reproducing formula for vector valued functions over Cartan domains, Esti. Tead. Akad. Toim. 42(1993), 213-221.

[P7]   J. Peetre, Hankel forms of arbitrary weights over a symmetric domain via the transvectant, Rocky Mountain J. Math 24 (1994).

[P8]   J. Peetre, Correspondence principle for quantized annulus Romanowski polynomials, Morse potential, J. Funct. Anal. 117(1993), 377-400.

[P9]   J. Peetre, The Berezin transform and Ha-plitz operators, J. Operator Theory 24(1990), 165-186.

[P]   V.V. Peller, Hankel operators in the class S_p and their implications (rational approximation, Gaussian processes, majorization problem for operators), Math. USSR Sbornik 41(1982), 443-479.

[Pe]   M. Peloso, Möbius invariant spaces on the unit ball, to appear in Michigan Math. J.

[Pi]   J.P. Pier, <u>Amenable Locally Compact Groups</u>, J. Wiley, New York, 1984.

[Py]   I.I. Pyaetskii–Shapiro, <u>Automorphic Functions and the Geometry of Classical Domains.</u> Gordon and Breach, New York, 1969.

[PZ1]   J. Peetre and Genkai Zhang, Harmonic analysis on the quantized Riemann sphere, Internat. J. Math. & Math. Sci. 16(1993), 225–244.

[PZ2]   J. Peetre and Genkai Zhang, A weighted Plancherel formula III. The Case of the hyperbolic matrix ball, Collect. Math. 43(1992), 273–301.

[R]   W. Rudin, <u>Function Theory in the Unit Ball of $\mathbb{C}^n$</u>, Springer–Verlag, New York, 1980.

[RV]   H. Rossi and M. Vergne, Analytic continuation of the holomorphic discrete series of a semisimple Lie group, Acta Math. 136(1976), 1–59.

[S]   W. Shmid, Die Randwerte holomorpher Functionen auf hermitesch symmetrischen Räumen, Inventiones Math. 9(1969), 61–80.

[Sa1]   S. Sahi, Explicit Hilbert Spaces for certain unipotent representations, Invent. Math. 110(1992) 409–418.

[Sa2]   S. Sahi, Unitary representation on the Shilov boundary of a symmetric tube domain, Representations of Groups and Algebras, Contemp. Math. 145(1993), 275–286, Amer. Math. Soc., Providence (1993).

[Sa3]   S. Sahi, The Capelli identity and unitary representation, Compositio Math. 81(1992), 247–260.

[U1]   H. Upmeier, <u>Symmetric Banach Manifolds and Jordan C*–algebras.</u> North–Holland Math. Studies No. 104, Elsevier, New York, 1985.

[U2]   H. Upmeier, <u>Jordan algebras in analysis, operator theory and quantum mechanics.</u> CBMS Regional Conference Series in Math., Amer. Math. Soc., Providence, RI No. 67, 1987.

[U3]   H. Upmeier, Jordan algebras and harmonic analysis on symmetric spaces, Amer. J. Math. 108(1986), 1–25.

[U4]   H. Upmeier, Weyl quantization of symmetric spaces; hyperbolic tube domains, J. Funct. Anal. 95(1991) 297–330.

[U5]   H. Upmeier, <u>Multivariable Toeplitz Operators and Index Theory</u>, to appear n Birkhauser–Verlag.

[UU1]   A. Unterberger and H. Upmeier, The Berezin transform and invariant differential operators, preprint (1993).

[UU2]   A. Unterberger and H. Upmeier, Pseudodifferential analysis on symmetric cones (1993).

[W]   N. Wallach, The analytic continuation of the discrete series, I. II, Trans. Amer. Math. Soc. 251(1979), 1–17, 19–37.

[Y1]   Z. Yan, Möbius transformations on bounded symmetric domains of tube type, preprint (1990).

[Y2]   Z. Yan, Invariant differential operators on holomorphic functions spaces, preprint (1991).

[Y3]   Z. Yan, Differential operators and function spaces, Contemporary Math. 142(1993), 121–142.

[Z1]   G. Zhang, A weighted Plancherel formula II. The case of a ball, Studia Math. 102(1992), 103–120.

[Z2]   G. Zhang, Tensor product sof Weighted Bergman spaces and invariant Ha–plitz operators, Math. Scand. 71(1992), 85–95.

[Z3]   G. Zhang, Jordan algebras and generalized principal series representations, preprint, Odense University (1994).

[Zhu1] K. Zhu, Möbius invariant Hilbert spaces of Holomorphic functions in the unit ball of $\mathbb{C}^n$, Trans. Amer. Math. Soc. 323(1991), 823–842.

[Zhu2] K. Zhu, Holomorphic Besov spaces on bounded symmetric domains, to appear in Quarterly J. Math.

[Zhu3] K. Zhu, Holomorphic Besov spaces on bounded symmetric domains II, preprint (1993).

[Zhu4] K. Zhu, Harmonic Analysis on Bounded Symmetric Domains, to appear in Harmonic Analysis in China.

Department of Mathematics
University of Haifa
Haifa 31905, ISRAEL

Electronic Address: jarazy@mathcs.haifa.ac.il

Contemporary Mathematics
Volume **185**, 1995

Homogeneous Operators and Systems of Imprimitivity

Bhaskar Bagchi and Gadadhar Misra [1]

Let Ω be a bounded symmetric domain in $\mathbb{C}^d$, and let $\mathcal{A}(\Omega)$ denote the normed linear space of functions holomorphic in some neighbourhood of $\bar{\Omega}$ with supremum on Ω as norm. Let G be the connected component of identity in the group of biholomorphic automorphisms of Ω. The domain Ω is a G - space with respect to the action $\omega \to \varphi(\omega)$ ([12, see p.158]). This action induces the map

$$f \to \varphi \cdot f \text{ for } f \in \mathcal{A}(\Omega) \text{ and } \varphi \in G, \text{ where } (\varphi \cdot f)(\omega) = f(\varphi^{-1}(\omega)).$$

DEFINITION 1 A bounded homomorphism $\varrho : \mathcal{A}(\Omega) \to \mathcal{L}(\mathcal{H})$ is said to be *homogeneous* if for each $\varphi \in G$, $\varrho(f)$ is unitarily equivalent to $\varrho(\varphi \cdot f)$ for all $f \in \mathcal{A}(\Omega)$ via a fixed unitary operator U_φ.

Here $\mathcal{H}$ is a separable Hilbert space and $\mathcal{L}(\mathcal{H})$ is the Banach space of bounded linear operators on $\mathcal{H}$ with operator norm.

Let $\mathcal{A}(\Omega) \otimes \mathcal{M}_k$ be the space of $k \times k$ matrices with entries from $\mathcal{A}(\Omega)$ with the norm $\|[f_{ij}]\| \overset{\text{def}}{=} \sup_{z \in \Omega} \|[f_{ij}(z)]\|_{\text{op}}$. Define $\varrho^{(k)} : A(\Omega) \otimes \mathcal{M}_k \to \mathcal{L}(\mathcal{H}) \otimes \mathcal{M}_k$ by $\varrho^{(k)}([f_{ij}]) = [\varrho(f_{ij})]$. It is clear that $\|\varrho^{(k)}\|$ is an increasing sequence and it need not be bounded even if the homomorphism ϱ is bounded. If $\sup_k \|\varrho^{(k)}\|$ is finite the homomorphism ϱ is said to be completely bounded.

Let $f_1, \ldots, f_d$ be the co - ordinate functions, and $T_1 = \varrho(f_1), \ldots, T_d = \varrho(f_d)$. The tuple $\boldsymbol{T} = (T_1 \ldots, T_d)$ is a commuting tuple of bounded operators. Further if p is any polynomial in d variables then we see that $p(\boldsymbol{T}) = \varrho(p)$ and hence

(1) $$\|p(\boldsymbol{T})\| \leq K \|p\|_\infty \quad \text{for all } p.$$

Conversely, if we have a commuting tuple of bounded operators $\boldsymbol{T}$ satisfying (1) then we obtain a homomorphism $\varrho : p \to p(\boldsymbol{T})$ which is bounded on the polynomials and therefore has a unique extension to the closure with respect to the supremum

[1]1991 Mathematics Subject Classification. Primary 47B 38, Secondary 22E 46, 32 H10
The second author was supported by a travel grant from NBHM

norm on Ω. Since this closure contains $\mathcal{A}(\Omega)$, we obtain a bounded homomorphism defined on $\mathcal{A}(\Omega)$.

It was shown in [10] that if a homogeneous homomorphism of $\mathcal{A}(\Omega)$ is irreducible (there is no common reducing subspace for $\varrho(f)$, $f \in \mathcal{A}(\Omega)$) then there exists a projective representation $\varphi \to U_\varphi$ of the group G on $\mathcal{H}$ such that

$$(2) \qquad (U_\varphi T_1 U_\varphi^*, \ldots, U_\varphi T_d U_\varphi^*) = \varphi(T_1, \ldots, T_d).$$

In the rest of this section and the next we assume that ϱ is irreducible. Thus to say that the homomorphism ϱ is homogeneous amounts to saying that the homomorphism ϱ of the subalgebra $\mathcal{A}(\Omega)$ of the C^* - algebra $C(\bar{\Omega})$ into $\mathcal{L}(\mathcal{H})$ and the projective representation $U : \varphi \to U_\varphi$ into the unitary operators on $\mathcal{H}$ satisfy the imprimitivity relation

$$(3) \qquad U_\varphi \, \varrho(f) \, U_\varphi^* = \varrho(\varphi \cdot f)$$

(cf. [12, Lemma 6.1]). This brings out a formal resemblance with Mackey systems of imprimitivity, where one has a homomorphism of the C^* - algebra $C(\bar{\Omega})$ and a group representation (projective) satisfying the imprimitivity condition. From the point of view of operator theory studying a Mackey system of imprimitivity amounts to studying the spectral behavior of a commuting tuple of normal operators. Indeed, let's say that a homomorphism ϱ is subnormal if $\{\varrho(f) : f \in \mathcal{A}(\Omega)\}$ is jointly subnormal. We showed in Theorem 5.1 of [2] that an irreducible homomorphism ϱ is subnormal if and only if the system in (3) embeds in a homogeneous Mackey system of imprimitivity on $C(\bar{\Omega})$. The "if" part is trivial while the "only if" part is an easy consequence of spectral theory for normal tuples. In view of this, it should be of great interest to Operator Theory to extend Mackey's theory to homogeneous homomorphisms in the absence of subnormality. In doing this, it will perhaps be useful to assume that the homomorphism is not merely bounded but completely contractive. In [3] this programme was carried out for single homogeneous contractions as opposed to tuples (the induced homomorphism in this case is automatically completely contractive). The explicit knowledge of the unitary dilation in terms of the given contraction was used in an essential manner. The theory of completely contractive homomorphisms as developed by Arveson gurantees that such a homomorphism dilates to a $*$ - homomorphism. If in the general case we start with a homogeneous completely contractive homomorphism then we should expect the dilation to be homogeneous as well (these are precisely the systems of imprimitivity of Mackey and are classified).

Our study of homogeneous homomorphisms has two parts. The first part is to obtain a large class of examples. The second part is to show that these homomorphisms are tractable.

1   Examples

It is clear that to obtain good examples one must have a manageable set of unitary invariants. Cowen and Douglas [4] have introduced some techniques from complex geometry which exhibit a certain real analytic function (curvature) as a complete unitary invariant for an operator tuple in the class $P_1(\Omega)$. The characteristic function of Sz.-Nagy and Foias is a complete unitary invariant for a completely non unitary contraction.

Cowen - Douglas operators

A commuting d - tuple of bounded operators $\boldsymbol{T} = (T_1, \ldots, T_d)$ is said to be in the Cowen - Douglas class $P_n(\Omega)$ if

(i) $\dim \cap \{\ker(T_k - \omega_k) : k = 1, 2, \ldots, d\} = n$ for all $\omega \in \Omega$,

(ii) The operator $T_\omega : \mathcal{H} \to \mathcal{H} \oplus \cdots \oplus \mathcal{H}$ defined by $T_\omega x = \oplus_{k=1}^{d}(T_k - \omega_k)x$ has closed range, and

(iii) the closed linear span of $\cap \{\ker(T_k - \omega_k) : k = 1, 2, \ldots, d\}$, $\omega \in \Omega$, is $\mathcal{H}$.

It was shown in [4] that each d - tuple $\boldsymbol{T}$ in $P_1(\Omega)$ determines a nonzero holomorphic map $\gamma : \Omega \to \mathcal{H}$ such that $\gamma(\omega) \in \cap\{\ker(T_k - \omega_k) : k = 1, 2, \ldots d\}$ for all $\omega \in \Omega$. Thus to each $\boldsymbol{T} \in P_1(\Omega)$, they associate a hermitian holomorphic vector bundle

$$E = \{(\omega, \gamma(\omega)) : T_k\gamma(\omega) = \omega_k\gamma(\omega)\} \xrightarrow{\pi} \Omega$$

over the domain Ω. It is then shown that two operator tuples in $P_1(\Omega)$ are unitarily equivalent if and only if the associated bundles are equivalent. Since the curvature

$$\mathcal{K}(\omega) = \left(\left(\frac{\partial^2}{\partial\omega_i\,\partial\omega_j}\log\|\gamma(\omega)\|^{-2}\right)\right).$$

is a complete invariant for line bundles it follows that two operator tuples in $P_1(\Omega)$ are unitarily equivalent if and only if their curvatures are equal.

Note that the operator tuple $\varphi(\boldsymbol{T})$ cresponds to the pullback bundle ψ^*E, where $\psi = \varphi^{-1} : \Omega \to \Omega$. Thus $\boldsymbol{T}$ is homogeneous if and only if E is unitarily equivalent to ψ^*E for each $\psi \in \Omega$. It follows that the corresponding curvature functions must be equal. Now consider the tangent bundle $T\Omega$ with the hermitian metric induced by

$$(4) \qquad \omega \to \left(\left(\frac{\partial^2}{\partial\omega_i\,\partial\omega_j}\log\|\gamma(\omega)\|^{-2}\right)\right).$$

The requirement that the curvatures for E and ψ^*E be the same for all $\psi \in G$ is the same thing as saying that the metric defined by (4) is G - invariant. However, upto scalar multiple there is only one such metric, namely the one induced by the Bergman kernel function $\gamma(\omega) = B(\omega, \omega)$ on Ω (cf. [7, Proposition 3.6, p.371]). It follows that an operator tuple $\boldsymbol{T}$ in $P_1(\Omega)$ is homogeneous if and only if

$$(5) \qquad \mathcal{K}(\omega) = \lambda\left(\left(\frac{\partial^2}{\partial\omega_i\,\partial\omega_j}\log\|B(\omega, \omega)\|^{-1}\right)\right),$$

for some positive constant λ.

We now exhibit operator tuples $\boldsymbol{T}_\lambda$ in $P_1(\Omega)$ with curvature as prescribed in (5). Let g be the genus (see [1]) of Ω, and let

$$\mathcal{W} = \{\lambda \geq 0 : B^{(\lambda)} \stackrel{\text{def}}{=} B^{\lambda/g} \text{ is nnd}\}$$

be the Wallach set (cf. [1]) and $\mathcal{H}^{(\lambda)}$ be the Hilbert space with reproducing kernel $B^{(\lambda)}$. Thus the only possible candidate for our homogeneous operator tuple is the adjoint of the tuple $\boldsymbol{M}^{(\lambda)} = (M_{z_1}, \ldots, M_{z_d})$ of multiplication by co-ordinate functions on the Hilbert space $\mathcal{H}^{(\lambda)}$.

THEOREM 1 *If a homogeneous tuple $\boldsymbol{T}$ is in $P_1(\Omega)$ then $\boldsymbol{T}^* = \boldsymbol{M}^{(\lambda)}$ for some $\lambda \in \mathcal{W}$.*

However, it is not clear if this tuple is bounded, let alone the question of its membership in $P_1(\Omega)$. It was shown in [10] that if $\lambda \geq g$ then $\boldsymbol{M}^{(\lambda)}$ is in $P_1(\Omega)$. In fact the argument in [10] together with a result from [6] shows that this is true for $\lambda > g - 1$. In section 2, we shall elaborate on this question and discuss our recent results from [2] in this connection.

What about operators in the Cowen - Douglas class $P_n(\Omega)$, $n \geq 2$? The difficulty in this situation stems from the fact that determining when two hermitian holomorphic bundles (not necessarily line bundles) are equivalent is a complicated affair. Recently D. Wilkins [14] has proved

THEOREM 2 *There is a one - one correspondence between isomorphism classes of homogeneous hermitian holomorphic bundles of rank n over the unit disk and unitary equivalence classes of pairs (F, A), where F and A are skew - Hermitian $n \times n$ matrices satisfying $[F, [F, A]] = -A$.*

Using this theorem he has been able to explicitly determine all homogeneous operators in $P_2(\mathbb{D})$. He has also studied homogeneous operators in $P_1(\mathbb{D})$, where the group is a Fuchsian subgroup of the Möbius group [13].

Sz.-Nagy and Foias Characteristic function

To each contraction operator T, Sz.-Nagy and Foias associate an operator valued holomorphic function Θ_T on the unit disk $\mathbb{D}$. To describe this function we need the following notation.

$$D_T = \sqrt{I - T^*T} \qquad D_{T^*} = \sqrt{I - TT^*}$$
$$\mathcal{D}_T = \operatorname{ran} D_T \qquad \mathcal{D}_{T^*} = \operatorname{ran} D_{T^*}$$
$$\Delta_T = \sqrt{I - \Theta_T \Theta_T^*} \qquad \mathcal{H} = H^2_{\mathcal{D}_{T^*}} \oplus \Delta_T L^2_{\mathcal{D}_T}$$
$$\mathcal{M} = \{(\Theta_T f, \Delta_T f) : f \in H^2_{\mathcal{D}_T}\} \qquad \mathcal{M}^\perp = \mathcal{H} \ominus \mathcal{M}.$$

The characteristic function Θ_T is defined as

$$\Theta_T(z) = -T + z D_{T^*}(I - zT^*)^{-1} D_T \in \mathcal{L}(\mathcal{D}_T, \mathcal{D}_{T^*}), \quad z \in \mathbb{D}.$$

It then follows that T is unitarily equivalent to the model operator $\mathcal{T} : (f, g) \to (zf, e^{it}g)$ on $\mathcal{H}$, compressed to $\mathcal{M}^\perp$. Sz.-Nagy and Foias prove that two completely non-unitary contractions T_1 and T_2 are unitarily equivalent if and only if there exists a unitary operator $U : \mathcal{D}_{T_1^*} \to \mathcal{D}_{T_2^*}$ and a unitary operator $V : \mathcal{D}_{T_2} \to \mathcal{D}_{T_1}$ such that $U\,\Theta_{T_1}(z)\,V = \Theta_{T_2}(z)$ for all $z \in \mathbb{D}$. Let φ be an automorphism of the unit disk $\mathbb{D}$. The fact that there exist two unitary operators U_φ and V_φ with

$$U_\varphi \Theta_{\varphi(T)}(z) V_\varphi = \Theta_T(\varphi^{-1}(z))$$

(cf. [11, p. 240]) immediately yields the following theorem (cf. [3]).

THEOREM 3 *Let T be a completely non-unitary contraction with the dimension of either $\mathcal{D}_T$ or $\mathcal{D}_{T^*}$ equal to 1. The operator T is homogeneous if and only if the characteristic function Θ_T is constant.*

It is easy to see that a completely non- unitary contraction which has a constant characteristic function must be homogeneous. However, if either dim $\mathcal{D}_T$ or dim $\mathcal{D}_{T^*}$ is strictly greater than 1 then such a contraction need not have a constant characteristic function. Examples of homogeneous contractions with non - constant characteristic functions were given in [3]. This paper also determined explicitly the projective representation of the automorphism group Aut($\mathbb{D}$) associated with a contraction whose characteristic function is constant. It was shown that the associated multiplier is nontrivial. However all these projective representations lift to linear representations of $SU(1,1)$. They are not necessarily irreducible. It will be an amusing problem to break up these representations into their irreducible parts and relate this to some spectral properties of the operator T.

Finally [3] makes the connection with the usual notion of a system of imprimitivity explicit in this context via dilations.

Any subspace which can be written as the difference of two invariant subspaces is said to be *semi-invaraint*. Let $\phi : C(\mathbb{T}) \to \mathcal{L}(\mathcal{H})$ be a $*$ - homomorphism and $U :$ Aut($\mathbb{D}$) $\to \mathcal{L}(\mathcal{H})$ be a unitary projective representation satisfying the imprimitivity relation. It is then easy to see that if $\mathcal{M}$ is a semi-invariant subspace for $\phi(\mathrm{id})$ and each U_φ leaves $\mathcal{M}$ invariant then the operator $T = P_\mathcal{M} \phi(\mathrm{id})\,|_\mathcal{M}$ is homogeneous with $U_\varphi T U_\varphi^* = \varphi(T)$. On the other hand, starting with an irreducible homogeneous operator T, we obtain a projective representation $\varphi \to V_\varphi$ satisfying $V_\varphi T V_\varphi^* = \varphi(T)$. Let W_T be the minimal unitary dilation for T on a Hilbert space $\mathcal{K}$ containing $\mathcal{H}$ as a semi-invariant space. Then there exists a projective representation $U :$ Aut($\mathbb{D}$) $\to \mathcal{L}(\mathcal{K})$ which leaves $\mathcal{H}$ invariant, and $U_\varphi W_T U_\varphi^* = \varphi(W_T)$. We thus obtain the usual system of imprimitivity.

In spite of all this we don't know how to parametrise the set of homogeneous contractions.

2 Wallach set and Subnormality

In this section Ω is a Cartan domain (i.e. irreducible bounded symmetric domain), with associated Wallach set $\mathcal{W}$, and $\lambda \in \mathcal{W}$ is fixed but arbitrary. We wish to examine the homomorphism $\varrho^{(\lambda)} : \mathcal{A}(\Omega) \to \mathcal{L}(\mathcal{H}^{(\lambda)})$ given by $\varrho^{(\lambda)}(f)(g) = f \cdot g$ (pointwise product), for $g \in \mathcal{H}^{(\lambda)}$, $f \in \mathcal{A}(\Omega)$ - provided it is defined. Equivalently, this is a study of the tuple $\boldsymbol{M}^{(\lambda)}$ of multiplication by coordinate functions on $\mathcal{H}^{(\lambda)}$.

Recall that according to Cartan's classification, the domain Ω is determined by five parameters d (dimension), g (genus), r (rank), a, b. Of these, d is the complex dimension of the ambient space, r is the number of orbits into which $G = \text{Aut}(\Omega)$ breaks $\partial\Omega$. The definition of the other parameters requires a Jordan theoretic background which we do not discuss; see Arazy [1]. These parameters are not independent, but are related by the two equations

$$(6) \qquad\qquad g \;=\; (r-1)a + b + 2$$

$$(7) \qquad\qquad g - 1 \;=\; (d/r) + (r-1)a/2.$$

Of these two equations, (6) is wellknown, while we have verified (7) by going through the Cartan classification case by case. Recall that the Wallach set $\mathcal{W}$ breaks up into two disjoint parts $\mathcal{W} = \mathcal{W}_d \cup \mathcal{W}_c$, where the discrete part $\mathcal{W}_d$ and the continuous part $\mathcal{W}_c$ are given by

$$\mathcal{W}_d \;=\; \{\, j\,a/2 : j = 0, 1, \ldots, r-1 \,\}$$
$$\mathcal{W}_c \;=\; \{\, \lambda : \lambda > (r-1)\,a/2 \,\}.$$

Thus (7) is saying that the r points in $(d/r) + \mathcal{W}_d$ partition the interval $[d/r, g-1] \subseteq \mathcal{W}_c$ into $r - 1$ equal parts. By Faraut and Koranyi [6], for $\lambda > g - 1$ - i.e., for $\lambda \in (d/r) + \mathcal{W}_c$ - the inner product on $\mathcal{H}^{(\lambda)}$ is given by integration against Lebesgue measure suitably weighted, so that $\varrho^{(\lambda)}$ is bounded and subnormal for $\lambda > g - 1$. On the other hand, for $\lambda \le (r-1)\,a/2$, i.e., for $\lambda \in \mathcal{W}_d$, the decomposition of $\mathcal{H}^{(\lambda)}$ into K-irreducibles (K = Isotropy of 0 in G) shows that not all analytic polynomials are in $\mathcal{H}^{(\lambda)}$ though $\mathcal{H}^{(\lambda)}$ contains the constant functions - whence $\varrho^{(\lambda)}$ is not every where defined for $\lambda \in \mathcal{W}_d$. Moreover, we shall see that for $\lambda < d/r$, $\varrho^{(\lambda)}$ is not a bounded homomorphism. Thus, from our point of view, the interval $[d/r, g-1]$ is the "mystery region", and (7) gells well with the subnormality conjecture given below.

In view of the above discussion and the Theorem 1.1 in [2], we feel confident to make the following

CONJECTURE 1 (BOUNDEDNESS) $\boldsymbol{M}^{(\lambda)}$ is bounded (equivalently $\varrho^{(\lambda)}$ is defined through out $\mathcal{A}(\Omega)$) if and only if $\lambda \in \mathcal{W}_c$.

By the preceeding discussion, the "only if" part is trivial and it is the "if" part which is of interest. In [2] this conjecture was proved for Ω of the type $I_{n,m}$, $m \ge n$, i.e., for the unit ball of the space of $n \times m$ complex matrices with the usual operator norm. Note that neither here nor in [2] do we address the question of whether the operator tuple $\boldsymbol{M}^{(\lambda)}$ corresponding to $\varrho^{(\lambda)}$ is in the Cowen - Douglas class. Probably it is, for $\lambda \in \mathcal{W}_c$. We are able to evade this question thanks to homogeneity of $\varrho^{(\lambda)}$ (when defined). Indeed, there is a naturally defined projective unitary representation of G on $\mathcal{H}^{(\lambda)}$, and the transformation rule for the reproducing kernel of $\mathcal{H}^{(\lambda)}$ under this action of G easily translates into homogeneity. Homogeneity of $\varrho^{(\lambda)}$ allows us to prove :

LEMMA 1 *For any fixed* $\lambda \in \mathcal{W}_c$, *the following are equivalent.*

(i) $\boldsymbol{M}^{(\lambda)}$ *is bounded (equivalently, $\varrho^{(\lambda)}$ is every where defined),*

(ii) $\varrho^{(\lambda)}$*(trace) is bounded,*

(iii) the kernel $(c^2 - \mathrm{trace}(z)\overline{\mathrm{trace}(\omega)})B^{(\lambda)}(z,\omega)$ on Ω is non-negative definite for some finite constant c.

Here $B^{(\lambda)} = B^{\lambda/g}$ is the reproducing kernel for $\mathcal{H}^{(\lambda)}$ and trace denotes the unique linear elementary spherical function (esf) (see [1]) on Ω. Of course, $(i) \Rightarrow (ii)$ is trivial. By the K - irreducibility of the space $\mathrm{Hom}\,(1)$ of linear (analytic) polynomials on Ω, this space is spanned by $\{\mathrm{trace} \circ k : k \in K\}$. So, by homogeneity, (ii) implies $\varrho^{(\lambda)}(f)$ is bounded for all f in $\mathrm{Hom}\,(1)$, in particular for all the co-ordinate functions. Hence $(ii) \Rightarrow (i)$. The equivalence $(ii) \Leftrightarrow (iii)$ is a standard fact from the theory of reproducing kernels.

In [2] we proved the boundedness conjecture for $\Omega = I_{n,m}$ by verifying (iii) of Lemma 1. This depends on our explicit knowledge (derived from Weil's character formula for $U(n)$) of how the pointwise product of an arbitrary esf and trace decomposes into a linear combination of esf's. It is only the absence of an equally explicit decomposition [2] for an arbitrary Cartan domain that stops us from proving the boundedness conjecture in full generality. In a private conversation with the second named author J. Arazy suggested that this conjecture may also be proved by an appeal to the theory of composition series of invariant Banach spaces (see [6]). Ignorance prevents us from commenting on this prospect.

¿From now on we assume that $\lambda \in \mathcal{W}_c$ and (hence ?) $\varrho^{(\lambda)}$ is defined. Recall that the Cartan domain Ω is the open unit ball with respect to a uniquely determined Banach space norm on $\mathbf{C}^d$. In the following, $\|\cdot\|$ refers to this norm.

LEMMA 2 *Let $z \in \mathbf{C}^d$ with $\|z\| = t > 1$. Then there is an element $\varphi \in G$ such that φ has an analytic continuation to $t\,\bar{\Omega}$ and $\|\varphi(z)\| > \|z\|$.*

Indeed, for a suitable choice of $\omega \in \Omega$, $\varphi = \varphi_\omega$, the unique involution in G interchanging 0 and ω, works. For the details of this computation in the case $\Omega = I_{n,m}$, see [2].

Using this lemma and homogeneity of $\varrho^{(\lambda)}$, it is easy to prove

THEOREM 4 *The joint Taylor spectrum of $\boldsymbol{M}^{(\lambda)}$ is $\bar{\Omega}$.*

Indeed, each point $\omega \in \Omega$ is an eigenvalue of the adjoint of $\boldsymbol{M}^{(\lambda)}$ with $B^{(\lambda)}(\cdot,\omega) \in \mathcal{H}^{(\lambda)}$ as the corresponding eigen vector. This shows that $\bar{\Omega}$ is contained in the Taylor spectrum. For the opposite inclusion take z in the spectrum maximising $\|z\|$. Say $\|z\| = t$. If the claimed inclusion was false we would have $t > 1$. Get hold of φ as in the Lemma. By the spectral mapping property $\varphi(z)$ is in the spectrum of $\varphi(\boldsymbol{M}^{(\lambda)})$. But by homogeneity, the latter is unitarily equivalent to $\boldsymbol{M}^{(\lambda)}$, so that $\varphi(z)$ is in the spectrum of $\boldsymbol{M}^{(\lambda)}$. But this contradicts the maximality of $\|z\|$.

---

[2]Note added during revision: G. Zhang has informed us that such a decomposition is now available in [15].

Notice how homogeneity allows us to avoid all computations of Koszul complexes in the above argument. Such computations would in all probability be impossible to execute anyway!

We would like to know all the values of λ for which $\varrho^{(\lambda)}$ is completely contractive. Subnormality of $\varrho^{(\lambda)}$ is an obvious sufficient condition for this. Here we only have conjectures inspired, again, by the case $I_{n,m}$.

CONJECTURE 2 (SUBNORMALITY) *For* $\lambda \in \mathcal{W}_c$ *the following are equivalent :*

(i) $\varrho^{(\lambda)}$ *is subnormal,*

(ii) $\lambda \in (d/r) + \mathcal{W}$,

(iii) The reproducing kernel $B^{(\lambda)}$ of $\mathcal{H}^{(\lambda)}$ has the Hardy kernel $B^{(d/r)}$ as a factor in the sense that $B^{(\lambda)}/B^{(d/r)}$ (pointwise ratio) is nonnegative definite on Ω.

Note that the equivalence $(ii) \Leftrightarrow (iii)$ is a trivial consequence of the definition of $\mathcal{W}$. We have included the statement (iii) since it highlights the central position of the Hardy space $\mathcal{H}^{(d/r)}$ and relates the subnormality question to the theory of module tensor products developed in [5]. Conjecturally, $\varrho^{(\lambda)}$ is subnormal if and only if $\mathcal{H}^{(\lambda)}$ has the Hardy space as a module factor. Also note that, as already remarked, we do have subnormality of $\varrho^{(\lambda)}$ for $\lambda > g-1$ i.e., for $\lambda \in (d/r) + \mathcal{W}_c$. Thus the conjecture is open only for $d/r < \lambda \leq g-1$. Arguing as in [2] it can be shown that $\varrho^{(\lambda)}$ is subnormal if and only if the inner product $\langle \cdot, \cdot \rangle_\lambda$ on $\mathcal{H}(\lambda)$ is "given by" a (necessarily unique) probability μ_λ supported inside $\bar{\Omega}$, in the sense that $\langle f, g \rangle_\lambda = \int f\bar{g}\,d\mu_\lambda$ for $f, g \in \mathcal{A}(\Omega)$. Plugging in this formula into the fact that we have a unitary representation (see 1.10 in [1]) of G on $\mathcal{H}^{(\lambda)}$, we obtain :

LEMMA 3 $\varrho^{(\lambda)}$ *is subnormal if and only if there is a probability μ_λ supported inside $\bar{\Omega}$ such that μ_λ is quasi - invariant with respect to the G - action, and $\left(\frac{d\mu_\lambda \circ \varphi}{d\mu_\lambda}\right)(z) = |J(\varphi, z)|^{2\lambda/g}$ for $z \in \bar{\Omega}$, $\varphi \in G$.*

In [2] we use Fourier analysis and the combinatorics of Schur polynomials to show that such a probability μ_λ exists for $\Omega = I_{n,m}$ if and only if $\lambda \in (d/r) + \mathcal{W}$, hence proving the subnormality conjecture in this case. Undoubtedly this Fourier analysis will generalise to prove the "only if" part of the conjecture in its complete generality. To prove the "if" part the requirement is a better understanding of the Jack polynomials associated with the Cartan domains.

Let det denote the esf on Ω corresponding to the signature $(1, \ldots, 1)$. It is not difficult to see that $\varrho^{(\lambda)}(\det)$ is a direct sum of weighted shifts for all λ in $\mathcal{W}_c$. Moreover, when Ω is tube like, det is almost K - invariant in the sense that $\det \circ k = \chi(k) \cdot \det$ for all $k \in K$. Using this invariance it is easy to compute the weights of the component shifts. Arguing as in [2], one then sees that :

THEOREM 5 *Let Ω be a tube like Cartan domain. Then*

(a) For $\lambda < d/r$, $\{\varrho^{(\lambda)}(\det^n) : n = 1, 2 \ldots\}$ is not bounded in norm, whence $\varrho^{(\lambda)}$ is not bounded for $\lambda < d/r$. In consequence, $\varrho^{(\lambda)}$ is not subnormal for $\lambda < d/r$.

(b) For $\lambda \geq d/r$, $\varrho^{(\lambda)}(\det)$ is subnormal.

When Ω is not tube like, we don't even have a conjecture as to the range of λ for which $\varrho^{(\lambda)}(\det)$ is subnormal.

Like subnormality, the question of boundedness of $\varrho^{(\lambda)}$ remains open for λ in the interval $d/r \leq \lambda \leq g-1$. However, we can prove :

THEOREM 6 *For* $\lambda \in (d/r) + \mathcal{W}$, $\varrho^{(\lambda)}$ *is completely contractive.*

Indeed, using the theory of reproducing kernels, one can see that $\varrho^{(\lambda)}$ is contractive if and only if for all $f \in \mathcal{A}(\Omega)$ with $\|f\| \leq 1$, $(1 - f(z)\overline{f(\omega)})B^{(\lambda)}(z, \omega)$ is an nnd kernel on Ω. However, for $\lambda \in (d/r) + \mathcal{W}$, we have the factorisation $B^{(\lambda)} = B^{(d/r)} A$, where $A(= B^{(\lambda - d/r)})$ is an nnd kernel. Since there is a probability (namely the unique K - invariant probability on the Shilov boundary of Ω) giving the inner product on $\mathcal{H}^{(d/r)}$, $\varrho^{(d/r)}$ is subnormal and hence contractive. It follows that $(1 - f(z)\overline{f(\omega)})B^{(d/r)}(z, \omega)$ is nnd. Since the product of two nnd kernels is nnd, this completes the proof that $\varrho^{(\lambda)}$ is contractive for $\lambda \in (d/r) + \mathcal{W}$. A similar argument proves complete contractivity.

Note that this last Theorem is an evidence in favour of our subnormality conjecture.

References

[1] J. Arazy, *A survey of invariant Hilbert spaces of analytic functions on bounded symmetric domains*, Mittag - Leffler preprint series, 1991.

[2] B. Bagchi and G. Misra, *On homogeneous tuples of multiplication operators on twisted Bergman spaces*, submitted.

[3] D.N. Clark and G. Misra, *On Homogeneous Contractions and Unitary Representation of* $SU(1,1)$, to appear, J. Operator Theory.

[4] M.J. Cowen and R.G. Douglas, *Operators possessing an open set of eigen values*, Fejer - Riesz conf., Budapest, 1980.

[5] R.G. Douglas and V.I. Paulsen, *Hilbert Modules over Function Algebras*, Longman Research Notes, 217, 1989.

[6] J. Faraut and A. Koranyi, *Function spaces and reproducing kernels on bounded symmetric domains*, J. Func. Anal., 88(1990), 64 -89.

[7] S. Helgason, *Differential Geometry, Lie Groups and Symmetric Spaces*, Academic Press, 1978.

[8] A.A. Kirillov, *Elements of the theory of representations*, Springer Verlag, 1976.

[9] G. Misra, *Curvature and discrete Series representation of* $SL_2(R)$, J. Integral Equations & Operator Theory, 9 (1986), 452-459.

[10] G. Misra and N.S.N. Sastry, *Homogeneous tuples of operators and holomorphic discrete series representation of some classical groups*, J. Operator Theory, 24 (1990), 23-32.

[11] B. Sz.-Nagy and C. Foias, *Harmonic analysis of Hilbert space operators*, North-Holland, 1970.

[12] V.S. Varadarajan, *Geometry of quantum theory*, second edition, Springer Verlag, 1985.

[13] D.R. Wilkins, *Operators, Fuchsian groups and automorphic vector bundles*, Math. Ann., 290, 405-424 (1991).

[14] D.R. Wilkins, *Homogeneous Vector Bundles and Cowen - Douglas Operators*, International J. of Math. Vol. 4, No. 3 (1993) 503-520.

[15] Genkai Zhang, *Some recurrence formulas for spherical polynomials on tube domains*, preprint, 1993.

Statistics and Mathematics Division
Indian Statistical Institute
Bangalore 560 059
India e-mail : bbagchi@isibang.ernet.in
e-mail : gm@isibang.ernet.in

Contemporary Mathematics
Volume **185**, 1995

DUALITY BETWEEN A^∞ AND $A^{-\infty}$ ON DOMAINS WITH NONDEGENERATE CORNERS

David E. Barrett

University of Michigan

ABSTRACT. We investigate the existence of a natural pairing between the algebra A^∞ of holomorphic functions smooth up to the boundary and the algebra $A^{-\infty}$ of holomorphic functions that blow up like a power of the distance to the boundary (or the corresponding spaces of $(n,0)$-forms). We show in particular that such a pairing exists when the boundary is piecewise smooth with complex-nondegenerate corners.

We give examples of domains for which this pairing is not a duality pairing. In particular we give a recipe for constructing functions in $A^{-\infty}$ that are orthogonal to A^∞. This recipe is shown in turn to be related to the existence of eigenvalues for certain compact non-self-adjoint operators connected with the $\overline{\partial}$ problem on bordered Riemann surfaces.

An alternate demonstration of failure of duality is provided by a biholomorphic mapping argument.

§1. Introduction

Let Ω be a relatively compact subdomain of a complex manifold M. Let $A^\infty_{(n,0)}(\Omega)$ denote the space of holomorphic $(n,0)$-forms on Ω whose coefficients extend C^∞-smoothly to $\overline{\Omega}$. Let $A^{-\infty}_{(n,0)}(\Omega)$ denote the space of holomorphic $(n,0)$-forms whose coefficients extend continuously to $\overline{\Omega}$ after being multiplied by a large enough power of the distance to the boundary.

We will say that Ω is a *Bell domain* if there is a separately continuous Hermitian pairing

$$\langle \cdot, \cdot \rangle : A^{-\infty}_{(n,0)}(\Omega) \times A^\infty_{(n,0)}(\Omega) \to \mathbb{C}$$

such that

$$\langle f, g \rangle = C_n \int_\Omega f \wedge \overline{g}$$

whenever $f \in A^\infty_{(n,0)}(\Omega)$, where $C_n = 2^{-n} i^{n(n-2)}$.

The topology on $A^\infty_{(n,0)}(\Omega)$ is induced from the usual Frechet topology on $C^\infty_{(n,0)}(\overline{\Omega})$; the locally convex space $A^{-\infty}_{(n,0)}(\Omega)$ is topologized as the inductive limit as $k \to \infty$ of Banach spaces of holomorphic forms with norm $\|(\text{dist}(\cdot, b\Omega))^k f\|_{L^\infty}$ [RR, V.2].

1991 *Mathematics Subject Classification.* 32H10.

Supported in part by a grant from the National Science Foundation.

When $M = \mathbb{C}^n$ we can of course strip the $dz_1 \wedge \cdots \wedge dz_n$ off of all of the forms and view $\langle \cdot, \cdot \rangle$ as a pairing between the corresponding function algebras $A^{-\infty}(\Omega)$ and $A^{\infty}(\Omega)$ extending the usual L^2-pairing on Ω.

In Section 2 we exhibit some sufficient conditions as well as some necessary conditions for a domain to be Bell. In particular, domains whose boundary is smooth except for complex-nondegenerate corners are Bell (Theorem 2.2).

Given a Bell domain Ω, it is natural to ask whether $A^{\infty}_{(n,0)}(\Omega)$ and $A^{-\infty}_{(n,0)}(\Omega)$ are in fact dual to each other. This is known to be the case for many important classes of domains (in particular for smooth strictly pseudoconvex domains and other domains with globally regular $\overline{\partial}$-Neumann operators; see [BB],[BS1]). On the other hand, if $A^{\infty}_{(n,0)}(\Omega)$ fails to be dense in $A^{-\infty}_{(n,0)}(\Omega)$ (see [Ba1], [BF] and the end of §2 below for examples) then an application of the Hahn-Banach theorem shows that there are linear functionals on $A^{-\infty}_{(n,0)}(\Omega)$ that are not represented by $A^{\infty}_{(n,0)}(\Omega)$.

In [Ki], Kiselman gave an example of a Bell domain on which $A^{\infty}_{(n,0)}(\Omega)$ is dense in $A^{-\infty}_{(n,0)}(\Omega)$ but which nevertheless admits non-zero forms in $A^{-\infty}_{(n,0)}(\Omega)$ that are orthogonal to $A^{\infty}_{(n,0)}(\Omega)$, so that duality again fails. In Section 3 we exhibit a method for constructing large families of such examples. The method depends on the existence of eigenvalues for certain non-self-adjoint compact operators; this eigenvalue problem is analyzed in Section 4.

Section 5 contains an alternate construction of functionals on $A^{-\infty}_{(n,0)}(\Omega)$ that are not represented by $A^{\infty}_{(n,0)}(\Omega)$ using a biholomorphic mapping argument.

A later paper will show how to obtain asymptotic expansions for the Bergman kernel on the domains discussed in Section 3; also, it will explain how these domains can be used as "model domains" (as in [Ba2]) for certain two-dimensional smooth bounded pseudoconvex domains. (In particular, it will turn out that for these domains there is a sort of partial converse to the results of Boas and Straube [BS2] to the effect when the infinite-type points of the boundary of a smooth bounded pseudoconvex domain in $\mathbb{C}^2$ form a bordered Riemann surface then the cohomology class they discuss does indeed "generically" obstruct the exact regularity of the Bergman projection.)

§2. BELL DOMAINS

Theorem 2.1 [Bel]. *(See also* [Str]*.) If $b\Omega$ is smooth then Ω is a Bell domain.*

Proof. First consider the Euclidean case $\Omega \subset\subset M = \mathbb{C}^n$.

Let r be a defining function for Ω, let T be a vector field of type $(0,1)$ such that $Tr \equiv 1$ in a neighborhood of $b\Omega$, and let $\chi \in C_0^{\infty}(\Omega)$ satisfy $Tr \equiv 1$ on $\mathrm{supp}(1 - \chi)$.

To define $\langle f, g \rangle$ pick s so big that $r^s f$ is continuous on $\overline{\Omega}$; then let

$$\langle f, g \rangle = \int_{\Omega} \frac{r^s}{s!} f \, \overline{(T^*)^s((1-\chi)g)} \, dV + \int_{\Omega} f \, \overline{\chi g} \, dV,$$

where $T^* = -\overline{(T + \mathrm{div}\, T)}$ is the formal adjoint of T. The integrals are convergent, and integration by parts (using $Tf = 0$ and noting that boundary terms vanish) reveals that the value of the first integral on the right hand side is independent of the choice of (sufficiently large) s; hence the pairing is well-defined and coincides with the L^2-pairing ($s = 0$) when $f \in A^{\infty}(\Omega)$.

The required continuity properties are easily checked. (The pairing fails to be jointly continuous, however.)

For a general ambient manifold M pick r, T, and χ as above and let $\mathcal{L}_T$ denote the Lie derivative with respect to T [KN, I.3]. Recalling the product rule

$$\mathcal{L}_T(\eta \wedge \omega) = (\mathcal{L}_T \eta) \wedge \omega + \eta \wedge (\mathcal{L}_T \omega)$$

we see that $\mathcal{L}_T$ is an "order zero" (multiplication) operator on holomorphic forms so that the integrals in

$$\langle f, g \rangle = C_n \left(\int_\Omega \frac{r^s}{s!} (\mathcal{L}_{-T})^s (f \wedge \overline{(1-\chi)g}) + \int_\Omega f \wedge \overline{\chi g} \right)$$

converge for s sufficiently large.

Verification proceeds as before, using Stokes' Theorem in the guise

$$\int_\Omega \mathcal{L}_T \omega = \int_{b\Omega} \iota_X \omega$$

to handle the integrations by parts. $\quad \square$

Theorem 2.2. *Let $\Omega \subset M$ be a relatively compact domain that may be written as an intersection $\bigcap\limits_{j=1}^{N} \Omega_j$ of smooth domains such that*

(i) *all intersections of the $b\Omega_j$ are transverse*

and

(ii) *for each $S \subset \{1, \ldots, N\}$ the intersection $B_S = \bigcap\limits_{j \in S} b\Omega_j$, if non-empty, is a CR-manifold of CR-dimension $n - |S|$.*

Then Ω is a Bell domain.

Note that the second hypothesis above guarantees in particular that every intersection of $n + 1$ distinct boundaries is empty.

Example. Products of smooth domains are Bell domains.

Proof of Thm. 2.2. Let r_k be a defining function for Ω_k. The hypotheses imply that the $(1,0)$-forms $\{\overline{\partial} r_k : r_k(p) = 0\}$ are linearly independent at p. Thus one can find vector fields $T_1^{(p)}, \ldots, T_N^{(p)}$ of type $(0,1)$ such that $T_j^{(p)} r_k \equiv \delta_{jk}$ in a neighborhood of p when $r_j(p) = r_k(p) = 0$. Using a partition of unity we can now pick vector fields T_j on M of type $(0,1)$ such that $T_j r_k \equiv \delta_{jk}$ on a neighborhood U_{jk} of $b\Omega_j \cap b\Omega_k$.

Let $\{\chi_S\}_{S \subset \{1, \ldots, N\}}$ be a partition of unity subordinate to the covering

$$\{U_S\}_{S \subset \{1, \ldots, N\}}$$

of M, where

$$U_S = \bigcap_{j, k \in S} U_{jk} \setminus \bigcup_{\ell \notin S} b\Omega_\ell.$$

To compute $\langle f, g \rangle$ pick a multi-index $s = (s_1, \ldots, s_N)$ so that $r^s f = r_1^{s_1} \ldots r_N^{s_N} f$ is continuous on $\overline{\Omega}$ and let

$$\langle f, g \rangle = \sum_{S \subset \{1, \ldots, N\}} \int_\Omega \left(\prod_{j \in S} \frac{r_j^{s_j}}{s_j!} \right) f \overline{\left(\left(\prod_{j \in S} (T_j^*)^{s_j} \right) \chi_S g \right)} \, dV \tag{2.1a}$$

or

$$C_n \sum_{S \subset \{1, \ldots, N\}} \int_\Omega \left(\prod_{j \in S} \frac{r_j^{s_j}}{s_j!} \right) \left(\prod_{j \in S} (\mathcal{L}_{-T_j})^{s_j} \right) (\chi_S f \wedge \overline{g}) . \tag{2.1b}$$

As before, if $f \in A^\infty$ then integration by parts shows that (2.1) agrees with the usual L^2 pairing.

It remains to be shown that our pairing is independent of s. This is obvious if A^∞ happens to be dense in $A^{-\infty}$. The general case can be treated by a computation involving commutators of the T_j, but it will be more convenient to use the following "local approximation" argument.

Note first that the integrals in (2.1) are still well-defined if $g \in C^\infty(\overline{\Omega})$; by a partition of unity argument it suffices now to show that they are independent of (large enough) s when g has small support.

In the Euclidean case we may assume in particular that there is a vector $v \in \mathbb{C}^n$ which points outward along each $b\Omega_j$ passing through the support of g. Thus

$$\langle f, g \rangle = \sum_{S \subset \{1, \ldots, N\}} \lim_{\epsilon \searrow 0} \int \left(\prod_{j \in S} \frac{r_j^{s_j}}{s_j!} \right) f(\cdot - \epsilon v) \overline{\left(\left(\prod_{j \in S} (T_j^*)^{s_j} \right) \chi_S g \right)} \, dV.$$

Integrating by parts as before and summing, we find that

$$\langle f, g \rangle = \lim_{\epsilon \searrow 0} \int_\Omega f(\cdot - \epsilon v) \overline{g} \, dV, \tag{2.2}$$

independent of s.

The case of a general ambient manifold is handled using the same technique in local coordinates. $\square$

We record the following consequence of the alternate description (2.2) of the pairing.

Addendum to Theorem 2.2. *The formula (2.1) actually defines a separately continuous Hermitian pairing on $A_{(n,0)}^{-\infty}(\Omega) \times C_{(n,0)}^\infty(\overline{\Omega})$ extending the usual L^2-pairing. This pairing is independent not only of s but of the choice of r_j, T_j, and χ_S.*

So it is clear that Bell domains need not have smooth boundaries. Nevertheless, the singularities of the boundaries of Bell domains are very restricted. In one dimension for example we have the following result.

Theorem 2.3. *Any quasidisk which is a Bell domain has a C^∞-smooth boundary.*

Proof. Let Ω be a quasidisk which is a Bell domain, let $\phi : \Omega \to \Delta$ be a Riemann mapping for Ω, and let $\psi = \phi^{-1}$.

The boundary distances in Ω and Δ are related by the estimate

$$\text{dist}(z, b\Omega) \leq C \,\text{dist}(\phi(z), b\Delta)^\epsilon$$

for suitable constants C and ϵ. (This follows in particular from the "hyperbolic bound property" of quasidisks [Jo], [Ge, 4.5].) Thus the pullback map for functions maps $A^{-\infty}(\Delta)$ to $A^{-\infty}(\Omega)$. Since $|\phi'(z)| \leq \text{dist}(z, b\Omega)^{-1}$ by Cauchy's estimate, the corresponding pullback map for (1,0)-forms maps $A^{-\infty}_{(1,0)}(\Delta)$ to $A^{-\infty}_{(1,0)}(\Omega)$.

For $g \in A^\infty_{(1,0)}(\Omega)$ we can thus define a functional

$$\lambda_g : A^{-\infty}_{(1,0)}(\Delta) \to \mathbb{C}$$
$$f \mapsto \langle \phi^* f, g \rangle_\Omega.$$

But it is known [Be2, Theorem 28.3] that any such functional is represented by pairing against a form $h \in A^\infty_{(1,0)}(\Delta)$; thus

$$\langle \phi^* f, g \rangle_\Omega = \langle f, h \rangle_\Delta$$

for $f \in A^{-\infty}_{(1,0)}(\Delta)$. Since this must hold in particular for all square-integrable f it is clear that $h = \psi^* g$. Taking $g = dz$ we find in particular that $\psi' \in A^\infty(\Delta)$. Since ψ is univalent, ψ' can have at most simple zeros on $b\Delta$. (This follows from geometric considerations or, alternatively, from the estimate $|S_\psi(z)| \leq 6/(1 - |z|^2)^2$ on the Schwarzian derivative $S_\psi = (\log \psi')'' - ((\log \psi')')^2/2$ of univalent functions [Le, II.1.20].) But if ψ' has a simple zero on $b\Delta$ then Ω has an inward-pointing cusp or slit, contrary to our assumption that Ω is a quasidisk [Ge, 3.6]. Hence ψ is a diffeomorphism and the boundary of Ω is smooth. $\square$

Note that if Ω is a piecewise-smooth domain with a simple corner or a jump discontinuity in some higher derivative at some point $p \in b\Omega$ then an alternate proof that Ω fails to be Bell is obtained by checking that

$$\varlimsup_{\Omega \not\ni q \to p} \int_\Omega (z - q)^{-k} dz \wedge \overline{dz} = \infty$$

for large enough k.

On the other hand, if $\Omega \subset \mathbb{C}$ is a domain with inward-pointing cusps or slits which is the image of the unit disk Δ under a biholomorphic mapping $\psi \in A^\infty(\Delta)$ then it is easy to check that the pull-back map ψ^* maps $A^{-\infty}_{(1,0)}(\Omega)$ to $A^{-\infty}_{(1,0)}(\Delta)$ and $A^\infty_{(1,0)}(\Omega)$ to $A^\infty_{(1,0)}(\Delta)$ so that Ω is a Bell (non-quasidisk) domain with pairing

$$\langle f, g \rangle_\Omega = \langle \psi^* f, \psi^* g \rangle_\Delta.$$

Also, if Ω is Bell and $p \in \Omega$ then $\Omega \setminus \{p\}$ is Bell. In fact, we can choose a projection Q from $A^{-\infty}(\Omega \setminus \{p\})$ onto the closed subspace $A^{-\infty}(\Omega)$ and set $\langle f, g \rangle_{\Omega \setminus \{p\}} = \langle Qf, g \rangle_\Omega$. Q is not unique, and neither is the pairing. As explained in the introduction, the failure of A^∞ to be dense in $A^{-\infty}$ keeps the duality map $A^\infty(\Omega \setminus \{p\}) \to (A^{-\infty}(\Omega \setminus \{p\}))^*$ from being surjective no matter which pairing is used.

§3 A FAMILY OF LOCALLY POLYHEDRAL DOMAINS

Let $R \subset\subset \hat{R}$ be a bordered Riemann surface and let h be real-valued and harmonic on $\hat{R}$. (We will assume that $\hat{R}$ retracts onto $\overline{R}$.) Let

$$\Omega = \Omega_{R,h} = \{(z, w) \in R \times \hat{\mathbb{C}} : \operatorname{Re} e^{ih(z)} w > 0\}.$$

(Here $\hat{\mathbb{C}}$ denotes the Riemann sphere.) Ω is thus relatively compact in $\hat{R} \times \hat{\mathbb{C}}$.

The boundary of Ω thus consists of two smooth pieces, $bR \times \hat{\mathbb{C}}$ (henceforth known as the *vertical boundary*) and $\{(z, w) \in \overline{R} \times \hat{\mathbb{C}} : \operatorname{Re} e^{ih(z)} w = 0\}$ (henceforth known as the *twisting boundary*), intersecting transversally in the totally real manifold $\{(z, w) \in bR \times \hat{\mathbb{C}} : \operatorname{Re} e^{ih(z)} w = 0\}$. Thus by Theorem 2.2 we have the following.

Theorem 3.1. $\Omega_{R,h}$ *is a Bell domain.*

Locally we can write h as the real part of a holomorphic function H so that the twisting boundary is defined by $\operatorname{Re} e^{iH(z)} w = 0$. Thus Ω is *locally* (but not, in general, globally) an analytic polyhedron. It is easy to check that $\overline{\Omega}$ is in fact locally biholomorphic to the closed bidisk $\overline{\Delta^2}$ and moreover that Ω is covered by Δ^2.

Theorem 3.2. *(Compare* [BF]*.)* $A_{(2,0)}^{\infty}(\Omega)$ *is dense in* $A_{(2,0)}^{-\infty}(\Omega)$.

Proof. Pick $\beta > 0$ small enough so that w^{β} (defined using a global branch of $\log w$) satisfies $|\arg(w^{\beta})| < \pi/4$ on Ω.

Fix $f = f(z, w) \, dz \wedge dw \in A_{(2,0)}^{-\infty}(\Omega)$ to be approximated. Since $e^{-\delta(w^{\beta} + w^{-\beta})} f$ has rapid decay at $w = 0, \infty$ and tends to f in $A_{(2,0)}^{-\infty}(\Omega)$ as $\delta \searrow 0$, we may assume that f itself decays rapidly at $w = 0, \infty$.

We may now invoke the Mellin inversion formula (see [Fol2, 7.6]) to assert that

$$f(z, w) = \frac{1}{2\pi} \int_{\mathbb{R}} w^{i\xi} \mathcal{M} f(z, \xi) \, d\xi \tag{3.1}$$

where the partial Mellin transform $\mathcal{M}f$ is defined by

$$\mathcal{M}f(z, \xi) = \int_{\substack{\arg w = \theta \text{ fixed} \\ -\pi/2 - h(z) < \theta < \pi/2 - h(z)}} w^{-i\xi - 1} f(z, w) dw$$

and decays rapidly as $\xi \to \pm\infty$.

Approximating the integral (3.1) by a Riemann sum we find that f can be approximated in $A_{(2,0)}^{-\infty}(\Omega)$ by a finite sum of the form

$$\sum_j c_j(z) w^{i\xi_j} \, dz \wedge dw$$

with $c_j \in A^{-\infty}(R)$.

The factors $w^{i\xi_j} \, dz \wedge dw$ can be tempered at $w = 0, \infty$ by multiplying by $e^{-\delta(w^{\beta} + w^{-\beta})}$ as before. It remains, then, to observe that by a small adaptation of [Be2, Lemma 28.2] the $c_j(z)$ can be approximated in $A^{-\infty}(R)$ by $c_j \in A^{\infty}(R)$. □

Theorem 3.3. *Let R, h, and $\Omega = \Omega_{R,h}$ be as above. Suppose that there exist a function $\psi \in C^\infty(\overline{R})$ and a complex number λ such that $\psi \equiv 0$ on $b\Omega$ and the (1,0) form $e^{i\lambda h} \partial \psi$ is holomorphic. Then the (2,0)-form $e^{i\lambda h} \partial \psi \wedge w^{\lambda/2-1}\, dw \in A_{(2,0)}^{-\infty}(\Omega)$ is orthogonal to $A_{(2,0)}^\infty(\Omega)$.*

Proof. $\hat{R}$ is non-compact and thus admits a nowhere-vanishing vector field Y of type (1,0) with holomorphic coefficient [For, Thm. 26.6]. Let η be the corresponding dual (1,0)-form; thus $\partial \psi = (Y\psi)\eta$ and $\mathcal{L}_Y \eta = 0$. Lift Y to a horizontal vector field (also called Y) on $\hat{R} \times \hat{\mathbb{C}}$; thus in particular $Yw = 0$.

Let X denote the vector field $Y - 2i(Yh)w\dfrac{\partial}{\partial w}$ on $\hat{R} \times \hat{\mathbb{C}}$. Recall that $\mathcal{L}_X df = d(Xf)$ and note that

(i) X is tangent to the twisting boundary since

$$X(\operatorname{Re} e^{ih(z)}w) = -i(Yh)(\operatorname{Re} e^{ih(z)}w);$$

(ii) $\mathcal{L}_X$ annihilates conjugate-holomorphic forms;

and

(iii)

$$\mathcal{L}_X\left(e^{i\lambda h(z)}\psi(z)\eta \wedge w^{\lambda/2-1}\, dw\right) = e^{i\lambda h(z)}\partial\psi \wedge w^{\lambda/2-1}\, dw \in A_{(2,0)}^{-\infty}(\Omega).$$

Pick β as in the proof of Theorem 3.2. Then

$$e^{-\delta(w^\beta + w^{-\beta})}e^{i\lambda h}\partial\psi \wedge w^{\lambda/2-1}\, dw \to e^{i\lambda h}\partial\psi \wedge w^{\lambda/2-1}\, dw$$

in $A_{(2,0)}^{-\infty}(\Omega)$ as $\delta \searrow 0$ so that

$$\langle e^{i\lambda h}\partial\psi \wedge w^{\lambda/2-1}\, dw, g\rangle$$

$$= \lim_{\delta\searrow 0}\langle e^{-\delta(w^\beta+w^{-\beta})}e^{i\lambda h}\partial\psi \wedge w^{\lambda/2-1}\, dw, g\rangle$$

$$= \lim_{\delta\searrow 0} C_2 \int_\Omega \mathcal{L}_X\left(e^{-\delta(w^\beta+w^{-\beta})}e^{i\lambda h(z)}\psi(z)\eta \wedge w^{\lambda/2-1}\, dw \wedge \overline{g}\right)$$

$$- \lim_{\delta\searrow 0} C_2 \int_\Omega X(e^{-\delta(w^\beta+w^{-\beta})}) \cdot e^{i\lambda h(z)}\psi(z)\eta \wedge w^{\lambda/2-1}\, dw \wedge \overline{g}.$$

The first term vanishes by Stokes' Theorem. (The boundary term vanishes since the quantity in parentheses vanishes on the vertical boundary where X fails to be tangential.) Since the integrand in the second term is holomorphic in w, we may interpret the second integral as a holomorphic pairing along w-fibers followed by an integration in the z-plane. But

$$X(e^{-\delta(w^\beta+w^{-\beta})}) \cdot w^{\lambda/2-1}\, dw \to 0$$

in $A^{-\infty}$ on each fiber as $\delta \searrow 0$, uniformly in z.

Thus $\langle e^{i\lambda h}\partial\psi \wedge w^{\lambda/2-1}\, dw, g\rangle = 0$, as required. $\square$

§4 AN EIGENVALUE PROBLEM

Let R and h be as in §3. Our goal in this section is to study the hypothesis of Theorem 3.3.

Let $T : L^2_{(1,1)}(R) \to L^2_{(1,0)}(R)$ be the L^2-minimizing inverse (also known as the "Kohn solution operator") to $\overline{\partial}$. That is, $T\eta$ is the unique square-integrable $(1,0)$-form orthogonal to $A^\infty_{(1,0)}(R)$ and solving $\overline{\partial}T\eta = \eta$ in the sense of distributions.

Let S_h denote the compact non-self-adjoint operator

$$L^2_{(1,1)}(R) \to L^2_{(1,1)}(R)$$

$$\eta \mapsto \overline{\partial}h \wedge T\eta.$$

Theorem 4.1. *A necessary and sufficient condition for the existence of a non-zero holomorphic $(1,0)$-form of the form $e^{i\lambda h}\partial\psi$ with $\psi \in C^\infty(\overline{R})$, $\psi \equiv 0$ on $b\Omega$ is that i/λ should be an eigenvalue for the operator S_h.*

Proof. If $e^{i\lambda h}\partial\psi$ is holomorphic and ψ vanishes on bR then $\overline{\partial}\partial\psi = -i\lambda\overline{\partial}h \wedge \partial\psi$ and $\partial\psi = T\overline{\partial}\partial\psi$ (since $\partial\psi$ is orthogonal to $A^\infty_{(1,0)}$), so $S_h\overline{\partial}\partial\psi = (i/\lambda)\overline{\partial}\partial\psi$. Also, $\overline{\partial}\partial\psi \equiv 0$ implies $\partial\psi \equiv 0$ since $\int_\Omega \partial\psi \wedge \overline{\partial\psi} = -\int_\Omega \psi\partial\overline{\partial\psi}$. (This also handles the case $\lambda = 0$.)

For the converse, we note that since T gains one degree of regularity on Sobolev spaces ([Be2, p. 61], [Fol2, 7.32]), any eigenform for S_h must be smooth on $\overline{\Omega}$. Thus if η is an eigenform for S then $T\eta$ is a smooth $(1,0)$ form orthogonal to $A^\infty_{(1,0)}$ and is thus of the form $\partial\psi$, where $\psi \in C^\infty(\overline{R})$ vanishes on bR [Be2, p. 61]. The eigenvalue equation now reads $\overline{\partial}h \wedge \partial\psi = \frac{i}{\lambda}\overline{\partial}\partial\psi$, which implies that $e^{i\lambda h}\partial\psi$ is holomorphic. Also, $e^{i\lambda h}\partial\psi \not\equiv 0$ since $\eta = \overline{\partial}\partial\psi \not\equiv 0$. $\square$

Since S_h is compact, $\mathrm{Spec}\, S_h \backslash \{0\}$ is discrete in $\mathbb{C} \backslash \{0\}$ and consists of eigenvalues; 0 itself is not an eigenvalue unless $\overline{\partial}h \equiv 0$ and hence h is constant.

Theorem 4.2. *If h_1 and h_2 have the same conjugate periods then $\mathrm{Spec}\, S_{h_1} = \mathrm{Spec}\, S_{h_2}$.*

In particular, if h is non-constant and has a single-valued harmonic conjugate then $\mathrm{Spec}\, S_h = \{0\}$ and (by the comment preceding the theorem) S_h has no eigenvalues.

Proof. It suffices to show $\mathrm{Spec}\, S_{h_1} \subset \mathrm{Spec}\, S_{h_2}$.

We may write $h_2 = h_1 + H + \overline{H}$, $H \in A^\infty(R)$. The result then follows from Theorem 4.1 by noting that if $e^{i\lambda h_1}\partial\psi$ is holomorphic then so is $e^{i\lambda H}e^{i\lambda h_1}\partial\psi = e^{i\lambda h_2}\partial(e^{-i\lambda\overline{H}}\psi)$. $\square$

Theorem 4.3. $\mathrm{Spec}\, S_h \cap i\mathbb{R} = \{0\}$.

Proof. If $e^{\mu h}\partial\psi$ is holomorphic with $\psi \equiv 0$ on bR and $\mu \in \mathbb{R}$ then $0 \le \int_R e^{\mu h}\partial\psi \wedge \overline{\partial\psi} = \int_R \overline{\psi}\partial(e^{\mu h}\partial\psi) = 0$ and thus $\partial\psi \equiv 0$. $\square$

Using the fundamental existence theorem for harmonic forms [Co,8-4] we can write $dh = \eta + *du$, where $\eta = \eta_h$ is a uniquely determined harmonic form with $\eta|_{bR} \equiv 0$ and u is a single-valued harmonic function on a neighborhood of $\overline{R}$.

Theorem 4.4. *If $\eta_h \not\equiv 0$ and $\int_\gamma \eta \in 2\pi\mathbb{Z}/k$, $(k \in \mathbb{R}$ fixed) for all closed curves γ in R as well as for all paths γ in $\overline{R}$ with endpoints in bR then $\dfrac{i}{2mk} \in \operatorname{Spec} S_h$ for all $m \in \mathbb{Z} \setminus \{0\}$.*

Proof. Since h is single-valued, the periods of $*du$ along any closed curve in R also lie in $2\pi\mathbb{Z}/k$. Thus letting v be a (multiple-valued) harmonic conjugate for u and fixing $z_0 \in bR$ we find that $e^{mk(u-iv)}(e^{i2mk \int_{z_0}^z \eta} - 1)$ is smooth and single-valued on $\overline{R}$ and vanishes on bR. Moreover,

$$e^{i2mkh}\partial\left(e^{mk(u-iv)}(e^{-i2mk \int_{z_0}^z \eta} - 1)\right) = Ce^{mk(u+iv)}\eta^{(1,0)}$$

is holomorphic (C a non-zero constant), so by Theorem 4.1 we are done. $\square$

The set V of real harmonic forms η with $\eta|_{bR} \equiv 0$ is a finite dimensional vector space; if we choose a basis $\gamma_1, \ldots, \gamma_k$ for $H^1(R, bR)$ and adjoin a collection $\gamma_{k+1}, \ldots, \gamma_N$ of paths joining a fixed component of bR to each of the remaining $N - k$ components of bR then the period map $P : V \to \mathbb{R}^N$, $\eta \mapsto (\int_{\gamma_1} \eta, \ldots, \int_{\gamma_N} \eta)$ is an isomorphism (apply [Co, 8-4] to the Schottky double).

By Theorem 4.2, $\eta_{h_1} = \eta_{h_2}$ implies $\operatorname{Spec} S_{h_1} = \operatorname{Spec} S_{h_2}$. Also, given $\eta \in V$ we can find real-valued harmonic functions h and u on a neighborhood of $\overline{R}$ so that $\overline{\partial}(h - iu) = \eta^{(0,1)}$ [For, Thm. 25.6]; thus $dh - *du = \eta$ and $\eta = \eta_h$.

With this in mind, we set $E = \{\eta \in V : \operatorname{Spec} S_h$ is non-trivial when $\eta_h = \eta\}$. Then E is open in W [Au,3.4.5]. Moreover, by Theorem 4.4, E contains $P^{-1}(\mathbb{Q}^N) \setminus \{0\}$.

By considering *complex*-valued harmonic functions we can easily construct a finite-dimensional complex vector space of operators corresponding to the complexification $V \otimes \mathbb{C}$ of V. Citing the results [Au, 7.1.10, 7.1.13] on spectra of holomorphic families of operators we have the following.

Theorem 4.5. *$V \setminus E$ is closed and pluripolar (and hence nowhere dense) in $V \otimes \mathbb{C}$.*

In particular, non-zero conjugate periods of h "usually" give rise to a non-trivial spectrum.

Question. *Is $E = V \setminus \{0\}$?*

§5 FUNCTIONALS INDUCED BY FORMS NOT IN A^∞

Keeping the notation of §3, let $\Omega_j = \Omega_{R_j, h_j}, j = 1, 2$, where

$$R_1 = \{z : e^{-\pi} < |z| < e^\pi\},$$
$$R_2 = \{z : e^{-\pi/2} < |z| < e^{\pi/2}\},$$
$$h_1(z) = \log|z|,$$

and

$$h_2(z) = -\log|z|/2.$$

Using a global branch of $\log w$ we can then define a biholomorphic map

$$\psi : \Omega_1 \to \Omega_2$$
$$(z, w) \mapsto (zw^{-i}, w^{1/2})$$

with inverse

$$\phi : \Omega_2 \to \Omega_1$$
$$(z, w) \mapsto (zw^{2i}, w^2).$$

(To check that $\psi(\Omega_1) = \Omega_2$ use that fact that $\operatorname{Re} \tau > 0$ if and only if $e^{-\pi/2} < |\tau^i| < e^{\pi/2}$.)

We note that the pull-back map

$$\phi^* : A_{(2,0)}^{-\infty}(\Omega_1) \to A_{(2,0)}^{-\infty}(\Omega_2)$$
$$f(z, w)dz \wedge dw \mapsto 2w^{1+2i} f(zw^{2i}, w^2)\, dz \wedge dw$$

is an isomorphism and that the inverse

$$\psi^* : A_{(2,0)}^{-\infty}(\Omega_2) \to A_{(2,0)}^{-\infty}(\Omega_1)$$
$$f(z, w)dz \wedge dw \mapsto \frac{1}{2} w^{-i-1/2} f(zw^{-i}, w^{1/2})\, dz \wedge dw$$

fails to map $A_{(2,0)}^{\infty}(\Omega_2)$ to $A_{(2,0)}^{\infty}(\Omega_1)$.

Nevertheless, forms in $\psi^*(A_{(2,0)}^{\infty}(\Omega_2))$ do induce linear functionals on $A_{(2,0)}^{-\infty}(\Omega_1)$ since we may set

$$\langle \psi^* f, g \rangle_{\Omega_1} = \langle f, \phi^* g \rangle_{\Omega_2}. \tag{5.1}$$

If $f \in A_{(2,0)}^{\infty}(\Omega_2)$ then $\psi^* f$ is the unique square-integrable form such that (5.1) holds for all holomorphic square-integrable g. Thus the functional on $A_{(2,0)}^{-\infty}(\Omega_1)$ induced by $\psi^* f$ cannot be induced by a form in $A_{(2,0)}^{\infty}(\Omega_1)$ unless $\psi^* f$ itself lies in $A_{(2,0)}^{\infty}(\Omega_1)$.

REFERENCES

[Au]   B. Aupetit, *A primer on spectral theory*, Springer-Verlag, New York, 1991.

[Ba1] D. Barrett, *Irregularity of the Bergman projection on a smooth bounded domain in $\mathbb{C}^2$*, Annals of Math. **119** (1984), 431-436.

[Ba2] _____, *Behavior of the Bergman projection on the Diederich-Fornæss worm*, Acta Math. **168** (1992), 1-10.

[BF]   D. Barrett and J. Fornæss, *Uniform approximation of holomorphic functions on bounded Hartogs domains in $\mathbb{C}^2$*, Math. Zeit. **191** (1986), 61-72.

[Be1] S. Bell, *Biholomorphic mappings and the $\bar{\partial}$-problem*, Annals of Math. **114** (1981), 103-113.

[Be2] S. Bell, *The Cauchy transform, potential theory, and conformal mapping*, CRC Press, Boca Raton, Fla., 1992.

[BB]   S. Bell and H. Boas, *Regularity of the Bergman projection and duality of holomorphic function spaces*, Math. Ann. **267** (1984), 473-478.

[BS1] H. Boas and E. Straube, *Equivalence of regularity for the Bergman projection and the $\bar{\partial}$-Neumann operator*, Manuscripta math. **67** (1990), 25-33.

[BS2]   H. Boas and E. Straube, *De Rham cohomology of manifolds containing the points of infinite type, and Sobolev estimates for the $\bar{\partial}$-Neumann problem*, Jour. Geometric Analysis **3** (1993), 225-235.

[Co]   H. Cohn, *Conformal mapping on Riemann surfaces*, Dover, New York, 1976.

[Fol1]   G. Folland, *Fourier analysis and its applications*, Wadsworth & Brooks/Cole, Pacific Grove, Calif., 1992.

[Fol2]   G. Folland, *Introduction to partial differential equations*, Princeton University Press, 1976.

[For]   O. Forster, *Fectures on Riemann surfaces*, New York, Springer-Verlag, 1981.

[Ge]   F. W. Gehring, *Characterstic properties of quasidisks*, Université de Montréal, 1982.

[Jo]   P. W. Jones, *Extension theorems for BMO*, Indiana Univ. Math. J. **29** (1980), 41–66.

[Ki]   C. Kiselman, *A study of the Bergman projection in certain Hartogs domains*, Proc. Symp. Pure Math. **52** (1991), 219-232.

[KN]   S. Kobayashi and K. Nomizu, *Foundations of differential geometry (vol. 1)*, Interscience, London, 1963.

[Le]   O. Lehto, *Univalent functions and Teichmüller spaces*, Springer-Verlag, New York, 1987.

[RR]   A. Robertson and W. Robertson, *Topological vector spaces (2nd ed.)*, Cambridge University Press, 1973.

[Str]   E. Straube, *Harmonic and analytic functions admitting a distribution boundary value*, Ann. Scuola Norm. Sup. Pisa Sci. Serie IV **10** (1984), 559-591.

DEPARTMENT OF MATHEMATICS, ANN ARBOR, MI 48109-1003 USA

E-mail address: barrett@math.lsa.umich.edu

Contemporary Mathematics
Volume 185, 1995

COMMUTATIVE SUBSPACE LATTICES, COMPLETE DISTRIBUTIVITY and APPROXIMATION

KENNETH R. DAVIDSON

ABSTRACT. This paper is a survey of certain aspects of CSL algebras, emphasizing the role of complete distributivity as a tool for doing various useful approximations both of the lattice and various operators in the algebra.

This paper was presented at an AMS summer seminar in Seattle in July 1993 on the topic of multivariate operator theory. It is the author's thesis that the study of interesting nonselfadjoint operator algebras is one important facet of multivariate operator theory. In this article, I am concerned with the class of reflexive algebras with *commutative subspace lattice*. That is, one begins with a lattice $\mathcal{L}$ of subspaces such the orthogonal projections onto these subspaces commute with each other. The algebra $\mathrm{Alg}(\mathcal{L})$ consists of all operators leaving each subspace in $\mathcal{L}$ invariant. Equivalently, these are reflexive operator algebras containing a masa (maximal abelian selfadjoint subalgebra). The goal of the subject is to classify various structural properties of the algebras, to the extent possible, by corresponding properties of the underlying lattice. This paper has some overlap with the survey article [21], but to a large extent, we are concentrating on subsequent developments.

To illustrate the ideas, we will need a few basic examples.

EXAMPLE 1. A *nest* is a complete chain of closed subspaces. The corresponding algebra of operators should be thought of as the operators with a given upper triangular form. These algebras are well understood, at least by comparison to more general CSL algebras. (See [9].) As such, they are a constant source of problems and possibilities in the general case.

1991 *Mathematics Subject Classification.* 47D25.
Partially supported by an NSERC grant.

One special nest to keep in mind is the *Volterra* nest $\mathcal{N}$ on $L^2(0,1)$ consisting of the subspaces

$$N_t = \{f \in L^2(0,1) \mid supp(f) \subset [0,t]\}$$

for $0 \leq t \leq 1$. The corresponding algebra will be denoted by $\mathcal{T}(\mathcal{N})$ rather than $\mathrm{Alg}(\mathcal{N})$ to emphasize its triangular nature. In this case, the nest is said to be *continuous* because no element of the nest has an immediate successor. The von Neumann algebra $\mathcal{N}'$ of operators commuting with the orthogonal projections $P(N_t)$ are precisely the multiplication operators by functions in $L^\infty(0,1)$, which is a nonatomic masa.

EXAMPLE 2. A CSL $\mathcal{L}$ which is also a Boolean algebra (under complementation) corresponds to a type I von Neumann algebra $\mathrm{Alg}(\mathcal{L}) = \mathcal{L}'$. In some sense, these algebras are the most different from the nest algebras of the previous example. The typical example is given by the collection of all subspaces of the form

$$L_E = \{f \in L^2(0,1) \mid f(x) = 0 \text{ a.e. } x \in (0,1) \setminus E\}$$

as E runs over all measurable subsets of $(0,1)$ modulo null sets. Then $\mathrm{Alg}(\mathcal{L}) = L^\infty(0,1)$.

EXAMPLE 3. Let X be a compact metric space. Suppose that m is a finite regular Borel measure on X. And suppose that there is a closed partial order $\leq$ on X, meaning that $\{(x,y) \in X^2 \mid y \leq x\}$ is closed. A subset E of X is *increasing* if $x \in E$ and $x \leq y$ implies that $y \in E$. The Hilbert space will be $L^2(m)$, and $\mathcal{L} = \mathcal{L}(X,m,\leq)$ consists of all subspaces of the form

$$L_E = \{f \in L^2(0,1) \mid f(x) = 0 \text{ a.e. } x \in (0,1) \setminus E\}$$

as E runs over all increasing Borel sets.

An important example to keep in mind, as it exhibits many of the difficulties, is to take X to be the unit square with planar Lebesgue measure and the standard partial order

$$(x_1, y_1) \leq (x_2, y_2) \quad \text{if and only if} \quad x_1 \leq x_2 \quad \text{and} \quad y_1 \leq y_2.$$

The increasing Borel sets are those subsets of the square bounded below by the graph of a decreasing function of $[0,1]$ into itself. The corresponding CSL algebra is, in fact, isomorphic to the tensor product of two copies of the Volterra nest algebra $\mathcal{T}(\mathcal{N})$. However, from the various points of view that we will pursue here, this algebra is much more complicated.

1. Arveson's work

The class of CSL algebras were introduced by Arveson in his seminal paper [5]. As the title of this paper indicates, the primary motivation was the study of invariant subspaces. In [3], Arveson showed that $\mathcal{B}(H)$ is the only weak-∗ closed algebra containing a masa with no proper invariant subspaces. Then in [33], Radjavi and Rosenthal showed that a WOT closed algebra containing a masa whose invariant subspaces formed a nest was necessarily the whole nest algebra. Arveson's paper settles the question of whether this is always the case for arbitrary algebras containing a masa. Surprizingly, the answer is no.

As the invariant subspace lattice of a masa is just the lattice of projections in the masa, the lattice of invariant subspaces of any superalgebra is a CSL. So consider a fixed commutative subspace lattice $\mathcal{L}$. The first ingredient of Arveson's analysis is to show that every CSL has a representation of the form $\mathcal{L} = \mathcal{L}(X, m, \leq)$ of Example 3 above.

This then permits the explicit construction of certain operators that belong to $\mathrm{Alg}(\mathcal{L})$. Indeed, if μ is a finite (complex) Borel measure on X^2, the two associated marginal measures are

$$\mu_1(E) = \mu(E \times X) \quad \text{and} \quad \mu_2(E) = \mu(X \times E).$$

When both μ_1 and μ_2 are dominated by some multiple of the measure m, one can define a *pseudo-integral* operator by the formula

$$(T_\mu f, g) = \iint_{X^2} f(y)\overline{g(x)} \, d\mu.$$

It isn't difficult to show that T_μ belongs to $\mathrm{Alg}(\mathcal{L})$ if and only if μ is supported on $\mathcal{G}(\leq) = \{(x, y) \in X^2 \mid y \leq x\}$. The difficulty lies in showing that $\mathcal{G}(\leq)$ actually supports enough such measures.

A simple, but non-trivial, example of such pseudo-integral operators are obtained by considering measures on the diagonal $D = \{(x, x) : x \in X\}$. If $f \in L^\infty(m)$, the measure μ obtained by transferring $f \, dm$ to D yields the operator $T_\mu = M_f$, the multiplication operator on $L^2(m)$ by f. Thus the masa belongs to the set of pseudo-integral operators.

Arveson shows that the set of operators of this form is very rich. Indeed, it is an algebra whose invariant subspaces are precisely $\mathcal{L}$. This shows that the algebra $\mathrm{Alg}(\mathcal{L})$ determines $\mathcal{L}$. This part of the theory now has a simple proof due to Haydon [18].

THEOREM 1.1. *Every commutative subspace lattice is reflexive:* $\mathrm{Lat}(\mathcal{L}) = \mathcal{L}$.

Arveson goes on to show that the algebra $\mathcal{A}_{\min}(\mathcal{L})$ obtained by closing the set of pseudo integral operators in $\mathrm{Alg}(\mathcal{L})$ in the weak-* topology has a remarkable property.

THEOREM 1.2. *Let $\mathcal{L}$ be a commutative subspace lattice, and let $\mathcal{A}_{min}(\mathcal{L})$ be the weak-* closed algebra generated by the pseudo-integral operators (with respect to some representation of $\mathcal{L}$). Then if $\mathcal{A}$ is any weak-* closed algebra containing a masa with $\mathrm{Lat}(\mathcal{A}) = \mathcal{L}$, then*

$$\mathcal{A}_{min} \subseteq \mathcal{A} \subseteq \mathrm{Alg}(\mathcal{L}).$$

What makes this result even more remarkable is that $\mathcal{A}_{\min}$ is therefore canonical even though the representation of $\mathcal{L}$ as $\mathcal{L}(X, m, \leq)$ is not (nor is the choice of a masa).

The coup de grace was Arveson's construction of a CSL for which $\mathcal{A}_{\min}$ is strictly smaller than $\mathrm{Alg}(\mathcal{L})$. This example was based on the failure of spectral synthesis in commutative harmonic analysis. Subsequently, this connection with spectral synthesis was strengthened by Froelich [17]. Froelich contructed a natural class of algebras associated to subsets of a compact abelian group generalizing Arveson's example. The lattice turns out to be synthetic precisely when the subset of the group is a set of spectral synthesis.

Consequently, Arveson coined the term *synthetic* lattice for those lattices satisfying the equality $\mathcal{A}_{\min}(\mathcal{L}) = \mathrm{Alg}(\mathcal{L})$. In particular, he showed that lattices generated by finitely many commuting nests (*finite width* lattices) are synthetic. Every Hilbert-Schmidt operator is a bona fide integral operator whenever the Hilbert space is represented as a function space. Thus all such operators in $\mathrm{Alg}(\mathcal{L})$ lie in $\mathcal{A}_{\min}(\mathcal{L})$. We shall shortly see that this yields another large class of synthetic lattices.

THEOREM 1.3. *There are commutative subspace lattices which are not synthetic. The class of synthetic lattices includes all finite width lattices, all completely distributive lattices, and all Boolean algebras.*

2. Complete Distributivity

In this section, we will examine the role of rank one operators in CSL algebras, and the connection with the lattice property complete distributivity. A rank one operator R can be written as $R = xy^*$ where x is a non-zero vector in the range of R and y^* is a functional annihilating the kernel of R. Whence

$$Rz = x(y^*z) = (z, y)x.$$

The prototypical result about the existence of many rank one operators in nest algebras is known as the Erdos Density Theorem [16].

THEOREM 2.1. *In any nest algebra $\mathcal{T}(\mathcal{N})$, there is a net of norm one finite rank operators in $\mathcal{T}(\mathcal{N})$ converging to the identity in the strong operator topology. Consequently, the finite rank operators in a nest algebra are weak-∗ dense in the whole algebra.*

Moreover, every rank n operator in a nest algebra is the sum of n rank one operators in the nest algebra. So the rank one operators are very plentiful. Now in a non-atomic von Neumann algebra like Example 2, there are no finite rank (or even compact) operators at all.

The following simple proposition due to Longstaff [25] sets the stage.

PROPOSITION 2.2. *Let $\mathcal{L}$ be a commutative subspace lattice, and let $R = xy^*$ be a rank 1 operator. Then $R \in \mathrm{Alg}(L)$ if and only if there is a subspace $L \in \mathcal{L}$ such that $x \in L$ and*

$$y \perp L_- := \bigvee \{M \in \mathcal{L} \mid M \not\geq L\}.$$

PROOF. Let $L = \overline{\mathrm{Alg}(\mathcal{L})x}$. This is the smallest invariant subspace of $\mathrm{Alg}(\mathcal{L})$ containing x. First suppose that $R \in \mathrm{Alg}(\mathcal{L})$. Now let $M \in \mathcal{L}$ such that $M \not\geq L$. Thus x does not belong to M. Yet for any $z \in M$,

$$M \ni Rz = (z, y)x.$$

Hence it follows that $(z, y) = 0$ for all $z \in M$, or equivalently, $y \perp M$. This holds for all $M \not\geq L$, and hence $y \perp L_-$.

Reversing this computation shows that if $y \perp L_-$, then R belongs to $\mathrm{Alg}(\mathcal{L})$. $\square$

Let $\mathcal{R}_1(\mathcal{L})$ denote the linear span of the rank one operators in $\mathrm{Alg}(\mathcal{L})$. We obtain the immediate consequence:

COROLLARY 2.3. *For $L \in \mathcal{L}$,*

$$\mathcal{R}_1(\mathcal{L})L = \bigvee \{M \in \mathcal{L} \mid M_- \not\geq L\} =: L_\#.$$

It follows that a necessary condition for $\mathcal{R}_1(\mathcal{L})$ to be weak-$*$ dense in $\mathrm{Alg}(\mathcal{L})$ is that $L_\# = L$ for all $L \in \mathcal{L}$. This property already occurred in the lattice theory literature of the 1950's.

A lattice is *distributive* if it satisfies the laws

$$L \vee (M \wedge N) = (L \vee M) \wedge (L \vee N) \tag{D}$$
$$L \wedge (M \vee N) = (L \wedge M) \vee (L \wedge N) \tag{D*}$$

For any lattice, the properties (D) and (D*) are equivalent. Every CSL satisfies both these laws, as is easily seen from the relations

$$P(L \wedge M) = P(L)P(M)$$

and

$$P(L \vee M) = P(L) + P(M) - P(L)P(M).$$

The property of *complete distributivity* is the corresponding law for distributing the join and meet of an arbitrary collection of sets of elements of the lattice. The rather horrible formula that results can be deduced by checking what works for sets. For the sake of complete disclosure, here it is. Let Λ be an index set. For each $\lambda \in \Lambda$, let

$$\mathcal{C}_\lambda = \{L_i^{(\lambda)} : i \in \mathcal{I}_\lambda\}$$

be a subset of $\mathcal{L}$ indexed by a set $\mathcal{I}_\lambda$. The complete distributive law becomes

$$\bigvee_{\lambda \in \Lambda} \left(\bigwedge_{i \in \mathcal{I}_\lambda} L_i^{(\lambda)} \right) = \bigwedge_{i \in \prod \mathcal{I}_\lambda} \left(\bigvee_{\lambda \in \Lambda} L_{i(\lambda)}^{(\lambda)} \right) \tag{CD}$$

Likewise, there is a law (CD*) dual to this one interchanging the role of meets and joins. A lattice satisfying both (CD) and (CD*) is called completely distributive.

It is easy to see that the lattice of all subsets of a given set satisfies these laws. Thus the same is true for any complete homomorphic image of a complete sublattice of sets. This includes all complete chains, for example.

However, the lattice of all measurable subsets of $(0, 1)$ modulo null sets (Example 2) is not completely distributive. Indeed, let E_n for $n \geq 1$ be a collection of independent subsets of $(0, 1)$ of measure $1/2$. Let $\mathcal{C}_n = \{E_n, E_n^c\}$. Consider (CD) for this collection. Clearly, the left hand side is the join of zero elements, and thus is 0. But the right hand side is the meet of many terms, each of which is the join of countably many independent subsets of measure $1/2$. Thus each of these terms has measure 1, and hence is the unit element 1 of $\mathcal{L}$. So the right hand side is 1.

The main result characterizing completely distributive lattices is due to Raney [34] in 1952.

THEOREM 2.4. *For a lattice $\mathcal{L}$, the following are equivalent:*
(1) *$\mathcal{L}$ satisfies* (CD).
(1') *$\mathcal{L}$ satisfies* (CD*).
(2) *$\mathcal{L}$ is a complete homomorphic image of a complete lattice of sets.*
(3) *$L = L_\# $ for every $L \in \mathcal{L}$.*

It can now be seen that the lattice on the unit square of Example 3 is also completely distributive. This can be verified by checking that $L = L_\#$ for all $L \in \mathcal{L}$.

So we now have a strong connection between complete distributivity of $\mathcal{L}$ and rank one operators in $\mathrm{Alg}(\mathcal{L})$. This is made more precise in the following theorem. It is this result that signalled the importance of complete distributivity in this subject.

THEOREM 2.5. *For a CSL lattice $\mathcal{L}$, the following are equivalent:*
(1) *$\mathcal{L}$ is completely distributive.*
(2) *$\mathcal{R}_1(\mathcal{L})$ is weak-$*$ dense in $\mathrm{Alg}(\mathcal{L})$.*
(3) *The Hilbert-Schmidt operators in $\mathrm{Alg}(\mathcal{L})$ are WOT dense in $\mathrm{Alg}(\mathcal{L})$.*

We have already observed that (2) implies (1) is elementary. The converse due to Laurie and Longstaff [24] relies significantly on the methods and results of Arveson. The equivalence of (1) and (3) is due to Hopenwasser, Laurie and Moore [22]. It is an immediate consequence that completely distributive lattices are synthetic. This theorem begs the following question, which has become an important open problem:

OPEN PROBLEM. *For which commutative subspace lattices $\mathcal{L}$ is the set of compact operators in $\mathrm{Alg}(\mathcal{L})$ weak-$*$ dense in the whole algebra?*

3. Semi-continuity

Every CSL algebra contains a masa, and every masa has faithful normal states. Indeed, a masa $\mathcal{M}$ can always be represented as an $L^\infty(\mu)$ space acting by multiplication on $L^2(\mu)$. The measure μ is unique up to equivalence of measures. Integration against any equivalent probability measure ν yields a faithful normal state. Hence the $L^1(\nu)$ norm yields a normal metric on $\mathcal{L}$:

$$d_\nu(L_1, L_2) := \|L_1 - L_2\|_{L^1(\nu)}.$$

All these metrics are equivalent to the order topology, and hence are also equivalent to the strong operator topology on the corresponding set of projections. So every CSL is a complete metric lattice.

A metric lattice $\mathcal{L}$ is *upper semi-continuous* if

$$\bigwedge_{\varepsilon > 0} \left(\bigvee \{ M \in \mathcal{L} \mid d_\nu(L, M) < \varepsilon \} \right) = L$$

for all $L \in \mathcal{L}$. This says that each element of the lattice is close to the span of all nearby elements. Lower semi-continuity is defined analogously.

Consider Example 3 on the unit square. It is not difficult to see that if L corresponds to an increasing set bounded by the graph of a decreasing function, then the span of all other sets within ε in the $L^1(m)$ metric (m is planar Lebesgue measure) lies within $2\sqrt{\varepsilon}$. Thus it follows that this lattice is upper (and lower) semi-continuous.

On the other hand, in Example 2, the non-atomic Boolean algebra has the property that the span of all subsets of measure ε is all of $[0,1]$ for every $\varepsilon > 0$. Thus this lattice is not upper semi-continuous.

The following theorem due to Pitts and the author [13] has proven to be a very powerful tool for approximating completely distributive lattices.

THEOREM 3.1. *For a commutative subspace lattice, the following are equivalent:*

(1) $\mathcal{L}$ *is completely distributive.*
(2) $\mathcal{L}$ *is upper semi-continuous.*
(3) $\mathcal{L}$ *is lower semi-continuous.*

The original motivation for this new characterization of complete distributivity was an attempt to solve the open question about the density of the compact operators in a CSL algebra. Wagner [40] had shown that when the compact operators are dense in $\mathrm{Alg}(\mathcal{L})$, the lattice is compact in its order topology. It was thought that perhaps compactness of the lattice implied complete distributivity (which would complete the circle). However, this is not the case, as shown by a curious example due to Haydon [19]. Pitts and the author were able to show that for a large class of lattices, these notions do coincide:

THEOREM 3.2. *If $\mathcal{L}$ is generated (as a CSL) by a completely distributive lattice and a finite width lattice, then the complete distributivity of $\mathcal{L}$, the compactness of $\mathcal{L}$ and the density of the compact operators in $\mathrm{Alg}(\mathcal{L})$ in the weak-$*$ topology are all equivalent.*

The natural conjecture still remains that the density of the compacts should imply complete distributivity of the lattice. This conjecture can be reformulated as a question about compact operators because of the implications of semi-continuity.

REFORMULATION. *Let K be a compact operator, and let $\mathcal{M}$ be a masa. Suppose that for every $\varepsilon > 0$, there are projections $P_1, \dots, P_n$ in $\mathcal{M}$ such that*

(i) $P_i \mathcal{H}$ *is invariant for K for $1 \leq i \leq n$.*
(ii) $\|P_i K\| < \varepsilon$ *for $1 \leq i \leq n$.*
(iii) $\bigvee_{i=1}^n P_i = I$.
Does this imply that $K = 0$?

4. Approximate Unitary Equivalence

It is a natural problem to try to classify commutative subspace lattices up to some appropriate equivalence. Three natural choices are unitary equivalence, approximate unitary equivalence, and similarity. One difficulty comes from the fact that the functional representation of $\mathcal{L}$ mentioned in the first section is not canonical. This has been remedied for completely distributive lattices by Orr and Power [**28**]. However, this does not seem to provide enough information to answer these questions.

Consider two CSL's $\mathcal{L}$ and $\mathcal{M}$ which are lattice isomorphic via an isomorphism $\theta : \mathcal{L} \to \mathcal{M}$. Say that θ is implemented by a unitary (or similarity) if there is a unitary (or invertible) operator U (or S) such that

$$UL = \theta(L) \quad \text{or} \quad SL = \theta(L) \quad \text{for all } L \in \mathcal{L}.$$

We also say that θ is implemented by a (*strong*) *approximate unitary equivalence* if there is a sequence U_n of unitary operators such that

$$\lim_{n \to \infty} \sup_{L \in \mathcal{L}} \|U_n P(L) U_n^* - P(\theta(L))\| = 0$$

(and the function $f(L) = U_n P(L) U_n^* - P(\theta(L))$ is norm continuous and compact valued).

A unitary equivalence between $\mathcal{L}$ and $\mathcal{M}$ implements a unitary equivalence of the type I von Neumann algebras $\mathcal{L}'$ and $\mathcal{M}'$ of operators commuting with the orthogonal projections onto $\mathcal{L}$ and $\mathcal{M}$ respectively. As in the well known theory of self-adjoint operators, there is a spectral measure associated to a CSL. We can think of $\mathcal{L}$ and $\mathcal{M}$ as two representations of the same lattice. The spectral theorem implies that there is a spectral measure which determines the representation. The unitary invariants are just the measure class and the multiplicity function determined by this spectral measure. This was done for nests by Erdos [**15**], and follows similarly for arbitrary CSL's.

In the nest case, the other two notions are more important; but also proved more resistant. One weaker spatial property implied by all these equivalences is the property of *preserving dimension* in the sense that

$$\text{for all } L_1 < L_2 \in \mathcal{L}, \qquad \dim \theta(L_2)/\theta(L_1) = \dim L_2/L_1.$$

The following theorem [**8**], known as the Similarity theorem for nests, is the prototype:

THEOREM 4.1. *Let* $\theta : \mathcal{N} \to \mathcal{M}$ *be an order isomorphism of two nests on a separable Hilbert space. Then the following are equivalent.*
(1) θ *preserves dimension.*
(2) θ *is implemented by a similarity*
(3) θ *is implemented by an approximate unitary equivalence.*
Moreover, when this occurs, the similarity may be chosen to be of the form $S = U + K$ *where* U *is unitary,* K *is compact, and* $\|K\| < \varepsilon$ *for any given* $\varepsilon > 0$.

An important tool for converting an approximate unitary equivalence to a similarity is Arveson's distance formula for nest algebras [**6**]: for any operator A and nest $\mathcal{N}$,

$$\text{dist}(A, \mathcal{T}(\mathcal{N})) = \sup_{N \in \mathcal{N}} \|P(N)^\perp A P(N)\|.$$

While this formula, strictly interpreted, cannot hold for other CSL algebras, there is the notion which is sufficient. A lattice $\mathcal{L}$ is called *hyper-reflexive* if there is a constant C so that for every operator $A \in \mathcal{B}(H)$,

$$\text{dist}(A, \text{Alg}(\mathcal{L})) \leq C \sup_{L \in \mathcal{L}} \|P(L)^{\perp} A P(L)\|.$$

Unfortunately, the collection of CSL's known to have this property is very limited.

The fact that any two continuous nests are approximately unitarily equivalent is due to Andersen [1]. This result is now understood to be closely related to Voiculescu's celebrated generalized Weyl–von Neumann Theorem [39] which gives useful criteria for determining when two representations of a separable C*-algebra are approximately unitarily equivalent. Arveson [7] extended this to apply to certain non-separable C*-algebras which were, in a certain precise sense, compactly generated. This succeeded in yielding both Voiculescu's theorem and Andersen's theorem as special cases. A fairly direct proof for nests in this spirit is contained in [9].

It turns out that Arveson's theorem can be applied to arbitrary completely distributive CSL's. This was done by Pitts and the author [14]. The result relies on a certain strong approximation property of completely distributive CSL's which is a consequence of semi-continuity:

LEMMA 4.2. *A commutative subspace lattice $\mathcal{L}$ with metric d_μ is completely distributive if and only if for every $\varepsilon > 0$, there is a finite sublattice $\mathcal{L}_0$ of $\mathcal{L}$ with the property that for every $L \in \mathcal{L}$, there are L_1 and L_2 in $\mathcal{L}_0$ such that $L_1 \leq L \leq L_2$ and $d_\mu(L_1, L_2) < \varepsilon$.*

This in turn allows certain uniform estimates required to apply Arveson's Weyl–von Neumann theorem. We obtain:

THEOREM 4.3. *If $\mathcal{L}$ is a completely distributive CSL, and $\theta : \mathcal{L} \to \mathcal{M}$ is a lattice isomorphism which preserves dimension, then θ is implemented by an approximate unitary equivalence. If, in addition, $\mathcal{L}$ is hyper-reflexive, then θ is implemented by a similarity.*

Notice that when $\mathcal{L}$ is not hyper-reflexive, the unitary which moves $\mathcal{L}$ uniformly close to $\mathcal{M}$ may not necessarily move $\text{Alg}(\mathcal{L})$ close to $\text{Alg}(\mathcal{M})$. Pitts [32] has recently shown that when (the unit balls of) two CSL algebras are close enough, the algebras are similar.

5. The Jacobson Radical

In the representation theory of Banach algebras, a lot of attention is paid to the Jacobson radical, the intersection of kernels of all irreducible representations. Again for CSL algebras, the prototype is the result for nests.

When the nest $\mathcal{F}$ is finite, as for the algebra of upper triangular matrices, the radical is the ideal of all strictly upper triangular operators $\mathcal{T}_0(\mathcal{F})$. Thus it follows easily that the operators which are strictly upper triangular with respect to some finite subnest of $\mathcal{N}$ are also in the radical of $\mathcal{T}(\mathcal{N})$. The radical is norm closed, so we also obtain limits of these elements in the radical. Such an operator

T has the property that for every $\varepsilon > 0$, there is a finite subset $\mathcal{F}$ of $\mathcal{N}$ so that the diagonal part $\Delta_{\mathcal{F}}(T)$ of the upper triangular form of T with respect to $\mathcal{F}$ has norm less than ε. That this in fact characterizes the radical is due to Ringrose [**35**].

THEOREM 5.1. *Let $\mathcal{N}$ be a nest. For $T \in \mathcal{T}(\mathcal{N})$, the following are equivalent*
(1) *T belongs to $\mathrm{rad}(\mathcal{T}(\mathcal{N}))$.*
(2) *T belongs to the closure of $\bigcup_{\mathcal{F} \subset \mathcal{N}} \mathcal{T}_0(\mathcal{F})$ as $\mathcal{F}$ runs over all finite subnests of $\mathcal{N}$.*
(3) *For every $\varepsilon > 0$, there is a finite subnest $\mathcal{F}$ of $\mathcal{N}$ so that $\|\Delta_{\mathcal{F}}(T)\| < \varepsilon$.*

It seems worthwhile to briefly sketch the proof, the difficulty being to show that T being in the radical implies condition (3), or equivalently, show that failing to satisfy (3) implies that T is not in the radical. To show that T is not in the radical, it suffices to find an operator A in the algebra so that AT satisfies $\|(AT)^n\| \geq 1$ for all $n > 0$, as this implies that AT has positive spectral radius, and thus is not quasinilpotent.

A technical device reduces (3) to a local version: for every $N \in \mathcal{N}$ and $\varepsilon > 0$, there are elements $N_1 < N < N_2$ in the nest so that

$$\|(P(N) - P(N_1))T(P(N) - P(N_1))\| < \varepsilon$$

and

$$\|(P(N_2) - P(N))T(P(N_2) - P(N))\| < \varepsilon.$$

So if (3) fails, we may suppose (after bookkeeping changes) that there is an element N of the nest so that

$$\|(P(N') - P(N))T(P(N') - P(N))\| > 1$$

for all $N' > N$. The lower semicontinuity of the norm in the strong operator topology allows us to extract a sequence

$$N_1 > N_2 > N_3 > \ldots > N$$

in the nest $\mathcal{N}$ so that

$$\|(P(N_i) - P(N_{i+1}))T(P(N_i) - P(N_{i+1}))\| > 1 \quad \text{for} \quad i \geq 1.$$

So we obtain unit vectors x_i and y_i in $N_i \ominus N_{i+1}$ so that

$$(Tx_i, y_i) > 1.$$

Now let

$$A = \sum_{i \geq 1} x_{i+1} y_i^*.$$

Since the rank one operators $x_{i+1} y_i^*$ shift "to the left", they all belong to $\mathcal{T}(\mathcal{N})$. As they all have norm one, and pairwise orthogonal domains and ranges, the norm of A is one. A routine computation shows that

$$((AT)^n x_1, x_{n+1}) > 1$$

for all n. Hence T is not in the radical.

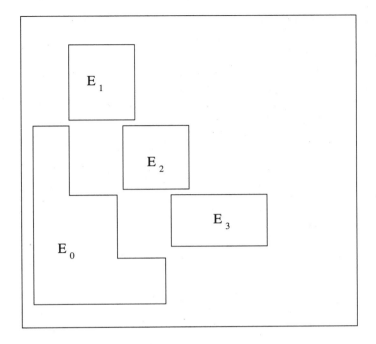

FIGURE 1. interval configurations

In arbitrary commutative subspace lattices, Hopenwasser [20] pointed out
that (2) and (3) are equivalent and imply membership in the radical, and asked
if this still characterized the radical. In the proof sketched above, it was easy
to construct A in $\mathcal{T}(\mathcal{N})$ which keeps the powers of AT large. As soon as one
considers a more complicated example such as the lattice on the unit square
(Example 3), it is no longer apparent how to do this.

Consider the figure below. Suppose that the compression of T to each interval
E_i has norm $\|E_i T E_i\| > 1$. Then it is possible to construct a norm one operator
A in $\mathrm{Alg}(\mathcal{L})$ so that $\|TAT\| > 1$. As before, choose unit vectors x_i and y_i in E_i
so that

$$(Tx_i, y_i) > 1.$$

It is no longer possible to map y_0 to x_i for any i. However, part of y_0 can
be mapped to each of the x_i for $i = 1, 2, 3$. That is, one can decompose the
interval E_0 into three subrectangles so that each is comparable to one of the E_i
for $i = 1, 2, 3$. This splits $y_0 = z_1 \oplus z_2 \oplus z_3$ so that

$$A = \sum_i x_i \|z_i\|^{-1} z_i^*$$

is a norm one operator in $\mathrm{Alg}(\mathcal{L})$. Then $\|TATx_0\| > 1$.

This more complicated kind of configuration can be handled whenever certain
combinatorial–geometric conditions apply. If inside each rectangle of the figure
above, there is a similar configuration, and another inside that for k stages, then

a norm one operator A can be constructed so that

$$\|(AT)^{2^k}\| > 1.$$

Piecing these together yields the desired operator.

In [**11**], Orr and the author solved this problem for all width two lattices. While this includes non-completely distributive lattices, it turns out that this holds because the non-CD parts of width two lattices are quite special, and can be handled. The real work comes for general completely distributive lattices. Complete distributivity allows us to use certain compactness arguments that yield a *finite* number of intervals with the right combinatorial configuration. Then we solve the interpolation problem to produce the desired operator A. One unfortunate hitch remains which so far has been insurmountable. Only in the width two case, did we succeed completely.

THEOREM 5.2. *For width two commutative subspace lattices, the Ringrose condition characterizes the Jacobson radical.*

One consequence of the radical conjecture is the uniformity of the radical in the sense of the theorem below. This has been established independently.

THEOREM 5.3. *Let $\mathcal{L}$ be a commutative subspace lattice, and let T belong to the radical of* $\mathrm{Alg}(\mathcal{L})$. *Then*

$$\lim_{n \to \infty} \sup_{\|A\| \le 1} \|(AT)^n\|^{1/n} = 0$$

where the sup *runs over the whole unit ball of* $\mathrm{Alg}(\mathcal{L})$.

This allows one to show that the radical satisfies another nice property. The point of interest is that while the radical is a norm closed ideal, there is no a priori reason that infinite sums should belong to the radical. This corollary shows that in a restricted, but important, case this is possible.

COROLLARY 5.4. *Suppose that T belongs to the radical of* $\mathrm{Alg}(\mathcal{L})$. *Let $\{E_i\}$, $i \ge 1$, be a countable family of pairwise orthogonal interval projections of $\mathcal{L}$. Then $\sum_{i \ge 1} E_i T E_i$ also belongs to the radical.*

These results, and similar results that are somewhat stronger, would be a consequence of the radical conjecture. As such, they can be taken as some positive evidence. This latter corollary is also extremely useful in the problem of constructing the operator A above. In the nest case, the linear order makes it easy to screen out extraneous terms that occur in the calculation of $(AT)^n$. This is no longer possible even in the width two case without something like this corollary.

6. Dilation theory

Another aspect of representation theory for operator algebras has its roots in an important theorem of single operator theory due to Sz. Nagy [36].

THEOREM 6.1. *If T is a bounded operator on a Hilbert space $\mathcal{H}$ with $\|T\| \leq 1$, then there is a larger Hilbert space $\mathcal{K} \supset \mathcal{H}$ and a unitary operator U on $\mathcal{K}$ such that*

$$T^n = P_{\mathcal{H}} U^n|_{\mathcal{H}} \qquad \text{for all } n \geq 0.$$

This unitary operator is called a unitary (power) dilation of T. Necessarily, the space $\mathcal{K}$ decomposes as a direct sum

$$\mathcal{K} = \mathcal{H}_+ \oplus \mathcal{H} \oplus \mathcal{H}_-$$

so that the matrix of the unitary operator U has the form

$$U = \begin{bmatrix} * & * & * \\ 0 & T & * \\ 0 & 0 & * \end{bmatrix}.$$

This result gave the first transparent explanation of the von Neumann inequality:

COROLLARY 6.2. *If T is a bounded operator on a Hilbert space $\mathcal{H}$ with $\|T\| \leq 1$, then*

$$\|p(T)\| \leq \|p\| := \sup_{|z| \leq 1} |p(z)|$$

for every polynomial p.

Arveson [4] put the representation of arbitrary subalgebras of C*-algebras into a similar context. Given a subalgebra $\mathcal{A}$ of a C*-algerba $\mathcal{B}$, say that a representation ρ of $\mathcal{A}$ on $\mathcal{H}$ has a ∗-dilation if there is a Hilbert space $\mathcal{K}$ containing $\mathcal{H}$ and a ∗-representation σ of $\mathcal{B}$ on $\mathcal{K}$ such that

$$\rho(a) = P_{\mathcal{H}} \sigma(a)|_{\mathcal{H}}$$

for all $a \in \mathcal{A}$.

To see why Sz. Nagy's theorem fits into this context, notice that a contraction T yields a representation ρ of the polynomials into $\mathcal{B}(H)$ by the formula

$$\rho(p) = p(T).$$

Von Neumann's inequality implies that ρ extends to a norm one (contractive) representation of the disk algebra $A(\mathbf{D})$. Sz. Nagy's unitary dilation yields a ∗-representation of the C*-algebra $C(\mathbf{T})$ of continuous functions on the unit circle into $\mathcal{B}(\mathcal{K})$ by

$$\sigma(f) = f(U).$$

Clearly, this is a ∗-dilation of ρ.

Arveson observed that a representation ρ of $\mathcal{A}$ determines a representation $\rho_n = \rho \otimes \mathrm{id}_n$ of the $n \times n$ matrices over $\mathcal{A}$ on the direct sum of n copies of $\mathcal{H}$. If ρ has a ∗-dilation, then the ∗-representation $\sigma_n = \sigma \otimes \mathrm{id}_n$ of the $n \times n$ matrices over $\mathcal{B}$ is a ∗-dilation of ρ_n. In particular, $\|\rho_n\| \leq 1$ since ∗-representations are contractive. Hence, in order for ρ to have a $*$-dilation, a necessary condition is that ρ be *completely contractive*:

$$\|\rho\|_{cb} := \sup_n \|\rho_n\| \le 1.$$

One of Arveson's main results is the converse.

THEOREM 6.3. *Suppose that ρ is a representation of a unital subalgebra $\mathcal{A}$ of a C^*-algebra $\mathcal{B}$. Then ρ has a $*$-dilation if and only if it is completely contractive.*

See Paulsen's book [**29**] for an overview of completely bounded maps.

We also note an additional feature of Sz. Nagy's result. It is only necessary to verify that the generator z is sent to a contraction $T = \rho(z)$ in order to deduce that the desired dilation exists. This is a sort of local contractivity that we wish to explore more in this section. In general, it is difficult to compute the norm of a representation of an algebra, even one as well understood as the disk algebra. In Sz. Nagy's Theorem, the fact that the representation is a contraction is deduced by first constructing the $*$-dilation using only the information $\|T\| \le 1$. Surprizingly, this is typical of almost every case in which $*$-dilations are known.

The first instance of this is the nest algebra case. First one must analyze representations of the $n \times n$ upper triangular matrices $\mathcal{T}_n$. This was done by McAsey and Muhly [**26**], but a more explicit dilation was given by Paulsen, Power and Ward [**31**] who were then able to extend the results to arbitrary nest algebras. Given a representation ρ of $\mathcal{T}_n$, a mild necessary condition for dilation is that each matrix unit be sent to a contraction. It follows that the diagonal matrix units are sent to pairwise orthogonal projections that decompose the Hilbert space into n orthogonal subspaces

$$\mathcal{H} = \mathcal{H}_1 \oplus \ldots \oplus \mathcal{H}_n.$$

The matrix units $E_{i,i+1}$ are sent to contractions T_i which map $\mathcal{H}_{i+1}$ into $\mathcal{H}_i$. One may use the Sz. Nagy unitary dilations of the T_i to write down an explicit $*$-dilation of ρ. In particular, it follows that the local conditions

$$\|\rho(E_{ij})\| \le 1 \quad \text{for} \quad 1 \le i \le j \le n$$

are sufficient to imply that ρ is completely contractive.

Then Paulsen, Power and Ward go on to show that every nest algebra can be approximated by subalgebras which are completely isometrically isomorphic to finite dimensional nest algebras. From this they obtain:

THEOREM 6.4. *Suppose that ρ is a weak-$*$ continuous contractive representation of a nest algebra $\mathcal{T}(N)$. Then ρ has a weak-$*$ continuous $*$-dilation.*

This suggests the following question.

OPEN PROBLEM. *Which CSL algebras have the property that every weak-$*$ continuous representation has a $*$-dilation?*

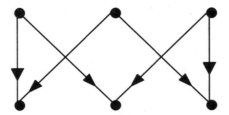

FIGURE 2. 6-cycle algebra

We are lead to consider representations of the finite dimensional CSL algebras which are *locally contractive* in the sense that each standard matrix unit in the algebra is sent to a contraction. A finite dimensional CSL algebra is a subalgebra of the $n \times n$ matrix algebra $\mathcal{M}_n$ which contains the diagonal algebra $\mathcal{D}_n$. Such an algebra is determined by the matrix units it contains. We can associate to the algebra $\mathcal{A}$ a directed graph $\mathcal{G}(\mathcal{A})$ with n vertices v_i corresponding to the standard basis vectors, and a directed edge from v_j to v_i when E_{ij} belongs to $\mathcal{A}$. Since $\mathcal{A}$ is an algebra, this becomes the graph of a (finite) transitive relation. One can often describe properties of the algebras in terms of properties of the graph. That is the case here.

The 6-cycle algebra represented by the graph above demonstrates an obstruction to the desired dilation of locally contractive representations. Consider the automorphism given by

$$
\rho\left(\left[\begin{array}{ccc|ccc}
a_1 & 0 & 0 & 0 & b & c \\
0 & a_2 & 0 & d & 0 & e \\
0 & 0 & a_3 & f & g & 0 \\
\hline
0 & 0 & 0 & a_4 & 0 & 0 \\
0 & 0 & 0 & 0 & a_5 & 0 \\
0 & 0 & 0 & 0 & 0 & a_6
\end{array}\right]\right) = \left[\begin{array}{ccc|ccc}
a_1 & 0 & 0 & 0 & b & -c \\
0 & a_2 & 0 & d & 0 & e \\
0 & 0 & a_3 & f & g & 0 \\
\hline
0 & 0 & 0 & a_4 & 0 & 0 \\
0 & 0 & 0 & 0 & a_5 & 0 \\
0 & 0 & 0 & 0 & 0 & a_6
\end{array}\right].
$$

This map is locally contractive because each matrix unit is sent to itself or to minus itself ($\rho(E_{16}) = -E_{16}$). However, it is not even a contractive map. Indeed, the matrix with all diagonal entries equal to 0, and

$$b = -c = d = e = f = g = 1$$

yields a matrix of norm $\sqrt{3}$ while the image with all non-negative entries has norm 2.

In fact, any cycle like this without "chords" (i.e. other edges of the between the vertices of the cycle) is an obstruction with one exception. The cycles can have any even length. A cycle of odd length must have two adjacent edges with the same direction, and this forces a chord because of transitivity. The exception is a 4-cycle with an *interpolating point* as shown below.

Notice that the four outer vertices form a 4-cycle, but the central point allows the edges of the four cycle to be factored through this single vertex. The existence of this interpolating point puts constraints on the allowable representations that rules out the examples of the type just given. This leads to the following class of directed graphs:

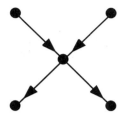

FIGURE 3. interpolating point

DEFINITION 6.5. An *interpolating graph* is a directed graph such that
  (i) every 4-cycle has an interpolating point, and
  (ii) every $2k$-cycle for $k \geq 3$ has a chord.

The main result for finite dimensional algebras is that this class of graphs completely determines the algebras we are looking for [**10**].

THEOREM 6.6. *For finite dimensional CSL algebras, the following are equivalent:*
  (1) *$\mathcal{A}$ has an interpolating graph.*
  (2) *Every locally contractive endomorphism of $\mathcal{A}$ is contractive.*
  (3) *Every locally contractive representation has a unitary $*$-dilation.*

For infinite dimensional algebras, it is necessary to find the appropriate analogue of the interpolating condition. For these algebras, rather than a graph, there is a measure theoretic type of representation. So it is natural that certain approximate versions of the conditions be used. An *interval* of a lattice is a projection of the form $P(L) - P(M)$ where $M < L$ are elements of the lattice. One says that intervals $E \succ F$ provided that $F\mathcal{B}(H)E$ is contained in $\mathrm{Alg}(\mathcal{L})$. The interpolating condition becomes:

DEFINITION 6.7. A completely distributive CSL algebra $\mathcal{A}$ with commutative subspace lattice $\mathcal{L}$ is called *interpolating* provided
  (i) If E_i and F_i are disjoint interval projections of $\mathcal{A}$ such that $E_i \succ F_j$ for $i, j \in \{1, 2\}$, then there exist non-zero subprojections $E_i' < E_i$ and $F_j' < F_j$ and an interpolating interval $G \neq 0$ so that $E_i' \succ G \succ F_j'$ for $i, j \in \{1, 2\}$.
  (ii) If there are disjoint intervals E_i and F_i for $1 \leq i \leq k$ and $k \geq 3$ so that $E_i \succ F_j$ if $j = i$ or $j = i + 1 \mod k$, then there is a partial chord, meaning that there are non-zero $E_{i_0}' < E_{i_0}$ and $F_{j_0}' < F_{j_0}$ where $j_0 \neq i_0, i_0 + 1 \mod k$ so that $E_{i_0}' \succ F_{j_0}'$.

The second condition actually implies that there is a full chord, meaning that it is not necessary to drop to sub-intervals. However, partial intervals are really needed for the interpolating condition. But with the assumption of complete distributivity, this can be improved. The semi-continuity formulation makes it possible to do the appropriate approximations so that the sub-intervals can be forced to have measure at least $1 - \varepsilon$ times the measure of the original intervals.

Complete distributivity plays another important role. We need to approximate the infinite dimensional CSL algebra by a sequence of subalgebras which are completely isometrically isomorphic to finite dimensional interpolating CSL algebras. It is much easier to approximate operators in completely distributive lattices because there are so many comparable intervals. Otherwise, it is very difficult to describe the operators in the algebra. Moreover, since we know that the span of the rank one operators is weakly dense, it suffices to approximate the rank one operators sufficiently well.

THEOREM 6.8. *If $\mathcal{L}$ is a completely distributive interpolating CSL, then there is a sequence of subalgebras $\mathcal{A}_n$ of $\mathrm{Alg}(\mathcal{L})$ such that each $\mathcal{A}_n$ is completely isometrically isomorphic to a finite dimensional interpolating CSL algebra, and the union of the unit balls of the $\mathcal{A}_n$ is norm dense in the unit ball of $\mathrm{Alg}(\mathcal{L}) \cap \mathcal{K}$ and weakly dense in the unit ball of $\mathrm{Alg}(\mathcal{L})$.*

As an immediate consequence of this and Arveson's theory, we obtain a dilation theorem.

THEOREM 6.9. *Let $\mathcal{A}$ be a completely distributive interpolating algebra, and let ρ be a contractive, weak-$*$ continuous representation of $\mathcal{A}$. The ρ is completely contractive, and thus has a $*$-dilation.*

The question of which lattices satisfy the conclusion of this theorem remains an important question. It is known that there are other lattices, such as the 4-cycle, that have this property [12].

Another important result of dilation theory is the Sz. Nagy–Foiaş commutant lifting theorem [37]:

THEOREM 6.10. *Suppose that T is a contraction in $\mathcal{B}(H)$ with a unitary dilation U in $\mathcal{B}(\mathcal{K})$. For each operator X commuting with T, there is an operator Y in $\mathcal{B}(\mathcal{K})$ commuting with U such that $\|Y\| = \|X\|$ and*

$$P_{\mathcal{H}} U^n Y^m|_{\mathcal{H}} = T^n X^m$$

for all non-negative integers n and m.

Two repetitions of this result yield a simultaneous dilation theorem due to Ando [2].

THEOREM 6.11. *Suppose that T and X are commuting contractions. Then there exist unitary dilations U and V of T and X respectively on the same Hilbert space $\mathcal{K}$ such that $UV = VU$ and*

$$P_{\mathcal{H}} U^n V^m|_{\mathcal{H}} = T^n X^m$$

for all non-negative integers n and m.

The nest algebra version is due to Paulsen and Power [30].

THEOREM 6.12. *Suppose that ρ is a weak-$*$ continuous contractive representation of a nest algebra $\mathcal{T}(\mathcal{N})$, and X is a contraction commuting with the range of ρ. Then there is a weak-$*$ continuous $*$-dilation σ of $\mathcal{B}(H)$ on a Hilbert space $\mathcal{K}$ and a unitary dilation U of X on $\mathcal{K}$ which commutes with the range of σ such that*

$$P_{\mathcal{H}} \sigma(T) U^m|_{\mathcal{H}} = \rho(T) X^m$$

for all $T \in \mathcal{T}(\mathcal{N})$ *and* $n \geq 0$.

This result can be extended to a class of algebras known as tree algebras [12] which are a subset of the interpolating algebras. The finite dimensional ones have digraphs which are generated by a bilateral tree (no cycles) as a transitive relation. A modified version applies in the infinite dimensional setting. Among the interpolating algebras, this theorem characterizes the tree algebras [10]. There is some evidence to indicate that this theorem characterizes tree algebras among all CSL algebras.

A recent result of Muhly and Solel [27] using a somewhat different version of commutant lifting determines a smaller class of algebras.

REFERENCES

1. Andersen, N.T., *Compact perturbations of reflexive algebras*, J. Functional Analysis 38 (1980), 366–400.
2. Ando, T., *On a pair of commuting contractions*, Acta. Sci. Math. (Szeged) 24 (1963), 88–90.
3. Arveson, W.B., *A density theorem for operator algebras*, Duke Math. J. 34 (1967), 635–647.
4. Arveson, W.B., *Subalgebras of C*-algebras*, Acta. Math. 123 (1969), 141–224.
5. Arveson, W.B., *Operator algebras and invariant subspaces*, Ann. Math. (2) 100 (1974), 433–532.
6. Arveson, W.B., *Interpolation problems in nest algebras*, J. Func. Anal. 3 (1975), 208–233.
7. Arveson, W.B., *Perturbation theory for groups and lattices*, J. Func. Anal. 53 (1983), 22–73.
8. Davidson, K.R., *Similarity and compact perturbations of nest algebras*, J. reine angew. Math. 348 (1984), 286–294.
9. Davidson, K.R., *Nest Algebras*, Pitman Research Notes in Mathematics Series, vol. 191, Longman Scientific and Technical Pub. Co., London, New York, 1988.
10. Davidson, K.R., *When locally contractive representations are completely contractive*, preprint, 1992.
11. Davidson, K.R., and Orr, J.L., *The Jacobson radical of a CSL algebra*, preprint, 1992.
12. Davidson, K.R., Paulsen, V.I. and Power, S.C., *Tree algebras, semi-discreteness, and dilation theory*, Proc. London Math. Soc., to appear.
13. Davidson, K.R., and Pitts, D.R., *Compactness and Complete Distributivity for Commutative Subspace Lattices*, J. London Math. Soc. (2) 42 (1990), 147–159.
14. Davidson, K.R., and Pitts, D.R., *Approximate unitary equivalence of completely distributive commutative subspace lattices*, preprint, 1993.
15. Erdos, J.A., *Unitary invariants for nests*, Pacific J. Math. 23 (1967), 229–256.
16. Erdos, J.A., *Operators of finite rank in nest algebras*, J. London Math. Soc. 43 (1968), 391–397.
17. Froelich, J., *Compact operators, invariant subspaces, and spectral synthesis*, J. Func. Anal. 81 (1988), 1–37.
18. Haydon, R., *Reflexivity of commutative subspace lattices*, Proc. A.M.S. 115 (1992), 1057–1060.
19. Haydon, R., *A compact lattice of projections*, J. London Math. Soc. (2) 46 (1992), 161–170.
20. Hopenwasser, A., *The radical of a reflexive operator algebra*, Pacific J. Math. 65 (1976), 375–392.
21. Hopenwasser, A., *Complete distributivity*, Proc. Symp. Pue Math., 51 (1), A.M.S., 1990.
22. Hopenwasser, A., Laurie, C. and Moore, R., *Reflexive algebras with completely distributive subspace lattices*, J. Operator Theory 11 (1984), 91–108.
23. Laurie, C., *On density of compact operators in reflexive algebras*, Indiana U. Math. J. 30 (1981), 1–16.
24. Laurie, C. and Longstaff, W., *A note on rank one operators in reflexive algebras*, Proc. Amer. Math. Soc. 89 (1983), 293–297.

25. Longstaff, W.E., *Operators of rank one in reflexive algebras*, Canad. J. Math. 28 (1976), 19–23.
26. McAsey, M.J. and Muhly, P.S., *Representations of nonselfadjoint crossed products*, Proc. London Math. Soc. (3) 47 (1983), 128–144.
27. Muhly, P.S. and Solel, B., *Hilbert modules over operator algebras*, preprint, 1993.
28. Orr, J.L. and Power, S.C., *Representations and refinement for reflexive operator algebras with completely distributive subspace lattices*, Indiana Univ. Math. J. 40 (1991), 617–638.
29. Paulsen, V.I., *Completely bounded maps and dilations*, Pitman Res. Notes Math. 146, Longman Sci. and Tech. 1986.
30. Paulsen, V.I. and Power, S.C., *Lifting theorems for nest algebras*, J. Operator Theory 20 (1988), 312–327.
31. Paulsen, V.I., Power, S.C. and Ward, J., *Semidiscreteness and dilation theory for nest algebras*, J. Func. Anal. 80 (1988), 76–87.
32. Pitts, D., *Close CSL algebras are similar*, preprint, 1993.
33. Radjavi, H. and Rosenthal, P., *On invariant subspaces and reflexive algebras*, Amer. J. Math. 91 (1969), 683–692.
34. Raney, G.N., *Completely distributive lattices*, Proc. Amer. Math. Soc. 3 (1952), 677–680.
35. Ringrose, J.R., *On some algebras of operators*, Proc. London Math. Soc. (3) 15 (1965), 61–83.
36. Sz. Nagy, B., *Unitary dilations of Hilbert space operators and related topics*, CBMS 19, Amer. Math. Soc., Providence, 1974.
37. Sz. Nagy, B. and Foiaş, C., *Dilatation des commutants d'opérateurs*, C.R. Acad. Sci. Paris, Ser. A 266 (1968), 493–495.
38. Sz. Nagy, B. and Foiaş, C., *Harmonic analysis of operators on Hilbert space*, American Elsevier, New York, 1970.
39. Voiculescu, D., *A non-commutative Weyl-von Neumann theorem*, Rev. Roum. Pures Appl. 21 (1976), 97–113.
40. Wagner, B., *Weak limits of projections and compactness of subspace lattices*, Trans. Amer. Math. Soc. 304 (1987), 515–535.

PURE MATH. DEPT., UNIVERSITY OF WATERLOO, WATERLOO, ONT. N2L–3G1, CANADA
E-mail address: krdavidson@math.uwaterloo.ca

Contemporary Mathematics
Volume 185, 1995

MODELS AND RESOLUTIONS FOR HILBERT MODULES

R. G. Douglas

0. Introduction

While operator theory has been studied for over a hundred years, the focus has been almost exclusively on the single operator case, even when the space is finite dimensional. Except for commuting collections of normal operators, no theory exists that provides a reasonable understanding of a multi–variate situation. In part, the difficulty stems from the higher level of sophistication necessary just to provide an appropriate context when more than one operator is being studied. An analogous phenomenon occurs in the transition from one to several variables in both algebra and complex analysis. Replacing a single operator by a commuting n–tuple of operators corresponds to what happens in that situation.

––––– – –––– – –––––

1991 Mathematical Subject Classification. Primary 46H25, 46J99, 46M20; Secondary 32A35, 46J15, 47B20, 47B38.

Research Supported in part by a grant from the National Science Foundation.

To appreciate the significance of this issue for multi–variate operator theory, one need only consider the definition of spectrum. For a single operator T, the definition is straight forward–the spectrum is the set of complex numbers for which $T-\lambda$ fails to be invertible. The fact that the spectrum is non–empty and compact as well as the relation of the spectrum to the functional calculus rests essentially on the recognition that the resolvent $(T-\lambda)^{-1}$ is an operator–valued holomorphic function. While various "elementary" notions of spectrum for (commuting) n–tuples exist, the definition due to J. Taylor [25], which is the one that has proved most useful, rests on the Koszul complex (and hence, implicitly on sheaf resolutions) and several complex variables including higher dimensional analogues of the Cauchy kernel. It is difficult to describe even the more recent approaches to this notion as being elementary.

Given the state of our understanding of multi–variate operator theory (M.O.T), it is difficult to know where and how to approach its development. While many authors have chosen to study operator n–tuples, usually commuting, I have opted for the context of Hilbert modules [12]. The assumption that there is a Hilbert space structure reflects the fact that I choose to study operators on Hilbert space. That I am considering modules, rather than n–tuples, means that no particular basis for the implicit algebra has been identified. In some instances the choice of a completion of $\mathbb{C}[z_1,...,z_n]$ is not important, but often even that fact needs to be kept in mind. I have chosen to consider Hilbert modules $\mathcal{M}$ over a "natural" functional algebra. The adjective "natural" is used in a descriptive, non–technical sense. The algebra is the supremum norm closure of some natural algebra of holomorphic functions on a "nice" domain. Although the algebra is usually topologically finitely generated, that is not assumed from the outset.

If Ω is a bounded open set in $\mathbb{C}^n$, then $A(\Omega)$ denotes the function algebra consisting of the uniform limits of functions holomorphic in some neighborhood of the closure of Ω. The simplest examples are obtained when Ω is the unit ball B^n or the unit polydisk D^n in $\mathbb{C}^n$. These examples will suffice to understand most of the topics discussed in this note but the results stated usually carry over to algebras defined for more general domains.

A first goal for M.O.T. is to provide enough examples to illustrate the various behavior or phenomena possible. If some family of examples is nice enough and sufficiently ample, it can be used to provide models for all possible examples of Hilbert modules over a particular algebra. In that situation, the study of the general case can be reduced to considering the examples as well as the mechanism that produces a model from a given module.

A second goal is to identify structure or general properties that a module might possess. To be useful one also needs effective methods for determining when a module has a particular property. One way to do this is by identifying invariants which are readily calculable and intrinsically related to the structure.

Finally, there is the general goal of relating the study of Hilbert modules to applications and other parts of mathematics. The most obvious kind of result here concerns examples of modules whose definition is based on concrete situations from other parts of mathematics. The most fruitful instances of the latter to date have involved commutative algebra and resulted in the rigidity theorems to be described later.

Before we proceed let me mention two dichotomies in the study of Hilbert modules. The first involves properties that depend on the boundary behavior of functions versus those that depend only on the behavior on the interior. The second concerns global properties of Hilbert modules versus local properties. All four kinds of properties are interesting and a well—developed multi—variate operator theory will encompass all. However, much of my effort has been directed to the study of interior behavior and local properties which can be investigated using commutative algebraic and algebraic geometric techniques. In the one variable case, the local theory is quite simple and leads to results which are mostly finite dimensional. That is not the case in several variables. Moreover, questions involving boundary behavior can be very technical and, perhaps, their consideration is best postponed until we have a better understanding of the local theory (but cf. [14], [7]). However, I should add that since zero sets in the several variables case are not isolated points, an easy division between local and global properties is not always possible.

1. Models for Hilbert Modules

Our purpose in this section is to recall some recent results about Hilbert modules relating these results to the overview just presented in the Introduction.

Let $H^2(\mathbb{D})$ denote the Hardy module in the one–dimensional case. If $\mathscr{M}$ is a non–empty (closed) submodule of $H^2(\mathbb{D})$, then one consequence of the Beurling Invariant Subspace Theorem [1] is that $\mathscr{M}$ is unitarily equivalent to $H^2(\mathbb{D})$. Hence, from the point of view of examples, nothing new has been obtained.

If $B^2(\mathbb{D})$ is the Bergman module and I is an ideal in $\mathbb{C}[z]$, then the closure [I] of I in $B^2(\mathbb{D})$ is not unitarily equivalent to [I] unless $[I]=B^2(\mathbb{D})$. However, if p(z) is a generator for the principal ideal I, then [I] is unitarily equivalent to the Bergman–like module obtained by taking the closure of $\mathbb{C}[z]$ in $L^2(|p|dA)$, where dA represents Lebesgue measure on $\mathbb{D}$. Thus, consider the family of Bergman–like modules $B^2(\nu)$ for the disk algebra $A(\mathbb{D})$ defined by forming the closure of $\mathbb{C}[z]$ in the spaces $L^2(\nu)$, where ν is a finite, positive measure on $\mathbb{D}$. The submodules obtained by taking the closure in $B^2(\nu)$ of ideals I in $\mathbb{C}[z]$, yield nothing new; that is, such a submodule is already unitarily equivalent to a member of the family.

There are submodules of $B^2(\mathbb{D})$ not unitarily equivalent to a member of this family but that phenomenon takes us outside the local–interior context. One way to obtain such an example is from certain ideals in the algebra of bounded holomorphic functions on $\mathbb{D}$. Another way for this phenomenon to happen is for the submodule to fail to be cyclic (cf. [19]) because such a submodule can't be unitarily equivalent to a member of the family.

In the several variables case the situation is very different. A rigidity phenomenon intervenes and shows that a family of models closed under submodules must be richer. If I is a principal ideal in $\mathbb{C}[z_1,..,z_n]$, then the closure [I] of I in $H^2(\mathbb{D}^n)$ is unitarily equivalent to a member of the appropriate family just as in the one–dimensional case. If $p(z_1,...,z_n)$ is a generator for I, then again [I] is unitarily equivalent to $H^2(|p|d\sigma_n)$, where σ_n is Lebesgue measure on T^n, the distinguished boundary of $\mathbb{D}^n$. (The analogous result is true for closures of principal ideals in $B^2(\mathbb{D}^n)$, $H^2(B^n)$ or $B^2(B^n)$.) However, for non–principal ideals something different happens.

<u>Theorem 1</u>:   Let I_1 and I_2 be ideals in $\mathbb{C}[z_1,...,z_n]$ such that the complex codimension of the zero set $\mathscr{Z}(I_i)$ is 2 or greater for $i = 1,2$ and assume that the algebraic components (zero sets of the ideals in the primary decomposition for I_i) of $\mathscr{Z}(I_i)$ intersect Ω. Then the two submodules $[I_1]$ and $[I_2]$ are quasi–similar if and only if $I_1 = I_2$.

This follows from the Rigidity Theorem ([12], [13]). (Quasi–similarity is implied by similarity or unitarily equivalence but is weaker). A slight variation of this result shows that the closure [I] of an ideal I in $B^2(\mathbb{D}^n)$ with $\text{codim}_{\mathbb{C}} \mathscr{Z}(I) > 1$ can't be similar to $B^2(\nu)$ for any measure ν on $\text{clos}\mathbb{D}^n$. Thus a full family of models must include not only all the $B^2(\nu)$ but also the closure [I] in $B^2(\nu)$ of each ideal I in $\mathbb{C}[z_1,...,z_n]$ for all measures ν, a result quite different from that for the one–variable case.

We are omitting the whole story, because, just as in the one–variable case, we did not consider all submodules of $B^2(\mathbb{D}^n)$ or $B^2(\nu)$. The same possibilities as in the one–variable case, and then some, exist in the higher dimensional situations.

Let me emphasize the point I am making. The Rigidity Phenomenon reveals that Bergman–like modules alone can not capture the full range of possibilities for Hilbert modules or M.O.T. in the case of several variables no matter how intricate the domains and measures considered. Another way to glean this truth is from the Koszul complex about which we will have more to say in later sections.

2.   Invariants for Hilbert Modules

The rigidity phenomenon is established using localization techniques, which can be approached via techniques that involve either local algebras or module tensor products and local modules. We will not provide a full presentation here but refer the reader to [12], [13] for complete details. However, we do describe how localization can be accomplished using the module tensor product by way of a simple but important example.

Recall for $\underline{z}$ in Ω that $\mathbb{C}_{\underline{z}}$ is the one–dimensional module over $A(\Omega)$ in which module multiplication is defined by $\varphi \times \lambda = \varphi(\underline{z})\lambda$ for φ in $A(\Omega)$ and λ in $\mathbb{C}_{\underline{z}}$. If we localize the Hardy module $H^2(\mathbb{D}^n)$ using $\mathbb{C}_{\underline{z}}$, we obtain

$$H^2(\mathbb{D}^n) \otimes_{A(\mathbb{D})} \mathbb{C}_{\underline{z}} \cong \mathbb{C}_{\underline{z}} \text{ for } \underline{z} \in \mathbb{D}^n.$$

However, if $H^2_{\underline{\Omega}}(\mathbb{D}^n)$ denotes the submodule of $H^2(\mathbb{D}^n)$ generated by the ideal $I_{\underline{0}} = \{p \in \mathbb{C}[z_1,...,z_n] \cdot p(\underline{0}) = 0\}$, then

$$H^2_{\underline{0}}(\mathbb{D}^n) \otimes_{A(\mathbb{D})} \mathbb{C}_{\underline{z}} \cong \begin{cases} \mathbb{C}_{\underline{z}} & \underline{z} \in \mathbb{D}^n \setminus \{\underline{0}\} \\ \\ \underbrace{\mathbb{C}_{\underline{0}} \oplus \ .. \ \oplus \ \mathbb{C}_{\underline{0}}}_{n \ \text{times}} & \underline{z} = \underline{0} \ . \end{cases}$$

Note that $I_{\underline{0}}$ is not principal for $n > 1$ and that the calculation shows that $H^2(\mathbb{D}^n)$ and $H^2_{\underline{0}}(\mathbb{D}^n)$ are not unitarily equivalent or even similar in this case.

We showed this latter fact by introducing a local invariant which is not equal for the two modules. Rigidity Theorems are proved by introducing enough local invariants to show that equivalence of the two Hilbert modules leads to a kind of strong algebraic equivalence of the associated algebraic modules, reducing the proof to earlier results of Zariski and Grothendieck in commutative algebra (cf. [28]).

There are other ways to obtain invariants for Hilbert modules. In [15] Yan and I introduced the notion of a Hilbert–Samuel polynomial for Hilbert modules. Again, I won't repeat that development in full but only enough of it to state the definitions and principal results.

A function algebra A is said to be finite at the maximal ideal m in the maximal ideal space M_A for A if $\dim_{\mathbb{C}} m/_{m^2} < \infty$. A key lemma (cf. Lemma 2, [15]) allows one to define the polynomial.

Lemma 1:   If the function algebra A is finite at m in M_A and the Hilbert module is finitely generated over A, then $\dim_{\mathbb{C}} \mathcal{H}/_{[m^k \ \mathcal{H}]} < \infty$ for $k = 1,2,3...$, where $[m^k \ \mathcal{H}]$ denotes the closure of $m^k \ \mathcal{H}$ in $\mathcal{H}$.

The fact that one is taking closures does not allow the standard results for Hilbert–Samuel polynomials to be applied directly, but suitable modifications and extensions of classical arguments enable us to establish (cf. Theorem 3, [15]):

Theorem 2:   With the same hypotheses as in the Lemma, there exists a polynomial $p_{m, \mathscr{H}}$ such that $p_{m, \mathscr{H}}(k) = \dim_{\mathbb{C}} \mathscr{H}/[m^k \mathscr{H}]$ for $k \gg 0$.

Moreover, the degree of p is less than or equal to $\dim_{\mathbb{C}} m/m_2$ and is related to the dimension of the module spectrum of $\mathscr{H}$ at m. Finally, if the polynomial is written in the form

$$\frac{\alpha_n}{n!} k^n + \alpha_{n-1} \binom{n}{1} k^{n-1} + \ldots + \alpha_0,$$

then the α_i are natural numbers and α_n is related to the multiplicity of $\mathscr{H}$ at m.

In some cases it is possible to calculate the Hilbert–Samuel polynomial for a Hilbert module directly. For example, for $H^2(\mathbb{D}^n)$ over $A(\mathbb{D}^n)$ one can use such an approach to show that

$$p_{\underline{z}_0, H^2(\mathbb{D}^n)}(k) = \binom{n+k}{k}$$

for $\underline{z}_0$ in $\mathbb{D}^n$. If $\mathscr{H}$ is the quotient module $H^2(\mathbb{D}^n)/[z_1 - z_2]$ over $A(\mathbb{D}^n)$, where $[z_1 - z_2]$ denotes the closure of the principal ideal in $H^2(\mathbb{D}^n)$ generated by the polynomial $z_1 - z_2$, then again one can calculate directly that

$$p_{\underline{z}_0, \mathscr{H}}(k) = \begin{cases} k & \underline{z}_0 \in V \\ 0 & \underline{z}_0 \notin V \end{cases}$$

where V is the intersection of $\mathscr{Z}(z_1 - z_2)$ and $\mathbb{D}^2$.

However, in general it is difficult to calculate the Hilbert–Samuel polynomial directly. In any case, one wants to connect the polynomial to other objects defined for the Hilbert module. For these reasons one seeks other approaches and other relationships for the Hilbert–Samuel polynomial. In [15] we offer one possibility, a relation of the polynomial to the final sheaf in the associated Koszul complex. Again we provide an overview and some additional comments.

For the Taylor approach to multi–variate spectral theory to apply we must assume that the algebra A is finitely generated and to fix n–generators $x_1, x_2,...,x_n$ for A.

The last stage of the Koszul complex $k^n(\underline{z})$ for the Hilbert module $\mathscr{H}$ defined relative to the commuting operator n–tuple $(T_1,...,T_n)$, where T_i is multiplication by x_i, is defined by

$$\overset{n}{\underset{\oplus}{}} \mathscr{H} \xrightarrow{\ d\ } \mathscr{H} \longrightarrow 0,$$

where d is defined by $d[(h_1, h_2 ,..., h_n)] = \sum_{i=1}^{n} (T_i - z_i)h_i$.

This complex is defined directly in terms of $\mathscr{H}$ and the module action of A on $\mathscr{H}$. There is a related complex defined less directly.

If $\mathcal{O}(\mathscr{H})$ denotes the sheaf of $\mathscr{H}$–valued analytic functions, then the last stage of this complex is defined by

$$\overset{n}{\underset{\oplus}{}} \mathcal{O}(\mathscr{H}) \xrightarrow{\ \tilde{d}\ } \mathcal{O}(\mathscr{H}) \longrightarrow 0,$$

where $\tilde{d}(f_1,...,f_n) = \sum_{i=1}^{n} (T_i - (\cdot)_i)f_i$. The quotient

$$\mathfrak{H} = \mathcal{O}(\mathscr{H})/\text{Im } \tilde{d}$$

is the last term of the cohomology sheaf for the "sheafified" Koszul complex (cf. [16]). Under appropriate hypotheses on A and $\mathscr{H}$, which are closely related to those invoked for the existence of the Hilbert–Samuel polynomial, the germ $\mathfrak{H}_{\underline{z}}$ of $\mathfrak{H}$ at $\underline{z}$ is a Noetherian module over the local ring $\mathcal{O}_{\underline{z}}$. A result from commutative algebra [28] establishes the existence of a polynomial $P_{\underline{z},\mathfrak{H}}$ that satisfies

$$P_{\underline{z},\mathfrak{H}}(k) = \dim {}^{\mathfrak{H}}\underline{z} / {}_{M_{\underline{z}}^k} \mathfrak{H}_{\underline{z}} ,$$

where $M_{\underline{z}}$ is the maximal ideal in $\mathcal{O}_{\underline{z}}$.

A fundamental question concerns the relation of $p_{\underline{z}_0, \mathscr{H}}$ and $P_{\underline{z}_0, \mathfrak{H}}$ when the two exist. In [15] we show in this case that $p_{\underline{z}_0, \mathscr{H}}(k) \geq P_{\underline{z}_0, \mathfrak{H}}(k)$ for $k \geq 2$. However, they are not always equal and the difference in the polynomials is closely related to the existence of nilpotents in $\mathscr{H}$. Under strong hypotheses that guarantee the absence of any nilpotents, we show the equality of the polynomials in [15]. Our hypothesis implies that $k^n(\underline{z})$ has constant sheaf dimension in a neighborhood of $\underline{z}_0$ which is clearly too strong but the method of proof we use depends on this hypothesis. One would expect it to be enough for the sheaf to be given by a holomorphic vector bundle supported on a subvariety for which $z_{\underline{0}}$ is a manifold point.

Why do we need two "Hilbert–Samuel polynomials" for a Hilbert module $\mathscr{H}$? The reason is that the two have different advantages and disadvantages. The main advantage of $p_{\underline{z}, \mathscr{H}}$ is that it is defined in terms of the action of A on $\mathscr{H}$ and thus can, in principle, be computed directly. Moreover, operator–theoretic properties of this action can be related concretely to it. The polynomial $P_{\underline{z},\mathfrak{H}}$ is related to the spectrum of $\mathscr{H}$ and its geometry but it is several steps removed from the Hilbert space and the operator action. The principal interest in both polynomials lies in their connection with the notion of multiplicity and this relation is key. Although there is a deep, well–established multiplicity theory in algebraic geometry and the notion of Hilbert–Samuel polynomial fits into it, it is not yet clear how useful that theory will be here, since one is seeking to understand multiplicity for the operator or module action. Along these lines there is the work of Nikolskii [21] which needs to be brought into the picture. In another direction there is the work of Carey–Pincus [5] and Vasilescu [26] and Putinar (cf. [16]) that plays an important role. There is much yet to do , both to connect these approaches and to obtain interesting results relating the operator action and the geometry..

3.   Modules and Resolutions

There is one very successful model theory, that for the case of contractive Hilbert modules over the disk algebra $A(\mathbb{D})$. A reinterpretation of the canonical model theory of Sz–Nagy and Foias [17] shows that a module resolution exists in this case [9] as follows. To simplify the exposition we restrict attention to contractive modules $\mathcal{M}$ over $A(\mathbb{D})$ for which the operator defined by multiplication by z is of C_{00}–class. Then there exist Hilbert spaces $\mathcal{E}$ and $\mathcal{E}_*$ such that

$$0 \longleftarrow \mathcal{M} \longleftarrow H^2_{\mathcal{E}}(\mathbb{D}) \longleftarrow H^2_{\mathcal{E}_*}(\mathbb{D}) \longleftarrow 0,$$

where the arrows represent partial isometric module maps. The notation $H^2_{\mathcal{E}}(\mathbb{D})$ and $H^2_{\mathcal{E}_*}(\mathbb{D})$ denotes Hardy modules of $\mathcal{E}$ and $\mathcal{E}_*$–valued functions respectively. Study and analysis of $\mathcal{M}$ can be reduced to considering these simple Hilbert modules and the module map connecting them which Sz.–Nagy and Foias show how to calculate explicitly in terms of the module action of $A(\mathbb{D})$ on $\mathscr{M}$.

What are the important properties of the modules $H^2_{\mathscr{E}}(\mathbb{D})$ and $H^2_{\mathscr{E}_*}(\mathbb{D})$ in this resolution? First, the modules are subnormal and the spectrum of the normal extension is contained in the Šilov boundary of $A(\mathbb{D})$ or they are what we have called <u>Šilov</u> <u>modules</u> [9]. Secondly, the spectral sheaf for them, which is equivalent to the sheaf defined by the last term of Koszul complex, is <u>locally</u> <u>free</u> or can be viewed as consisting of holomorphic sections of a vector bundle. Thirdly, these modules are <u>acyclic</u> or all the Koszul cohomology for them except for the last group vanishes at each point of $\mathbb{D}$.

Although resolutions by Šilov modules were sought originally (cf. [12]), it is now clear that there are problems with such a program, at least for modules over function algebras. (See [20] for results which suggest a different conclusion for Hilbert modules over noncommutative algebras.) In general, it is not known if Šilov resolutions always exist. However, the more critical issue is that even if a Šilov resolution exists it will not necessarily enable one to reduce the study of the given module to that of modules in the resolution and the connecting maps. In particular, a Šilov module need not be locally free or acylic. For example, the Hilbert module $H^2_{\Omega}(\mathbb{D}^2)$ over $A(\mathbb{D}^2)$ is a Šilov module since it is subnormal and its normal spectrum is T^2, the distinguished boundary of $\mathbb{D}^2$. This follows easily from the fact that z_1 and z_2 define commuting isometries on $H^2_{\Omega}(\mathbb{D}^2)$ with a joint normal extension given by multiplication by z_1 and z_2 on $L^2(T^2)$. The calculation in section 1 shows that $H^2_\Omega(\mathbb{D}^2)$ is not locally free at $\underline{0}$ since the dimension of the sheaf fiber is 2 at $\underline{0}$ and 1 on $\mathbb{D}^2 \backslash \{\underline{0}\}$. It is not acylic at $\underline{0}$ either. To see this one can calculate directly or argue as follows. The Euler index of the cohomology of the complex is locally constant at Fredholm points at $\underline{z}$ in $\mathbb{D}^2$. It equals -1, the same as that for $H^2(\mathbb{D}^2)$. Since the dimension of the final group for $H^2_\Omega(\mathbb{D}^2)$ is just the dimension of the cokernel, and that is 2 at $\underline{0}$, it follows that the middle cohomology group for $H^2_\Omega(\mathbb{D}^2)$ at $\underline{0}$ must be non zero.

Considerations of $H^2(\mathbb{D}^2)$ and $H^2_\Omega(\mathbb{D}^2)$ also shed light on a question about decomposable n—tuples. Over the last several decades the notion of spectral operator due to Bishop [2] was developed by many authors (cf. [26]). We do not review that notion here except to make a remark regarding condition β derived from Bishop's work [2]. The latter is a one—variable concept but an appropriate analogue was introduced into the multi—variate context by Frunza [18]. At first it was thought that this condition was implied by subnormality. However, the module $H^2_\Omega(\mathbb{D}^2)$ or essentially the closure of any non—principal ideal provides examples showing it does not, since condition β implies acyclicity.

Before continuing I want to raise a couple of questions and make an observation. Despite the fact that the Taylor spectrum is defined using the Koszul complex and that requires identifying a fixed set of generators for the algebra, it is still possible to define a module spectrum [12] which is a compact subset of M_A, the maximal ideal space of the algebra. For a module $\mathcal{H}$ and a finite subset of A having k elements, one can define a Taylor spectrum for the commuting k–tuple of these elements acting on $\mathscr{H}$. This spectrum is a compact subset of $\mathbb{C}^k$. The Gelfand transform enables one to pull this subset back to M_A. Standard functorial properties of the Taylor spectrum show that these compact subsets of M_A, indexed by the partially ordered collection of finite subsets of A, fit together to yield an inverse limit which can be identified as a compact subset of M_A. The module spectrum $\mathrm{Spec}_A(\mathscr{H})$ of $\mathscr{H}$ is defined to be this subset of M_A. One can also try to define the module spectrum directly using derived functors (cf. [23]). In neither approach is there any locality to the definition even though the spectrum is, in principle, a local notion. My first question asks whether that must be the case.

<u>Question 1:</u>   Is there a local definition of module spectrum?

Since this problem may seem somewhat vague, let me ask two more precise test questions that should be implied by an affirmative response to Question 1. Consider the Hilbert module $\mathscr{H}_\Omega$ over $A(\mathbb{D}^n)$ obtained from the closure of $\mathbb{C}[\underline{z}]$ in $L^2(\Omega)$, where Ω is an open set in $\mathbb{D}^n$ and the measure is volume measure. It would seem that whether a point $\underline{z}_0$ in $\mathbb{D}^n$ is in $\mathrm{Spec}_{A(\mathbb{D}^n)}(\mathscr{H}_\Omega)$ should depend only on the nature of $\Omega \cap \Delta_{\underline{z}_0}$, where $\Delta_{\underline{z}_0}$ is an open neighborhood of $\underline{z}_0$ in $\mathbb{D}^n$. However, no definition of Taylor spectrum makes this apparent but a positive response to the Question 1 should.

Another test question concerns a possible relation between the module spectrum of a module tensor product and the intersection of spectra for the two factors. In particular, is it true that $\mathrm{Spec}_A(\mathscr{M} \otimes_A \mathscr{N}) = \mathrm{Spec}_A(\mathscr{M}) \cap \mathrm{Spec}_A(\mathscr{N})$? What if we place appropriate finiteness conditions on A, $\mathcal{M}$ and $\mathcal{N}$? In this case, this is a theorem in the algebraic case.

The second question concerns two local module properties we discussed earlier: acyclicity and local freeness.

<u>Question 2</u>:   In the presence of appropriate finiteness conditions on a function algebra and a Hilbert module, is there a relation between a module being locally free at a point and it's being acyclic? If not, what if we assume the properties to hold on an open set?

In the algebraic case, such a relation is valid and depends on the fact that the vanishing of the first Koszul cohomology group implies the existence of "good coordinates" at a manifold point of the variety. However, the technique of using a "regular sequence" does not seem to carry over.

It seems clear that one would like module resolutions by modules that have both properties. Acyclicity is necessary for homological methods to be effective. Moreover, it would seem to have an algebraic character. Its local/global nature depends on the solution to the first problem. Local freeness is definitely a local property and is geometrical. Further, obtaining invariant metric properties from a resolution will be much simpler if one is dealing with spaces consisting of sections of hermitian holomorphic vector bundles and that is what local freeness means. In particular, in such a situation one can apply the complex geometric techniques introduced by M. Cowen and me [8] as was shown by Chen and me [6]. Under appropriate hypotheses equivalent to assuming local freeness and that the fibers generate the

Hilbert module, we show that local metric invariants are sufficient to characterize such Hilbert modules up to unitary equivalence.

The remark I wanted to make before continuing is that for a resolution to be useful it is not necessary that the connecting homomorphisms be partially isometric. In fact, the resolution could be useful even if it is only topologically exact, that is, if each kernel of a map equals the <u>closure</u> of the range of the preceding map. The reason is that much of the information will be obtained from localizations of the resolution and in the presence of appropriate finiteness conditions such a localization will consist of finite dimensional modules for which topological and algebraic exactness coincide.

Maps that don't preserve the metric on subsequent Hilbert modules are not quite so easily dealt with. One must construct classes that measure the change in metrics. This problem was solved in the hermitian vector bundle case by Bott and Chern [4] and such classes are now referred to as Bott–Chern classes. To consider Hilbert modules more globally leads one to Arakelov theory and the work of Bismuit–Gillet–Soulé [3]. We do not comment further on the connection here but it is basic and multi–variate operator theory provides interesting examples for this intricate and deep program in hermitian algebraic geometry.

There is work by K. Yan [27] related to the preceding discussion. We do not attempt to summarize it but only point out that it contains a development of the notion of hermitian holomorphic sheaf. The appropriate definition, at least for the context of Hilbert modules, is not obvious and the extension of algebraic and geometric machinery to this context can be most challenging.

4.   Existence of Resolutions

A basic question in the study of Hilbert modules concerns the existence of "nice resolutions."   As we indicated in the last section, "nice" should probably mean locally free and acyclic on $M_A \backslash \partial_A$, where ∂_A denotes the Šilov boundary for A.   Under appropriate finiteness conditions on an algebra and a Hilbert module, the spectral sheaf for $\mathcal{H}$, that is, the sheaf obtained via localization by $\mathbb{C}_{\underline{z}}$ or equivalently, the sheaf obtained from the first Koszul cohomology group, can be shown to be coherent on the Fredholm spectrum.   This requires showing first that $\text{Spec}_A \, \mathcal{H}$ is an algebraic variety.   Coherence means essentially that locally a sheaf resolution exists where the sheaves are sections of vector bundles.   However, there is no guarantee that a resolution exists over all of $M_A \backslash \partial_A$ but, more important, there is no connection implied between the terms of this resolution and Hilbert modules or multi–variate operator actions on Hilbert space.   That is, the sheaves in such a resolution need not come from Hilbert modules or even be equipped with a metric.   While there may be a use for "unnatural but nice sheaves", constructions involving natural or interesting objects would seem much preferable.

Question 3:   Under appropriate finiteness conditions on A and $\mathcal{H}$, does there exist a resolution of $\mathcal{H}$ by locally free, acyclic Hilbert modules?

While we do not know if such resolutions always exist, we can show using spectral sequence techniques that the length of some resolutions would have to be long enough to correspond to the homological dimension of the algebra.

<u>Theorem 3:</u>   There exists a finitely generated Hilbert module $\mathcal{H}$ over $A(\mathbb{D}^n)$ such that any resolution by acyclic Hilbert modules has length at least n.

One does not expect resolutions to be unique and one of the principal challenges is to extract invariant properties from any given resolutions. Solving this problem in the algebraic case was one of the early goals of homological algebra. Some of the results developed to accomplish this will carry over but many will not. Ultimately one will have to consider extensions of techniques in sheaf theory, especially homology theory, spectral sequences and more, to understand the multi–variate case. Extensive results including much foundational work has been obtained toward the study of resolutions by Putinar (cf. [16]) although his point of view is somewhat different.

There are some situations in which nice resolutions are known to exist and uniqueness holds. If I and J are ideals in $\mathbb{C}[\underline{z}]$ satisfying $J \subset I$, then there is a natural resolution for $[I]/[J]$. However, it will be "nice" only when both I and J are principal. Otherwise, [I] and [J] will be neither locally free nor acyclic. Thus, if we assume that $[I] = H^2(\mathbb{D}^n)$ and $[J] \subseteq H^2(\mathbb{D}^n)$ with J principal, then

$$0 \longleftarrow H^2(\mathbb{D}^n)/{[J]} \longleftarrow H^2(\mathbb{D}^n) \longleftarrow [J] \longleftarrow 0$$

is a nice resolution for $\mathcal{Q}_J = H^2(\mathbb{D}^n)/[J]$. Are there others and can we extract invariants for $\mathscr{L}_J$? We can answer the first question in part using a result of Foias and mine [10].

Theorem 4:   If $\mathscr{M}_1$, $\mathscr{M}_2$ are submodules of $H^2(\mathbb{D}^n)$ for which $H^2(\mathbb{D}^n)/\mathscr{M}_1$ and $H^2(\mathbb{D}^n)/\mathscr{M}_2$ are unitarily equivalent $A(\mathbb{D}^n)$ modules, then $\mathscr{M}_1 = \mathscr{M}_2$.

One would expect that $\mathrm{Spec}_{A(\mathbb{D}^n)}(\mathscr{L}_J) = \mathscr{Z}(J) \cap (\mathbb{D}^n)$ and that $\mathrm{Spec}^e_{A(\mathbb{D}^n)}(\mathscr{L}_J) = \mathscr{Z}(J) \cap \partial(\mathbb{D}^n)$, where the latter is the essential module spectrum. In general, proving these statements requires confronting

the notion of analytic transversality. Attempts to establish the necessary results have been partially successful [24], but the general case remains open.

One issue that is left unmentioned in the foregoing discussion concerns the nilpotents that arise in $\mathscr{L}_J$ because J may not be prime. Dealing with them and, in particular, understanding them operator—theoretically is another challenge. Again, to be more precise consider the examples [11] obtained from $J = z_1\mathbb{C}[\underline{z}]$ and $J^2 = z_1^2\mathbb{C}[\underline{z}]$. The action of z_1 on $\mathscr{L}_J$ is just 0 while on $\mathscr{L}_{J^2}$ it is the nilpotent operator $\begin{bmatrix} 0 & 1 \\ 0 & 0 \end{bmatrix} \otimes I_{H^2(\mathbb{D}^{n-1})}$.

Paulsen considers the uniqueness issue for certain Hilbert modules over the annulus algebra [22]. In this case all possible resolutions can be calculated and co—ordinates introduced for them. His results show how complicated even the simplest situations can be when resolutions are not unique.

As this note tries to make clear, the study of multi—variate operator theory is intimately related to algebraic geometry, commutative algebra, complex geometry and homological algebra as well as to operator theory and functional analysis. It is an exciting time in the development of M.O.T. since we are just now coming to grips with some of the fundamental problems and have identified important examples. The subject has achieved a certain richness and we are poised to achieve significant understanding in the near future.

Bibliography

1.   A. Beurling, On two problems concerning linear transformations in Hilbert space, Acta Math. **81** (1949) 239–255.

2.   E. Bishop, A duality theorem for an arbitrary operator, Pacific J. Math **9** (1959) 379–397.

3.   J.–M. Bismut, H. Gillet, and C. Soulé, Bott–Chern currents and complex immersions, Duke Math. J. **60** (1990) 255–284.

4.   R. Bott and S. S. Chern, Hermitian vector bundles and the equidistribution of the zeros of their holomorphic sections, Acta Math. **114** (1968) 71–112.

5.   R. Carey and J. Pincus, Principal functions, index theory, geometric measure theory and function algebras, Int. Eq. Op. Theory **2** (1979) 441–483.

6.   X. Chen and R. G. Douglas, Localization of Hilbert modules, Michigan Math. J. **39** (1992) 443–454.

7.   X. Chen and R. G. Douglas, Rigidity of Hardy submodules on the unit ball, Houston Math. J. **18** (1992) 117–125.

8.   M. Cowen and R. G. Douglas, Complex geometry and operator theory, Acta Math. **141** (1976) 187–261.

9.   R. G. Douglas, On Šilov resolution of Hilbert modules, in Operator Theory: Advances and Applications, vol. 28, Birkhâuser Verlag, Basel, 1988, pp 51–60.

10.   R. G. Douglas and C. Foias, Uniqueness of multi–variate canonical models, Acta Sci. Math. (Szeged) **57** (1993) 79–81.

11.   R. G. Douglas and G. Misra, Some calculations for Hilbert modules, Preprint, 1986.

12.   R. G. Douglas and V. I. Paulsen, <u>Hilbert Modules over Function Algebras</u>, Pitman Res. Notes Math. Ser., Vol. 219, Longman, London, 1989.

13.   R. G. Douglas, V.I. Paulsen, K. Yan and H. Sah, Algebraic reduction and rigidity for Hilbert modules, Amer. J. Math. (to appear).

14.   R. G. Douglas and K. Yan, On the rigidity of Hardy submodules, Int. Eq. Op. Theory **13** (1990) 350–363.

15.   R. G. Douglas and K. Yan, Hilbert–Samuel Polynomials for Hilbert Modules, Indiana U. Math. J. **42** (1993) 811–820.

16.   J. Eschmeier and M. Putinar, <u>Spectral Decompositions and Analytic Sheaves</u>, to appear.

17.   C. Foias and B. Sz–Nagy, <u>Harmonic analysis of operators on Hilbert space</u>, North Holland, Amsterdam, 1970.

18   S. Frunza, The Taylor spectrum and spectral decompositions, J. Funct. Anal. **19** (1975) 390–421.

19.   H. Hedenmalm, B. Korenblum and K. Zhu, Beurling Type Invariant Subspaces of the Bergman Spaces, preprint.

20.   P. Muhly and B. Solel, in preparation.

21.   N. K. Nikolskii, <u>Operators and Function Theory</u> (Proc. NATO Adv. Study Inst., Lancaster, 1984) Reidel, 1985, pp. 87–137.

22.   V. I. Paulsen, Resolutions of Hilbert modules, preprint.

23.   M. Putinar, Operator Theory and Sheaf Theory II, Math. Zeit. **192** (1986) 478–490.

24.   M. Putinar and N. Salinas, Analytic transversality and Nullstellensatz in Bergman spaces, Contemp. Math. **137** (1992) 367–381.

25.   J. L. Taylor, A joint spectrum for several commuting operators, J. Funct. Anal. **6** (1970) 172–191.

26.   F. H. Vasilescu, <u>Analytic functional calculus and spectral decompositions</u>, Reidel Publ. Co., Dordrecht, 1982.

27.   K. Yan, Hermitian Sheaves and Operator Theory, preprint, 1992.

28.   O. Zariski and O. Samuel, <u>Commutative Algebra I, II</u>, van Nostrand, Princeton, 1958.

State University of New York
Stony Brook, NY 11794

Contemporary Mathematics
Volume **185**, 1995

POSITIVITY, EXTENSIONS AND THE
TRUNCATED COMPLEX MOMENT PROBLEM

LAWRENCE FIALKOW

1. Introduction.

Given a closed subset $K \subset \mathbf{C}$ and a doubly indexed finite sequence of complex numbers

$$(1.1) \quad \gamma : \gamma_{00}, \ \gamma_{01}, \ \gamma_{10}, \ \gamma_{02}, \ \gamma_{11}, \gamma_{20}, ..., \ \gamma_{0,2n}, \ \gamma_{1,2n-1}, ..., \ \gamma_{2n-1,1}, \ \gamma_{2n,0},$$

where $\gamma_{00} > 0$ and $\gamma_{ji} = \bar{\gamma}_{ij}$,

the *truncated K complex moment problem* entails finding a positive Borel measure μ such that

$$(1.2) \qquad \gamma_{ij} = \int \bar{z}^i z^j \ d\mu \ (0 \le i + j \le 2n)$$

and

$$(1.3) \qquad supp \ \mu \subset K.$$

Any sequence γ as in (1.1) is a *truncated moment sequence* and any measure μ as in (1.2)-(1.3) is a *representing measure* for γ. In the present note we study the truncated complex moment problem using an approach based on matrix positivity and extension. Apart from its intrinsic interest, the truncated complex moment problem is significant due to its proximity to several moment problems which have recently been applied to questions in the theory of subnormal and hyponormal Hilbert space operators (as detailed below; see [Cu] for operator theoretic terminology).

By analogy with (1.1)-(1.3), we also consider the *full K moment problem*, in which the finite sequence of (1.1) is replaced by the *full moment sequence* $\gamma \equiv (\gamma_{ij})_{i,j \ge 0}$, $\gamma_{00} > 0$, $\gamma_{ji} = \bar{\gamma}_{ij}$. The classical theory shows that γ has a representing measure if and only if the associated functional F_γ acting on $\mathbf{C}[z, \bar{z}]$ is *strongly positive* [Sch] [ShT]. This condition is difficult to verify, so various authors have

1991 *Mathematics Subject Classification*. Primary 44A60; Secondary 47A20.
This paper is in final form and no version of it will be submitted for publication elsewhere.
Research supported by grants from The National Science Foundation and the NYS/UUP Professional Development and Quality of Working Life Committee

studied more concrete criteria for the existence of representing measures in special cases.

For $K = \mathbf{D}$ (the closed unit disk in $\mathbf{C}$), the full complex moment problem was solved by Atzmon [Atz] as follows: a natural positivity hypothesis on a kernel function associated to γ induces a Hilbert space norm on the polynomial space $\mathbf{C}[z]$, and an additional positivity hypothesis on γ forces M_z (multiplication by z) to be a subnormal contraction on this space; the spectral measure of the minimal normal extension then induces a representing measure for γ. (A closely related moment problem was solved by M. Putinar [P1] using the theory of hyponormal operators.) Given a *truncated* moment sequence (as in (1.1)), it is only possible to define Atzmon's norm on the space of polynomials of degree at most n, but since this space is not invariant for M_z, the preceding argument is obstructed; perhaps for this reason, there appears to be little published literature on the truncated complex moment problem, although the problem is well-known to specialists.

In [Atz] Atzmon used the solution of the full complex moment problem to recover some classical results concerning the following K *power moment problem* for a complex sequence $\beta \equiv (\beta_j)_{j \geq 0}$:

$$(1.4) \qquad\qquad \beta_j = \int z^j \, d\mu \ (j \geq 0); \ supp \, \mu \subset K.$$

Results of Stieltjes, Toeplitz, Hamburger, and Hausdorff provide necessary and sufficient conditions for solving the power moment problem in the cases $K = [0, +\infty)$, $K = \mathbf{T} \equiv \{z : |z| = 1\}$, $K = \mathbf{R}$, and $K = [a, b]$ $(a, b \in \mathbf{R})$, respectively [AhK] [Ak] [KrN] [Lan] [ShT]. For example, Hamburger's theorem for $K = \mathbf{R}$ implies that there exists a representing measure for $\beta \iff$ for every $n \geq 0$, the *Hankel matrix* $H(n) \equiv (\beta_{i+j})_{0 \leq i,j \leq n}$ is positive semidefinite. The classical theory also provides uniqueness theorems and concrete parameterizations of representing measures; moreover, the theory has been partially extended to multidimensional moment problems [B] [F] (indeed, the complex moment problem is equivalent to a power moment problem in two *real* variables).

Techniques from a wide variety of subjects have been brought to bear on moment problems, including real analysis, analytic function theory, continued fractions, extensions of positive functionals on convex cones in function spaces, and operator theory. Concerning the latter, the above-mentioned results of Atzmon and Putinar illustrate the use of operator theory in constructing representing measures; the operator theoretic approach originated in a development of Hamburger's theorem using the theory of self-adjoint extensions of symmetric operators [Sar] [St].

The interplay between operator theory and moment problems goes in both directions [P2]. Indeed, Atzmon's results on the complex moment problem were motivated by a reformulation of the invariant subspace problem for subnormal operators [Atz]. Further, in [CF1] Curto and the author solved Stampfli's subnormal completion problem by first solving the truncated Stieltjes moment problem (as described below), and [CF1] motivated our study of truncated moment problems in [CF3], [CF4] and in the present work. Subsequent to [CF1], Curto and Putinar [CP1] [CP2] used techniques from the theory of moments to solve a long-standing problem in operator theory by proving the existence of an operator that is *polynomially hyponormal* but not subnormal.

Let $\{e_n\}_{n\geq 0}$ denote the canonical orthonormal basis for $l_2(\mathbf{Z}_+)$; for a bounded sequence of positive numbers $\alpha \equiv \{\alpha_n\}$, we define the *unilateral weighted shift* operator W_α by $W_\alpha e_n = \alpha_n e_{n+1}$; we define the *moments* of W_α by $\beta_0 = 1$, $\beta_{n+1} = \beta_n \alpha_n^2$ ($n \geq 0$). The *subnormal completion problem* (even case) [S] asks for conditions on an initial segment of positive weights α: $\alpha_0, ..., \alpha_{2k}$ such that α can be completed to the weight sequence $\{\alpha_n\}_{n\geq 0}$ of a *subnormal* unilateral weighted shift on $l_2(\mathbf{Z}_+)$. The *truncated Stieltjes moment problem* asks for conditions on a sequence of positive numbers β: $\beta_0, ..., \beta_{2k+1}$ such that there exists a positive Borel measure μ supported in $[0, +\infty)$ satisfying $\beta_j = \int t^j \, d\mu$ ($0 \leq j \leq 2k + 1$). The main result of [CF1] describes the equivalence of these two problems as follows. Given an initial segment of weights α: $\alpha_0, ..., \alpha_{2k}$, we can define moments $\beta \equiv \beta(\alpha)$: $\beta_0, ..., \beta_{2k+1}$ as above. We then define Hankel matrices

$$(1.5) \qquad A(k) \equiv (\beta_{i+j})_{0\leq i,j\leq k}, \ B(k) \equiv (\beta_{i+j+1})_{0\leq i,j\leq k},$$

and we denote the rightmost column of $B(k)$ by v_{k+1} .

Theorem 1.1. *[CF1] (Subnormal completion problem, even case) For an initial weight segment α: $\alpha_0, ..., \alpha_{2k}$, the following are equivalent:*

i) *α has a subnormal completion;*
ii) *α has a $(k + 1)$-hyponormal completion (cf. [Cu]);*
iii) *The truncated Stieltjes moment problem for $\beta(\alpha)$ has a representing measure;*
iv) *$A(k) \geq 0, B(k) \geq 0$, and $v_{k+1} \in Ran \, A(k)$.*

Theorem 1.1-iv) reduces the subnormal completion problem to elementary linear algebra; the operator matrix techniques of [CF1], particularly the range condition of Theorem 1.1-iv), appear to be new in the literature of moment problems, and Theorem 1.1 appears to give the first complete solution of the truncated Stieltjes moment problem. The proof of Theorem 1.1 in [CF1] uses the spectral theorem to construct a representing measure, but in [CF3] we subsequently found a "finite" proof. The main results of [CF3] lead to elementary algorithms for solving the *truncated K power moment problem* for $K = [0, +\infty), \mathbf{R}, [a, b]$, and $\mathbf{T}$. These algorithms entail only basic linear algebra, matrix positivity, finite dimensional operator theory, Lagrange interpolation, and the calculation of roots of polynomials. In particular, we are able to circumvent the Riesz functional extension theorem [Akh], the infinite dimensional spectral theorem, Gaussian quadrature, and the theory of orthogonal polynomials, all of which play large roles in standard approaches to moment problems.

The basic result of [CF3] solves the *truncated Hamburger moment problem*. Let β: $\beta_0, ..., \beta_{2k}$ be real numbers, $\beta_0 > 0$. We seek a positive Borel measure μ such that $\beta_j = \int t^j \, d\mu$ ($0 \leq j \leq 2k$) and $supp \, \mu \subset \mathbf{R}$. Associated to β we have the Hankel matrix $H(k) \equiv (\beta_{i+j})_{0\leq i,j\leq k}$, with columns $v_0, ..., v_k$. If $H(k)$ is singular, there exists a minimal r, $1 \leq r \leq k$, such that $v_r \in \langle v_0, ..., v_{r-1}\rangle$; thus there exist unique real scalars $\phi_0, ..., \phi_{r-1}$ such that $v_r = \phi_0 v_0 + \cdots + \phi_{r-1} v_{r-1}$.

Theorem 1.2. *[CF3] The following are equivalent for β: $\beta_0, ..., \beta_{2k}$:*

i) *β has a representing measure;*
ii) *β has a compactly supported representing measure;*

iii) β *has a finitely atomic representing measure;*

iv) β *has a rank* $H(k)$-*atomic representing measure;*

v) $H(k)$ *admits an extension to* $H(k+1) \geq 0$;

vi) $H(k) \geq 0$ *and there exists an extension* $H(k+1)$ *such that*
    rank $H(k+1) = $ *rank* $H(k)$;

vii) $H(k) \geq 0$, *and either* $H(k)$ *is nonsingular, or* $H(k)$ *is singular and*

$$\beta_n = \phi_0\beta_{n-r} + \cdots + \phi_{r-1}\beta_{n-1} \ (r \leq n \leq 2k)$$

$$(equivalently, v_j = \phi_0 v_{j-r} + \cdots + \phi_{r-1}v_{j-1} \ (r \leq j \leq k)).$$

The structure theorem for a positive singular Hankel matrix $H(k)$ states that the preceding recursive relation for β_n *always* holds for $r \leq n \leq 2k - 1$ [CF3], so the issue in Theorem 1.2-vii) is simply whether the relation also holds for $n = 2k$. If vii) holds and $H(k)$ is singular, there is a unique representing measure, whose support consists of the r distinct real roots of the polynomial $t^r - (\phi_0 + \cdots + \phi_{r-1}t^{r-1}$). If $H(k)$ is nonsingular, then for each choice of β_{2k+1}, there exists $c \equiv c(\beta; \beta_{2k+1})$ (obtained using $H(k)^{-1}$) such that $\beta_{2k+2} \geq c \Longleftrightarrow$ the corresponding Hankel extension $H(k+1)$ is positive; moreover, $H(k+1)$ is singular $\Longleftrightarrow \beta_{2k+2} = c$. Thus if $H(k)$ is nonsingular and $H(k+1)$ is singular, the corresponding moment problem for $H(k+1)$ has a unique solution (as described above), which also produces a distinct representing measure for β parameterized by the choice of β_{2k+1}. If $H(k+1)$ is nonsingular, the above procedure may be repeated to yield an extension $H(k+2) \geq 0$. By continuing in this way, we have a method for producing all finitely atomic representing measures for β (parameterized by the successive choices of $\beta_{2k+1}, \beta_{2k+2}, \beta_{2k+3}, ...$). [CF3] also contains analogous results for the truncated power moment problems for $[0, +\infty)$, $[a, b]$, and **T**. For future reference we note that in the latter case, there exists a representing measure for β: $\beta_{-k}, ..., \beta_0, ..., \beta_k$, where $\beta_{-i} = \bar{\beta}_i \Longleftrightarrow$ the Toeplitz matrix $T(k)(\beta) \equiv (\beta_{j-i})_{0 \leq i,j \leq k}$ is positive.

In the present note we study a conjecture concerning an analogue of Theorem 1.2 for the truncated complex moment problem. Our main tool in studying this problem is the *moment matrix* $M(n) \equiv M(n)(\gamma)$ that we associate to the truncated moment sequence γ; this matrix plays a central role in Conjecture 1.3 that we present below. $M(n)$ is designed as the analogue of Hankel or Toeplitz matrix appropriate for the complex moment problem.

For $n \geq 1$, let $m \equiv m(n) = (n+1)(n+2)/2$. For $A \in M_m(\mathbf{C})$ (the m x m complex matrices), we denote the successive rows and columns according to the following lexicographic-functional ordering: $1, z, \bar{z}, ..., z^n, z^{n-1}\bar{z}, ..., z\bar{z}^{n-1}, \bar{z}^n$. For $0 \leq i + j \leq n$, $0 \leq l + k \leq n$, we denote the entry in row $\bar{z}^l z^k$, column $\bar{z}^i z^j$ by $A_{(l,k)(i,j)}$. (In notation such as $0 \leq i + j \leq n$, it is always implicit that $i, j \geq 0$.) We now define $M(n)(\gamma) \in M_m(\mathbf{C})$ as follows: for $0 \leq i + j \leq n$, $0 \leq l + k \leq n$,

(1.6) $$M(n)_{(l,k)(i,j)} = \gamma_{i+k,j+l}.$$

For example, with $n = 1$, we associate to the *quadratic moment problem* for γ: $\gamma_{00}, \gamma_{01}, \gamma_{10}, \gamma_{02}, \gamma_{11}, \gamma_{20}$ the moment matrix

$$M(1) = \begin{pmatrix} \gamma_{00} & \gamma_{01} & \gamma_{10} \\ \gamma_{10} & \gamma_{11} & \gamma_{20} \\ \gamma_{01} & \gamma_{02} & \gamma_{11} \end{pmatrix}.$$

(To study *multidimensional* truncated moment problems in several real or complex variables, we can define moment matrices subordinate to lexicographic orderings of the several variables; most of the results from [CF4] concerning $M(n)$ that we cite below extend to these more general moment matrices.) The use of functional headings for the columns of $M(n)$ is deliberately suggestive; as we show below, a necessary condition for the existence of representing measures is that the columns of $M(n)$ behave like the functions naming them (in a sense we will make precise). This perspective, which seems to be new in moment problems, is perhaps our main contribution, and is exploited systematically below and in [CF4]. To further explain the significance of $M(n)$ we require additional notation. Let P_n denote the vector space of polynomials in $z, \bar{z}$ of total degree $\leq n$. Each $p \in P_n$ has a unique representation $p \equiv p(z, \bar{z}) = \sum_{0 \leq i+j \leq n} a_{ij} \bar{z}^i z^j$ $(a_{ij} \in \mathbf{C})$; let $\hat{p}$ denote the coefficient vector $(a_{00}, a_{01}, a_{10}, ..., a_{0,n}, ..., a_{n,0})$. For $A \in M_m(\mathbf{C})$ we define a sesquilinear form $\langle \cdot, \cdot \rangle_A$ on P_n by $\langle p, q \rangle_A = (A\hat{p}, \hat{q})$.

The fundamental connection between $M(n)(\gamma)$ and any representing measure μ for γ is provided by the identity

(1.7) $\int f\bar{g}d\mu = \langle f, g \rangle_{M(n)} = (M(n)\hat{f}, \hat{g})$ $(f, g \in P_n)$ [CF4].

For $f = g$, we thus have $0 \leq \int |f|^2 d\mu = (M(n)\hat{f}, \hat{f})$, which implies the following basic necessary condition for the existence of representing measures:

(1.8)   *If γ has a representing measure, then $M(n)(\gamma) \geq 0$.*

Let $\mathbf{1}, \mathbf{Z}, \bar{\mathbf{Z}}, ..., \mathbf{Z^n}, \mathbf{Z^{n-1}\bar{Z}}, ..., \mathbf{Z\bar{Z}^{n-1}}, \bar{\mathbf{Z}}^\mathbf{n}$ denote the successive columns of $M(n)$ and let $C_{M(n)}$ denote the column space of $M(n)$ in $\mathbf{C}^m$; we refer to the linear span of $\mathbf{1}, \mathbf{Z}, ..., \mathbf{Z^i}, ..., \mathbf{Z^n}$ in $\mathbf{C}^m$ as the *analytic space* of γ. For $p \in P_n$, $p(z, \bar{z}) \equiv \sum a_{ij} \bar{z}^i z^j$, let $\mathbf{p}(\mathbf{Z}, \bar{\mathbf{Z}}) \equiv \sum a_{ij} \bar{\mathbf{Z}}^\mathbf{i} \mathbf{Z}^\mathbf{j} \in C_{M(n)}$. Dependence relations in the columns of $M(n)$ can be used to help locate the support of any representing measure for γ:

(1.9)   *If μ is a representing measure for γ and $p \in P_n$, then*

$\mathbf{p}(\mathbf{Z}, \bar{\mathbf{Z}}) = 0 \Longleftrightarrow \text{supp } \mu \subset Z(p) \equiv \{z \in \mathbf{C} : p(z, \bar{z}) = 0\}$ [CF4].

In the sequel we study the following conjecture.

Conjecture 1.3. *For a truncated moment sequence γ, the following are equivalent:*
  i) γ has a representing measure;
  ii) γ has a compactly supported representing measure;
  iii) γ has a finitely atomic representing measure;
  iv) γ has a rank $M(n)$-atomic representing measure;
  v) $M(n)$ admits an extension $M(n+1) \geq 0$;

$vi)$ $M(n) \geq 0$ admits an extension $M(n+1)$ satisfying
rank $M(n+1) =$ rank $M(n)$.

In section 3 we affirm this conjecture in the case when $\mathbf{Z} = \bar{\mathbf{Z}}$ in $C_{M(n)}$, i.e., when there exists a sequence β: $\beta_0, ..., \beta_{2n}$ such that $\gamma_{ij} = \beta_{i+j}$. We reduce the $K \equiv \mathbf{R}$ truncated complex moment problem for γ to the truncated Hamburger moment problem for β by proving that $M(n)(\gamma) \geq 0 \iff H(n)(\beta) \geq 0$ (Proposition 3.1). Thus, for $K = \mathbf{R}$, Conjecture 1.3 is equivalent to Theorem 1.2. In section 4 we show that the equivalence of i), ii), iii), and v) of Conjecture 1.3 is also true when $n > 1$ and $\mathbf{Z}\bar{\mathbf{Z}} = \mathbf{1}$ in $C_{M(n)}$, i.e., $\gamma_{ij} = \beta_{j-i}$ for some auxiliary sequence β. We prove in Proposition 4.1 that in this case $M(n)(\gamma) \geq 0 \iff T(2n)(\beta) \geq 0$; thus, for $K = \mathbf{T}$, Conjecture 1.3 reduces to the truncated Toeplitz (or *trigonometric*) moment problem.

The main result of [CF4] provides additional significant evidence in support of Conjecture 1.3:

Theorem 1.4. *[CF4] The following are equivalent for a truncated complex moment sequence γ:*

i) γ has a rank $M(n)(\gamma)$-atomic representing measure;

ii) $M(n) \geq 0$ admits an extension $M(n+1)$ satisfying rank $M(n+1) =$ rank $M(n)$.

Theorem 1.4 is used in [CF4] to prove Conjecture 1.3 for n=1, and to thereby solve the quadratic moment problem for γ: γ_{00}, γ_{01}, γ_{10}, γ_{02}, γ_{11}, γ_{20}. As a further application of Theorem 1.4, [CF4] contains a solution of the truncated complex moment problem in the case when the *analytic columns* of $M(n)$, $\mathbf{1}, \mathbf{Z}, ..., \mathbf{Z^i}, ..., \mathbf{Z^n}$, are dependent. In this case there exists a minimal r, $1 \leq r \leq n$, such that $\mathbf{Z^r} \in \langle \mathbf{1}, ..., \mathbf{Z^{r-1}} \rangle$; thus there exist unique scalars ϕ_j such that $\mathbf{Z^r} = \phi_0\mathbf{1} + \cdots + \phi_{r-1}\mathbf{Z^{r-1}}$.

Theorem 1.5. *[CF4] Suppose the analytic columns of $M(n)$ are dependent. Then γ has a representing measure if and only if $M(n) \geq 0$ and $\{\mathbf{1}, ..., \mathbf{Z^{r-1}}\}$ spans $C_{M(n)}$; in this case γ has a unique representing measure, $\mu = \sum \rho_i \delta_{z_i}$, whose support $\{z_0, ..., z_{r-1}\}$ consists of the r distinct roots of $z^r - (\phi_0 + \cdots + \phi_{r-1}z^{r-1})$ and whose densities ρ_j satisfy the Vandermonde equation*

$$V(z_0, ..., z_{r-1}) (\rho_0, ..., \rho_{r-1})^T = (\gamma_{00}, ..., \gamma_{0,r-1})^T.$$

The case of the truncated complex moment problem in which the analytic columns are *independent* is unsolved and, in particular, the case when $M(n)$ is positive and invertible ($M(n) > 0$) is open for $n > 1$. (If $M(1)(\gamma) > 0$, then γ has infinitely many 3-atomic representing measures [CF4].) In section 5 we consider a particular example of $M(2) \geq 0$ in which $\{\mathbf{1}, \mathbf{Z}, \mathbf{Z^2}\}$ is a basis for $C_{M(2)}$. For this example we use our function-column techniques to determine a parameterization of all 3-atomic representing measures, in accordance with Conjecture 1.3 (and in agreement with Conjecture 2.6 below).

Section 2 contains preliminary results on positive operator matrices. We prove in Proposition 2.4 the following extension principle for positive matrices, which is used frequently in [CF4]: a dependence relation in the columns of an orthogonal compression of a positive matrix extends to a corresponding dependence relation in the full columns of the matrix. In section 2 we also discuss the Extension Problem for positive moment matrices and its relation to Conjecture 1.3.

2. Positive matrices and extensions.

Let $H \equiv H_1 \oplus H_2$ denote an orthogonal direct sum decomposition of a Hilbert space H into a sum of closed subspaces. For bounded linear operators $A: H_1 \mapsto H_1$, $B: H_2 \mapsto H_1$, $C: H_2 \mapsto H_2$, let $\tilde{A}: H \mapsto H$ denote the operator with matrix

$$(2.1) \qquad \begin{pmatrix} A & B \\ B^* & C \end{pmatrix}.$$

We begin with several characterizations of positivity for $\tilde{A}$.

Proposition 2.1. $\tilde{A} \geq 0$ if and only if $A \geq 0$ and there exists a bounded operator $V: H_2 \mapsto H_1$ such that $B = A^{1/2}V$ and $C \geq V^*V$.

Proposition 2.1 may easily be derived from an equivalent characterization, due to Smul'jan [Smu], which requires $Ran\ V \subset H_1 \ominus ker\ A$ (cf. [CF1, Proposition 2.2]). Smul'jan also proved that $\tilde{A} \geq 0 \iff A \geq 0$, $C \geq 0$, and there exists a linear contraction $E: H_2 \mapsto H_1$ such that $B = A^{1/2}EC^{1/2}$. Still another characterization of positivity is that $A \geq 0$, $C \geq 0$, and $|\langle Bx, y \rangle|^2 \leq \langle Ay, y \rangle \langle Cx, x \rangle$ for every $x \in H_2$, $y \in H_1$ [Pa, page 39]. We also have a partial analogue of Proposition 2.1 which does not explicitly refer to $A^{1/2}$.

Proposition 2.2. If $A \geq 0$ and there exists a bounded operator $W: H_2 \mapsto H_1$ such that $B = AW$ and $C \geq W^*AW$, then $\tilde{A} \geq 0$. The converse is true $\iff Ran\ B \subset Ran\ A$.

Proof. Assume $A \geq 0$, $B = AW$, and $C \geq W^*AW$. Let $V = A^{1/2}W$, so that $A^{1/2}V = AW = B$ and $V^*V = W^*AW \leq C$; Proposition 2.1 implies that $\tilde{A} \geq 0$.

For the converse to hold, clearly $Ran\ B = Ran\ AW \subset Ran\ A$. Now suppose $\tilde{A} \geq 0$ and $Ran\ B \subset Ran\ A$. Proposition 2.1 implies that there exists $V: H_2 \mapsto H_1$ satisfying $B = A^{1/2}V$ and $C \geq V^*V$. Let P denote the orthogonal projection of H_1 onto $init\ A \equiv H_1 \ominus ker\ A$. Then $A^{1/2}V = A^{1/2}(PV)$ and $V^*V \geq (PV)^*(PV)$. We may thus assume that $Ran\ V \subset init\ A$. Since $Ran\ B \subset Ran\ A$, [D] implies that there exists a bounded operator $W: H_2 \mapsto H_1$ such that $B = AW$; replacing W by PW, we may assume $Ran\ W \subset init\ A$. Since $A^{1/2}(A^{1/2}W - V) = 0$ and $Ran(A^{1/2}W - V) \subset init\ A$, then $V = A^{1/2}W$, whence $W^*AW = V^*V \leq C$. $\square$

Recall that a positive operator A has closed range if and only if $Ran\ A^{1/2} \subset Ran\ A$ [FW, page 259]. In this case, the "converse" in Proposition 2.2 always holds, since from Proposition 2.1, $Ran\ B \subset Ran\ A^{1/2} \subset Ran\ A$. If A has nonclosed range, the "converse" implication in Proposition 2.2 holds in some cases but not in others. If $A \geq 0$ and $B = C = A$, then $\tilde{A} \geq 0$ by Proposition 2.2 (with $W = 1$) whether or not A has closed range. On the other hand, if $A \geq 0$, A has nonclosed range, $B = A^{1/2}$, and $C = 1$, then $\tilde{A} \geq 0$ by Proposition 2.1 (with $V = 1$), but there is no bounded operator W such that $A^{1/2} = AW$.

We next focus on the case when H is finite dimensional and A, B, C are complex matrices of compatible sizes. Note that if $A \geq 0$ and $B = AW$, $B = AV$ for matrices W, V, then $W^*AW = V^*AV$.

Corollary 2.3. *(cf. [CF1, Proposition 2.3]) i) $\tilde{A} \geq 0 \iff A \geq 0$, Ran $B \subset$ Ran A (so that $B = AW$ for some matrix W), and $C \geq W^*AW$ for some (every) W satisfying $B = AW$; ii) If $A \geq 0$ and $B = AW$ for some matrix W, then $C \equiv W^*AW$ is the unique choice of C such that rank $\tilde{A} = $ rank A.*

In the sequel we refer to a matrix of the form (2.1) as an *extension* of A; this is *not* the usual notion of extension for an operator A. An extension $\tilde{A}$ satisfying *rank $\tilde{A} = $ rank A* is said to be a *flat extension*; it is clear from Corollary 2.3 that if A is positive and B is specified such that *Ran $B \subset$ Ran A*, then there is a unique flat extension of A using B. If A is known to be in a special class, such as a positive Hankel, Toeplitz, or moment matrix, we are interested in conditions on A which insure the existence of (flat) extensions $\tilde{A}$ in the same class; we will return to this extension problem below. The next result is our fundamental observation concerning extensions. Although this result is an easy consequence of Smul'jan's theorem, it does not seem to appear explicitly in the literature, and its important consequences for matrix extension problems seem not to have been previously noticed. Let $M_k(\mathbf{C})$ denote the $k \times k$ complex matrices. Let $A \in M_{k+1}(\mathbf{C})$, $A \equiv (a_{ij})_{0 \leq i,j \leq k}$, and let $v_0, ..., v_k$ denote the columns of A. For $0 \leq p \leq k$, let $A(p) \equiv (a_{ij})_{0 \leq i,j \leq p} \in M_{p+1}(\mathbf{C})$; let $v(0,p)$, ..., $v(p,p)$ denote the columns of $A(p)$.

Proposition 2.4. *(Extension Principle) If $A \equiv A(k) \geq 0$ and there exist r, $0 \leq r \leq k$, and scalars c_0, ..., c_r such that $c_0 v(0,r) + \cdots + c_r v(r,r) = 0$, then $c_0 v_0 + \cdots + c_r v_r = 0$.*

Proof. Consider the following orthogonal partitioning of A:

$$A \equiv \begin{pmatrix} A(r) & B_r \\ B_r^* & C_r \end{pmatrix}.$$

Corollary 2.3 implies *Ran $B_r \subset$ Ran $A(r)$*, whence *ker $A(r) \subseteq$ ker B_r^** . Let $v \equiv (c_0, ..., c_r) \in \mathbf{C}^{r+1}$ and let $w \equiv (c_0, ..., c_r, 0, ..., 0) \in \mathbf{C}^{k+1}$. The hypothesis implies $A(r)v = 0$; thus $B_r^* v = 0$, and so $Aw = 0$, i.e., $c_0 v_0 + \cdots + c_r v_r = 0$. $\square$

The Extension Principle is used in [CF4] to prove the following structure theorem for positive moment matrices; this provides an important tool for the proofs of Theorems 1.4 and 1.5 in [CF4].

Theorem 2.5. *[CF4] Suppose $M(n)(\gamma) \geq 0$. If p, q, $pq \in P_{n-1}$ and $p(\mathbf{Z}, \bar{\mathbf{Z}}) = 0$, then $(pq)(\mathbf{Z}, \bar{\mathbf{Z}}) = 0$.*

Theorem 2.5 can be used to recover the "recursive" structure theorems for positive singular Hankel matrices and positive singular Toeplitz matrices (see sections 3 and 4). More importantly, it provides the following necessary condition for the existence of a positive moment matrix extension $M(n + 1)$:

(J)   *If p, q, $pq \in P_n$ and $p(\mathbf{Z}, \bar{\mathbf{Z}}) = 0$, then $(pq)(\mathbf{Z}, \bar{\mathbf{Z}}) = 0$.*

Condition (J) requires that $J_\gamma \equiv \{p \in P_n : p(\mathbf{Z}, \bar{\mathbf{Z}}) = 0\}$ behave like an "ideal" in the polynomial space P_n. Note that condition (J) can be verified using only elementary linear algebra.

Conjecture 2.6. *For $M(n)(\gamma) \geq 0$, the following are equivalent:*
i) *There exists an extension $M(n + 1) \geq 0$;*
ii) *There exists a flat extension $M(n + 1)$;*
iii) *There exist flat extensions $M(n + k)$ for every $k \geq 1$;*
iv) *$M(n)$ satisfies (J).*

An affirmation of this conjecture would provide (J) as a "concrete" condition equivalent to v) of Conjecture 1.3 (and analogous to condition vii) of Theorem 1.2); together with (1.8), (1.9), Theorem 1.4, and Theorem 2.5, this would completely prove Conjecture 1.3. Indeed, to prove Conjecture 1.3, it suffices to prove that if $M(n) \geq 0$ satisfies (J), then there exists a flat extension $M(n + 1)$. The results of section 3 and Theorem 1.2 imply that Conjecture 2.6 is true in the case when $\mathbf{Z} = \bar{\mathbf{Z}}$. Conjecture 2.6 is also true if $rank\ M(n) = rank\ M(n - 1)$ [CF4], or if n = 1 (the quadratic moment problem), but it remains unresolved for the case when $rank\ M(n)$ is maximal; note that if $M(n)$ is invertible, then (J) is satisfied vacuously.

Conjecture 2.7. *If $M(n)$ is positive and invertible, then $M(n)$ has a flat extension $M(n + 1)$.*

Conjecture 2.7 is true for $n = 1$ [CF4].

3. The $K \equiv R$ truncated complex moment problem.

We begin by defining a useful block decomposition of $M(n)$. For $0 \leq i, j \leq n$, let B_{ij} denote the $i + 1 \times j + 1$ matrix of the form

(3.1) $(\gamma_{i+r-s, j-r+s})_{0 \leq s \leq i, 0 \leq r \leq j}$;

for example, $B_{2,3}$ has the form

$$\begin{pmatrix} \gamma_{23} & \gamma_{32} & \gamma_{41} & \gamma_{50} \\ \gamma_{14} & \gamma_{23} & \gamma_{32} & \gamma_{41} \\ \gamma_{05} & \gamma_{14} & \gamma_{23} & \gamma_{32} \end{pmatrix}.$$

$M(n)$ admits the block decomposition $M(n) = (B_{ij})_{0 \leq i, j \leq n}$. Note that B_{ij} has the Toeplitz-like property of being constant on diagonals; moreover, $M(n)$ has the Hankel-like property that the cross diagonal blocks .... $B_{i+1, j-1}$, B_{ij}, $B_{i-1, j+1}$, ... all contain the same elements (those γ_{pq} with $p + q = i + j$), though arranged in differing patterns.

In this section we consider the case in which each block B_{ij} is constant; we assume there exists a sequence β: β_0, ..., β_{2n} (of necessarily *real* scalars) such each entry of B_{ij} is equal to $\gamma_{ij} \equiv \beta_{i+j}$ ($i, j \geq 0$, $0 \leq i+j \leq 2n$). Observe that $M(n)$

block-constant $\Longrightarrow \mathbf{Z} = \bar{\mathbf{Z}}$ in $C_{M(n)} \Longrightarrow$ any representing measure μ for γ satisfies $supp \, \mu \subset \mathbf{R}$ (1.9) $\Longrightarrow M(n)$ block-constant. We will show that Conjectures 1.3, 2.6, and 2.7 are true in this case and that the block-constant case of the truncated complex moment problem for γ is equivalent to the truncated Hamburger moment problem for β. With γ, β as above, let $H(n) \equiv H(n)(\beta)$ denote the Hankel matrix $(\beta_{i+j})_{0 \leq i,j \leq n}$ and let $HM(n)$ denote the block-constant $M(n)(\gamma)$.

Proposition 3.1. $HM(n) \geq 0 \Longleftrightarrow H(n) \geq 0$.

Proof. Since $H(n)$ is the compression of $HM(n)$ to the analytic space, if $HM(n) \geq 0$, then $H(n) \geq 0$. For the converse, consider the decomposition of $HM(n)$ as

$$HM(n) = \begin{pmatrix} A & B \\ B^* & C \end{pmatrix},$$

where $A = HM(n-1)$,

$$B = \begin{pmatrix} \beta_n & \cdots & \beta_n \\ \beta_{n+1} & \cdots & \beta_{n+1} \\ \beta_{n+1} & \cdots & \beta_{n+1} \\ \vdots & & \vdots \\ \beta_{2n-1} & \cdots & \beta_{2n-1} \\ \vdots & & \vdots \\ \beta_{2n-1} & \cdots & \beta_{2n-1} \end{pmatrix} \quad n(n+1)/2 \times (n+1),$$

$$D = \begin{pmatrix} 1 & \cdots & 1 \\ \vdots & & \vdots \\ 1 & \cdots & 1 \end{pmatrix} (n+1) \times (n+1),$$

and $C = \beta_{2n} D$.

Since $H(n) \geq 0$, Corollary 2.3 implies there exist scalars $\phi_0, \ldots, \phi_{n-1}$ such that

(3.2) $\beta_{n+i} = \phi_0 \beta_i + \cdots + \phi_{n-1} \beta_{n-1+i} \quad (0 \leq i \leq n-1)$

and

(3.3) $\phi_0 \beta_n + \cdots + \phi_{n-1} \beta_{2n-1} \leq \beta_{2n}.$

Let

$$W = \begin{pmatrix} \phi_0 & \cdots & \phi_0 \\ \phi_1 & \cdots & \phi_1 \\ 0 & \cdots & 0 \\ \phi_2 & \cdots & \phi_2 \\ 0 & \cdots & 0 \\ 0 & \cdots & 0 \\ \vdots & & \vdots \\ \phi_{n-1} & \cdots & \phi_{n-1} \\ 0 & \cdots & 0 \\ \vdots & & \vdots \\ 0 & \cdots & 0 \end{pmatrix} \quad n(n+1)/2 \times (n+1).$$

Now $AW = B$ by (3.2) and $W^*AW = (\phi_0\beta_n + \cdots + \phi_{n-1}\beta_{2n-1})D \leq C$ by (3.3). Thus $HM(n) \geq 0$ by Corollary 2.3. $\square$

We can now use Theorem 2.5 to recover the structure theorem for positive singular Hankel matrices. In the sequel we denote the successive columns of $H(n)(\beta)$ by $\mathbf{1}, \mathbf{t}, \ldots, \mathbf{t^n}$. Assume $H(n)$ is singular and let r, $1 \leq r \leq n$, be minimal with the property that $\mathbf{t^r} \in \langle \mathbf{1}, \ldots, \mathbf{t^{r-1}} \rangle$; thus there exist unique real scalars $\phi(\beta) : \phi_0, \ldots, \phi_{r-1}$ such that $\mathbf{t^r} = \phi_0\mathbf{1} + \cdots + \phi_{r-1}\mathbf{t^{r-1}}$.

Proposition 3.2. *[CF3] If $H(n)(\beta)$ is positive and singular, and $r < n$, then $\mathbf{t^{r+s}} = \phi_0\mathbf{t^s} + \cdots + \phi_{r-1}\mathbf{t^{r+s-1}}$ $(0 \leq s \leq n-r-1)$; equivalently, $\beta_{r+s} = \phi_0\beta_s + \cdots + \phi_{r-1}\beta_{r+s-1}$ $(0 \leq s \leq 2n-r-1)$ and $\beta_{2n} \geq \phi_0\beta_{2n-r} + \cdots + \phi_{r-1}\beta_{2n-1}$.*

Proof. Since $H(n) \geq 0$, Proposition 3.1 implies $HM(n) \geq 0$, and in $C_{HM(n)}$ we have $\mathbf{Z^r} = \phi_0\mathbf{1} + \cdots + \phi_{r-1}\mathbf{Z^{r-1}}$. Theorem 2.5 implies that for $0 \leq s \leq n-r-1$, $\mathbf{Z^{r+s}} = \phi_0\mathbf{Z^s} + \cdots + \phi_{r-1}\mathbf{Z^{r+s-1}}$, whence the identity for $\mathbf{t^{r+s}}$ follows by compression; the inequality for β_{2n} follows from Corollary 2.3. $\square$

We next solve the block-constant moment problem by showing its equivalence to the truncated Hamburger moment problem. Let γ, β be as above, i.e., $\gamma_{ij} = \beta_{i+j}$ $(0 \leq i+j \leq 2n)$.

Proposition 3.3. *Suppose $\mathbf{Z} = \bar{\mathbf{Z}}$, i.e., $\gamma_{ij} = \beta_{i+j}$ $(0 \leq i+j \leq 2n)$. The following are equivalent:*

  i) *γ has a representing measure;*
  ii) *$HM(n)$ admits a positive (flat) extension $HM(n+1)$;*
  iii) *$H(n)$ admits a positive (flat) extension $H(n+1)$;*
  iv) *$H(n) \geq 0$, and either $H(n)$ is invertible, or $H(n)$ is singular and*
       $\phi(\beta)$ satisfies $\mathbf{t^n} = \phi_0\mathbf{t^{n-r}} + \cdots + \phi_{r-1}\mathbf{t^{n-1}}$;
  v) *The truncated Hamburger moment problem for β has a representing measure;*
  vi) *γ has a rank $HM(n)$-atomic representing measure supported in $\mathbf{R}$.*

Proof. The equivalence of ii) and iii) follows from Proposition 3.1 and the fact that $rank\ HM(p) = rank\ H(p)$ for $0 \leq p \leq n$. The equivalence of iii), iv) and v) is part of Theorem 1.2. If μ is a representing measure for β, then $\int \bar{z}^i z^j d\mu = \int t^{i+j} d\mu = \beta_{i+j} = \gamma_{ij}$, so μ is a representing measure for γ; thus v)

$\Rightarrow$ i). Suppose i) holds and let ν be a representing measure for γ. Since $\mathbf{Z} = \bar{\mathbf{Z}}$ in $C_{HM(n)}$, (1.9) implies $supp\ \nu \subset \mathbf{R}$, whence ν is a representing measure for β; thus v) holds. The equivalence of i) and vi) follows from the equivalence of i) and v), from Theorem 1.2, and from the fact that $rank\ HM(n) = rank\ H(n)$. $\square$

4. The $K \equiv T$ truncated complex moment problem.

For $n > 1$, the moment problem for $K \equiv \mathbf{T}$ corresponds to the case in which $M(n)$ satisfies $\mathbf{Z}\bar{\mathbf{Z}} = 1$. This is the case in which there exists a sequence $\alpha: \alpha_{-2n},\ ...,\ \alpha_0,\ ...,\ \alpha_{2n}$ such that $\gamma_{ij} = \alpha_{j-i}$ $(0 \le i+j \le 2n)$ (so that $\alpha_{-p} = \bar{\alpha}_p$). In this case we denote $M(n)(\gamma)$ by $TM(n)(\gamma)$ and we let $T(2n)(\alpha)$ denote the Toeplitz matrix $(\alpha_{j-i})_{0 \le i,j \le 2n}$. For example, with $n = 2$ and $\gamma_{ij} = \alpha_{j-i}$ $(0 \le i+j \le 4)$, $TM(2)$ and $T(4)$ are given by

$$\begin{pmatrix} \alpha_0 & \alpha_1 & \alpha_{-1} & \alpha_2 & \alpha_0 & \alpha_{-2} \\ \alpha_{-1} & \alpha_0 & \alpha_{-2} & \alpha_1 & \alpha_{-1} & \alpha_{-3} \\ \alpha_1 & \alpha_2 & \alpha_0 & \alpha_3 & \alpha_1 & \alpha_{-1} \\ \alpha_{-2} & \alpha_{-1} & \alpha_{-3} & \alpha_0 & \alpha_{-2} & \alpha_{-4} \\ \alpha_0 & \alpha_1 & \alpha_{-1} & \alpha_2 & \alpha_0 & \alpha_{-2} \\ \alpha_2 & \alpha_3 & \alpha_1 & \alpha_4 & \alpha_2 & \alpha_0 \end{pmatrix}$$

and

$$\begin{pmatrix} \alpha_0 & \alpha_1 & \alpha_2 & \alpha_3 & \alpha_4 \\ \alpha_{-1} & \alpha_0 & \alpha_1 & \alpha_2 & \alpha_3 \\ \alpha_{-2} & \alpha_{-1} & \alpha_0 & \alpha_1 & \alpha_2 \\ \alpha_{-3} & \alpha_{-2} & \alpha_{-1} & \alpha_0 & \alpha_1 \\ \alpha_{-4} & \alpha_{-3} & \alpha_{-2} & \alpha_{-1} & \alpha_0 \end{pmatrix} .$$

Although it is possible to use Theorem 2.5 to recover the "recursive" structure theorem for positive singular Toeplitz matrices [CF3], the proof is lengthy, so we omit this. For $n > 1$, the following result reduces the truncated $\mathbf{T}$ moment problem for γ to the *truncated trigonometric moment problem*

$$(4.1) \qquad \alpha_k = \int z^k\ d\mu\ (0 \le k \le 2n);\ \mu \ge 0;\ supp\ \mu \subset \mathbf{T}.$$

Proposition 4.1. *For* $n \ge 0$, $TM(n) \ge 0 \Longleftrightarrow T(2n) \ge 0$.

Proof. We prove by induction on n that $TM(n) \ge 0 \Longrightarrow T(2n) \ge 0$, and for $n = 0$ we have $TM(0) = T(0) = (\alpha_0)$. Assume $TM(n) \ge 0$ and consider the orthogonal decomposition

$$TM(n) \equiv \begin{pmatrix} A_n & B_n \\ B_n^* & C_n \end{pmatrix},$$

where $A_n \equiv TM(n-1)$. In $TM(n)$ we denote the first 1-step extension of $TM(n-1)$ by

$$P \equiv \begin{pmatrix} Q & w_n \\ w_n^* & \alpha_0 \end{pmatrix},$$

where $Q \equiv TM(n-1)$ and $w_n = (\alpha_n,\ \alpha_{n-1},\ \alpha_{n+1},\ \alpha_{n-2},\ \alpha_n,\ \alpha_{n+2},\ ...,\ \alpha_1,\ \alpha_3,\ ...,\ \alpha_{2n-1})^t$. Since $P \ge 0$, $w_n \in Ran\ Q$, so there exists h_n such that

$$(4.2) \qquad Qh_n = w_n \text{ and } \alpha_0 \ge h_n^* w_n.$$

For $i = 0$, $j = 1$, ..., $n - 1$ and $j = 0$, $i = 0$, ..., n, let $\mathbf{L^{i-j}}$ denote the truncation of column $\mathbf{\bar{Z}^i Z^j}$ of $TM(n)$ through level $\mathbf{Z^n}$. Since columns $\mathbf{1, L}$, ..., $\mathbf{L^{n-1}}$, $\mathbf{L^{-1}}$, ..., $\mathbf{L^{1-n}}$ span the columns of $TM(n-1)$, there exist scalars ϕ_0, ..., ϕ_{2n-2} such that (4.2) may be realized as

$$(4.3) \qquad \sum_{0 \leq i \leq n-1} \phi_{i+n-1} \mathbf{L^i} + \sum_{1 \leq j \leq n-1} \phi_{-j+n-1} \mathbf{L^{-j}} = \mathbf{L^n}$$

and

$$(4.4) \qquad \alpha_0 \geq \sum_{0 \leq i \leq 2n-2} \phi_i \alpha_{i-2n+1}.$$

Now consider $T(2n-1)$ as the 1-step extension of $T(2(n-1))$ in $T(2n)$:

$$T(2n-1) \equiv \begin{pmatrix} R & u_n \\ u_n^* & \alpha_0 \end{pmatrix},$$

where $R \equiv T(2(n-1))$ and $u_n \equiv (\alpha_{2n-1}, ..., \alpha_1)^t$; let $\mathbf{1}$, $\mathbf{w}$, ..., $\mathbf{w^{2n-2}}$ denote the columns of R. Since $TM(n) \geq 0$, so is $TM(n-1)$, and by induction we may assume that $R \geq 0$. (4.3) readily implies

$$\sum_{0 \leq i \leq 2n-2} \phi_i \mathbf{w}^i = u_n,$$

and this relation and (4.4) imply that there exists g_n such that $R g_n = u_n$ and $g_n^* u_n \leq \alpha_0$; thus $T(2n-1) \geq 0$ by Corollary 2.3.

Next we consider the decompositions

$$TM(n) \equiv \begin{pmatrix} S & v_n \\ v_n^* & \alpha_0 \end{pmatrix},$$

where $v_n \equiv (\alpha_{-n}, ..., \alpha_{-2})^t$, and

$$T(2n) \equiv \begin{pmatrix} U & y_n \\ y_n^* & \alpha_0 \end{pmatrix},$$

where $U \equiv T(2n-1)$ and $y_n \equiv (\alpha_{2n}, ..., \alpha_1)^t$. The positivity of $TM(n)$ implies that there exists r_n such that $S r_n = v_n$ and $r_n^* v_n \leq \alpha_0$; a detailed analysis of these relations (using the same method as above) implies that there exists t_n such that $T(2n-1) t_n = y_n$ and $t_n^* y_n \leq \alpha_0$; since $T(2n-1) \geq 0$, it follows that $T(2n) \geq 0$.

We will omit a detailed proof of the converse. We note only that to prove $T(2n) \geq 0 \Longrightarrow TM(n) \geq 0$, we use techniques similar to those above, but the argument is lengthier, because we must consider $n+1$ successive 1-step extensions of $TM(n-1)$ to build up to the positivity of $TM(n)$. $\quad\square$

Proposition 4.2. *Let $n > 1$ and suppose $\mathbf{Z\bar{Z}} = \mathbf{1}$, i.e., $\gamma_{ij} = \alpha_{j-i} \, (0 \leq i+j \leq 2n)$. The following are equivalent:*

   i) γ has a representing measure;

   ii) $TM(n)(\gamma) \geq 0$;

iii) $T(2n)(\alpha) \geq 0$;

iv) The truncated trigonometric moment problem for α has a rank $T(2n)(\alpha)-$ atomic representing measure which is also a representing measure for γ;

v) $TM(n)(\gamma)$ admits an extension $TM(n+1) \geq 0$.

Proof. i) $\Rightarrow$ ii) is clear and Proposition 4.1 gives ii) $\Rightarrow$ iii). [CF3] shows that if iii) holds, then the truncated trigonometric problem for α has a *rank $T(2n)(\alpha)$*-atomic representing measure μ; observe that since *supp* $\mu \subset \mathbf{T}$ and $\mu \geq 0$, then

$$\gamma_{ij} = \alpha_{j-i} = \int z^{j-i} \, d\mu = \int z^j \bar{z}^i \, d\mu,$$

so μ is a representing measure for γ, whence iv) holds. iv) $\Rightarrow$ i) is trivial. The preceding equivalences and Proposition 4.1 imply the equivalence v) $\Leftrightarrow$ ii). $\square$

Example 4.3. For $n = 1$, the existence of representing measures for the $TM(1)$ moment problem is still equivalent to the positivity of $TM(n)$, but the result no longer follows from Proposition 4.2, but rather from the solution of the quadratic moment problem [CF4]. Indeed, in this case *supp* μ need not be contained in $\mathbf{T}$.

To see this, for $0 < x < 1$, let $\gamma_{00} = 1$, $\gamma_{01} = \gamma_{10} = x$, $\gamma_{02} = \gamma_{11} = \gamma_{20} = 1$, and let $\alpha_{-2} = \alpha_2 = 1$, $\alpha_{-1} = \alpha_1 = x$, $\alpha_0 = 1$. Then $\gamma_{ij} = \alpha_{j-i}$ $(0 \leq i + j \leq 2)$ and

$$TM(1) \equiv \begin{pmatrix} 1 & x & x \\ x & 1 & 1 \\ x & 1 & 1 \end{pmatrix}.$$

It is easy to verify that $\mu \equiv (1-x^2)\delta_{2x} + x^2 \delta_{(2x^2-1)/x}$ is a representing measure for γ, and *supp* $\mu \subset \mathbf{T}$ only if $x = 1/2$. $\square$

5. An example of the function-column method.

In this section we illustrate how the identification of columns with functions can be used to solve a moment problem not covered by our previous results or those of [CF4]. This example provides evidence supporting Conjectures 1.3 and 2.6 in a case in which the analytic columns are independent. Consider the following truncated moment problem given implicitly via $M(2)(\gamma)$:

	1	$\mathbf{Z}$	$\bar{\mathbf{Z}}$	$\mathbf{Z}^2$	$\mathbf{Z}\bar{\mathbf{Z}}$	$\bar{\mathbf{Z}}^2$
1	1	$1+i$	$1-i$	$2/3+2i$	$8/3$	$2/3-2i$
	$1-i$	$8/3$	$2/3-2i$	$4+8i/3$	$4-8i/3$	$-4i$
	$1+i$	$2/3+2i$	$8/3$	$4i$	$4+8i/3$	$4-8i/3$
	$2/3-2i$	$4-8i/3$	$-4i$	10	$14/3-8i$	$-10/3-8i$
	$8/3$	$4+8i/3$	$4-8i/3$	$14/3+8i$	10	$14/3-8i$
	$2/3+2i$	$4i$	$4+8i/3$	$-10/3+8i$	$14/3+8i$	10

Straightforward calculations yield

(5.1) $\bar{\mathbf{Z}} = \alpha\mathbf{1} + \beta\mathbf{Z}$, where $\alpha = -2i$, $\beta = 1$;

(5.2)
$$\mathbf{Z}\bar{\mathbf{Z}} = \alpha\mathbf{Z} + \beta\mathbf{Z}^2$$

(5.3)
$$\bar{\mathbf{Z}}^2 = \alpha\bar{\mathbf{Z}} + \beta\mathbf{Z}\bar{\mathbf{Z}};$$

(5.4)
$$\{\mathbf{1},\ \mathbf{Z},\ \mathbf{Z}^2\}\ \text{is a basis for}\ C_{M(2)}.$$

These calculations imply that γ satisfies property (J). We next verify that $M(2)(\gamma) \geq 0$, and to this end we let $P(i)$ denote the successive upper left-hand corners of $M(2)$ ($0 \leq i \leq 5$). Since $det\ P(0) > 0$ and $det\ P(1) > 0$, and since (5.1) implies $rank\ P(2) = rank\ P(1)$, it follows from Corollary 2.3 that $P(2) \geq 0$. Let

$$b = \begin{pmatrix} 2/3 + 2i \\ 4 + 8i/3 \\ 4i \end{pmatrix} \text{ and } w = \begin{pmatrix} 2/3 - 2i \\ 2 + 2i \\ 0 \end{pmatrix}.$$

Since $P(2)w = b$ and $b^*w = 88/9 < 10$, Corollary 2.3 implies $P(3) \geq 0$; now (5.4) implies $rank\ P(5) = rank\ P(4) = rank\ P(3)$, whence Corollary 2.3 implies $M(2)(\gamma) \equiv P(5) \geq 0$.

In accordance with Conjectures 1.3 and 2.6 we will show that γ has a 3-atomic representing measure; indeed, we will give a complete parameterization of the 3-atomic representing measures. If μ is a representing measure, then (5.1) and (1.9) imply that $supp\ \mu \subset L \equiv \{z \in \mathbf{C} : \bar{z} = \alpha + \beta z\} = \{r + i : r \in \mathbf{R}\}$. If, additionally, μ is 3-atomic, then $L^2(\mu) = \langle 1, z, z^2 \rangle$, so there exist unique scalars a, b, c such that $z^3 \equiv a + bz + cz^2 \in L^2(\mu)$. (1.9) now implies that

(5.5)
$$\mathbf{Z}^3 = a\mathbf{1} + b\mathbf{Z} + c\mathbf{Z}^2 \text{ in the column space of } M(3)[\mu].$$

Let N denote the first three rows of the portion of $M(3)[\mu]$ inherited from $M(2)(\gamma)$ in columns $\mathbf{1}$, $\mathbf{Z}$, $\mathbf{Z}^2$, $\mathbf{Z}\bar{\mathbf{Z}}$, $\bar{\mathbf{Z}}^2$, $\mathbf{Z}^3$, $\mathbf{Z}^2\bar{\mathbf{Z}}$, $\mathbf{Z}\bar{\mathbf{Z}}^2$, $\bar{\mathbf{Z}}^3$; the submatrix of N corresponding to columns of degree three is thus of the form

$$\begin{pmatrix} 4i & 4 + 8i/3 & 4 - 8i/3 & -4i \\ 14/3 + 8i & 10 & 14/3 - 8i & -10/3 - 8i \\ -10/3 + 8i & 14/3 + 8i & 10 & 14/3 - 8i \end{pmatrix}.$$

Specializing (5.5), we have

(5.6)
$$\mathbf{Z}^3 = a\mathbf{1} + b\mathbf{Z} + c\mathbf{Z}^2 \text{ in the column space of } N.$$

We will next show, conversely, that to each choice of a, b, $c \in \mathbf{C}$ satisfying (5.6) and such that $p(z) \equiv z^3 - (a + bz + cz^2)$ has three distinct roots on line L, there corresponds a 3-atomic representing measure for γ; we will parameterize these measures by an algebraic curve. Suppose a, b, c satisfy the preceding conditions. Note that in the column space of N we have

(5.7)
$$\bar{\mathbf{Z}}\mathbf{Z}^2 = \alpha\mathbf{Z}^2 + \beta\mathbf{Z}^3;$$

(5.8)
$$\bar{\mathbf{Z}}^2\mathbf{Z} = \alpha\bar{\mathbf{Z}}\mathbf{Z} + \beta\bar{\mathbf{Z}}\mathbf{Z}^2;$$

(5.9) $$\bar{Z}^3 = \alpha\bar{Z}^2 + \beta Z\bar{Z}^2.$$

Let z_0, z_1, z_2 denote the distinct roots of $p(z)$ and define ρ_0, ρ_1, ρ_2 as the unique solution of the Vandermonde equation

$$\begin{pmatrix} 1 & 1 & 1 \\ z_0 & z_1 & z_2 \\ z_0^2 & z_1^2 & z_2^2 \end{pmatrix} \begin{pmatrix} \rho_0 \\ \rho_1 \\ \rho_2 \end{pmatrix} = \begin{pmatrix} \gamma_{00} \\ \gamma_{01} \\ \gamma_{02} \end{pmatrix}.$$

Let $\mu \equiv \rho_0\delta_{z_0} + \rho_1\delta_{z_1} + \rho_2\delta_{z_2}$; clearly, $\int z^i \, d\mu = \gamma_{0i}$, $i = 0, 1, 2$. We next show that $\int \bar{z}^i z^j \, d\mu = \gamma_{ij}$ ($0 \le i + j \le 4$). We have, successively,

$\int \bar{z} \, d\mu = \int (\alpha + \beta z) d\mu = \alpha\gamma_{00} + \beta\gamma_{01} = \gamma_{10}$ (5.1);

$\int z\bar{z} \, d\mu = \int (\alpha z + \beta z^2) \, d\mu = \alpha\gamma_{01} + \beta\gamma_{02} = \gamma_{11}$ (5.2);

$\int \bar{z}^2 \, d\mu = \int (\alpha\bar{z} + \beta z\bar{z}) \, d\mu = \alpha\gamma_{10} + \beta\gamma_{11} = \gamma_{20}$ (5.3);

$\int z^3 \, d\mu = \int (a + bz + cz^2) d\mu = a\gamma_{00} + b\gamma_{01} + c\gamma_{02} = \gamma_{03}$ (5.6);

$\int \bar{z}z^2 \, d\mu = \int (\alpha z^2 + \beta z^3) \, d\mu = \alpha\gamma_{02} + \beta\gamma_{03} = \gamma_{12}$ (5.7);

$\int z\bar{z}^2 \, d\mu = \int (\alpha z\bar{z} + \beta z^2\bar{z}) d\mu = \alpha\gamma_{11} + \beta\gamma_{12} = \gamma_{21}$ (5.8);

$\int \bar{z}^3 \, d\mu = \int (\alpha\bar{z}^2 + \beta z\bar{z}^2) d\mu = \alpha\gamma_{20} + \beta\gamma_{21} = \gamma_{30}$ (5.9)

$\int z^4 \, d\mu = \int (az + bz^2 + cz^3) d\mu = a\gamma_{01} + b\gamma_{02} + c\gamma_{03} = \gamma_{04}$ (5.6);

$\int \bar{z}z^3 \, d\mu = \int (a\bar{z} + bz\bar{z} + cz^2\bar{z}) d\mu = a\gamma_{10} + b\gamma_{11} + c\gamma_{12} = \gamma_{13}$ (5.6);

$\int \bar{z}^2z^2 \, d\mu = \int (\alpha\bar{z}z^2 + \beta\bar{z}z^3) d\mu = \alpha\gamma_{12} + \beta\gamma_{13} = \gamma_{22}$ (5.7);

$\int \bar{z}^3z \, d\mu = \int (\alpha\bar{z}^2z + \beta z^2\bar{z}^2) \, d\mu = \alpha\gamma_{21} + \beta\gamma_{22} = \gamma_{31}$ (5.9);

$\int \bar{z}^4 \, d\mu = \int (\alpha\bar{z}^3 + \beta z\bar{z}^3) d\mu = \alpha\gamma_{30} + \beta\gamma_{31} = \gamma_{40}$ (5.9).

Thus μ interpolates γ, and since $M(2)(\gamma) \ge 0$, [CF4] implies $\mu \ge 0$ (cf. [CF3, pp. 614-615]; thus μ is a representing measure for γ.

It remains to prove the existence of scalars a, b, c satisfying the above requirements, and to parameterize the resulting 3-atomic measures. Let r, s, t be distinct real numbers and let $z_0 = r + i$, $z_1 = s + i$, and $z_2 = t + i$. Then $p(z) \equiv (z - z_0)(z - z_1)(z - z_2) = z^3 - (a + bz + cz^2)$, where $a = (rs - 1)t - r - s + ((rs - 1) + tr + ts)i$, $b = 3 - rs - rt - st - 2(r + s + t)i$, and $c = r + s + t + 3i$. The general solution of $A1 + BZ + CZ^2 = Z^3$ in C_N is

$$\begin{pmatrix} A \\ B \\ C \end{pmatrix} = \begin{pmatrix} 5 - 3i \\ 1 + 6i \\ 0 \end{pmatrix} + \eta \begin{pmatrix} -2/3 + 2i \\ -2 - 2i \\ 1 \end{pmatrix}, \ \eta \in \mathbf{C};$$

thus we require

(5.10) $\qquad (rs - 1)t - r - s + ((rs - 1) + tr + ts)i = 5 - 3i + \eta(-2/3 + 2i);$

(5.11) $\qquad 3 - rs - rt - st - 2(r + s + t)i = 1 + 6i + \eta(-2 - 2i);$

(5.12) $\qquad r + s + t + 3i = \eta.$

Substituting from (5.12) into (5.10) and (5.11) yields the equivalent system

(5.13) $\qquad 6rst + 6 = 2(r + s + t) = 4 + rs + t(r + s).$

If $r + s = 2$, the solutions of (5.13) are $t = 1$ and $r = 2$, $s = 0$ or $r = 0$, $s = 2$. If $r + s = 6rs$, the solutions are $r = 6 + 34^{1/2}$ or $r = 6 - 34^{1/2}$, $s = 2/r$, $t = 9/5$. If the latter two cases do not hold, then (5.13) is equivalent to the system

(5.14) $\qquad (-2 + rs)/(6rs - r - s) = (4 + rs - 2r - 2s)/(2 - r - s);$

$\qquad t = (-2 + rs)/(6rs - r - s).$

This system may be parameterized by $s \in \mathbf{R}$ as follows:

$u(s) \equiv 3 - 13s + 6s^2 \pm (1 - 18s + 105s^2 - 96s^3 + 24s^4)^{1/2};$

$r \equiv r(s) = u(s)/(2 - 12s + 6s^2);$

$t = (-2 + r(s)s)/(6r(s)s - r(s) - s).$

(Various values of s must be excluded to insure that r, s, and t are distinct.)

Added in proof.

In recent work with R. Curto, we learned of additional evidence concerning the conjectures discussed in this paper. Using [Sch], we can exhibit an example of $M(3)(\gamma)$ that is positive and invertible (and hence satisfies property (J)), but such that γ admits no representing measure. In particular (via Theorem 1.4), $M(3)(\gamma)$ admits no flat extension $M(4)$. Thus Conjectures 1.3, 2.6, and 2.7 are false as stated. We continue to conjecture that i)-iv) and vi) of Conjecture 1.3 are equivalent. Conjecture 2.6 is true in case $\bar{\mathbf{Z}} = \alpha\mathbf{1} + \beta\mathbf{Z}$ or $\mathbf{Z}^k \in \langle \bar{\mathbf{Z}}^i \mathbf{Z}^j \rangle_{0 \le i+j \le k-1}$ where $k \le [n/2] + 1$. The above-mentioned example also shows that Conjecture 2.7 is false for $n = 3$. For $n = 2$, [BCJ] implies that if $M(2)(\gamma) \ge 0$, then γ has a representing measure; Conjecture 2.6 $(iv) \implies ii)$) and Conjecture 2.7 remain open for n = 2. For additional recent related results see [P3] [P4] [StSz].

REFERENCES

[AhK] N.I. Ahiezer and M.G. Krein, *Some Questions in the Theory of Moments*, Trans. Math. Monographs **2** (1962), Amer. Math. Soc., Providence.

[Ak] N.I. Akhiezer, *The Classical Moment Problem*, Hafner Publ. Co., New York.

[Atz] A. Atzmon, *A moment problem for positive measures on the unit disk*, Pac. J. Math. **59** (1975), 317-325.

[B] C. Berg, *The multidimensional moment problem and semigroups*, Proc. Symposia Appl. Math. **37** (1987), 110-124.

[BCJ] C. Berg, J. P. R. Christensen, and C. U. Jensen, *A remark on the multidimensioanl moment problem*, Math. Ann. **243** (1979), 163-169.

[Cu] R.E. Curto, *Joint hyponormality: A bridge between hyponormality and subnormality*, Proc. Symposia Pure Math. **51** (1990), 69-91.

[CF1] R.E. Curto and L.A. Fialkow, *Recursively generated weighted shifts and the subnormal completion problem*, Integral Eq. Operator Thy. **17** (1993), 1-42.

[CF2] R.E. Curto and L.A. Fialkow, *Recursively generated weighted shifts and the subnormal completion problem II*, Integral Eq. Operator Thy. **18** (1994), 369-426.

[CF3] R.E. Curto and L.A. Fialkow, *Recursiveness, positivity, and truncated moment problems*, Houston J. Math. **17** (1991), 603-635.

[CF4] R.E. Curto and L.A. Fialkow, *The truncated complex moment problem*, Memoirs Amer. Math. Soc. (to appear).

[CP1] R.E. Curto and M. Putinar, *Existence of nonsubnormal polynomially hyponormal operators*, Bull. Amer. Math. Soc. **25** (1991), 373-378.

[CP2] R.E. Curto and M. Putinar, *Nearly subnormal operators and moment problems*, J. Funct. Anal. **115** (1993), 480-497.

[D] R.G. Douglas, *On majorization, factorization and range inclusion of operators in Hilbert space*, Proc. Amer. Math. Soc. **17** (1966), 413-415.

[FW] P.A. Fillmore and J.P. Williams, *On operator ranges*, Advances in Math. **7** (1971), 254-281.

[Fu] B. Fuglede, *The multidimensional moment problem*, Expo. Math. **1** (1983), 47-65.

[KrN] M. Krein and A. Nudel'man, *The Markov Moment Problem and Extremal Problems*, Transl. Math. Monographs **50** (1977), Amer. Math. Soc., Providence.

[L] H. Landau, *Classical background of the moment problem*, Proc. Symposia Appl. Math. 37 (1987), Moments in Mathematics, Amer. Math. Soc., Providence.

[Pa] V.I. Paulsen, *Completely bounded maps and dilations*, Pitman Res. Notes Math. **146** (1986), Longman Scientific & Technical, Essex, UK.

[P1] M. Putinar, *A two-dimensional moment problem*, J. Funct. Anal. **80** (1988), 1-8.

[P2] M. Putinar, *Moment problems and operators*, Lecture notes, GPOTS.

[P3] M. Putinar, *Positive polynomials on compact semi-algebraic sets*, preprint.

[P4] M. Putinar, *Extremal solutions of the two-dimensional L-problem of moments*, preprint.

[Sar] D. Sarason, *Moment problems and operators in Hilbert space*, Proc. Symposia Appl. Math.37 (1987), 54-70, Moments in Mathematics, Amer. Math. Soc., Providence.

[Sch] K. Schmudgen, *An example of a positive polynomial which is not a sum of squares of polynomials- a positive, but not strongly positive functional*, Math. Nachr. **88** (1979), 385-390.

[ShT] J. Shohat and J. Tamarkin, *The Problem of Moments*, Math. Surveys **1** (1943), Amer. Math. Soc., Providence.

[Smu] J.L. Smul'jan, *An operator Hellinger integral*, Mat. Sb. **91** (1959), 381-430. (Russian)

[S] J. Stampfli, *Which weighted shifts are subnormal*, Pac. J. Math. **17** (1966), 367-379.

[St] M.H. Stone, *Linear Transformations in Hilbert Space*, Amer. Math. Soc., New York, 1932.

[StSz] J. Stochel and F. Szafraniec, *Algebraic operators and moments on algebraic sets*, Portugaliae Math. **51** (1994), 1-21.

DEPARTMENT OF MATHEMATICS AND COMPUTER SCIENCE, SUNY COLLEGE AT NEW PALTZ, NEW PALTZ, NY 12561

E-mail: fialkowl@snynewvm.bitnet

Contemporary Mathematics
Volume **185**, 1995

Torsion Invariants for Finite von Neumann Algebras

Donggeng Gong * and Joel Pincus

Abstract

We define a numerical torsion invariant for an n-tuple T of commuting elements in a finite von Neumann algebra $\mathcal{A}$ and study properties of this torsion. New spectral invariants supported on the spectrum $\sigma(T)$ of T are also discussed and are related to the torsion invariant as well as certain invariants of Novikov-Shubin. We use the torsion invariant to get a homotopy invariant for a compact oriented Riemannian manifold and a torsion-index for a family of elliptic operators.

1 Introduction

This paper studies a numerical torsion invariant for an n-tuples of commuting elements in $\mathcal{A}$. The definition makes use of the Fuglede- Kadison determinant. Since this determinant may vanish for some nonregular operators, we restrict our attention to n-tuples of commuting elements

---

* Research at MSRI Supported in part by NSF grant # DMS 9022140 while on leave of absence from Dept. of Math., Univ. of Chicago.

AMS (MSC) subject classifications (1991) 47B20,46A67,46L45

Key words and phrases:torsion,commuting operator n-tuple- homotopy invariant,von Neumann algebra

whose restricted Laplace operators have nonzero determinants, and introduce spectral invariants supported on the Taylor joint spectrum $\sigma(T)$ of an n-tuple T of commuting elements in $\mathcal{A}$. We then define the torsion invariant of T via the Koszul complex and the restricted Laplace operators associated with T. We use this torsion invariant to define a family of maps from the K-theory group $K_0(\mathcal{A})$ to C and from these we get the so called torsion-index for a family of elliptic operators on a compact Riemannian manifold. Then, using the Hilsum-Skandalis theorem [HS], we produce a homotopy invariant for M. This can be considered as an application of multivariable operator theory to obtain specifically geometric results.

Our work is related to certain other studies by the authors. In the thesis [Go] a weak K-theory valued torsion invariant for n-tuples of bounded commuting $\mathcal{A}$-operators on a finitely generated Hilbert $\mathcal{A}$-module was defined. The main properties of this invariant were studied and in the single operator case it was shown that it is both determined by and determines the absolute value of the adjoint of the operator (up to conjugation by isomorphism and a direct sum factor of weak isomorphisms). In [Car-Pin] a new F-valued "joint-torsion" was introduced for pairs A, B of commuting endomorphisms on a vector space $\mathcal{H}$ over an arbitrary field F. This joint torsion, $\tau(A, B; \mathcal{H})$, is defined purely in terms of the Koszul complex $K_*(A, B) \otimes \mathcal{H}$ but is then identified in terms of Steinberg symbols i.e. as an object of algebraic K-Theory. It was found that the resulting construction gives a canonical 'index' factorization to the Milnor excision map in algebraic K-theory. Thus let R be a subring of the endomorphisms on a vector space $\mathcal{H}$ over a field F which contains the two sided ideal $\mathcal{F}$ of finite rank operators and which is such that $R/\mathcal{F}$ is commutative. Then for elements $a, b \in GL_n(R/\mathcal{F})$ which have commutative lifts A, B to $M_n(R)$ there exist maps $\delta_{a;b}$ and $\delta_{b;a}$ so that $\tau(A, B; \mathcal{H}) = -[\partial_1 a; \partial(B) + \delta_{a;b}] + [\partial_1 b; \partial(A) + \delta_{b;a}] = \partial_2\{a, b\}$

as a class in $K_1(R, \mathcal{F})$. Here $\partial_{2,1}$ are, respectively, the boundary map $\partial_2 : K_2(R/\mathcal{F}) \to K_1(F)$ and $\partial_1 : K_1(R/\mathcal{F}) \to K_0(\mathcal{F})$ which occur in the Milnor exact sequence of K groups. The maps $\delta_{a;b}$ and $\delta_{b;a}$ are obtained from the boundary maps of the long exact sequence in homology associated to the subcomplexes $(\mathcal{E}_A; \bar{\partial}(B)) = K_*(A) \otimes \mathcal{H}$ and $(\mathcal{E}^B; \bar{\partial}(A)) = K_*(B) \otimes \mathcal{H}$ of $K_*(A, B) \otimes \mathcal{H}$.

The joint torsion is thus a bicharacter which is invariant under perturbations by elements of $\mathcal{F}$. It is explicitly computed in many examples in [Car-Pin], where also if $\mathcal{H}$ is a Hilbert space the metric permitted the definition of a joint analytic torsion $\tau_a(A, B; \mathcal{H}) \equiv \frac{\tau_a(\mathcal{E}_A; \bar{\partial}(B))}{\tau_a(\mathcal{E}^B; \bar{\partial}(A))}$ in terms of analytic torsions of the indicated complexes. For a pair of elements in a type II_∞ von Neumann algebra there is a similar construction in [Car-Pin] using the Fuglede Kadison determinant which was related to the measure defined by L. Brown [Br] in his generalization of the Lidskii trace formula.

The present results deal with the corresponding situation in which there may be more than two operators. The exposition is arranged as follows. We define in Section 2 spectral invariants $\theta_j(T)$ for an n-tuple T and study its properties.

It is interesting that the $\theta_j(T)$ are similarity invariants.

The finiteness of $\theta_j(T)$ guarantees the nonvanishing of the determinants of the restricted Laplace operators associated with T; and this enables us to define the torsion invariant of T with $\theta_j(T) < \infty$ in Section 3. In Section 4 we give torsion formulas for adjoint operators and tensor products of two n-tuples. We also find a formula for the torsions of two similar n- tuples. In Section 5 we concentrate on the single operator case. In that case the torsion is related to the Brown measure. We obtain a mapping formula for the torsion under an additional condition. We also obtain in Section 6 a formula for computing the torsion of single operator when $\mathcal{A}$ is abelian. In particular, the torsion of an

$m \times m$ matrix is just the product of the absolute values of the nonzero eigenvalues of the matrix counted according to multiplicity. Finally, we treat the torsion for some special tuples in Section 7 and use it to get a family of maps from the K-theory group $K_0(\mathcal{A})$ to C. Hence we produce a torsion-index for a family of elliptic operators on a compact Riemannian manifold M and a homotopy invariant for M. Finally we state several open questions.

2   Spectral Invariants

Let $\mathcal{A}$ be a finite von Neumann algebra. This means that there is a finite faithful and normal trace τ on $\mathcal{A}$. Denote by H the completion of $\mathcal{A}$ under the inner product $< a, b >= \tau(b^*a)$. Let $T = (T_1, T_2, \ldots, T_n)$ be an n-tuple of commuting elements in $\mathcal{A}$. Thus, $T_iT_j = T_jT_i, i, j = 1, 2, \ldots, n$. Let $\sigma = (\sigma_1, \sigma_2, \ldots, \sigma_n)$ be n indeterminants and $\wedge^k(\sigma)$ the exterior space of σ with base $\{\sigma_{j_1}, \ldots, \sigma_{j_k} : 1 \leq j_1 < \ldots < j_k \leq n\}, k = 0, 1, \ldots, n$. We form a standard Koszul complex $\{C_*, d_*\}$ associated with T and H by setting

$$C_k = H \otimes \wedge^k(\sigma)$$

with differential $d_k : C_k \rightarrow C_{k+1}$ given by $d_k = \sum_{j=1}^n T_j S_j$, where $S_j : C_k \rightarrow C_{k+1}$ is defined as

$$S_j(\xi) = \sigma_j \wedge \xi, \quad \xi \in C_k, \quad j = 1, 2, \ldots, n.$$

The adjoint S_j^* of S_j is $S_j^*(\xi_1 + \sigma_j \wedge \xi_2) = \xi_2$ for $\xi_1 + \sigma_j \xi_2 \in C_{k+1}$ with $\sigma_j \wedge \xi_1$ and $\sigma_j \wedge \xi_2 \neq 0$. We have

$$S_iS_j + S_jS_i = 0, \quad i, j = 1, 2, \ldots, n.$$

$$(2.1)$$

$$S_iS_j^* + S_j^*S_i = \begin{cases} 1, & i = j, \\ 0, & i \neq j. \end{cases}$$

Using (2.1) and the commutativity of T_j, we get $d_{j+1}d_j = 0$. Define the Laplace operator Δ_j of d_j as

$$\Delta_j = d_j^* d_j + d_{j-1} d_{j-1}^*, j = 1, 2, \ldots, n-1,$$

and $\Delta_0 = d_0^* d_0, \Delta_n = d_{n-1} d_{n-1}^*$, where $d_j^* = \sum_{i=1} T_i^* S_i^*$ is the adjoint of d_j on C_{j+1}, $j = 0, 1, \ldots, n-1$. Δ_j is an $m_j \times m_j$ matrix with entries in $\mathcal{A}$, where m_j is the dimension of the space $\wedge^j(\sigma)$. Δ_j is nonnegative selfadjoint operator on C_j. We can thus talk about the spectral representation of Δ_j

$$\Delta_j = \int \lambda dE_j(\lambda),$$

where $E_j(\lambda)$ is the projection-valued spectral measure associated with Δ_j.

Let $P_j : C_j \rightarrow Ker\Delta_j$ be the orthogonal projection and $\Delta_j' = (1-P_j)\Delta_j(1-P_j)$. Δ_j' belongs to $\mathcal{A}_j = (1-P_j)M_{m_j}(\mathcal{A})(1-P_j)$, which is a finite von Neumann algebra [D ix]. Without confusion, we will not distinguish the algebras $\mathcal{A}$, $M_{m_j}(\mathcal{A})$ and $\mathcal{A}_j$ and their traces.

With the above notation, the Fuglede-Kadison determinant of Δ_j is defined as

$$\Delta(\Delta_j) = e^{\int \ln \lambda dN_j(\lambda)} \neq 0,$$

provided Δ_j is invertible, where $N_j(\lambda) = \tau(E_j(\lambda))$. In general, one can still define the determinant

$$\Delta(\Delta_j') = e^{\int \ln \lambda d(N_j(\lambda)-b_j)}$$

(2.2)

for $b_j = \tau(P_j)$. But $\Delta(\Delta_j')$ may be zero. Since the nonvanishing of $\Delta(\Delta_j')$ is important for our definition of the torsion invariant, we will have to investigate when $\Delta(\Delta_j') \neq 0$. This is the goal of the present section.

We see by (2.2) that $\Delta(\Delta_j') \neq 0$ iff $\int \ln \lambda d(N_j(\lambda) - b_j) > -\infty$, and that if $N_j(\lambda) - b_j$ approaches 0 like $\lambda^{\alpha/2}$ for some $\alpha > 0$ and λ near 0, then $\Delta(\Delta_j') \neq 0$. This motivates us to set

$$\alpha_j(T) = \sup\{\beta : N_j(\lambda) - b_j = O(\lambda^{\beta/2}) \ as \ ; \lambda \to 0^+\}.$$

$$(2.3)$$

Here, $\beta = \infty$ is accepted, which means $N_j(\lambda) - b_j = 0$ for λ near 0. Thus for $\Delta_j \equiv 0$, we take $\alpha_j(T) = \infty$. In case β in (2.3) does not exist, we take $\alpha_j(T) = 0$.

Definition 2.1: Let $\theta_j(T) = \frac{1}{\alpha_j(T)}$. The numbers $\{\theta_0(T), \ldots, \theta_n(T)\}$ are called the spectral invariants of the n-tuple T at zero.

Remark 2.2: (1) $\{\alpha_j(T)\}_{j=0}^n$ are the analogue of the Novikov-Shubin invariants [NS] in the operator case. Novikov and Shubin defined α_j' s for the Laplace operators on a smooth Riemannian manifold and the definition was extended by Lott [Lo] to topological manifolds.

(2) Another way to study (2.2) is via the singular numbers for finite von Neumann algebras [Fa]. Recall that for $B \in \mathcal{A}$ with spectral representation $|B| = |B^*B|^{\frac{1}{2}} = \int_0^\infty \lambda dE^B(\lambda)$, the s-numbers are defined as

$$S_t(B) = \min\{\lambda \geq 0 : \tau(1 - E^B(\lambda)) \leq t\}, \ \ t \geq 0.$$

$S_t(B)$ is a decreasing function on $(0, \infty]$. Then

$$\Delta_{\tau(1)}(B) = \exp\{\int_0^{\tau(1)} \log S_t(B) dt\}$$

is the Fuglede-Kadison determinant for B invertible. In our situation, $\Delta_{\tau(1)}(\Delta_j') = \Delta(\Delta_j')$. Thus the convergence of the integral $\int_0^{\tau(1)} \log S_t(\Delta_j') dt$ guarantees the nonvanishing of $\Delta(\Delta_j')$.

(3) If $0 \in C \setminus \sigma(T)$, then each Δ_j is invertible. $N_j(\lambda)$ is zero in a neighborhood of $\lambda = 0$. We have $\theta_j(T) = 0, j = 0, \ldots, n$. Thus by the spectral mapping theorem, if one of the T_j' s is invertible, then $\theta_j(T) = 0$.

Now we state certain properties of the spectral invariants in the following theorems.

Theorem 2.3: Let $T = (T_1, \ldots, T_n)$ be an n-tuple of commuting elements in $\mathcal{A}$. Then

1. $\theta_j(zT) = \theta_j(T), z \in C, z \neq 0$,

2. $\theta_{n-j}(T^*) = \theta_j(T), j = 0, 1, \ldots, n$,

3. $\{\theta_j(T)\}$ are unitary invariants, namely, $\theta_j(U^*TU) = \theta_j(T)$ for unitary element U in $\mathcal{A}$.

Proof: (1) Let $\Delta_j(zT)$ and $\Delta_j(T)$ be the Laplace operators associated with zT and T respectively, and $E_j^{zT}(\lambda)$ and $E_j^T(\lambda)$ be the spectral families of $\Delta_j(zT)$ and $\Delta_j(T)$, respectively. Then $\Delta_j(zT) = |z|^2 \Delta_j(T)$. Let $\Delta_j(T) - \lambda = U_j^T(\lambda)|\Delta_j(T) - \lambda|$ be the polar decomposition. We have

$$\Delta_j(zT) - z^2\lambda^2 = |z|^2 U^{zT}(z^2\lambda)|\Delta_j(T) - \lambda| = |z|^2 U_j^T(\lambda)|\Delta_j(T) - \lambda|.$$

Hence, $U_j^{zT}(z^2\lambda) = U_j^T(\lambda)$. This proves

$$E_j^{zT}(z^2\lambda) = 1 - \frac{U_j^{zT}(z^2\lambda)^2 + U_j^{zT}(z^2\lambda)}{2} = E_j^T(\lambda),$$

and then $N_j^{zT}(z^2\lambda) = N_j^T(\lambda)$. Since $Ker(\Delta_j(zT)) = Ker(\Delta_j(T))$, we get $\alpha_j(zT) = \alpha_j(T)$.

(2) Define a unitary operator $\sharp : C_* \to C_*$ by

$$\sharp(\xi) = i^{n(n-1)/2} \sum_{j_1 < \ldots < j_p} S_{j_1}^* \ldots S_{j_p}^* S_1 \ldots S_n x_{j_1 \ldots j_p} \in C_{n-p}$$

for $\xi = \sum_{j_1 < \ldots < j_p} S_{j_1} \ldots S_{j_p} x_{j_1 \ldots j_p} \in C_p$ [Vas]. Then as shown in [Go],

$$\Delta_j(T) = \sharp^* \Delta_{n-j}(T^*)\sharp,$$
$$|\Delta_j(T) - \lambda| = ((\Delta_j(T)^* - \lambda^*)(\Delta_j(T) - \lambda))^{\frac{1}{2}} = \sharp^*|\Delta_{n-j}(T^*) - \lambda|\sharp,$$
$$U_j^T(\lambda) = \sharp^* U_{n-j}^{T^*}(\lambda)\sharp,$$

and

$$E_j^T(\lambda) = 1 - \frac{U_j^T(\lambda)^2 + U_j^T(\lambda)}{2} = \sharp^* E_{n-j}^{T^*}(\lambda)\sharp.$$

Hence, $N_j^T(\lambda) = \tau(E_j^T(\lambda)) = N_{n-j}^{T^*}(\lambda)$. This implies that $\theta_j(T^*) = \theta_{n-j}(T^*)$, since $b_j(T) = b_{n-j}(T^*)$.

The proof of (3) is clear. $Q.E.D.$

To prove the next theorem, we need the following lemma. Let

$$\alpha_j'(T) = \sup\{\beta : \tau(e^{-t\Delta_j'(T)}) = O(t^{-\beta/2}) \ as \ t \to \infty\}.$$

Lemma 2.4 [ES]: $\alpha_j(T) = \alpha_j'(T), j = 0, 1, \ldots, n$.

Theorem 2.5: Let $T^{(k)} = (T_1^{(k)}, \ldots, T_n^{(k)})$ be two n-tuples of commuting elements in $\mathcal{A}$, $k = 1, 2$. Then

1. $\theta_j(T^{(1)} \oplus T^{(2)}) = \max\{\theta_j(T^{(1)}), \theta_j(T^{(2)})\}, j = 0, 1, \ldots, n,$

2. $\theta_j(T^{(1)} \otimes 1 + 1 \otimes T^{(2)}) = \max_{p+q=j}\{\theta_p(T^{(1)}), \theta_q(T^{(2)})\}.$

Proof: (1) We have $\Delta_j(T^{(1)} \oplus T^{(2)}) = \Delta_j(T^{(1)}) \oplus \Delta_j(T^{(2)})$ and $U_j^{T^{(1)} \oplus T^{(2)}}(\lambda) = U_j^{T^{(1)}}(\lambda) \oplus U_j^{T^{(2)}}(\lambda)$. Furthermore, $b_j(T^{(1)} \oplus T^{(2)}) = b_j(T^{(1)}) + b_j(T^{(2)})$. We obtain that $\alpha_j(T^{(1)} \oplus T^{(2)}) = \min\{\alpha_j(T^{(1)}), \alpha_j(T^{(2)})\}$ and then the required result.

(2) Let $T = T^{(1)} \otimes 1 + 1 \otimes T^{(2)}$. As shown in [Go],

$$\Delta_j(T) = \bigoplus_{p+q=j} (\Delta_p(T^{(1)}) \otimes 1 + 1 \otimes \Delta_q(T^{(2)})),$$

$$Ker\Delta_j(T) = \bigoplus_{p+q=j} (Ker(\Delta_p(T^{(1)})) \otimes Ker(\Delta_q(T^{(2)})))$$

and

$$Ker\Delta_j(T)^\perp = \bigoplus_{p+q=j} \{(Ker(\Delta_p(T^{(1)}))^\perp$$

$$\otimes Ker(\Delta_q(T^{(2)}))) \oplus (Ker(\Delta_p(T^{(1)})) \otimes Ker(\Delta_q(T^{(2)}))^\perp)$$

$$\oplus (Ker(\Delta_p(T^{(1)}))^\perp \otimes Ker(\Delta_q(T^{(2)}))^\perp\}.$$

Hence,

$$\tau(e^{-t\Delta'_j(T)}) \;=\; \sum_{p+q=j} \left(\tau(e^{-t\Delta'_p(T^{(1)})})\tau(e^{-t\Delta'_q(T^{(2)})}) + \right.$$

$$\left. + \;\; \tau(P_p(T^{(1)}))\tau(e^{-t\Delta'_q(T^{(2)})}) + \tau(P_q(T^{(2)}))\tau(e^{-t\Delta'_p(T^{(1)})}) \right)$$

By Lemma 2.4, we get

$$\alpha_j(T) = \min_{p+q=j}\{\alpha_p(T^{(1)}) + \alpha_q(T^{(2)}), \alpha_p(T^{(1)}), \alpha_q(T^{(2)})\}.$$

The proof is complete. $Q.E.D.$

One of the important properties of the spectral invariant $\theta_j(T)$ is the homotopy invariance in the sense of the following lemma.

Lemma 2.6 [GS]: Let $\{C^{(k)}_*, d^{(k)}_*\}$ be two bounded complexes of Hilbert $\mathcal{A}$-modules, $k = 1, 2$. Suppose $U : C^{(1)}_* \to C^{(2)}_*$ and $V : C^{(2)}_* \to C^{(1)}_*$ are two bounded $\mathcal{A}$-homomorphisms of the complexes. If UV and VU are cochain homotopic to Id, respectively, then $\alpha_j(C^{(1)}_*) = \alpha_j(C^{(2)}_*)$.

Recall that a complex $\{C_*, d_*\}$ is bounded if $C_k = 0$ for $|k| > k_0$ with some $k_0 > 0$. The invariant $\alpha_j(C_*)$ is defined in the same way as $\alpha_j(T)$ in Definition 2.1.

The proof of Lemma 2.6 is the same as in [GS] except that we state the result for general finite von Neumann algebras.

As an application of Lemma 2.6, we have

Theorem 2.7 Let $T^{(k)} = (T^{(k)}_1, \ldots, T^{(k)}_n)$ be two n-tuples of commuting elements in $\mathcal{A}$, $k = 1, 2$. Suppose $U, V \in \mathcal{A}$ satisfy that $UT^{(1)}_j = T^{(2)}_j U, VT^{(2)}_j = T^{(1)}_j V$, $j = 1, 2, \ldots, n$. If there are elements $h^{(k)}_j \in \mathcal{A}$ such that $h^{(k)}_j$ commute with $T^{(k)}_j$ and

$$VU - 1 = \sum_{j=1}^{n} h^{(1)}_j T^{(1)}_j$$

and

$$UV - 1 = \sum_{j=1}^{n} h^{(2)}_j T^{(2)},$$

then $\theta_j(T^{(1)}) = \theta_j(T^{(2)}), j = 0, 1, \ldots, n$.

Proof: Since U and V interwine $T^{(k)}$, U and V induce morphisms of the complexes $\{C_*^{(k)}, d_*^{(k)}\}$ associated with $T^{(k)}$. By Lemma 2.6, it suffices to show that UV and VU are homotopic to the identity map, respectively.

Let $\tilde{h}_j^{(k)} : C_j^{(k)} \to C_{j-1}^{(k)}$ be defined by

$$\tilde{h}_j^{(k)} = \sum_{i=1}^n h_i^{(k)} S_i^*.$$

By (2.1), we obtain

$$\tilde{h}_{j+1}^{(k)} d_j^{(k)} + d_{j-1}^{(k)} \tilde{h}_j^{(k)} = \sum_{i,l=1}^n h_i^{(k)} T_l^{(k)} (S_i^* S_l + S_l S_i^*) = \sum_{i=1}^n h_i^{(k)} T_i^{(k)}.$$

This proves by assumption that $VU - 1 = \tilde{h}^{(1)} d^{(1)} + d^{(1)} \tilde{h}^{(1)}$ and $UV - 1 = \tilde{h}^{(2)} d^{(2)} + d^{(2)} \tilde{h}^{(2)}$.   *Q.E.D.*

Corollary 2.8: With the notation of Theorem 2.7, if $T^{(1)}$ and $T^{(2)}$ are similar, then $\theta_j(T^{(1)}) = \theta_j(T^{(2)}), j = 0, 1, \ldots, n$.

Proof: Let $U \in \mathcal{A}$ be an invertible element such that $UT^{(1)} = T^{(2)}U$. Then $T^{(1)}U^{-1} = U^{-1}T^{(2)}$. Since $U^{-1}U = 1$, we can take $h_j^{(k)} = 0$ in Theorem 2.7 and get the result.   *Q.E.D.*

We close this section with an example:

Example 2.9: Let $\mathcal{A} = L^\infty([0,1])$ be a finite abelian von Neumann algebra with trace $\tau(\varphi) = \int_0^1 \varphi(x)dx$. We have

$$\tau(e^{-t\Delta_0'(\varphi-\lambda)}) = \tau(e^{-t\Delta_0(\varphi-\lambda)}) - \tau(\chi_{\{\varphi=\lambda\}})$$
$$= \int_{\{\varphi\neq\lambda\}} e^{-t|\varphi-\lambda|^2} dx,$$

where χ_X denotes the characteristic function of the set X. Thus for $\varphi(x) = \sqrt{\ln(1+x)}$ on $[0,1]$,

$$\theta_j(\varphi - \lambda) = \begin{cases} \frac{1}{2}, & \lambda = 0, \\ \leq 1, & \lambda = \sqrt{\ln(1+s)}, s \in (0,1], \\ 0, & \lambda \neq \sqrt{\ln(1+s)}, s \in [0,1]. \end{cases}$$

This example shows that the spectral invariant $\theta_j(T)$ is not a continuous function on T in general.

3 Determinants and Torsion Invariants

The torsion invariant defined in [Go] has values in the weak K-theory group $K_1^w(\mathcal{A})$ of a finite von Neumann algebra $\mathcal{A}$. We know that the determinant on $\mathcal{A}$ induces a map from the K-theory group $K_1(\mathcal{A})$ to C. If the determinant induces also a homomorphism from $K_1^w(\mathcal{A})$ to C, then we would have obtained a numerical torsion invariant by applying the determinant to the weak K-theory valued torsion.But, such a map is not well defined as we explained in the introduction.

We begin with the Fuglede-Kadison determinant on finite von Neumann algebra $\mathcal{A}$. As we noted in (2.2), the determinant $\Delta(B)$ of $B \in \mathcal{A}$ i s defined as

$$\Delta(B) = e^{\tau(\log|B|)} = e^{\int \log \lambda dN^B(\lambda)},$$

where $|B| = (B^*B)^{\frac{1}{2}} = \int \lambda dE^B(\lambda)$ is the spectral representation and $N^B(\lambda) = \tau(E^B(\lambda))$. $\Delta(B)$ enjoys the following properties.

Lemma 3.1 [FK] [Br]: Let $B, B_1, B_2 \in \mathcal{A}$.

1. $\Delta(\lambda B) = |\lambda|\Delta(B), \lambda \in C, \lambda \neq 0,$

2. $\Delta(B^*) = \Delta(B) = \Delta(|B|),$

3. $\Delta(B_1 B_2) = \Delta(B_1)\Delta(B_2),$

4. $\Delta(UBU^{-1}) = \Delta(B)$ for invertible $U \in \mathcal{A},$

5. $\Delta(B) = \lim_{\varepsilon \to 0} \Delta(B + \varepsilon I)$ for $B > 0,$

6. $\Delta(f(B)) = \exp\{\int \log|f(\lambda)|dN^B(\lambda)\}$ for a normal operator B and a continuous function f such that $f \neq 0$ on $\sigma(B),$

7. More generally, if f is holomorphic in a neighborhood of $\sigma(B) \cup \{0\}$ and $f(0) = 1$, then

$$\Delta(f(B)) = \exp\{\int \log|f(\lambda)|d\mu^B(\lambda)\},$$

where $\mu^B(\lambda)$ is the measure defined by L. Brown [Br].

Using this determinant, we can define our basic notion. Let $[\mathcal{D}_n(\mathcal{A}) = \{T = (T_1, \ldots, T_n) : T_iT_j = T_jT_i, T_i \in \mathcal{A}, \theta_j(T) < \infty, j = 0, 1, \ldots, n\}$.

Recall that $\Delta_j(T)$ is the Laplace operator associated with T and $\Delta_j'(T)$ is the restriction of $\Delta_j(T)$ to $(Ker\Delta_j(T))^\perp$. We noted already in Section 2 that $\Delta(\Delta_j'(T)) \neq 0$ iff $\int \log \lambda d(N_j^T(\lambda) - b_j) > -\infty$. A sufficient condition f or $\Delta(\Delta_j'(T)) \neq 0$ is that $\theta_j(T) < \infty$ for $j = 0, 1, \ldots, n$.

Definition 3.2: Let $T \in \mathcal{D}_n(\mathcal{A})$. The torsion invariant $\tau(T)$ of T is defined by

$$\log \tau(T) = \frac{1}{2}\sum_{j=0}^{n}(-1)^{j+1}j\log(\Delta(\Delta_j'(T))).$$

Remark 3.3: (1) By convention, we take $\log\Delta(\Delta_j'(T)) = 0$ for $\Delta_j(T) \equiv 0$.

(2) By Lemma 3.1(4), $\tau(T)$ is a unitary invariant, namely, $\tau(T) = \tau(UTU^*)$ for unitary element $U \in \mathcal{A}$.

(3) If λ is not in the Taylor spectrum $\sigma(T)$ of T, then each $\Delta_j(T-\lambda)$ is invertible. In that case, $T - \lambda \in \mathcal{D}_n(\mathcal{A})$ and $\tau(T - \lambda)$ is well defined.

(4) If $\mathcal{A} = M_m(C)$ is the algebra of $m \times m$ matrices over C, then $\mathcal{D}_n(\mathcal{A})$ is the space of all n-tuples of commuting matrices, since for $\Delta_j(T) \neq 0$, $\Delta_j'(T)$ is invertible in this case.

Lemma 3.4: Let $T \in \mathcal{D}_n(\mathcal{A})$. Let

$$\zeta^T(s) = \int_0^\infty \sum_{j=0}^{n}(-1)^j j\frac{1}{\Gamma(s)}t^{s-1}\tau(e^{-t\Delta_j'(T)})dt, s \gg 0, \qquad (3.1)$$

where $\Gamma(s)$ is the Γ-function defined by $\Gamma(s) = \int_0^\infty t^{s-1}e^{-ts}dt$. Then

$$\log \tau(T) = \frac{1}{2}\frac{d}{ds}(\zeta^T(s))|_{s=0}.$$

$$(3.2)$$

Proof: Since $\theta_j(T) < \infty$, $\zeta_j^T(s)$ extends to be a holomorphic function for s near 0. We have $\dfrac{1}{2}\dfrac{d}{ds}(\zeta^T(s))|_{s=0} =$

$$\frac{1}{2}\frac{d}{ds}\Big(\frac{1}{\Gamma(s)}\int_0^\infty \sum_{j=0}^n (-1)^j j t^{s-1}\int d\lambda e^{-\lambda t} d(N_j^T(\lambda) - b_j)dt\Big)|_{s=0}$$

$$= \frac{1}{2}\frac{d}{ds}\Big(\int \sum_{j=0}^n (-1)^j j \lambda^{-s} d(N_j^T(\lambda) - b_j)\Big)|_{s=0}$$

$$= \frac{1}{2}\int \sum_{j=0}^n (-1)^{j+1} j \ln \lambda \, d(N_j^T(\lambda) - b_j) = \log \tau(T).$$

Here the interchange of the integrals is validated by assumption.

$$Q.E.D.$$

The advantage of (3.2) is that sometimes certain of the integrals $\int_0^\infty t^{s-1}\tau(e^{-t\Delta_j'(T)})dt$ may not be well defined, but their alternating sum makes sense via the cancellation.

We now come to the relative torsion. Let $T^{(k)}$ be two n-tuples of commuting elements in $\mathcal{A}$, $k = 1, 2$. Suppose $U \in \mathcal{A}$ interwines $T^{(k)}$: $T_j^{(2)}U = UT_j^{(1)}$. U induces a morphism between two Koszul complexes $C_*^{(k)}$ associated with $T^{(k)}$. Define a mapping cone $\{C_*(U), d_*(U)\}$ of U as

$$C_j(U) = C_{j+1}^{(1)} \oplus C_k^{(2)}, d_k(U) = \begin{pmatrix} -d_{j+1}^{(1)} & 0 \\ U & d_j^{(2)} \end{pmatrix}, j = -1, 0, 1, \ldots, n.$$

Here, $d_j^{(k)} = 0$ for $j < 0$. Let $\Delta_j(U) = d_j^*(U)d_j(U) + d_{j-1}(U)d_{j-1}^*(U)$ be its Laplace operator and $\Delta_j'(U)$ the restriction of $\Delta_j(U)$ to $(Ker\Delta_j(U))^\perp$.

Definition 3.5: (1) The torsion $\tau(U)$ of U relative to $T^{(k)}$ is

$$\tau(U, T^{(k)}) = \frac{1}{2}\sum_{j=0}^n (-1)^{j+1} j \log(\Delta(\Delta_j'(U))),$$

provided $\Delta(\Delta_j'(U)) \neq 0$.

(2) U is said to be a weak cohomology equivalence relative to $T^{(k)}$ if the cohomology $H^*(C_*(U))$ of the mapping cone $C_*(U)$ is zero. If U is

a weak cohomology equivalence relative to $T^{(k)}$ and $\log \tau(U, T^{(k)}) = 0$, then U is called a weakly simple cohomology equivalence relative to $T^{(k)}$.

The Laplace operator $\Delta_j(U)$ can be written as a matrix

$$\Delta_j(U) = \begin{pmatrix} \Delta_{j+1}(T^{(1)}) + U^*U & U^*d_j^{(2)} - d_j^{(1)}U^* \\ (d_j^{(2)})^*U - U(d_j^{(1)})^* & \Delta_j(T^{(2)}) + UU^* \end{pmatrix}.$$

Using this formula and the notation $\Delta_j^{(k)} = \Delta_j(T^{(k)})$, we can prove the following

Theorem 3.6: Assume $U \in \mathcal{A}$ is invertible and interwines both $T^{(k)}$ and $(T^{(k)})^*$. Suppose $T^{(k)} \in \mathcal{D}_n(\mathcal{A})$ and

$$\frac{1}{2} \sum_{j=0}^{n} (-1)^{j+1} (\log \Delta((\Delta_j^{(1)} + U^*U)|_{(Ker\Delta_j^{(1)})^\perp}) + \log \Delta(U^*U|_{Ker\Delta_j^{(1)}})) = 0.$$

$$(3.3)$$

Then U is a weakly simple cohomology equivalence relative to $T^{(k)}$.

Proof: Since U interwines both $T^{(k)}$ and $(T^{(k)})^*$ and is injective, $\Delta_j(U)$ and $\Delta_j'(U)$ can be simplified as

$$\Delta_j(U) = (\Delta_{j+1}^{(1)} + U^*U) \oplus (\Delta_j^{(2)} + UU^*),$$

$$\begin{aligned} \Delta_j'(U) = &((\Delta_{j+1}^{(1)} + U^*U)|_{(Ker\Delta_{j+1}^{(1)})^\perp} \oplus (U^*U)|_{Ker\Delta_{j+1}^{(1)}}) \\ &\oplus ((\Delta_j^{(2)} + UU^*)|_{(Ker\Delta_j^{(2)})^\perp} \oplus (UU^*)|_{Ker\Delta_j^{(2)}}). \end{aligned}$$

We have

$$\begin{aligned} \tau(e^{-t\Delta_j'(U)}) = &\tau(e^{-t(\Delta_{j+1}^{(1)} + U^*U)'}) + \tau(e^{-t(U^*U)'}) \\ &+ \tau(e^{-t(\Delta_j^{(2)} + UU^*)'}) + \tau(e^{-t(UU^*)'}). \end{aligned}$$

Here for simplicity, $(B)'$ means the restriction of B to an appropriate subspace.

$$\sum_{j=0}^{n} (-1)^j j \tau(e^{-t\Delta_j'(U)}) =$$

$$\sum_{j=0}^{n}(-1)^{j-1}(j-1)(\tau(e^{-t(\Delta_{j+1}^{(1)}+U^*U)'})+\tau(e^{-t(U^*U)'}))$$

$$+\sum_{j=0}^{n}(-1)^{j}j(\tau(e^{-t(\Delta_{j}^{(2)}+UU^*)'})+\tau(e^{-t(UU^*)'}))$$

$$=\sum_{j=0}^{n}(-1)^{j}j(\tau(e^{-t(\Delta_{j}^{(1)}+U^*U)'})-\tau(e^{-t(\Delta_{j}^{(2)}+UU^*)'})+\tau(e^{-t(U^*U)'})-$$
$$\tau(e^{-t(UU^*)'}))+\sum_{j=0}^{n}(-1)^{j}(\tau(e^{-t(\Delta_{j}^{(1)}+U^*U)'})+\tau(e^{-t(U^*U)'})). \qquad (3.4)$$

We want to prove that
$$\tau(e^{-t(\Delta_{j}^{(1)}+U^*U)'})=\tau(e^{-t(\Delta_{j}^{(2)}+UU^*)'}) \text{ and } \tau(e^{-t(U^*U)'})=\tau(e^{-t(UU^*)'}).$$
In fact , using $Ud_{j}^{(1)}=d_{j}^{(2)}U$ and $U(d_{j}^{(1)})^*=(d_{j}^{(2)})^*U$, we have $U\Delta_{j}^{(1)}=\Delta_{j}^{(2)}U$. Thus, $U:(Ker\Delta_{j}^{(1)})^{\pm}\to(Ker\Delta_{j}^{(2)})^{\pm}$ is invertible. It follows that $U(\Delta_{j}^{(1)}+U^*U)')U^{-1}=(\Delta_{j}^{(2)}+UU^*)'$ and $U(U^*U)'U^{-1}=UU^*$. we get

$$\tau(e^{-t(\Delta_{j}^{(2)}+UU^*)'})=\tau(Ue^{-t(\Delta_{j}^{(1)}+U^*U)'}U^{-1})=\tau(e^{-t(\Delta_{j}^{(1)}+U^*U)'})$$

and

$$\tau(e^{-t(UU^*)'})=\tau(Ue^{-t(U^*U)'}U^{-1})=\tau(e^{-t(U^*U)'}).$$

Therefore, (3.4) reduces to

$$\sum_{j=0}^{n}(-1)^{j}j\tau(e^{-t\Delta_{j}'(U)})=\sum_{j=0}^{n}(-1)^{j}(\tau(e^{-t(\Delta_{j}^{(1)}+U^*U)'})+\tau(e^{-t(U^*U)'})).$$

$$\begin{aligned}\zeta^{U}(s)&=\int_{0}^{\infty}\sum_{j=0}^{n}(-1)^{j}j\frac{1}{\Gamma(s)}t^{s-1}\tau(e^{-t\Delta_{j}'(U)})dt\\&=\int_{0}^{\infty}\sum_{j=0}^{n}(-1)^{j}\frac{1}{\Gamma(s)}t^{s-1}\tau(e^{-t(\Delta_{j}^{(1)}+U^*U)'})dt+\\&+\int_{0}^{\infty}\sum_{j=0}^{n}(-1)^{j}\frac{1}{\Gamma(s)}t^{s-1}\tau(e^{-t(U^*U)|_{Ker\Delta_{j}^{(1)}}}).\end{aligned}$$

$$(3.5)$$

The second integral on the right hand side of (3.5) is convergent since

$U^*U|_{Ker\Delta_j^{(1)}}$ is invertible. To estimate the first integral in the right hand side of (3.5), we use the fact that $(\Delta_j^{(1)})'(U^*U)' = (U^*U)'(\Delta_j^{(1)})'$ and then $e^{-t(\Delta_j^{(1)}+U^*U)'} = e^{-t(\Delta_j^{(1)})'}e^{-t(U^*U)'}$. Hence

$$|\tau(e^{-t(\Delta_j^{(1)}+U^*U)'})| \leq Const \cdot |\tau(e^{-t(\Delta_j^{(1)})'})|.$$

By assumption, the first integral of (3.5) is thus absolutely convergent. We obtain by (3.4)

$$\tau(U, T^{(k)}) = \frac{1}{2}\frac{d}{ds}(\zeta^U(s))|_{s=0} =$$

$$\tfrac{1}{2}\sum_{j=0}^n(-1)^{j+1}(\log(\Delta((\Delta_j^{(1)}+U^*U)|_{(Ker\Delta_j^{(1)})^\perp})+\log\Delta(U^*U|_{Ker\Delta_j^{(1)}}))$$

$$= 0.$$

It remains to check that the cohomology of the mapping cone $\{C_*(U), d_*(U)\}$ is zero. This can be easily verified. See [Go]. Q.E.D.

Remark 3.7: Condition (3.3) in Theorem 3.6 is satisfied if for $n > 1$, $T^{(k)} = (T_1^{(k)}, \ldots, T_n^{(k)})$ have the property that $[T_i^{(k)}, (T_j^{(k)})^*] = 0$ and $T^{(k)} \in \mathcal{D}_n(\mathcal{A})$.

In fact, we get for such $T^{(k)}$,

$$\Delta_j(T^{(k)}) = \sum_{i=0}^n T_i^{(k)}(T_i^{(k)})^*,$$

$$Ker\Delta_j(T^{(k)}) = Ker(\sum_{i=0}^n T_i^{(k)}(T_i^{(k)})^*) \otimes \wedge^j(\sigma).$$

Hence

$$\log(\Delta((\Delta_j^{(1)} + U^*U)|_{(Ker\Delta_j^{(1)})^\perp}) =$$

$$\binom{n}{j}\log\Delta((\sum_{i=0}^n T_i^{(1)}(T_i^{(1)})^* + U^*U)|_{(Ker\sum_{i=0}^n T_i^{(k)}(T_i^{(k)})^*)^\perp}). \quad \text{and}$$

$$\log\Delta(U^*U)|_{(Ker\sum_{i=0}^n T_i^{(k)}(T_i^{(k)})^*)\otimes\wedge^j(\sigma))}$$

$$= \binom{n}{j} \log \Delta(U^*U)|_{(Ker \sum_{i=0}^{n} T_i^{(k)}(T_i^{(k)})^*)}.$$

Here $m_j = \binom{n}{j}$ is the dimension of $\wedge^j(\sigma)$. Adding these identities together, we get

$$\sum_{j=0}^{n}(-1)^j (\log(\Delta((\Delta_j^{(1)} + U^*U)|_{(Ker\Delta_j^{(1)})^\perp}) + \log \Delta(U^*U|_{Ker\Delta_j^{(1)}})))$$

$$= \sum_{j=0}^{n}(-1)^j \log \Delta((\sum_{i=0}^{n} T_i^{(1)}(T_i^{(1)})^* + U^*U)|_{(Ker \sum_{i=0}^{n} T_i^{(k)}(T_i^{(k)})^*)^\perp})$$

$$+ \quad \log \Delta(U^*U)|_{(Ker(\sum_{i=0}^{n} T_i^{(k)}(T_i^{(k)})^*))} = 0,$$

since $\sum_{j=0}^{n}(-1)^j \binom{n}{j} = 0$ for $n > 1$.

In particular, (3.3) holds for $T^{(k)} \in \mathcal{D}_n(\mathcal{A})$ with selfadjoint $T_j^{(k)}$.

4   Properties of the Torsion Invariants

We begin with the following basic lemma.

Lemma 4.1: Let $T \in \mathcal{D}_n(\mathcal{A})$. Then

$$\sum_{j=0}^{n}(-1)^j \log \Delta(\Delta_j'(T)) = 0.$$

Proof : Since $C_j(T)$ can be decomposed as $C_j(T) = \overline{Im\, d_{j-1}} \oplus Ker\Delta_j \oplus \overline{Im\, d_j^*}$, $\Delta_j'(T) = \Delta_j|_{(Ker\Delta_j)^\perp} = d_j^* d_j|_{\overline{Im\, d_j^*}} \oplus d_{j-1} d_{j-1}^*|_{\overline{Im\, d_{j-1}}}$.

Accordingly we have

$$\tau(e^{-td_{j-1}^* d_{j-1}|_{\overline{Im\, d_{j-1}^*}}}) = \int_t^\infty \frac{d}{dt} \tau(e^{-td_{j-1}^* d_{j-1}|_{\overline{Im\, d_{j-1}^*}}}) dt$$

$$= \int_t^\infty \tau(-d_{j-1}^* e^{-td_{j-1} d_{j-1}^*|_{\overline{Im\, d_{j-1}}}} d_{j-1}) dt$$

$$= \int_t^\infty \tau(-d_{j-1} d_{j-1}^* e^{-td_{j-1} d_{j-1}^*|_{\overline{Im d_{j-1}}}}) dt = \tau(e^{-td_{j-1} d_{j-1}^*|_{\overline{Im\, d_{j-1}}}}).$$

Hence,

$$\sum_{j=0}^{n}(-1)^j\tau(e^{-t\Delta'_j(T)}) = \sum_{j=0}^{n}(-1)^j(\tau(e^{-td_{j-1}d^*_{j-1}|\overline{Imd_{j-1}}}) + \tau(e^{-td^*_j d_j|\overline{Im\,d^*_j}}))$$

$$= \sum_{j=0}^{n-1}(-1)^{j+1}\tau(e^{-td^*_j d_j|\overline{Im\,d^*_j}}) + \sum_{j=0}^{n-1}(-1)^j\tau(e^{-td^*_j d_j|\overline{Im\,d^*_j}}) = 0.$$

We obtain

$$\sum_{j=0}^{n}(-1)^j\log\Delta(\Delta'_j(T)) =$$

$$-\frac{d}{ds}\left(\frac{1}{\Gamma(s)}\int_0^\infty t^{s-1}\sum_{j=0}^{n}(-1)^j\tau(e^{-t\Delta'_j(T)})dt\right)|_{s=0} = 0 \qquad Q.E.D.$$

Basic properties of the torsion invariant are stated in the following theorems.

Theorem 4.2: Let $T \in \mathcal{D}_n(\mathcal{A})$.

1. $\log\tau(\lambda T) = \log\tau(T) + \sum_{j=0}^{n}(-1)^{j+1}j\ln|\lambda|, \lambda \in C, \lambda \neq 0$,

2. If $\tilde{T} = (T_{\nu(1)},\ldots,T_{\nu(n)})$ is obtained by permuting $(T_1,\ldots,T_n)$, then $\log\tau(\tilde{T}) = \log\tau(T)$,

3. $\log\tau(T) = (-1)^{n+1}\log\tau(T^*)$.

Proof: (1) We have $\Delta_j(\lambda T) = |\lambda|^2\Delta_j(T), \Delta'_j(\lambda T) = |\lambda|^2\Delta'_j(T)$. By Lemma 3.1(1), we get

$$\log\tau(\lambda T) = \frac{1}{2}\sum_{j=0}^{n}(-1)^{j+1}j\log\Delta(|\lambda|^2\Delta'_j(T))$$

$$= \frac{1}{2}\sum_{j=0}^{n}(-1)^{j+1}j(\log\Delta(\Delta'_j(T)) + \log|\lambda|^2)$$

$$= \log\tau(T) + \sum_{j=0}^{n}(-1)^{j+1}j\log|\lambda|.$$

(2) The result follows from the fact that the Koszul complexes associated with $\tilde{T}$ and T are unitarily equivalent.

(3) We have by the proof of Theorem 2.3 that $\Delta_j(T) = \sharp^*\Delta_{n-j}(T^*)\sharp$ and $\Delta'_j(T) = \sharp^*\Delta'_{n-j}(T^*)\sharp$, where $\sharp : C_* \to C_*$ is a unitary operator.

Therefore, by Lemma 3.1(4) and Lemma 4.1, we obtain

$$\log \tau(T) = \frac{1}{2} \sum_{j=0}^{n} (-1)^{j+1} j \log \Delta(\Delta'_{n-j}(T^*))$$

$$= \frac{1}{2} \sum_{j=0}^{n} (-1)^{n-j+1} (n-j) \log \Delta(\Delta'_j(T^*)) = (-1)^{n+1} \log \tau(T^*).$$

$$Q.E.D.$$

Theorem 4.3: Let $T^{(k)} = (T_1^{(k)}, \ldots, T_n^{(k)}) \in \mathcal{D}_n(\mathcal{A}), k = 1, 2.$

1. $\log \tau(T^{(1)} \oplus T^{(2)}) = \log \tau(T^{(1)}) + \log \tau(T^{(2)}),$

2. $\log \tau(T^{(1)} \otimes 1 + 1 \otimes T^{(2)})$

   $= \chi(C_*(T^{(1)})) \log \tau(T^{(2)}) + \chi(C_*(T^{(2)})) \log \tau(T^{(1)}),$

   where $C_*(T^{(k)})$ is the Koszul complex associated with $T^{(k)},$
   $P_j(T^{(k)}) : C_j(T^{(k)}) \to Ker\Delta_j(T^{(k)})$ is the orthogonal projection
   and $\chi(C_*(T^{(k)})) = \sum_{j=0}^{n} (-1)^j \tau(P_j(T^{(k)})).$

Proof: (1) Let $T = T^{(1)} \oplus T^{(2)}$. We have $\Delta_j(T) = \Delta_j(T^{(1)}) \oplus \Delta_j(T^{(2)})$
and $\Delta'_j(T) = \Delta'_j(T^{(1)}) \oplus \Delta'_j(T^{(2)})$. By assumption,
$\Delta(\Delta'_j(T)) = \Delta(\Delta'_j(T^{(1)}))\Delta(\Delta'_j(T^{(2)})) \neq 0.$ Hence,

$$\tau(T) = \frac{1}{2} \sum_{j=0}^{n} (-1)^{j+1} j (\log \Delta(\Delta'_j(T^{(1)})) + \log \Delta(\Delta'_j(T^{(2)})))$$

$$= \log \tau(T^{(1)}) + \log \tau(T^{(2)}).$$

(2) Let $T = T^{(1)} \otimes 1 + 1 \otimes T^{(2)}$. We note that

$$\Delta_j(T) = \bigoplus_{p+q=j} (\Delta_p(T^{(1)}) \otimes 1 + 1 \otimes \Delta_q(T^{(2)}))$$

and

$$(Ker\Delta_j(T))^{\perp} =$$

$$\bigoplus_{p+q=j} ((Ker\Delta_p(T^{(1)})^{\perp} \otimes Ker\Delta_q(T^{(2)})^{\perp}) \oplus (Ker\Delta_p(T^{(1)}) \otimes Ker\Delta_q(T^{(2)})^{\perp})$$

$$\oplus (Ker\Delta_p(T^{(1)})^{\perp} \otimes Ker\Delta_q(T^{(2)}))).$$

It follows that

$$\tau(e^{-t\Delta'_j(T)}) = \sum_{p+q=j} (\tau(e^{-t\Delta'_p(T^{(1)})})\tau(e^{-t\Delta'_q(T^{(2)})}) + \tau(P_p(T^{(1)}))\tau(e^{-t\Delta'_q(T^{(2)})})$$
$$+ \tau(P_q(T^{(2)})\tau(e^{-t\Delta'_p(T^{(1)})})).$$

In view of Lemma 4.1, we get

$$\sum_{j=0}^{n}(-1)^j j\tau(e^{-t\Delta'_j(T)}) = \sum_{p,q=0}^{n} (-1)^{p+q}(p+q)(\tau(e^{-t\Delta'_p(T^{(1)})})\tau(e^{-t\Delta'_q(T^{(2)})})$$
$$+ \tau(P_p(T^{(1)}))\tau(e^{-t\Delta'_q(T^{(2)})}) + \tau(P_q(T^{(2)})\tau(e^{-t\Delta'_p(T^{(1)})}))$$
$$= \chi(C_*(T^{(1)})) \sum_{q=0}^{n}(-1)^q q\tau(e^{-t\Delta'_q(T^{(2)})})$$
$$+ \chi(C_*(T^{(2)})) \sum_{p=0}^{n}(-1)^p p\tau(e^{-t\Delta'_p(T^{(1)})}).$$

Therefore,

$$\log\tau(T) = \frac{1}{2}\frac{d}{ds}(\frac{1}{\Gamma(s)}\int_0^\infty t^{s-1}\sum_{j=0}^{n}(-1)^j j\tau(e^{-t\Delta'_j(T)})dt)_{s=0}$$
$$= \log\tau(T^{(1)})\chi(C_*(T^{(2)})) + \log\tau(T^{(2)})\chi(C_*(T^{(1)})).$$

Q.E.D.

Let $T^{(k)} = (T_1^{(k)}, \ldots, T_n^{(k)})$ be two n-tuples of commuting elements in $\mathcal{A}$. We say that $T^{(1)}$ and $T^{(2)}$ are similar if there is an invertible element $U \in \mathcal{A}$ such that $T_j^{(2)} = UT_j^{(1)}U^{-1}$. The following theorem gives a torsion formula for similar n-tuples.

Theorem 4.4: Suppose $T^{(1)}$ and $T^{(2)}$ are similar n-tuples via an invertible element $U \in \mathcal{A}$. If $\theta_j(T^{(k)}) < \infty$ and $H^*(C_*(T^{(k)})) = 0, k = 1, 2, j = 0, 1, \ldots, n$, then

$$\log\tau(T^{(2)}) =$$

$$\log\tau(T^{(1)}) + \frac{1}{2}\sum_{j=0}^{n}(-1)^{j+1}(\frac{1}{\Gamma(s)}\int_0^\infty dt\, t^{s-1}\int_0^1 du\, \tau(\nu_u e^{-t\Delta_j(T_u)}))_{s=0}, \quad (4.1)$$

where $T_u = T^{(2)}$ and ν_u is given by (4.4) below.

Proof: We divide the proof into four steps.

Step 1. Define a new inner product $< \cdot >_1$ on H by

$$< x, y >_2 = < U^{-1}x, U^{-1}y >$$

for $x, y \in H$. This product is equivalent to the old one on H. With respect to this new inner product on H, $U : (H, < \cdot >) \to (H, < \cdot >_2)$ is a unitary operator. In view of of the unitary invariance of the torsion, $\tau(T^{(1)}) = \tau(T^{(2)}, H, < \cdot >_2)$, where the right hand side is the torsion of $T^{(2)}$ with respect to the new inner product $< \cdot >_2$. As a result, we need only to prove (4.1) for an n-tuple $T \in \mathcal{D}_n(\mathcal{A})$ wi th $\theta_j(T) < \infty$ and the Hilbert space H with two equivalent inner products $< \cdot >_k, k = 1, 2$.

Step 2. Define a smooth path of inner products $< \cdot >_u$ on H by

$$< x, y >_u = u < x, y >_1 + (1 - u) < x, y >_2, \quad x, y \in H.$$

Let $i_u : (H, < \cdot >_1) \to (H, < \cdot >_u)$ be the identity map, and let i_u^* be its adjoint, $< i_u x, y >_u = < x, i_u^* y >_1$. The adjoint T_u^* of T in $(H, < \cdot >_u)$ is given by $(T_j^*)_u = (i_u^*)^{-1} T_j^* i_u^*$, where T_j^* is the adjoint of T_j in $(H, < \cdot >_1)$. Let $\Delta_j(T_u)$ be the Laplace operators associated with $T_u = T$ on $(H, < \cdot >_u)$.

$$\Delta_j(T_u) = (i_u^*)^{-1} d_j^* i_u^* d_j + d_{j-1} (i_u^*)^{-1} d_{j-1}^* i_u^*,$$

where d_j is the differential of the Koszul complex $C_*(T)$ associated wi th T on $(H, < \cdot >_1)$. Let $\nu_u = (i_u^*)^{-1} \frac{d}{du}(i_u^*)$. Since $(i_u^*)^{-1}(i_u^*) = Id$, $\frac{d}{du}((i_u^*)^{-1}) = -(i_u^*)^{-1} \frac{d}{du}(i_u^*)(i_u^*)^{-1} = -\nu_u(i_u^*)^{-1}$. We obtain

$$
\begin{aligned}
\frac{d}{du}(\Delta_j(T_u)) &= -\nu_u(i_u^*)^{-1} d_j^*(i_u^*) d_j + (i_u^*)^{-1} d_j^*(i_u^*)\nu_u d_j \\
&\quad - d_{j-1}\nu_u(i_u^*)^{-1} d_{j-1}^*(i_u^*) + d_{j-1}(i_u^*)^{-1} d_{j-1}^*(i_u^*)\nu_u \\
&= -\nu_u d_j^*(u) d_j(u) + d_j^* \nu_u d_j(u) \\
&\quad - d_j(u)\nu_u d_{j-1}^*(u) + d_{j-1}(u) d_{j-1}^*(u)\nu_u.
\end{aligned}
$$

Step 3. Since $\Delta_j(T_u)$ is bounded, we have that $\frac{d}{du}(\tau(e^{-t\Delta_j(T_u)})) = -t\tau(\frac{d}{du}(\Delta_j(T_u))e^{-t\Delta_j(T_u)})$. Note that assumption implies that the cohomology of the Koszul complex associated with T_u is trivial. By the formula in Step 2,

$$\frac{d}{du}\tau(e^{-t\Delta_j(T_u)}) =$$
$$-t(\tau(-\nu_u d_j^*(u)d_j(u)e^{-t\Delta_j(T_u)}) + \tau(d_j^*\nu_u d_j(u)e^{-t\Delta_j(T_u)})$$
$$+ \tau(-d_j(u)\nu_u d_{j-1}^*(u)e^{-t\Delta_j(T_u)}) + \tau(d_{j-1}(u)d_{j-1}^*(u)\nu_u e^{-t\Delta_j(T_u)}))$$
$$= -t(-\tau(\nu_u d_j^*(u)d_j(u)e^{-t\Delta_j(T_u)}) + \tau(\nu_u d_j(u)d_j^*(u)e^{-t\Delta_{j+1}(T_u)})$$
$$- \tau(\nu_u d_{j-1}^*(u)d_{j-1}(u)e^{-t\Delta_{j-1}(T_u)}) + \tau(\nu_u d_{j-1}(u)d_{j-1}^*(u)e^{-t\Delta_j(T_u)})).$$

$$\begin{aligned}
\frac{d}{du}(\zeta^{T_u}(s)) &= \frac{1}{\Gamma(s)}\sum_{j=0}^{n}(-1)^j j\int_0^\infty t^{s-1}\frac{d}{du}\tau(e^{-t\Delta_j(T_u)})dt \\
&= \frac{1}{\Gamma(s)}\int_0^\infty dt\, t^{s-1}t(\sum_{j=0}^{n-1}(-1)^{j+1}\tau(\nu_u d_j^*(u)d_j(u)e^{-t\Delta_j(T_u)}) \\
&\quad + \sum_{j=1}^{n}(-1)^{j+1}\tau(\nu_u d_{j-1}(u)d_{j-1}^*(u)e^{-t\Delta_j(T_u)})) \\
&= \frac{1}{\Gamma(s)}\int_0^\infty dt\, t^{s-1}t\sum_{j=0}^{n}(-1)^{j+1}\tau(\nu_u\Delta_j(T_u)e^{-t\Delta_j(T_u)}) \\
&= \sum_{j=0}^{n}(-1)^j\frac{1}{\Gamma(s)}\int_0^\infty dt\, t^{s-1}t\frac{d}{dt}\tau(\nu_u e^{-t\Delta_j(T_u)}) \\
&= \sum_{j=0}^{n}(-1)^{j+1}\frac{s}{\Gamma(s)}\int_0^\infty t^{s-1}\tau(\nu_u e^{-t\Delta_j(T_u)})dt
\end{aligned}$$

$$(4.2)$$

for s near 0. Here we have used the fact that $\theta_j(T_u) < \infty$ by Corollary 2.8, which implies that $t^s\tau(\nu_u e^{-t\Delta_j(T_u)}) \to 0$ as $t \to \infty$. Also $\frac{1}{\Gamma(s)}\int_0^\infty t^{s-1}\tau(\nu_u e^{-t\Delta_j(T_u)})dt$ extends to a holomorphic function for s near 0. Consequently,

$$\begin{aligned}
\frac{d}{du}\log\tau(T_u, <\cdot>_u) &= \frac{1}{2}\frac{d}{ds}(\frac{d}{du}\zeta^{T_u}(s))_{s=0} \\
&= \frac{1}{2}\frac{d}{ds}(\sum_{j=0}^{n}(-1)^{j+1}\frac{s}{\Gamma(s)}\int_0^\infty t^{s-1}\tau(\nu_u e^{-t\Delta_j(T_u)})dt)_{s=0} \\
&= \frac{1}{2}\sum_{j=0}^{n}(-1)^{j+1}(\frac{1}{\Gamma(s)}\int_0^\infty t^{s-1}\tau(\nu_u e^{-t\Delta_j(T_u)})dt)_{s=0}.
\end{aligned}$$

Therefore,

$$\log\tau(T, <\cdot>_1) - \log\tau(T, <\cdot>_2) =$$

$$\frac{1}{2}\sum_{j=0}^{n}(-1)^{j+1}(\frac{1}{\Gamma(s)}\int_0^\infty dt t^{s-1}\int_0^1 du\tau(\nu_u e^{-t\Delta_j(T_u)}))_{s=0}.$$

$$(4.3)$$

Step 4. Finally, we analyze ν_u in (4.3) in terms of the original operators. In Step 3, we have $< x,y>_1=< x,y>, < x,y>_2=< U^{-1}x, U^{-1}y>, x,y \in H$. Then

$$
\begin{aligned}
< x,y >_u &= u < x,y >_1 +(1-u) < x,y >_2 \\
&=< x, uy + (1-u)(U^{-1})^* U^{-1}y > \\
&= < x, uy + (1-u)(U^{-1})^* U^{-1}y >_1,
\end{aligned}
$$

and

$$< i_u x, y >_u=< x,y >_u=< x, i_u^* y >_1 .$$

We get from these two identities that $i_u^* = u+(1-u)(U^{-1})^* U^{-1}$. Hence, $(i_u^*)^{-1} = UU^*, u = 0; Id, u = 1;$ and $\frac{1}{u}UU^*(UU^* + \frac{1-u}{u})^{-1}, 0 < u < 1.$ Also $\frac{d}{du}(i_u^*) = 1 - (U^{-1})^* U^{-1}$. Combining these identities , we obtain

$$
\begin{aligned}
\nu_u &= (i_u^*)^{-1}\frac{d}{du}(i_u^*) = (i_u^*)^{-1}(1 - (U^*)^{-1}U^{-1}) \\
&= \begin{cases}
UU^* - 1, & u = 0, \\
1 - (U^*)^{-1}U^{-1}, & u = 1, \\
\frac{1}{u}(UU^* + \frac{1-u}{u})^{-1}(UU^* - 1), & 0 < u < 1.
\end{cases}
\end{aligned}
$$

$$(4.4)$$

The proof is complete.   Q.E.D.

We note out that Formula (4.1) is different from the variation formula of the Ray-Singer torsion [RS]. Our formula has an extra term due to the possible nonvanishing of the meromorphic extension of $\frac{1}{\Gamma(s)}\int_0^\infty t^{s-1}\tau(\nu_u e^{-t\Delta_j(T_u)})dt$ at $s = 0$. But the proof of Step 3 above is similar to a corresponding proof in [RS].

As a corollary of the proof, we have the following.

Corollary 4.5: Let $T^{(k)} \in \mathcal{D}_n(\mathcal{A}), k = 1, 2$. Suppose $T^{(1)}$ and $T^{(2)}$ are similar by means of an invertible element $U \in \mathcal{A}$. Then

$$\log \tau(T^{(1)}) - \log \tau(T^{(2)}) = \frac{1}{2}\int_0^1 \sum_{j=0}^n (-1)^j \tau(\nu_u P_j(T_u))du +$$

$$+ \frac{1}{2}\sum_{j=0}^{n}(-1)^{j+1}(\frac{1}{\Gamma(s)}\int_0^\infty dt t^{s-1}\int_0^1 du\tau(\nu_u e^{-t\Delta_j(T_u)+P_j(T_u)}))_{s=0},$$

$$(4.5)$$

where ν_u is given by (4.4) and $P_j(T_u) : C_j(T_u) \to Ker\Delta_j(T_u)$ is the orthogonal projection.

Proof: we use the same notation as in the proof of Theorem 4.4. By (4.2) we have

$$\frac{d}{du}\log\tau(T_u,<\cdot>_u) = \frac{1}{2}\frac{d}{ds}(\sum_{j=0}^{n}(-1)^{j+1}\frac{s}{\Gamma(s)}\int_0^\infty t^{s-1}\tau(\nu_u e^{-t\Delta_j'(T_u)})dt)_{s=0} =$$

$$\frac{1}{2}\frac{d}{ds}(\sum_{j=0}^{n}(-1)^{j+1}\frac{s}{\Gamma(s)}\int_0^\infty t^{s-1}\tau(\nu_u e^{-t(\Delta_j(T_u)+P_j(T_u))})dt)_{s=0}$$

$$+\frac{1}{2}\frac{d}{ds}(\sum_{j=0}^{n}(-1)^{j}\frac{s}{\Gamma(s)}\int_0^\infty t^{s-1}\tau(\nu_u P_j(T_u)e^{-t})dt)_{s=0}$$

$$= \frac{1}{2}\sum_{j=0}^{n}(-1)^{j+1}(\frac{1}{\Gamma(s)}\int_0^\infty t^{s-1}\tau(\nu_u e^{-t(\Delta_j(T_u)+P_j(T_u))})dt)_{s=0}$$

$$+\frac{1}{2}\sum_{j=0}^{n}(-1)^{j}\tau(\nu_u P_j(T_u)).$$

$$(4.6)$$

Taking the integral on both sides of (4.6), we get (4.5).   *Q.E.D.*

We know that the analytic torsion of an even dimensional compact manifold is zero. The following proposition gives a much stronger vanishing result in the operator case.

Proposition 4.6: Suppose $T \in \mathcal{D}_n(\mathcal{A})$ and $[T_i, T_j^*] = 0, i, j = 1, 2, \ldots, n$. Then for $n > 1$,

$$\log\tau(T) = 0.$$

Proof: It is immediate that the Laplace operator $\Delta_j(T)$ associated with T has the following form

$$\Delta_j(T) = \sum_{i=1}^{n} T_i T_i^*, j = 0, 1, \ldots, n.$$

We have $(Ker\Delta_j(T))^\perp = (Ker \sum_{i=1}^n T_iT_i^*)^\perp \otimes \wedge^j(\sigma)$.

$$\begin{aligned}
\log \tau(T) &= \frac{1}{2}\sum_{j=0}^n (-1)^{j+1} j \log \Delta((\sum_{i=1}^n T_iT_i^*)|_{(Ker \sum_{i=1}^n T_iT_i^*)^\perp \wedge^j(\sigma)}) \\
&= \frac{1}{2}\sum_{j=0}^n (-1)^{j+1} j \binom{n}{j} \log \Delta(\sum_{i=1}^n T_iT_i^*|_{(Ker \sum_{i=1}^n T_iT_i^*) \ perp}) \\
&= 0,
\end{aligned}$$

since $\sum_{j=0}^n (-1)^j j \binom{n}{j} = 0$ for $n > 1$. $\qquad\qquad$ Q.E.D.

5   The Torsion of a single operator

We will now concentrate on the case $n = 1$, namely the case of a single operator. Let T be an element in $\mathcal{A}$. The Laplace operators $\Delta_j(T)$ associated with T are $\Delta_0(T) = T^*TS_1 \simeq T^*T$ and $\Delta_j(T) = TT^*S_1^* \simeq TT^*$. Thus the torsion T is

$$\log \tau(T) = \frac{1}{2}\log \Delta(\Delta_1'(T)) = \frac{1}{2}\log \Delta(TT^*|_{(KerTT^*)^\perp}),$$

$$(5.1)$$

provided $T \in \mathcal{D}_1(\mathcal{A})$. If $KerTT^* = KerT^* = 0$, then $\log \tau(T) = \frac{1}{2}\log \Delta(TT^*) = \log \Delta(T)$. Thus, the torsion is then just the determinant of the operator.

We saw in Section 2 that the finiteness of the spectral invariants $\theta_j(T)$ guarantees a meaningful torsion. We give another such sufficient condition by considering the measure $d\mu^T(\lambda)$ defined by L. Brown [Br] in connection with his generalization of the trace theorem of Lidskii. Given an element $T \in \mathcal{A}$, $u(z) = \log \Delta(1 - zT)$ is a subharmonic function of $z \in C$. Let $d\mu_0^T(z)$ be the Riesz measure associated with $u(z)$, $d\mu_0^T(z) = \frac{1}{2\pi}\nabla^2 u(z)$, where ∇^2 is the usual Laplacian on C. The measure $d\mu^T(\lambda)$ is defined as $d\mu^T(\lambda) = d\mu_0^T(\frac{1}{\lambda})$. $d\mu^T(\lambda)$ is supported on $\sigma(T) \setminus \{0\}$. It has the following remarkable properties.

(1) $u(z) = \int \log(1 - zw)d\mu^T(w)$,   (5.2)

More generally, for any holomorphic function f in a neighborhood of $\sigma(T)$ such that $f(0) = 0$ if $0 \in \sigma(T)$, then

$\log \Delta(f(T)) = \int \log |f(w)|d\mu^T(w)$,   (5.3)

(2) $d\mu^{f(T)} = f_* d\mu^T$, i.e., $\mu^{f(T)}(F) = \mu^T(f^{-1}(F))$ for Borel set $F \subset \sigma(f(T))$,

(3) $d\mu^T$ can be extended to the point 0 by setting

$\mu^T(\{0\}) = \tau(1) - \mu(\sigma(T) \setminus \{0\})$. The measure $d\mu^T$ is finite on any compact set in $\sigma(T) \setminus \{0\}$,

(4) Given $T = \begin{pmatrix} T_1 & T_3 \\ 0 & T_2 \end{pmatrix}$, $d\mu^T = d\mu^{T_1} + d\mu^{T_2}$.

(5) $\mu^{ST} = \mu^{TS}, \forall S, T \in \mathcal{A}$ and $d\mu^{T^*}(\lambda) = d\mu^T(\bar{\lambda})$.

We are most interested in (5.3). Taking $f(\lambda) = \lambda$, we have

$$\log \Delta(T) = \int \log |\lambda| d\mu^T(\lambda).$$

This implies that $\log \Delta(T)$ is finite if $\mu^T(\lambda)$ behaves like $|\lambda|^\alpha$ as $\lambda \to 0$ for some $\alpha > 0$.

Definition 5.1: Let $\lambda \in \sigma(T)$. We define $\tilde{\alpha}(T, \lambda)$ as

$$\tilde{\alpha}(T, \lambda) = \sup\{\beta \geq 0 : \mu^T(B_w(\lambda)) = O(|w|^\beta), w \to 0\},$$

where $B_w(\lambda)$ is the ball of radius w in C with center at λ.

We see $\tilde{\alpha}(T, \lambda) = \frac{1}{4}\alpha_1(T - \lambda)$ for a normal operator T and noneigen-value λ of T, since in this case, $\mu^T(\lambda) = N_1^T(\lambda^{\frac{1}{2}})$. This follows from the fact that $\mu^T(\lambda)$ and $N_1^T(\lambda)$ are the traces of the spectral measures of T and TT^*, respectively. For an n-tuple $T = (T_1, \ldots, T_n) \in \mathcal{D}_n(\mathcal{A})$, we have $\tilde{\alpha}(\Delta'_j(T), 0) = \frac{1}{2}\alpha_j(T), j = 0, 1, \ldots, n$. $\tilde{\alpha}(T, \lambda)$ is also a similarity invariant as shown below.

Lemma 5.1: Let T_k be two elements in $\mathcal{A}$, $k = 1, 2$. Suppose T_1 and T_2 are similar. Then for any $\lambda \in \sigma(T_1)$,

$$\tilde{\alpha}(T_1, \lambda) = \tilde{\alpha}(T_2, \lambda).$$

Proof: It suffices to show $\mu^{T_1}(\lambda) = \mu^{T_2}(\lambda)$. In fact, these measures are determined by (5.2). Let $U \in \mathcal{A}$ be an invertible element such that $T_2 = U^{-1}T_1U$. Then by Lemma 3.1(4),

$$\log \Delta(1 - zT_2) = \log \Delta(U^{-1}(1 - zT_1)U) = \log \Delta(1 - zT_1).$$

Hence, by the uniqueness of μ^T, $\mu^{T_1}(\lambda) = \mu^{T_1}(\lambda)$. The result also follows directly from (5) above. $\hfill Q.E.D.$

Let T be an element in $\mathcal{A}$. Suppose f is holomorphic in a neighborhood of $\sigma(T)$. We are going to relate the torsion of $f(T)$ to the measure $d\mu^T$. First let $f(z) \neq 0$ on $\sigma(T)$. Then $f(T)$ is invertible by the spectral mapping theorem. Using (5.1), we have

$$\log \tau(f(T)) = \log \Delta(f(T)) = \int \log |f(w)| d\mu^T(w).$$

$$(5.4)$$

Let now $f(z_i) = 0$ for some $z_i \in \sigma(T)$. But we still assume that $f(T)$ is injective. Then (5.4) is also true. The question is when the integral in (5.4) is finite. One sufficient condition is that $\tilde{\alpha}(T, z_i) > 0$. Indeed, let $f(z) = \prod_{i=1}^{m}(z - z_i)^{k_i}g(z)$, $g(z) \neq 0$ on $\sigma(T)$. Then

$$\int \log |f(w)| d\mu^T(w) = \sum_{i=1}^{m} k_i \log |w| d\mu^T(w + z_i) + \int log|g(w)| d\mu^T(w)$$
$$> -\infty,$$

since $\mu^T(B_w(z_i)) \sim O(|w|^\beta)$ for w near 0 and $\beta > 0$.

More generally, we assume $f(z_i) = 0$ for $z_i \in \sigma(T)$, $i = 1, \ldots, m$ and $f(T)$ is not injective. Let $f(z) = \prod_{i=1}^{m}(z - z_i)^{k_i}g(z)$. Then

$$Kerf(T) = \overline{span}\{Ker(T - z_i)^{k_i}, i = 1, \ldots, m\}.$$

Let $H = Kerf(T) \oplus H_f$ be the orthogonal decomposition. Since T preserves $Kerf(T)$, we get a corresponding decomposition for T,

$$T = \begin{pmatrix} T_1 & * \\ 0 & T_2 \end{pmatrix}. \text{ Then}$$

$$f(T) = \begin{pmatrix} f(T_1) & * \\ 0 & f(T_2) \end{pmatrix} = \begin{pmatrix} 0 & T_0 \\ 0 & f(T_2) \end{pmatrix}$$

$$(5.5)$$

is the decomposition of $f(T)$ on H.

Theorem 5.3: Let T be an element in $\mathcal{A}$. Suppose $f(z) = \prod_{i=1}^{m}(z - z_i)^{k_i} g(z)$ is a holomorphic function in a neighborhood of $\sigma(T)$ with $g(z) \neq 0$ on $\sigma(T)$ and $z_i \in \sigma(T)$, $k_i \geq 0$ integers.

(1) If $f(T)$ is injective and $\tilde{\alpha}(T, z_i) > 0$ with $k_i > 0$, then $\log \tau(f(T))$ is finite and

$$\log \tau(f(T)) = \int \log |f(w)| d\mu^T(w),$$

(2) If $f(T)$ is not injective, but $f(T_2)$ in (5.4) is invertible, then

$$\log \tau(f(T)) = \int \log |f(w)| d\mu^{T_2}(w)$$
$$+ \frac{1}{2} \sum_{j=0}^{1} (-1)^{j+1} \left(\frac{1}{\Gamma(s)} \int_0^\infty dt t^{s-1} \int_0^1 du \tau(\nu_u^{(2)} e^{-t(\Delta_j(T_u^{(2)}) + P_j(T_u^{(2)}))}) \right)_{s=0}$$
$$+ \frac{1}{2} \sum_{j=0}^{1} (-1)^j \int_0^1 du \tau(\nu_u^{(2)} P_j(T_u^{(2)})),$$

$$(5.6)$$

where $T^{(2)} = f(T)$, $P_j(T_u^{(2)})$ is the projection from $C_*(T^{(2)})$ onto $Ker\Delta_j(T_u^{(2)})$, $\nu_u^{(2)} = ((i_u^{(2)})^*)^{-1} \frac{d}{du}((i_u^{(2)})^*)$, $(i_u^{(2)})^* = u + (1-u)(S^*)^{-1} S^{-1}$,

$$S = \begin{pmatrix} 1 & T_0 f(T_2)^{-1} \\ 0 & 1 \end{pmatrix}, \quad T_u^{(2)} = T^{(2)} \text{ on } (H, < \cdot >_u) \text{ with inner}$$

products $< x, y >_u = u < x, y > + (1-u) < S^{-1}x, S^{-1}y >$.

Proof: From the above discussion, it suffices to prove (5.6). We want to show that $f(T)$ and $0 \oplus f(T_2)$ are similar on H. Indeed, $S(0 \oplus f(T_2))S^{-1} = f(T)$. Using Corollary 4.5 and part (1) for T_2, we get the result.   *Q.E.D.*

Remark 5.4: In general, by the proof of Lemma 4.1, we have

$$\log \tau(f(T)) = \frac{1}{2} \log \Delta(f(T)f(T)^*|_{(Ker\, f(T)f(T)^*)^\perp})$$

$$= \frac{1}{2} \log \Delta(f(T)^* f(T)|_{(Ker\, f(T)^* f(T))^\perp}) = \frac{1}{2} \log \Delta(T_0^* T_0 + f(T)^* f(T)),$$

provided $(T_0^* T_0 + f(T)^* f(T)) \in \mathcal{D}_1(\mathcal{A})$.

6 Abelian Case: Examples

In this section we compute the torsion invariant for a single operator when $\mathcal{A}$ is an abelian von Neumann algebra.

Let X be a compact second countable set with a bounded measure ν. Let $\mathcal{A}$ be the algebra $L^\infty(X, \nu)$ of all bounded measurable functions on X and $H = L^2(X, \nu)$. Suppose $\varphi_1, \ldots, \varphi_n \in \mathcal{A}$. By Proposition 4.6, the torsion of the n-tuple $T = (\varphi_1, \ldots, \varphi_n)$ is zero for $n > 1$. Let us focus on the case $n = 1$ and assume $\varphi \in \mathcal{A}$. Then

$$(Ker\varphi)^\perp = \{f \in H : f = 0 \ \ on \ \ \{\varphi = 0\}\}.$$

Thus the projection P_φ from H onto $Ker\, \varphi$ is equal to χ_φ, the characteristic function of $\{\varphi = 0\}$. Suppose that for m sufficiently large, $\nu(\{x \in X : |\varphi(x)|^2 < \frac{1}{m}\}) \leq \frac{a_0}{m^{b_0}}$ for some fixed $a_0, b_0 > 0$. Then for $m = t^k, 0 < k < 1$,

$$\tau(e^{-t\Delta_0'(\varphi)}) = \int_{\varphi \neq 0} e^{-t|\varphi|^2} d\nu \leq e^{-t^{1-k}} \nu(X) + \frac{a_0}{t^k}.$$

On the other hand, for $|\varphi|^2 \leq M_0$, we have

$$\tau(e^{-t\Delta_0'(\varphi)}) = \int_{\varphi \neq 0} e^{-t|\varphi|^2} d\nu \geq e^{-tM_0} \nu\{\varphi \neq 0\}.$$

Thus we have shown that $\alpha_j(\varphi) \neq 0$, i.e., $\theta_j(\varphi) < \infty, j = 0, 1$. We can then see that

$$\log \tau(\varphi) = \frac{1}{2} \log \Delta(\varphi\varphi^*|_{(Ker\varphi\varphi^*)^\perp})$$

$$= \frac{1}{2}\frac{d}{ds}(-\frac{1}{\Gamma(s)}\int_0^\infty dt t^{s-1}\tau(e^{-t(\varphi\varphi^*)|_{(Ker\varphi\varphi^*)^\perp}}))_{s=0}$$

$$= \frac{1}{2}\frac{d}{ds}(-\frac{1}{\Gamma(s)}\int_0^\infty dt t^{s-1}\int_{\varphi\neq 0}e^{-t|\varphi|^2}d\nu)_{s=0}$$

$$= -\frac{1}{2}\frac{d}{ds}(\int_{\varphi\neq 0}\frac{d\nu}{|\varphi|^{2s}})_{s=0}.$$

To summarize, we have

Proposition 6.1: Let $\mathcal{A}$ be an abelian von Neumann algebra $L^\infty(X,\nu)$ with X compact and second countable and ν a bounded measure on X. Suppose $\varphi \in \mathcal{A}$ is such that that there are $a_0, b_0 > 0$ such that for m sufficiently large, $\nu\{x \in X : |\varphi(x)|^2 < \frac{1}{m}\} < \frac{a_0}{m^{b_0}}$. Then $\theta_j(\varphi) < \infty, j = 0, 1$, and

$$\log \tau(\varphi) = -\frac{1}{2}\frac{d}{ds}(\int_{\varphi\neq 0}\frac{d\nu}{|\varphi|^{2s}})_{s=0}$$

(6.1)

Example 6.2: Let $X = [0,1]$ and $\varphi(x) = x^k$ in Proposition 6.1. We get by (6.1)

$$\log \tau(\varphi) = -\frac{1}{2}\frac{d}{ds}(\int_0^1 \frac{dx}{|x|^{2ks}})_{s=0} = k.$$

Example 6.3: Let $\mathcal{A} = M_k(C)$. Suppose $T = (a_{ij})_{i,j=1}^k \in \mathcal{A}$ is an $m \times m$ matrix. Then

$$\frac{1}{\Gamma(s)}\int_0^\infty dt t^{s-1}\tau(e^{-t(TT^*)|_{KerTT^*)^\perp}}) = \sum_{\lambda_j>0}(\lambda_j)^{-s},$$

where $\{\lambda_j\}$ are the nonzero eigenvalues of TT^*. Therefore,

$$\log \tau(T) = -\frac{1}{2}\frac{d}{ds}(\sum_{\lambda_j>0}(\lambda_j)^{-s})_{s=0} = \ln(\prod_{\xi_j\neq 0}|\xi_j|),$$

where $\{\xi_j\}$ are the nonzero eigenvalues of T. We conclude $\tau(T) = \prod_{\xi_j\neq 0}|\xi_j|$.

The computation of the torsion invariant for n-tuples of commuting $m \times m$ matrices is quite complicated. Computer aided calculations show that the torsion of a 3-tuple $T = (T_1, T_1, T_1)$ for $T_1 = (a_{ij})_{i,j=1}^2$ is zero if $\Delta_j(T)$ is invertible. But the torsion for $T = (T_1, T_1, T_2)$ can be seen not to vanish in general.

7   Invariants of Compact Manifolds

We now use the torsion invariant for n-tuples of commuting elements to get a homotopy invariant for a compact oriented Riemannian manifold and a torsion-index for elliptic operators.

Let $\mathcal{A}$ be a finite von Neumann algebra. Suppose $P \in M_k(\mathcal{A})$ is a projection. Let

$$\mathcal{A}_n(P) = \{T(P) = (T_1, \ldots, T_n) : T_i T_j = T_j T_i,$$

$$\mathrm{T}_i = (\alpha_{kl}^{(i)} + \beta_{kl}^{(i)} P)_{k,l=1}^2, \alpha_{12}^{(i)} = \alpha_{21}^{(i)} = 0, \}, n > 1$$

and let

$$\mathcal{A}_1(P) = \{(\beta_{kl}^{(1)} P)_{k,l=1}^2, \beta_{kl}^{(1)} \in C\}.$$

The Laplace operator $\Delta_j(T(P))$ associated with $T(P)$ is a matrix with entries given by $\alpha_{kl} + \beta_{kl} P$. Hence $\Delta_j(T(P)) = (a_{kl}^{(j)})P + (b_{kl}^{(j)})(1 - P)$, where $(a_{kl}^{(j)})$ and $(b_{kl}^{(j)})$ are matrices with $a_{kl}^{(j)}$ and $b_{kl}^{(j)}$ in C. We have $\Delta_j'(T(P)) = (a_{kl}^{(j)})|_{(Ker(a_{kl}^{(j)}))^\perp} P \oplus (b_{kl}^{(j)})|_{(Ker(b_{kl}^{(j)}))^\perp} (1-P)$. We thus reduce the problem essentially to the finite dimensional case. This implies that $T(P) \in \mathcal{D}_n(\mathcal{A}), \forall T(P) \in \mathcal{A}_n(P)$. The torsion $\log \tau(T(P))$ is thus well defined.

Theorem 7.1: $\log \tau(T(P))$ defines a map from the K-theory group $K_0(\mathcal{A})$ to C.

Proof: We refer to [Bl] for the definition of $K_0(\mathcal{A})$. If P_1 and P_2 are two unitarily equivalent projections in $M_k(\mathcal{A})$, i .e., $P_2 = UP_1U^*$ with $U \in M_k(\mathcal{A})$ unitary. Then $T(P_2) = UT(P_1)U^*$. By Remark 3.3(3), $\log \tau(T(P_2)) = \log \tau(T(P_1))$. Clearly, $T(P_1 \oplus P_2) = T(P_1) \oplus T(P_2)$. Hence, $\log \tau(T(P_1 \oplus P_2)) = \log \tau(T(P_1)) + \log \tau(T(P_2))$. To prove $\log \tau(T(P \oplus 0)) = \log \tau(T(P))$, it suffices to check $\log \tau(T(0)) = 0$. In fact, $T(0) = 0$ for $n = 1$ and $T(0)$ is an n-tuple of diagonal matrices. By Propo sition 4.6, we get $\log \tau(T(0)) = 0$.   Q.E.D.

We have thus defined a family $\log \tau_n(T(P))$ of functions of several variables given by $\log \tau_n(T(P)) = \log \tau(T(P))$ for $T(P) = (T_1, \ldots, T_n)$,

$n \geq 1$. Each T_i is a function of α_{ij}, β_{ij} and P. Note that for $\Delta_j(T(P)) = (a_{kl}^{(j)})P + (b_{kl}^{(j)})(1 - P)$,

$$\log \tau(T(P)) = \tau(P)\frac{1}{2}\sum_{j=0}^{n}(-1)^{j+1}j \log \Delta((a_{kl}^{(j)})|_{(Ker(a_{kl}^{(j)}))^{\perp}}),$$

$$(7.1)$$

since by $\log \tau(T(0)) = 0$,

$$\tau(1 - P)\frac{1}{2}\sum_{j=0}^{n}(-1)^{j+1}j \log \Delta((b_{kl}^{(j)})|_{(Ker(b_{kl}^{(j)}))^{\perp}}) = 0.$$

The most interesting case is perhaps when $\alpha_{kl}^{(i)}, \beta_{kl}^{(i)}$ are related to P (e.g., $\tau(P)$).

To consider the application of the above map to index theory, let us assume that E_0 is a finitely generated (projective) Hilbert $\mathcal{A}$-module. Suppose M is a compact Riemannian manifold of even dimension and E is an $\mathcal{A}$-bundle over M whose fiber is E_0. One can define elliptic operators with coefficients in E. For such an operator D, there is an analytic index $Ind_{\mathcal{A}}^a(D)$ and a topological index $Ind_{\mathcal{A}}^t(D)$ in $K_0(\mathcal{A})$ [MF].

Definition 7.2: The analytic (resp. topological) torsion-index of D is defined by

$$\tau_n - Ind_{\mathcal{A}}^a(D) = \log \tau_n(T(Ind_{\mathcal{A}}^a(D)))$$

$$(resp. \ \tau_n - Ind_{\mathcal{A}}^t(D) = \log \tau_n(T(Ind_{\mathcal{A}}^t(D)))).$$

By the Miscenko-Fomenko theorem [MF], these two torsion-indices are equal. We can also consider the torsion-index for a family D_X of elliptic operators parameterized by a compact topological space X [AS]. Indeed, the index $Ind(D_X)$ of D_X lies in $K_0(C(X))$. Using the inclusion map $J : C(X) \rightarrow L^{\infty}(X) = \mathcal{A}$, we can define a torsion-index of D_X as $\log \tau_n(T(J_*(Ind(D_X))))$.

Let us in particular consider the signature operator. Let $\tilde{M}$ be the universal covering space of M and let Γ be the fundamental group of

M. Assume $\mathcal{A}$ is the finite von Neumann algebra generated by $R(\Gamma)$ in $B(l^2(\Gamma))$, where R is the right regular representation of Γ on $l^2(\Gamma)$. Let $E = \mathcal{A} \times_\Gamma \tilde{M}$. E is a flat $\mathcal{A}$-bundle over M. Let D be the signature operator with coefficients in E and $P = Ind_{\mathcal{A}}(D)$. Define for even dimensional M,

$$\tau_n(M) = \sup_{T(P) \in \mathcal{A}_n(P), |\alpha_{kl}^{(i)}| \leq 1, |\beta_{kl}^{(i)}| \leq 1} \{\log \tau(T(P))\},$$

and if M is odd dimensional , then set $\tau_n(M) = \tau_n(M \times S^1)$.

Prosposition 7.3: $\tau(M)$ is a homotopy invariant of the compact oriented Riemannian manifold M.

Proof: By (7.1), $\tau(M)$ is well-defined. By the Hilsum-Skandalis theorem [HS] (see [Go] for the equivariant case), $Ind_{\mathcal{A}}(D)$ is a homotopy invariant. Hence $\tau_n(M)$ is also a homotopy invariant.   Q.E.D.

Finally, we discuss some open questions.

Remark 7.4: (1) We know that the analytic torsion of a compact Riemannian manifold is not in general a homotopy invariant. Here we have used the finite von Neumann algebra and signature element to produce a homotopy invariant. To compute it, one needs to know the torsion invariant for n-tuples of commuting elements in $M_k(C)$. One also wants to know an analytic formula for the torsion-index. One may further take the coefficients $\alpha_{kl}^{(i)}, \beta_{kl}^{(i)}$ of $T(P)$ in $L^\infty(X, \nu)$ for some compact second countable space X with bounded measure ν.

(2) There are two sequences of commuting projections in Jones' basic construction [Jo]. One would like to know the torsion invariants of commuting 2×2 matrices coming from these two sequences and its relation to knots.

(3) The relationship between the torsion invariants associated to an n-tuple and its sub-tuples is not yet clear. A partial result was obtained in [Go].

(4) It is important to study the torsion invariant for a noncommut-

ing tuple. The definition of an appropriate spectrum and functional calculus for noncommuting tuples is a fundamental problem in multivariable operator theory. One can sometimes handle the spectral problem. (Certain deformation results of Carey and Pincus allow one to pass from Koszul systems to complexes. See the remark in [Car-Pin].) But the appropriate definition of a functional calculus is in general lacking.

(5) Another important problem is to consider when the spectral invariants are finite. This is related to the Nishio-Sunada-Lott-Luck conjecture that the Novikov - Shubin invariants are positive [NiS] [LL]. In other words, if $\mathcal{A}$ is the von Neumann algebra generated by a finitely generated discrete group Γ and $T = (a_{ij})_{ij=1}^k$ with entries in $Z\Gamma$, then $\theta_j(T) < \infty, j = 0, 1$. We should point out that $\theta_j(T)$ are not finite in general even for an abelian von Neumann algebra.

References

[AS] M. Atiyah and I. Singer, The index of elliptic operators: IV, Ann. Math, 93 (197 1), 119-138.

[Bl] B. Blackadar, K-Theory for Operator Algebras, MSRI Publ, No. 5, Springer, NY, 1986.

[Br] L. Brown, Lidskii's theorem in the type II case, Geometric Method in Operator Algebras, 1-35, Ed H. Araki & E. Effros, Longman Scientific Technical, NY, 1986.

[AM] A. Carey and V. Mathai, L^2-torsion invariants, J. Funct. Anal. 107 (1992), 369-386.

[Dix] J. Dixmier, von Neumann Algebras, North-Holland, NY, 1981.

[Car-Pin] R. Carey and J. D. Pincus, Steinberg Symbols and Torsion,Preprint 1993

[Fa] T. Fack, Sur la notion de valeur caractéristique, J. Oper. Theory, 7 (1982), 307-333.

[ES] D. Efremov and M. Shubin, Spectrum distribution function and variational principle for automorphic operators on hyperbolic space, Seminaire Equations aux Derivees Partielles, École Polytechnique, Centre de mathematiques, 1988-1989 , Expose VIII, P.19.

[FK] B. Fuglede and R. Kadison, Determinant theory in finite factors, Ann. Math. 55 (1952), 520-530.

[Go] D. Gong, L^2-analytic torsions, equivariant cyclic cohomology and the Novikov conjecture, Ph. D. thesis, SUNY, StonyBrook 1992.

[GS] M. Gromov and M. Shubin, von Neumann spectra near zero, Geom. Funct. Anal. 1 (1991), 375-404.

[HS] M. Hilsum and G. Skandalis, Invariance par homotopie de la signature à coefficients dans un fibré presque plat, J. reine. angew. Math. 423 (1 992), 73-99.

[Jo] V. F. R. Jones, Index of subfactors, Invent. Math. 72 (1983), 1-25.

[Lo] J. Lott, Heat kernels on covering spaces and topological invariants, J. Diff. Geom. 35 (1992), 471-510.

[LL] J. Lott and W. Lück, L^2-topological invariants of 3-manifolds, Preprint, 1992.

[LR] W. Lück and M. Rothenberg, Reidemeister torsion and the K-theory of von Neumann algebras, K-theory, 5 (1991), 213-264.

[MF] A. S. Miscenko and A. Fomenko, The index of elliptic operators over C^*-algebras, Izv. Akad. Nauk. SSSR. Ser. Math. 43 (1979), 831-859.

[NS] S. Novikov and M. Shubin, Morse theory and von Neumann II_1-factors, Dokl. Akad. Nauk. SSSR. 289 (1986), 289-292.

[NiS] M. Nishio and T. Sunada, Trace formula in spectral geometry, Proc. IMC. Vol 1, 5 77-586, Kyoto, 1990.

[RS] D. Ray and I. Singer, R-torsion and the Laplacian on Riemannian manifolds, Adv . Math. 7 (1971), 145-210.

[Va] F. Vasilescu, Analytic Functional Calculus and Spectral Decomposition, Ed Academei & D. Reidel. Co. Bucharest and Dordrecht, 1982.

D. Gong
Department of Mathematics
University of Chicago, Chicago,Illinois

J. Pincus
Department of Mathematics
SUNY at Stony Brook, NY 11794

Contemporary Mathematics
Volume 185, 1995

Algebraic K-theory Invariants for Operator Theory

JEROME KAMINKER

ABSTRACT. We present a modern version of an invariant first introduced by Larry Brown in his study of the Universal Coefficient Theorem for Ext(X)[1].

1. Introduction

In this paper we will show how the invariant introduced by Larry Brown in [1] can be related to the Connes-Karoubi pairing [3]. This will be applied to the study of finitely summable extensions as in [5], [4], [9], [2]. If a finitely summable extension, τ satisfies $\mathrm{Ind}(\tau) = 0$, it may still not be trivial. One needs an additonal invariant to determine this. The universal coefficient theorem of L. G. Brown, [1], provides this, but the invariant used there–the short exact sequence obtained by applying K_0 to the extension–is difficult to determine explicitely. In [1] there was a second invariant described which involved algebraic K-theory and is the one we will elaborate on here. It has the advantage of being directly tied to secondary characteristic classes in the same way that Index is tied to the Chern character.

In the present note we will introduce the invariant and give some elementary properties. In a later paper we will discuss its relation to the original invariant of Brown as well as to spectral flow and the $\mathbb{Z}/k$ index theorem of Freed and Melrose [6].

1.1. Basic definitions. We will work with $p + 1$-summable Fredholm modules over a $*$-algebra $\mathcal{A}$, and the associated group $Ell^p(\mathcal{A})$ defined by Connes and Karoubi. We first recall some definitions from [3].

1991 *Mathematics Subject Classification.* Primary 46L87, 19K33; Secondary 19D55.
The author was supported in part by NSF Grant DMS-9104636.

DEFINITION 1.1. *An odd Fredholm module, $(F, \sigma, \mathcal{H})$, over the $*$-algebra $\mathcal{A}$ consists of a Hilbert space $\mathcal{H}$, a representation $\sigma : \mathcal{A} \to \mathcal{L}(\mathcal{H})$, and a self-adjoint operator F satisfying $F^2 = 1$ and $[F, \sigma(a)] \in \mathcal{K}(\mathcal{H})$ for all $a \in \mathcal{A}$.*

An even $p + 1$-summable Fredholm module, $(F, \sigma, \mathcal{H})$, consists of the same ingredients and in addition $\mathcal{H}$ is a graded Hilbert space, with grading operator ε, satisfying σ is a grading preserving homomorphism and $F\varepsilon = -\varepsilon F$

The Schatten ideal, $\mathcal{L}^p(\mathcal{H})$, is defined to be $\{T \in \mathcal{K}(\mathcal{H}) : Tr(|T|^p) < \infty\}$.

DEFINITION 1.2. *A Fredholm module is p-summable if $[F, \sigma(a)] \in \mathcal{L}^p(\mathcal{H})$ for all $a \in \mathcal{A}$*

Following [2] we will assume that $\mathcal{A}$ is isomorphic to a holomorphically closed subalgebra of $\mathcal{L}(\mathcal{H})$.

In certain circumstances an unbounded operator can yield a $p + 1$-summable Fredholm module. Let D be a closed, unbounded operator which satisfies

(i) $[D, \sigma(a)]$ is densely defined and extends to a bounded operator
(ii) the resolvants, $(D - \lambda)^{-1} \in \mathcal{L}^{p+1}$.

Given an operator D satisfying these conditions, one can show that the phase of D, (i.e. the partial isometry part of the polar decomposition of D) satisfies the conditions necessary to be a $p + 1$-summable Fredholm module. For all of this see [2].

In the odd case, a $p + 1$-summable Fredholm module naturally leads to a C^*-algebra extension. We recall how this occurs.

Let $\mathcal{A}$ be a C^*-algebra and let $(F, \sigma, \mathcal{H})$ be an odd Fredholm module over $\mathcal{A}$. Let $\mathcal{A}^\infty = \{a \in \mathcal{A} : [a, F] \in \mathcal{L}^{p+1}\}$ and assume that this is a dense subalgebra of $\mathcal{A}$. Then $(F, \sigma, \mathcal{H})$ is a $p + 1$-summable Fredholm module over the algebra $\mathcal{A}^\infty$. Moreover, $\mathcal{A}^\infty$ is closed under the holomorphic functional calculus.

Let $P = (I + F)/2$ be the non-negative spectral projection for F. Define the $*$-homomorphism $\tau : \mathcal{A} \to \mathcal{Q}(\mathcal{H})$ by $\tau(a) = \pi(PaP)$. Then, by a pull-back construction, τ induces an extension

$$(1.1) \qquad\qquad 0 \to \mathcal{K} \to \mathcal{E} \to \mathcal{A} \to 0.$$

Since the Fredholm module is $p + 1$-summable the map τ actually factors through $\mathcal{L}(\mathcal{H})/\mathcal{L}^{p+1}(\mathcal{H})$ and induces another extension of algebras

$$(1.2) \qquad\qquad 0 \to \mathcal{L}^{p+1} \to \tilde{\mathcal{E}} \to \mathcal{A}^\infty \to 0.$$

This yields the diagram

$$(1.3)$$

$$
\begin{array}{ccccccccc}
0 & \longrightarrow & \mathcal{K} & \longrightarrow & \mathcal{E} & \longrightarrow & \mathcal{A} & \longrightarrow & 0 \\
 & & \uparrow & & \uparrow & & \uparrow & & \\
0 & \longrightarrow & \mathcal{L}^{p+1} & \longrightarrow & \tilde{\mathcal{E}} & \longrightarrow & \mathcal{A}^\infty & \longrightarrow & 0.
\end{array}
$$

1.2. The group $Ell^p(\mathcal{A})$. Let us consider the set of $p + 1$-summable Fredholm modules over $\mathcal{A}$ which have the same parity as p. In [**3**] a natural notion of equivalence and direct sum is defined on this set. The set of equivalence classes can be made into a group which is denoted $Ell^p(\mathcal{A})$. For later use we recall that there is a "classifying algebra" for $Ell^p(\mathcal{A})$, denoted $\mathcal{M}^p$. It has the property that there is a canonical element $\tau_p \in Ell^p(\mathcal{M}^p)$ such that for any $\tau \in Ell^p(\mathcal{A})$, there is a homomorphism, $\rho \colon \mathcal{A} \to \mathcal{M}^p$ satisfying $\rho^*(\tau_p) = \tau$.

To each Fredholm module is associated an element of the K-homology group of the C^*-algebraic closure of $\mathcal{A}$, $\bar{\mathcal{A}}$. This establishes a map

$$Ell^p(\mathcal{A}) \to K^p(\mathcal{A})$$

and if $\tau \in Ell^p(\mathcal{A})$ we denote its image by $\bar{\tau} \in K^p(\bar{\mathcal{A}})$.

Connes has defined a Chern character $ch : Ell^p(\mathcal{A}) \to HC^p(\mathcal{A})$. If $\mathcal{A}$ is commutative then $ch \otimes \mathbb{C}$ is an isomorphism.

1.3. The Connes-Karoubi pairing. Our main interest is in a pairing which was defined in [**3**] between $Ell^p(\mathcal{A})$ and the algebraic K-theory group, $K^{alg}_{p+1}(\mathcal{A})$.

$$(1.4) \qquad\qquad Ell^p(\mathcal{A}) \times K^{alg}_{p+1}(\mathcal{A}) \to \mathbb{C}^\times.$$

To that end we recall informally several constructions. The group $K^{alg}_{p+1}(\mathcal{A})$, the *algebraic K-theory of* $\mathcal{A}$, is by definition $\pi_{p+1}(BGL(\mathcal{A})^+)$ where $GL(\mathcal{A}) = \lim GL_N(\mathcal{A})$, $GL_N(\mathcal{A})$ has the discrete topology, and $BGL(\mathcal{A})^+$ is the Quillen plus construction applied to the classifying space of $GL(\mathcal{A})$. There is a fibration

$$(1.5) \qquad\qquad F(\mathcal{A}) \to BGL(\mathcal{A})^+ \to BGL^{top}(\mathcal{A})$$

where $BGL^{top}(\mathcal{A})$ is the classifying space of $GL^{top}(\mathcal{A})$. Then $\pi_{p+1}(BGL(\mathcal{A})^{top}) = K^{top}_{p+1}(\mathcal{A})$ is the *topological K-theory* of $\mathcal{A}$. The *relative K-theory* of $\mathcal{A}$ is defined to be $K^{rel}_{p+1}(\mathcal{A}) = \pi_{p+1}(F(\mathcal{A}))$. The homotopy exact sequence of the fibration (1.5) yields the exact sequence

$$(1.6) \qquad\qquad \cdots \to K^{rel}_{p+1}(\mathcal{A}) \to K^{alg}_{p+1}(\mathcal{A}) \to K^{top}_{p+1}(\mathcal{A}) \to \cdots .$$

This can be compared with the sequence

$$(1.7) \qquad\qquad \cdots \xrightarrow{S} HC_p(\mathcal{A}) \xrightarrow{B} HH_{p+1}(\mathcal{A}) \xrightarrow{I} HC_{p+1}(\mathcal{A}) \to \cdots$$

via Chern character maps [**7**]. On $K^{top}_{p+1}(\mathcal{A})$ one uses the usual Chern character. On $K^{alg}_{p+1}(\mathcal{A})$ one uses the Chern character defined by Karoubi, [**7**]. One takes the relative Chern character $ch^{rel} \colon K^{rel}_{p+1}(\mathcal{A}) \to HC_p(\mathcal{A})$ on relative K-theory. It can be viewed as a secondary characteristic class map. There is a third exact sequence which can be inserted between (1.6) and (1.7), and one obtains a diagram we will have need of later

(1.8)

$$
\begin{array}{ccccccc}
\cdots \xrightarrow{\ \lambda\ } & K_{p+1}^{rel}(\mathcal{A}) & \xrightarrow{\ \mu\ } & K_{p+1}(\mathcal{A})) & \longrightarrow & K_{p+1}^{top}(\mathcal{A}) & \longrightarrow \cdots \\
& \downarrow{\scriptstyle Ch^{rel}} & & \downarrow{\scriptstyle Ch^{alg}} & & \downarrow{\scriptstyle =} & \\
\cdots \longrightarrow & HC_p(\mathcal{A}) & \longrightarrow & MK_{p+1}(\mathcal{A}) & \longrightarrow & K_{p+1}^{top}(\mathcal{A}) & \longrightarrow \cdots \\
& \downarrow{\scriptstyle =} & & \downarrow{\scriptstyle \kappa} & & \downarrow{\scriptstyle Ch^{top}} & \\
\cdots \xrightarrow{\ S\ } & HC_p(\mathcal{A}) & \xrightarrow{\ B\ } & HH_{p+1}(\mathcal{A}) & \xrightarrow{\ I\ } & HC_{p+1}(\mathcal{A}) & \longrightarrow \cdots .
\end{array}
$$

Here $MK_{p+1}(\mathcal{A})$ is Karoubi's multiplicative K-theory. For this and the diagram (1.8) we refer to [**7**], [**8**].

Now we are ready to discuss the definition of the pairing. Let $x \in K_{p+1}(\mathcal{A})$ and $\tau \in Ell^p(\mathcal{A})$. First suppose x lifts back to $x' \in K_{p+1}^{rel}(\mathcal{A})$. Then one can pair $Ch^{rel}(x') \in HC_p(\mathcal{A})$ and $ch(\tau) \in HC^p(\mathcal{A})$. The result will be a complex number

$$< Ch^{rel}(x'), ch(\tau) > \in \mathbb{C}.$$

It may depend on the choice of the lift x', but the image in $\mathbb{C}/((2\pi i)^{[\frac{p}{2}]+1})\mathbb{Z}$ does not. Thus, one may define

(1.9) $$ck(\tau)(x) = < x, \tau > = \overline{< Ch^{rel}(x'), ch(\tau) >} \in \mathbb{C}^{\times}.$$

Here we have identified $\mathbb{C}/((2\pi i)^{[\frac{p}{2}]+1})\mathbb{Z}$ with $\mathbb{C}^{\times}$, the non-zero complex numbers.

If x does not lift back, then one uses the algebra $\mathcal{M}^p$. For any $\tau \in Ell^p(\mathcal{A})$ let $\rho \colon \mathcal{A} \to \mathcal{M}^p$ be the associated homomorphism. Each element in $K_{p+1}^{alg}(\mathcal{M}^p)$, such as $\rho_*(x)$, lifts back so one may define

$$ck(\tau)(x) = < \rho_*(x), \tau_p > .$$

This definition is consistent with the former case. It is useful to view this in terms of the following diagram.

(1.10)

$$
\begin{array}{ccccccccc}
\longrightarrow & K_{p+2}^{top}(\mathcal{A}) & \longrightarrow & K_{p+1}^{rel}(\mathcal{A}) & \longrightarrow & K_{p+1}(\mathcal{A}) & \xrightarrow{\ \pi\ } & \\
& \downarrow{\scriptstyle \mathrm{Ind}(\tau)} & & \downarrow{\scriptstyle ck^{rel}(\tau)} & & \downarrow{\scriptstyle ck(\tau)} & & \\
0 \longrightarrow & \mathbb{Z} & \longrightarrow & \mathbb{C} & \longrightarrow & \mathbb{C}^{\times} & \longrightarrow & 0.
\end{array}
$$

where $ck(\tau)$ is obtained as above and $ck^{rel}(\tau)$ is defined via

(1.11) $$ck^{rel}(\tau)(x') = < Ch^{rel}(x'), ch(\tau) > .$$

We will refer to this as the *Connes-Karoubi pairing*, but also view it as a homomorphism

(1.12) $$ck(\tau) \colon K_{p+1}^{alg}(\mathcal{A}) \to \mathbb{C}^{\times}$$

For $p = 1$ and $p = 2$ there are interesting interpretations of the pairings in terms of determinants and central extensions of loop groups. [**3**]

2. The definition of the invariant

In this section we will define the invariant generalizing Brown's.

2.1. $p+1$-summable Fredholm modules with trivial index. We first recall the universal coefficient theorem.

$$(2.1) \qquad 0 \to \mathrm{Ext}(K^{top}_{p+1}(\mathcal{A}), \mathbb{Z}) \to K^p(\bar{\mathcal{A}}) \xrightarrow{Ind} \hom(K^{top}_p(\mathcal{A}), \mathbb{Z}) \to 0.$$

If $\tau \in Ell^p(\mathcal{A})$ then its image in $K^p(\bar{\mathcal{A}})$ will be denoted $\bar{\tau}$. We say that $\tau \in Ell^p(\mathcal{A})$ has *trivial index* if it satisfies $Ind(\bar{\tau}) = 0$. It follows from this that $ch(\bar{\tau}) = 0$ and, hence, that $\bar{\tau}$ is of finite order. The main goal here is to describe the corresponding element of $\mathrm{Ext}(K^{top}_{p+1}(\mathcal{A}), \mathbb{Z})$ using the Connes-Karoubi pairing. Note that this can give an effective way to study if an element of $K^1(\bar{\mathcal{A}})$ is trivial.

2.2. Algebraic preliminaries. Let $\tau \in Ell^p(\mathcal{A})$. In this section we will set up the diagrams to define our invariant. The goal is to extract the torsion information in τ. First, we will need to refer to the diagram (1.4) which defines the Connes-Karoubi pairing.

$$(2.2)$$

$$
\begin{array}{ccccccc}
\cdots \longrightarrow & K^{top}_{p+2}(\mathcal{A}) & \xrightarrow{\lambda} & K^{rel}_{p+1}(\mathcal{A}) & \xrightarrow{\mu} & K^{alg}_{p+1}(\mathcal{A}) & \longrightarrow \cdots \\
& \downarrow{Ind(\bar{\tau})} & & \downarrow{ck^{rel}(\tau)} & & \downarrow{ck(\tau)} & \\
0 \longrightarrow & \mathbb{Z} & \xrightarrow{\alpha} & \mathbb{C} & \xrightarrow{\beta} & \mathbb{C}^\times & \longrightarrow 0.
\end{array}
$$

where $ck^{rel}(\tau)$ is defined in (1.11).

We next set up an algebraic problem whose obstruction to solution will be the required invariant.

PROPOSITION 2.1. *If $\tau \in Ell^p(\mathcal{A})$, with $Ind(\bar{\tau}) = 0$, then there exists a homomorphism $\varphi \colon K^{alg}_{p+1}(\mathcal{A}) \to \mathbb{C}$ such that $\varphi\mu = ck^{rel}(\tau)$.*

PROOF. One first finds a homomorphism $\tilde{\varphi} \colon \mathrm{im}\,\mu \to \mathbb{C}$ with $\tilde{\varphi}\mu = ck^{rel}(\tau)$. For φ to exist it is necessary that $ck^{rel}(\tau)$ descend to $\mathrm{im}\,\mu \cong K^{rel}_{p+1}(\mathcal{A})/\mathrm{im}\,\lambda$. This requires $ck^{rel}(\tau)$ to vanish on $\mathrm{im}\,\lambda$. But $ck^{rel}(\tau)\lambda = \alpha\,Ind(\bar{\tau}) = 0$, so this is satisfied and $\tilde{\varphi}$ is defined. Thus, we have

$$(2.3)$$

$$
\begin{array}{ccc}
0 \longrightarrow \mathrm{im}\,\mu & \longrightarrow & K^{alg}_{p+1}(\mathcal{A}) \\
\downarrow{\tilde{\varphi}} & & \\
\mathbb{C} & &
\end{array}
$$

and since $\mathbb{C}$ is divisible, there is a homomorphism $\varphi \colon K^{alg}_{p+1}(\mathcal{A}) \to \mathbb{C}$, extending $\tilde{\varphi}$, and satisfying $\varphi\mu = ck^{rel}(\tau)$. $\square$

Now, the map φ extending $ck^{rel}(\tau)$ is not unique. In particular, the one chosen need not satisfy $\alpha\varphi = ck(\tau)$. In fact, there is an obstruction to choosing φ to satisfy this condition. We formulate this in a general way.

PROPOSITION 2.2. *Consider the diagram*

$$
\begin{array}{ccccccccc}
0 & \longrightarrow & D & \overset{i}{\longrightarrow} & B & \overset{j}{\longrightarrow} & A & \longrightarrow & 0 \\
& & & & \downarrow{\scriptstyle g} & & \downarrow{\scriptstyle h} & & \\
0 & \longrightarrow & \mathbb{Z} & \overset{\alpha}{\longrightarrow} & \mathbb{C} & \overset{\beta}{\longrightarrow} & \mathbb{C}^{\times} & \longrightarrow & 0
\end{array}
$$

(2.4)

There is an obstruction, $\mathrm{ob}(g,h) \in \mathrm{Ext}(A,\mathbb{Z})$ with the property that a homomorphism $\varphi\colon B \to \mathbb{C}$ exists with $\beta\varphi = h$ and $\varphi i = h$ if and only if $\mathrm{ob}(g,h) = 0$.

PROOF. This is essentially a diagram chase. Using the fact that $\mathbb{C}$ is divisible one first chooses an arbitrary $\varphi\colon B \to \mathbb{C}$ satisfying $\varphi i = g$. We want to modify φ so that $\beta\varphi = h$ also holds. There is a $\psi\colon A \to \mathbb{C}^{\times}$ with $\psi j = h - \beta\varphi$. Indeed, define $\psi(x) = h(x') - \beta\varphi(x')$ where x' is any element of B with $j(x') = x$. There is a canonical map $p\colon \hom(A,\mathbb{C}^{\times}) \to \mathrm{Ext}(A,\mathbb{Z})$ and we define

$$\mathrm{ob}(g,h) = p(\psi).$$

We must show that $\mathrm{ob}(g,h)$ is independent of the choices made and satisfies the necessary conditions. The former is direct and we will show the latter. Clearly, if φ can be chosen so that $h = \beta\varphi$, then $\mathrm{ob}(g,h) = 0$. Conversely, if $\mathrm{ob}(g,h) = 0$, then from the sequence

$$
(2.5) \qquad \cdots \to \hom(A,\mathbb{C}) \overset{\beta_*}{\longrightarrow} \hom(A,\mathbb{C}^{\times}) \overset{p}{\longrightarrow} \mathrm{Ext}(A,\mathbb{Z}) \to 0
$$

we see that there is a $\hat{\psi}\colon A \to \mathbb{C}$ such that $\beta\hat{\psi} = \psi$. Let $\hat{\varphi} = \varphi - \hat{\psi}j$. Then $h = \beta\hat{\varphi}$ and $\hat{\varphi}i = g$ $\square$

We apply this to

$$
\begin{array}{ccccccccc}
0 & \longrightarrow & \mathrm{im}\,\mu & \longrightarrow & K_{p+1}^{alg}(\mathcal{A}) & \longrightarrow & \mathrm{im}\,\pi & \longrightarrow & 0 \\
& & & & \downarrow{\scriptstyle ck^{rel}(\tau)} & & \downarrow{\scriptstyle ck(\tau} & & \\
0 & \longrightarrow & \mathbb{Z} & \overset{\alpha}{\longrightarrow} & \mathbb{C} & \overset{\beta}{\longrightarrow} & \mathbb{C}^{\times} & \longrightarrow & 0
\end{array}
$$

(2.6)

where $\pi\colon K_{p+1}^{alg}(\mathcal{A}) \to K_{p+1}^{top}(\mathcal{A})$ and $ck^{rel}(\tau)$ was shown to be defined on $\mathrm{im}\,\mu$ in Proposition 2.2. This yields the element

$$\mathrm{ob}(ck^{rel}(\tau), ck(\tau)) \in \mathrm{Ext}(\mathrm{im}\,\pi, \mathbb{Z}).$$

To complete the definition of the invariant we must relate $\mathrm{Ext}(\mathrm{im}\,\pi, \mathbb{Z})$ to $\mathrm{Ext}(K_0^{\langle}(\mathcal{A})\mathcal{A}), \mathbb{Z})$. For this we bring in the middle row of diagram (1.8). The first observation is that the map $ck(\tau)\colon K_{p+1}^{alg}(\mathcal{A}) \to \mathbb{C}^{\times}$ actually factors through $MK_{p+1}(\mathcal{A})$.

PROPOSITION 2.3. *There is a homomorphism* $\tilde{ck}\colon MK_{p+1}(\mathcal{A}) \to \mathbb{C}^\times$ *so that the following diagram commutes.*

(2.7)

$$
\begin{array}{ccc}
K_{p+1}^{alg}(\mathcal{A}) & \xrightarrow{\;ck\;} & \mathbb{C}^\times \\
\;\;\downarrow{\scriptstyle Ch^{alg}} & & \| \\
MK_{p+1}(\mathcal{A}) & \xrightarrow{\;\tilde{ck}\;} & \mathbb{C}^\times
\end{array}
$$

PROOF. We will define $\tilde{ck}$ in the same way that ck was defined earlier, using diagram (1.8). Thus, $\tilde{ck}(\tau)(z) = \overline{< z', ch(\tau) >} \in \mathbb{C}^\times$ where $z' \in HC_p(\mathcal{A})$ is a lift of $z \in MK_{p+1}(\mathcal{A})$. As before, one uses the algebra $\mathcal{M}^p$ to carry out the lifting if necessary. This is well-defined by a similar argument as before. But one now has that $\tilde{ck}(\tau)(Ch^{alg}(x)) = < x'', ch(\tau) >$ where x'' lifts $Ch^{alg}(x)$. Since $Ch^{rel}(x')$ can be taken to be x'' the result follows. $\square$

This shows that the invariant could have been obtained from the diagram

(2.8)

$$
\begin{array}{ccccccccc}
\cdots & \longrightarrow & K_{p+2}^{top}(\mathcal{A}) & \xrightarrow{\;Ch\;} & HC_p(\mathcal{A}) & \xrightarrow{\;\hat{\mu}\;} & MK_{p+1}(\mathcal{A}) & \xrightarrow{\;\hat{\pi}\;} & \cdots \\
& & \downarrow{\scriptstyle Ind(\tau)} & & \downarrow{\scriptstyle <ch(\tau),\,>} & & \downarrow{\scriptstyle \tilde{ck}(\tau)} & & \\
0 & \longrightarrow & \mathbb{Z} & \longrightarrow & \mathbb{C} & \longrightarrow & \mathbb{C}^\times & \longrightarrow & 0
\end{array}
$$

rather than (1.4). Thus, $ob(ck^{rel}(\tau), ck(\tau)) \in \mathrm{Ext}(\mathrm{im}\,\hat{\pi}, \mathbb{Z})$. Our next goal is to identify $\mathrm{Ext}(\mathrm{im}\,\hat{\pi}, \mathbb{Z})$.

PROPOSITION 2.4. $\mathrm{Ext}(\mathrm{im}\,\hat{\pi}, \mathbb{Z}) \cong \mathrm{Ext}(K_{p+1}^{top}(\mathcal{A}), \mathbb{Z})$

PROOF. Consider the exact sequence

$$0 \to \ker \hat{\pi} \to MK_{p+1}(\mathcal{A}) \to \mathrm{im}\,\hat{\pi} \to 0.$$

From (2.8) one sees that $\mathrm{im}\,\hat{\pi} = \ker\mathrm{Ch}$ and this contains the torsion subgroup of $K_{p+1}^{top}(\mathcal{A})$. This implies that the two Ext groups are isomorphic. $\square$

Finally, to define our invariant we use Bott periodicity to identify $K_{p+1}^{top}(\mathcal{A}) \cong K_0^{top}(\mathcal{A})$.

DEFINITION 2.1. *Let* $[\tau] \in Ell^p(\mathcal{A})$ *satisfy* $Ind(\bar{\tau}) = 0$. *Then* $\tilde{ck}_p(\tau) \in \mathrm{Ext}(K_0^{top}(\mathcal{A}), \mathbb{Z})$ *is defined to be the image under Bott periodicity of*

$$ob(ck(\tau), ck^{rel}(\tau)).$$

It is easy to check the following naturality property of the invariant.

PROPOSITION 2.5. *Let* $f\colon \mathcal{A} \to \mathcal{B}$ *be a homomorphism. Let* $\tau \in Ell^p(\mathcal{A})$. *Then* $f_*(\tau) \in Ell^p(\mathcal{B})$ *and* $\tilde{ck}(f_*(\tau)) = f^*(\tilde{ck}(\tau))$.

REMARK 2.1. If $\mathcal{A} = C^\infty(M)$ and if $\tau \in Ell^p(\mathcal{A})$, then $ch(\tau) \in HC^p(\mathcal{A}) = Z_{DR}^p(M) \oplus H_{DR}^{p-2}(M) \oplus \cdots$. If $\mathrm{Ind}(\tau) = 0$, then $ch(\tau)$ lies in the component $Z_{DR}^p(M)$. Thus, our invariant is actually determined by only this unstable piece of the relative Chern character. This gives a fairly precise hold on the invariant.

REFERENCES

1. L.G. Brown, *Operator algebras and algebraic K-theory*, Bull. Amer. Math. Soc., (1975), 1119-1121
2. A. Connes, *Non-commutative Differential Geometry*, Academic Press, 1994
3. A. Connes and M. Karoubi, *Caractère multiplicative d'un module de Fredholm*, K-theory, 2(1988), 431-463
4. R. G. Douglas, *On the smoothness of elements of Ext*, 1980
5. R. G. Douglas and D. Voiculescu, *On the smoothness of sphere extensions* J. Operator Theory, 6(1981), 103-115
6. D.S. Freed and R.B. Melrose, *A mod k index theorem*, Invent. Math., 107(1992), 283-299
7. M. Karoubi, *Homologie Cyclique et K-Théorie*, Asterisque, 149(1987)
8. M. Karoubi, *K-théory multiplicative et homologie cyclique* C. R. Acad. Sci. Paris Sér. A-B, 303(1986), 507-510
9. X. Wang, *Smooth K-homology, Connes-Chern character and a non-commutative Sobolev lemma*, 1993

DEPARTMENT OF MATHEMATICAL SCIENCES, IUPUI, INDIANAPOLIS, INDIANA 46202-3216
E-mail address: kaminker@math.iupui.edu

Contemporary Mathematics
Volume **185**, 1995

Fundamentals of Harmonic Analysis on Domains in Complex Space

STEVEN G. KRANTZ

ABSTRACT. A survey of Hardy space theory in both one and several complex variables is provided. Special emphasis is placed on the real variable point of view, particularly the interaction with the theory of singular integrals. The paper culminates with a discussion of Hardy spaces of several complex variables, emphasizing the geometric point of view in the subject.

0. Prehistory

In the days of Euler and Bernoulli, the subject that we now call Fourier analysis concerned itself with the summing of certain series. Indeed the fundamental question was whether any given function (chosen from some reasonable class) can be represented as a convergent Fourier series. Bear in mind that the notion of convergence of a series had, in the late eighteenth century, not yet been given a precise definition (this was later done by Dirichlet, around 1815). And it would not be known for quite some time that usually Taylor series don't converge, and even when they do it is often *not* to the function that generated the series.

Amusingly, Euler argued that the initial data for a vibrating string could be something like

$$f(x) = \begin{cases} x & \text{if } 0 \le x \le 1 \\ 1 - \ln x / \ln 2 & \text{if } 1 < x \le 2. \end{cases}$$

Considered as a physical problem, a string in this configuration, when released, would certainly vibrate in some predictable Newtonian fashion. However a Fourier series, Euler reasoned, is a sum of functions $\sin kx$ and $\cos kx$, and these

1991 *Mathematics Subject Classification.* Primary 31B20, 31B99; Secondary 32A35, 42B20.

Key words and phrases. Fourier analysis, Hardy space, bounded mean oscillation, singular integral, space of homogeneous type.

The author was supported in part by grants #DMS8800523 and #DMS9101104 from the National Science Foundation.

This is the final version of this paper; it will be not be published elsewhere.

are *single functions.* Our initial data function f, by contrast, is the concatenation of *two* functions. So f cannot have a series representation in sines and cosines (see [**LAN**] for an extensive treatment of both Euler's and Bernoulli's ideas).

Fourier himself, in the early nineteenth century, gave an algorithm for calculating the Fourier coefficients of any given function. Although his reasoning was specious, he came up with the correct formulas for Fourier coefficients that we use today. His classic *Treatise on the Theory of Heat* presented Fourier's derivation, together with several chapters each carefully computing the series development of a specific function.

In the early twentieth century the point of view in the subject of Fourier series changed. Let us sketch why. Suppose that to any L^1 function f on $[0, 2\pi)$ we associate the Fourier coefficients

$$c_j = \frac{1}{2\pi} \int_0^{2\pi} f(t) e^{-int} \, dt$$

and the series

$$S = \sum_{j=-\infty}^{\infty} c_j e^{ij\theta}.$$

At first the series S is just a formal object. It may or may not converge, and its limit may or may not be f. We write $f \sim S$.

Define the partial sums of S by

$$S_N f(\theta) = \sum_{j=-N}^{N} c_j e^{ij\theta}.$$

The series is said to converge to f if $S_N f \to f$. [It is a curious but not well known fact that if the partial sums are defined in virtually any fashion, say $\tilde{S}_N = \sum_{j=-\sqrt{N}}^{\ln(N+3)}$, then an equivalent theory results. This assertion may be derived from the considerations that follow.] Let us consider convergence in the L^p topology.

Clearly if f is a trigonometric polynomial (i.e. a finite linear combination of exponentials) then $S_N f \to f$ in any topology; for the S_N's stabilize on f. An elementary approximation argument (using the Uniform Boundedness Principle) then shows that $S_N f \to f$ for all f in L^p if and only if

$$\|S_N\|_{(L^p, L^p)} \leq C.$$

Define $\tau_k g(\theta) = e^{ik\theta} g(\theta)$. Then the operator $T_N = \tau_N \circ S_N \circ \tau_{-N}$ sums the Fourier series from index 0 to $2N$. The operators τ_N have norm 1 in any rearrangement invariant topology, so we see that $S_N f \to f$ for all f in L^p if and only if

$$\|T_N\|_{(L^p, L^p)} \leq C.$$

But now a simple limiting argument shows that $S_N f \to f$ for all f in L^p if and only if

$$\|T\|_{(L^p, L^p)} \leq C,$$

where T is the operator that assigns to f the sum of the "half series"

$$\sum_{j=0}^{\infty} c_j e^{ij\theta}.$$

The remarkable fact here is that the problem of summing a Fourier series has been reduced to determining the boundedness of *just one operator*. In fact this operator can be calculated: formally, the operator T consists of convolution with the kernel

$$K(\theta) = \sum_{j=0}^{\infty} e^{ij\theta},$$

and this last is just a geometric series. The result is that the operator T is essentially

$$Tf = f * \cot(\theta/2). \tag{$*$}$$

The difference $\cot \theta/2 - 2/\theta$ is absolutely integrable, so studying $(*)$ is, modulo operators that are trivial to estimate, equivalent to studying the operator

$$Hf = f * (2/\theta).$$

The convolutions that we have discussed up to this point have been convolutions on $[0, 2\pi)$, or on the circle group. However, using a process of Poisson summation, the problem can be transferred to the real line. The resulting problem, after some normalizations, is to study the operator

$$\mathcal{H} : f \mapsto \int_{\mathbb{R}} \frac{f(t)}{x - t} \, dt.$$

To repeat, the L^p summability of Fourier series is equivalent to the L^p boundedness of $\mathcal{H}$.

The operator $\mathcal{H}$ is arguably the most important operator of classical analysis. It is called the *Hilbert transform*, and is the only singular integral operator (in the sense of Calderón and Zygmund) in dimension one.

In general, a C-Z singular integral operator is an operator on functions on $\mathbb{R}^N$ given by convolution with a kernel $K(x) = \Omega(x)/|x|^N$ such that Ω is smooth away from 0 and

(1) $\Omega : \mathbb{R}^N \setminus \{0\} \to \mathbb{R}$ is homogeneous of degree zero;

(2) $\displaystyle\int_{\Sigma} \Omega(x) \, d\sigma(x) = 0$

In the last condition, Σ denotes the unit sphere in $\mathbb{R}^N$ and $d\sigma$ is rotationally invariant area measure on the sphere. In this article we will treat only translation invariant C-Z operators; for a more general treatment, we refer the reader to [**CHR**].

We are abusing language a bit to speak of convolving with such a kernel K, because K is not absolutely integrable. Indeed, singular integrals must be interpreted in the *Cauchy principal value* sense:

$$T_K f(x) = K * f(x) = \lim_{\epsilon \to 0^+} \int_{|x-t|>\epsilon} f(t) K(x-t) \, dV(t). \qquad (**)$$

Of course dV is Lebesgue volume measure. It can be shown that, for a kernel K homogeneous of degree $-N$, the equation $(**)$ can define a distribution testing functions f (evaluate $(**)$ at $x = 0$) only if $K(x) = \Omega(x)/|x|^N$ and Ω satisfies the mean value condition specified in (2) above.

Returning to the operator $\mathcal{H}$ on the space $\mathbb{R}$, we see that $1/t = (\mathrm{sgn}\, t)/|t|$, hence $\mathcal{H}$ is a Calderón-Zygmund kernel—indeed the unique one up to constant multiples on $\mathbb{R}^1$. Thus the problem of norm convergence of Fourier series has given birth to the theory of singular integrals.

1. Singular Integral Operators

Applying the operator $\mathcal{H}$ to a characteristic function $\chi_{[a,b]}$ of an interval, we find that the outcome is a logarithmic function. Thus $\mathcal{H}$ is not bounded on L^∞. Also $\mathcal{H}$ is, up to a minus sign, self-adjoint. It follows that $\mathcal{H}$ cannot be bounded on L^1.

Since the kernel of $\mathcal{H}$ is homogeneous of degree "minus the dimension," and since that kernel induces a distribution, we may consider the Fourier transform of the kernel and determine that it is homogeneous of degree zero. In particular it is bounded. Any L^∞ Fourier multiplier is bounded on L^2. We conclude that norm convergence for L^2 functions is valid.

M. Riesz [**RIE**] extended the result of the last paragraph to other L^p spaces by devising a special argument that works for *even p*. Thus $\mathcal{H}$ is bounded on L^2, L^4, L^6, etc. He then, in effect, invented the complex method of interpolation and applied it to prove that $\mathcal{H}$ is bounded on L^p, $2 \le p < \infty$. Self-adjointness of $\mathcal{H}$ and duality then implied boundedness for $1 < p < 2$. Thus norm convergence of Fourier series is valid in L^p, $1 < p < \infty$. Details of these arguments appear in [**KAT**].

The modern method of handling these matters, in all dimensions simultaneously, and for all C-Z operators, is as follows: The Fourier transform method for getting L^2 boundedness is still valid, for conditions (1) and (2) imply that K induces a distribution and its Fourier transform must be bounded. Then one uses delicate techniques of Calderón and Zygmund to see that T_K is bounded on L^1 in a certain weak sense:

$$m\{x \in \mathbb{R}^N : |T_K f(x)| > \lambda\} \le C \cdot \frac{\|f\|_{L^1}}{\lambda} \qquad \text{for all } \lambda > 0.$$

This is called a *weak type* (1,1) inequality. One can then interpolate between the (strong) L^2 result and the weak (1,1) result and obtain L^p boundedness of the

Calderón-Zygmund operator on L^p, $1 < p \leq 2$. The result for $2 < p < \infty$ then follows by duality.

The L^p boundedness of any Calderón-Zygmund operator does not speak directly to the norm convergence of multiple Fourier series. But it has important applications in partial differential equations and other parts of harmonic analysis.

2. Ideas of Hardy

It is desirable to find a natural substitute for the spaces L^1 and L^∞, substitutes on which the Calderón-Zygmund operators act naturally. We begin by recalling the *Hardy spaces*:

DEFINITION 2.1. Let f be holomorphic on the unit disc D and $0 < p < \infty$. We say that $f \in H^p(D)$ if

$$\sup_{0<r<1} \left[\int_0^{2\pi} |f(re^{it})|^p \, dt \right]^{1/p} \equiv \|f\|_{H^p(D)} < \infty.$$

We say that $f \in H^\infty(D)$ if f is bounded, and we endow H^∞ with the supremum norm.

The most fundamental result about Hardy spaces is the following:

THEOREM 2.2. *Let* $f \in H^p(D)$, $0 < p \leq \infty$. *Then for almost every* $\theta \in [0, 2\pi)$ *it holds that*

$$\lim_{r \to 1^-} f(re^{i\theta})$$

exists. We call the limit function f^*. *Plainly* f^* *is* p^{th} *power integrable.*

We shall not prove this result, but refer the reader to [**KRA1**] for a thorough discussion.

Now consider the case $p = 1$. If $f \in H^1$ then the boundary function f^* lies in L^1. Write $f^* = u^* + iv^*$. Then u^*, v^* are real valued L^1 functions. Moreover, the operator $\mathcal{H}$ that we derived from Fourier series considerations is also the operator that mediates between functions on the boundary of the disc that are boundary limits of harmonic functions that are harmonic conjugates. This is so, at least philosophically, because the Fourier series operation

$$\sum_{j=-\infty}^{\infty} c_j e^{ij\theta} \longmapsto \sum_{j=0}^{\infty} c_j e^{ij\theta}$$

corresponds to the operation on harmonic functions given by

$$\sum_{j=-\infty}^{-1} c_j \bar{z}^{|j|} + \sum_{j=0}^{\infty} c_j z^j \longmapsto \sum_{j=0}^{\infty} c_j z^j.$$

We leave details to the interested reader, or see [**KAT**].

In any event, v^* is essentially the image under $\mathcal{H}$ of u^*. Thus u^* is a special L^1 function—it is one that the Hilbert transform $\mathcal{H}$ acts on boundedly. This suggest a way of thinking about H^1 that makes no reference to complex analysis. Namely, we define

$$H^1_{\mathrm{Re}} = \{f \in L^1([0, 2\pi)) : \mathcal{H}f \in L^1\}.$$

We could make a similar definition for values of p other than $p = 1$.

Now it is virtually a tautology that $\mathcal{H}$ maps H^1_{Re} to itself, once one notices that $\mathcal{H} \circ \mathcal{H}$ is minus the identity. So we have in effect discovered a subspace of L^1, at least in one dimension, on which the singular integral $\mathcal{H}$ acts naturally.

G. H. Hardy proved the following theorem in the early part of this century:

THEOREM 2.3. *Let* $f \in H^1$. *Then the Fourier series* $\sum_j c_j e^{ij\theta}$ *of* f *satisfies* $\sum_{j \neq 0} |c_j|/|j| < \infty$.

Of course, by linearity over $\mathbb{R}$, the theorem applies to H^1_{Re} as well. Let us put this result in context. Traditionally, there was great interest in the class of functions with absolutely convergent Fourier series. Usually some sort of smoothness assumption on f is a standard hypothesis to guarantee that its Fourier series converge absolutely. It is known that absolute continuity is *not* sufficient to guarantee absolute convergence. However the theorem says that if the derivative of f is in $\mathcal{H}^1$ then the Fourier series of f converges absolutely.

Thus we see that H^1_{Re} is natural both from the point of view of the operator H and from the point of view of Fourier series.

3. Ideas of Stein and Weiss

In the early 1960's ([**STW1**], [**STE1**]), Stein and Weiss considered extending the ideas of the last section to several dimensions, independent of complex analysis or Fourier series considerations. For $j = 1, \ldots, N$ we consider the operators R_j having kernels

$$K_j = \frac{x_j}{|x|^{N+1}}$$

(here we are omitting certain constant coefficients that depend on the dimension and are of no interest for us). These are the *Riesz transforms*. Each R_j is a Calderón-Zygmund operator. The N-tuple $(R_1, \ldots, R_N)$ behaves naturally under dilations, rotations, and translations (the three groups that act naturally on Euclidean space). It is in fact uniquely determined by that property (see [**STE2**] for details).

The idea now is to define

$$H^1_{\mathrm{Re}}(\mathbb{R}^N) = \{f \in L^1 : R_j f \in L^1, j = 1, \ldots, N\}.$$

From the point of view of singular integral operators, H^1_{Re} turns out to be the right substitute for L^1, although this is not so easy to see. In particular we shall find that any Calderón-Zygmund operator is bounded on H^1_{Re}. In order to see this, we must develop some auxiliary machinery.

4. Maximal Functions

In the last section we rewrote history a bit. Stein and Weiss's original point of view was to consider so-called *Cauchy Riemann systems*. This is an $(N+1)$-tuple of functions $(u_0, u_1, \ldots, u_N)$ on $\mathbb{R}_+^{N+1} \equiv \{(x_0, x_1, \ldots, x_N) : x_0 > 0\}$ satisfying the relations

$$\frac{\partial u_j}{\partial x_k} = \frac{\partial u_k}{\partial x_j}, \qquad j, k = 0, \ldots, N,$$

and

$$\sum_{j=0}^{N} \frac{\partial u_j}{\partial x_j} = 0.$$

Note that when $N = 1$ these equations reduce to the classical Cauchy-Riemann equations on the upper half plane. Stein and Weiss required that each of the functions u_j be the Poisson integral of an L^1 function f_j on the boundary $\mathbb{R}^N$. Then it turns out that each f_j is the jth Riesz transform of f_0. Thus f_0 determines the entire $(N + 1)$-tuple with which we began. [Note here that we confine our remarks to the Hardy space H^1. This results in certain simplifications which are useful in exposition. But in fact there is an analogous theory for $0 < p < 1$.]

The modern point of view is to suppress the upper half space and work in the context of the intrinsic harmonic analysis of the boundary. However let us momentarily return to the original point of view. If $f \in L^1(\mathbb{R}^N)$ then let u be its Poisson integral to $\mathbb{R}_+^{N+1}$. Define the maximal function

$$f^+(x) = \sup_{y>0} |u(x, y)|, \qquad x \in \mathbb{R}^N.$$

In principle, f^+ could be identically infinity. However it turns out (see [**BGS**] for the one dimensional result and [**FS**] for the N-dimensional result) that f is in H_{Re}^1 if and only if f^+ lies in L^1.

In fact a stronger result is true. Fix a number $\alpha > 0$. Define, for $f \in L^1(\mathbb{R}^N)$,

$$f^*(x) = \sup_{\{(t,y):y>0, |x-t|<\alpha y\}} |u(t, y)|.$$

Obviously $f^*(x) \geq f^+(x)$. Yet (see [**FS**]) it turns out that $f \in H_{\text{Re}}^1$ if and only if $f^* \in L^1(\mathbb{R}^N)$.

If one wishes to use the real-variable Hardy space theory (that we have been describing) in other contexts, then it is desirable to eliminate the dependence on harmonic function theory, in particular on the Poisson integral. Thus the following result of Fefferman and Stein [**FS**] was a real breakthrough:

THEOREM 4.1. *Fix a $\phi \in C_c^\infty(\mathbb{R}^N)$ and set $\phi_\epsilon(x) = \epsilon^{-N}\phi(x/\epsilon)$. Let $f \in L^1(\mathbb{R}^N)$. Define*

$$M_\phi f(x) = \sup_{\epsilon>0} |f * \phi_\epsilon(x)|.$$

Then $f \in H_{\text{Re}}^1$ if and only if $M_\phi f \in L^1$.

It is an elementary exercise to see that the Poisson kernel $P(x, y)$ for the upper half space is equal to (up to a constant multiple) $\Phi_y(x)$, where $\Phi(x) = 1/(1 + |x|^2)^{(N+1)/2}$. Of course Φ is not compactly supported, as in the theorem, but it decays nicely at infinity so suffices, in its role in defining f^+, for characterizing H^1_{Re}. The last theorem is the natural generalization of the f^+ characterization of the Hardy space.

Following the paradigm set by the characterization of Hardy spaces by f^*, Fefferman and Stein also proved:

THEOREM 4.2. *Let* $\alpha > 0$. *With* $f \in L^1$ *and* ϕ_ϵ *as above, define*

$$M_\phi^\alpha f(x) = \sup_{\{(t,\epsilon):\epsilon>0, |t-x|<\alpha\epsilon\}} |u(x,t)|.$$

Then $f \in H^1_{\mathrm{Re}}$ *if and only if* $M_\phi^\alpha \in L^1(\mathbb{R}^N)$.

The last two theorems spawned what is now known as the intrinsic real variable theory of Hardy spaces. Although we have not given a careful definition of $H^p_{\mathrm{Re}}, 0 < p < 1$, we note that H^p_{Re} can be characterized by the property that any of f^+ or f^* or $M_\phi f$ or $M_\phi^\alpha f$ be p^{th} power integrable. For $1 < p < \infty$, the Riesz transforms of an L^p function are always in L^p (by Calderón-Zygmund theory) and the maximal function of an L^p function is always L^p (just because the much simpler Hardy-Littlewood maximal function is). So we see that the Hardy spaces and the L^p spaces form a unified continuum, with a unified characterization, for $0 < p < \infty$.

It is possible to use (a slight variant of) the last characterization of Hardy spaces to prove the next remarkable result. First a definition, which we will state for all p.

DEFINITION 4.3. *Let* $0 < p \leq 1$. *Let* a *be a measurable function on* $\mathbb{R}^N$. *We say that* a *is a* p-*atom if*

(1) *The function* a *is supported in a ball* $B(P,r)$;
(2) *We have*

$$|a(x)| \leq \frac{1}{|B(P,r)|^{1/p}};$$

   [here $|B(P,r)|$ *denotes the Lebesgue volume measure of the ball].*
(3) *The function* a *satisfies the moment conditions*

$$\int a(x)x^\alpha \, dx = 0 \qquad \text{for all multi-indices} |\alpha| \leq \left[N\left(\frac{1}{p}-1\right)\right].$$

Now we have

THEOREM 4.4. *Let* $0 < p \leq 1$. *Let* f *be a* p^{th} *power integrable function on* $\mathbb{R}^N$. *Then* $f \in H^p$ *if and only if there are* p-*atoms* a_j *and coefficients* λ_j *with*

$$f = \sum_j \lambda_j a_j.$$

The sequence $\{\lambda_j\}$ can be taken to lie in ℓ^p, and its $\|\ \ \|_{\ell^p}$ norm to be comparable to the H^p_{Re} norm of f. In case $p = 1$ then the convergence of the series is in H^1 norm. For $p < 1$ the convergence is in the sense of distributions.

The upshot of this theorem is that one can define real variable Hardy spaces in any context where it makes sense to define balls, a measure, and moment conditions (for these three attributes are needed to define atoms). Before we consider that possibility, we need to introduce one more player into the discussion.

5. Bounded Mean Oscillation

In the early 1960's, Moser, De Giorgi, and others were considering certain quasi-linear partial differential equations in divergence form. Many of these equations arose from minimal surface considerations and had rough coefficients. As a result, a new technique needed to be developed for studying regularity of solutions. This was the Moser iteration technique, and it gave rise to spaces of functions defined by conditions of the following form:

Fix the dimension N. Let $\alpha > 0$. Let $\mathcal{P}_k$ denote the polynomials of degree not exceeding k. Set

$$\|f\| = \sup_Q \inf_{p \in \mathcal{P}_k} \frac{1}{|Q|^{1+\alpha}} \int_Q |f(x) - p(x)| \, dV(x).$$

Here Q is any cube with sides parallel to the coordinate axes. It turns out (see [**CAMP**], [**RA3**]) that a function with this norm finite is locally Lipschitz of some order. These spaces of functions are called Campanato-Morrey spaces.

Fix attention on the case $k = 0$. It became a matter of interest to consider what sort of functions arose when $\alpha \to 0^+$. Thus in 1961 John and Nirenberg [**JON**] gave the following definition:

DEFINITION 5.1. Let f be a locally integrable function on $\mathbb{R}^N$. We say that f is of bounded mean oscillation, and write $f \in BMO$, if

$$\|f\|_* = \sup_Q \inf_{c \in \mathbb{C}} \frac{1}{|Q|} \int_Q |f(x) - c| \, dV(x).$$

Note that $\|\ \ \|_*$ annihilates constant functions so that, properly speaking, this is a norm on the space (BMO/constants). This norm is delicate, in that it measures both size and some slight smoothness. For instance, in $\mathbb{R}^1$, $f(x) = \log|x|$ is BMO yet $g(x) = \operatorname{sgn} x \cdot \log|x|$ is not. On the other hand, any L^∞ function is BMO.

It is not difficult to see that any element of the dual of H^1_{Re} can be represented by integration against a BMO function. To verify this assertion, think of an H^1_{Re} function as an N-tuple $F = (f_0, f_1, \ldots, f_N) \in L^1 \times \cdots \times L^1$ with $f_j = R_j f_0$, $j = 1, \ldots, N$. If $\mu \in (H^1)^*$ then, by the Hahn-Banach theorem, it can be continued as a linear functional on *all* of $L^1 \times \cdots \times L^1$. Thus μ is represented by integration against some $(h_0, h_1, \ldots, h_N) \in L^\infty \times \cdots \times L^\infty$. We write

$$\mu F = \sum_{j=0}^{N} \int_{\mathbb{R}^N} h_j f_j$$

$$= \int_{\mathbb{R}^N} h_0 f_0 + \sum_{j=1}^{N} \int_{\mathbb{R}^N} h_j R_j f_0$$

$$= \int_{\mathbb{R}^N} h_0 f_0 - \sum_{j=1}^{N} \int_{\mathbb{R}^N} (R_j h_j) f_0.$$

But one can calculate that any singular integral operator maps L^∞ to BMO. Thus $k = f_0 - \sum_j R_j h_j$ is a BMO function. We have succeeded in representing μ by integration against a BMO function.

The converse direction is much more difficult and is known as Fefferman's theorem. Let $f \in H^1_{\mathrm{Re}}$ and $\phi \in BMO$. Identify each function with its Poisson extension to the half space. Consider the integral $\int_{\mathbb{R}^N} f\phi \, dV$. Pass to an integral on the half space using Green's theorem. The various terms that arise are controlled using (i) the maximal function characterizations of H^1, (ii) the fact that $x_0 \cdot |\nabla\phi|^2$ is a Carleson measure (see [**FS**]), and (iii) the fact that H^1_{Re} functions integrate nicely against Carleson measures. For more on this development, see [**HOR**], [**FS**].

Now that we know that $(H^1_{\mathrm{Re}})^* = BMO$, we can learn about mapping properties of H^1 by studying BMO. The space BMO is defined by a fairly simple geometric condition, and is relatively easier to treat than is H^1_{Re}. For example, any singular integral can be calculated to map BMO to BMO. Invoking duality, we then have

THEOREM 5.2. *Let T be a Calderón-Zygmund operator. Then T maps H^1_{Re} to H^1_{Re}.*

With greater effort, one can show that a Calderón-Zygmund operator maps H^p_{Re} to H^p_{Re}, $0 < p < 1$. The result for $1 < p < \infty$ is just the classical Calderón-Zygmund theorem. We summarize our results as follows:

THEOREM 5.3. *Let T be a Calderón-Zygmund operator. Then T maps H^p_{Re} to H^p_{Re}, $0 < p < \infty$. It also maps BMO to BMO.*

Next we turn to the development of all these ideas in more general settings.

6. Spaces of Homogeneous Type

The notion of a space of homogeneous type finds its genesis in ideas of Hörmander [**HOR**] and K. T. Smith [**SMI**]. More recent variants are due to Christ [**CHR**], Krantz/Li [**KRL3**], and others. But the idea was crystalized and developed into a decisive tool of harmonic analysis by Coifman and Weiss ([**COW1**], [**COW2**]).

DEFINITION 6.1. Let $(X, d\mu)$ be a measure space and a topological space and let
$\{B(x, r)\}_{x \in X, r \in \mathbb{R}^+}$ be a family of distinguished (and usually, but not always, open) subsets of X which we call *balls*. We say that these constitute a *space of homogeneous type* if

(1) Each ball satisfies $0 < \mu(B(x, r)) < \infty$.
(2) If $r < s$ and $x \in X$ then $B(x, r) \subseteq B(x, s)$.
(3) There is a constant $C_1 > 0$ such that $\mu(B(x, 2r)) \leq C_1 \mu(B(x, r))$ for all balls $B(x, r)$.
(4) There is a constant $C_2 > 0$ such that if $B(x, r) \cap B(y, s) \neq \emptyset$ and if $s \geq r$ then $B(y, C_2 s) \supseteq B(x, r)$.

It is easy to see that Euclidean space, equipped with the balls arising from the usual metric and with Lebesgue measure, is a space of homogeneous type. But also, for example, the boundary of a smoothly bounded domain in $\mathbb{R}^N$, equipped with the metric inherited from Euclidean space (and the corresponding balls) and endowed with $(N-1)$-dimensional surface measure, is a space of homogeneous type. The constant C_1 will depend on the curvatures of the boundary; the constant C_2 depends only on the metric and can therefore be taken to be 3, just as in the Euclidean case. It is proved in [**MAS**] that, even when the natural structure does not induce balls that are open, there is an equivalent structure that produces open balls; thus there is no loss of generality to assume that the balls in our space of homogeneous type are open.

One of the main motivations for defining spaces of homogeneous type is that they have the minimal structure necessary for proving a covering lemma and the Hardy-Littlewood maximal theorem. We begin with the covering lemma:

LEMMA 6.2. *Let the topological space X, with measure $d\mu$ and balls $B(x, r)$, be a space of homogeneous type. Let K be a compact set of X. Suppose that the balls $\{B(x_\alpha, r_\alpha)\}_{\alpha \in A}$ cover K. Then there is a pairwise disjoint subcollection $\{B(x_{\alpha_1}, r_{\alpha_1}), \ldots, B(x_{\alpha_k}, r_{\alpha_k})\}$ such that the dilated balls $\{B(x_{\alpha_1}, C_2 r_{\alpha_1}), \ldots, B(x_{\alpha_k}, C_2 r_{\alpha_k})\}$ still cover K.*

PROOF. Since K is compact, we may at the outset suppose that we are working with a finite collection $\{B(x_\ell, r_\ell)\}_{\ell=1}^m$. Now select $B(x_{\ell_1}, r_{\ell_1})$ to be the (possibly non-unique) ball among these with greatest radius.

Select $B(x_{\ell_2}, r_{\ell_2})$ to be, of the remaining balls disjoint from $B(x_{\ell_1}, r_{\ell_1})$, that which has greatest radius. Then the third ball is chosen to be the largest remaining ball that is disjoint from the first two. And so on. Continue in this fashion.

Of course the process must stop. We claim that the selected balls $B(x_{\ell_1}, r_{\ell_1})$, $\ldots, B(x_{\ell_k}, r_{\ell_k})$ satisfy the desired conclusions.

The selected balls are clearly pairwise disjoint by construction. To see that their dilates by the factor of C_2 still cover K, it suffices to see that these dilates still cover each of the original balls $B(x_\ell, r_\ell)$. Pick one of the original balls—call it $B = B(x, r)$. If B is also one of the selected balls, then of course it is covered. If it is not a selected ball, then there is a first selected ball $B(x_{\ell_p}, r_{\ell_p})$ that intersects B. But, by the way we chose the selected balls, it must be that $r_{\ell_p} \geq r$. Therefore $B(x_{\ell_p}, C_2 r_{\ell_p}) \supseteq B$ by axiom (4) for a space of homogeneous type. That is what we wished to establish. $\square$

Now the Hardy-Littlewood theorem is this:

THEOREM 6.3. *Let the topological space X, with measure $d\mu$ and balls $B(x,r)$, be a space of homogeneous type. Assume that $d\mu$ is inner regular. If f is integrable on X then define*

$$Mf(x) = \sup_{r>0} \frac{1}{\mu(B(x,r))} \int_{B(x,r)} |f(t)| \, d\mu(t).$$

Then the operator M is weak type (1,1) and (strongly) bounded on L^p for $1 < p \leq \infty$.

PROOF. The operator M is trivially bounded on L^∞. If we can show that M is weak type (1,1) then the result follows from the Marcinkiewicz interpolation theorem (for which see [**STW2**]).

Therefore fix $\lambda > 0$. Let

$$S_\lambda = \{x \in X : Mf(x) > \lambda\}.$$

We wish to measure S_λ, and in order to do so it suffices to measure any compact subset K of S_λ. For each point $x \in K$ there is, by the definition of M, a number $r_x > 0$ such that

$$\frac{1}{\mu(B(x,r_x))} \int_{B(x,r_x)} |f(t)| \, d\mu(t) > \lambda.$$

In other words,

$$\frac{1}{\lambda} \int_{B(x,r_x)} |f(t)| \, d\mu(t) > \mu(B(x,r_x)).$$

Of course the balls $B(x, r_x)$ cover K. Use the covering lemma to extract a pairwise disjoint subcollection $B(x_1, r_{x_1}), \ldots, B(x_k, r_{x_k})$ whose dilates by a factor of C_2 still cover K. Then we have

$$\mu(K) \leq \mu\left(\bigcup_j B(x_j, C_2 r_{x_j})\right)$$

$$\leq \sum_j \mu\left(B(x_j, C_2 r_{x_j})\right)$$

$$\leq \sum_j (C_1)^{(\log_2 C_2)+1} \mu\left(B(x_j, r_{x_j})\right)$$

$$\leq (C_1)^{(\log_2 C_2)+1} \sum_j \frac{1}{\lambda} \int_{B(x_j, r_j)} |f(t)| \, d\mu(t)$$

$$\leq (C_1)^{(\log_2 C_2)+1} \frac{1}{\lambda} \int_X |f(t)| \, d\mu(t)$$

$$\equiv C' \frac{\|f\|_{L^1(X, d\mu)}}{\lambda}.$$

This is what we wished to establish. $\qquad\square$

The Hardy-Littlewood theorem is just what is needed to prove the Lebesgue differentiation theorem for integrals (see [**STE2**]). Coifman and Weiss proved that one can in fact study singular integral theory on a space of homogeneous type (see [**COW1**], [**COW2**]).

For our purposes, spaces of homogeneous type are the right context in which to define "atomic" real-variable Hardy spaces. For simplicity let us restrict attention to the case $p = 1$.

DEFINITION 6.4. Let X be a topological space equipped with a measure $d\mu$ and balls $B(x, r)$ that make it a space of homogeneous type. We define a 1-atom a to be a measurable function satisfying the following conditions:

(1) The function a is supported in a ball $B(P, r)$;
(2) We have
$$|a(x)| \leq \frac{1}{\mu(B(P, r))};$$
(3) The function a satisfies the moment conditions
$$\int a(x) \, d\mu(x) = 0.$$

We say that a measurable function f on X is in the real variable Hardy space H^1_{Re} if f can be decomposed in the form $f = \sum \lambda_j a_j$ with each a_j a 1-atom and the sequence $\{\lambda_j\}$ lying in ℓ^1. Of course BMO is easy to define once one has balls and a measure. One can show (see [**COW1**], [**COW2**]) that $(H^1_{\mathrm{Re}})^* = BMO$, that singular integrals preserve BMO, and thus that singular integrals preserve H^1_{Re}.

We note in passing that the simple moment condition $\int a(x) \, d\mu(x) = 0$ can be used on any space of homogeneous type to define Hardy spaces for range of p that is less than, but close to, one (see [**COW2**]). It is not at all clear how to define higher order moment conditions as in Definition 4.3, (3). New ways to define moment conditions in abstract contexts are explored in [**KRA4**] and [**KRL1**].

At long last we have set the stage for the questions that we wish to ask, and answer, about the harmonic analysis of several complex variables.

7. Harmonic Analysis in Several Complex Variables

Let $\Omega \subseteq \mathbb{C}^n$ be a smoothly bounded domain. We assume that Ω is given by a defining function:

$$\Omega = \{z \in \mathbb{C}^n : \rho(z) < 0\},$$

with $\nabla \rho \neq 0$ on $\partial\Omega$. For $\epsilon > 0$ small, say $\epsilon_0 > \epsilon > 0$, we let

$$\partial\Omega_\epsilon = \{z \in \Omega : \rho(z) = -\epsilon\}.$$

Equip $\partial\Omega_\epsilon$ with $(2n-1)$ dimensional area measure, and call this measure $d\sigma_\epsilon$. Define, for $0 < p < \infty$, the *holomorphic Hardy space* $\mathcal{H}^p(\Omega)$ to be

$$\mathcal{H}^p(\Omega) = \{f \text{ holomorphic on } \Omega : \sup_{0 < \epsilon < \epsilon_0} \int_{\partial\Omega_\epsilon} |f(\zeta)|^p \, d\sigma_\epsilon(\zeta)^{1/p} \equiv \|f\|_{\mathcal{H}^p} < \infty\}.$$

This definition turns out to be independent of the choice of defining function ρ—that is to say, the norm changes in size but the space of functions remains the same and any two such norms are comparable provided only that the defining functions are C^2. We note that $\mathcal{H}^\infty(\Omega)$ is defined to be the space of bounded holomorphic functions on Ω with the sup norm topology. [Details of these matters may be found in [**KRA1**].] The basic theorem then is that modeled on the ideas of Fatou:

THEOREM 7.1. *Let $\Omega \subseteq \mathbb{C}^n$ have C^2 boundary. Let $0 < p \leq \infty$. Let $f \in \mathcal{H}^p(\Omega)$. Then for almost every point $P \in \partial\Omega$ it hold, with ν_P denoting the unit outward normal to $\partial\Omega$ at P, that*

$$\lim_{t \to 0^+} f(P - t\nu_P)$$

exists and is finite.

If we denote the resulting almost everywhere defined function on the boundary by f^*, then the original function f may be recovered from f^* either by Poisson integration, or Poisson-Szegö integration, or by Cauchy-Szegö integration.

We note that, for $p < 1$, the integrations alluded to in the last paragraph of the theorem must be interpreted in the sense of distributions. A detailed treatment of holomorphic Hardy spaces may be found in [**KRA1, Ch. 8**].

The interesting feature of harmonic analysis in several complex variables is that there are now two Hardy space theories and they are really distinct. [Recall that in one complex variable the real Hardy space was just the set of real parts of holomorphic functions in the classical Hardy spaces, so the distinction between the two theories amounted to a nicety.] The second Hardy space theory is obtained by endowing $\partial\Omega$ with the structure of a space of homogeneous type and then constructing an atomic Hardy space. The interesting question is then to determine what is the relationship between the real-variable atomic Hardy space and the holomorphic Hardy space. The remainder of this article is concerned with the answer to this question.

8. Homogeneous Type Structures on the Boundary of a Domain in $\mathbb{C}^n$

At this point in time we do not know the correct homogeneous type structure for the boundary of absolutely any smoothly bounded domain in several complex variables. The literature is confusing in this wise, for some writings would lead us to believe that we do in fact know the answer. Therefore some preliminary discussion is in order.

One possible homogeneous type structure on the boundary of $\Omega \subseteq \mathbb{C}^n$ is obtained by using the forgetful functor: discard the complex structure and think of Ω as a subset of $\mathbb{R}^{2n}$. Equip $\partial\Omega$ with the isotropic balls $B(P, r) \cap \partial\Omega$ inherited from the standard Euclidean structure of space and endow the boundary with standard $(2n - 1)$-dimensional area measure. Then $\partial\Omega$ is indeed a space of homogeneous type, as previously noted, and an atomic Hardy space theory can be constructed. Where do these Hardy spaces fit into the mathematical firmament?

¿From the point of view of harmonic analysis, Euclidean balls arise naturally in the context that they are sublevel sets for the fundamental solution $\Gamma(x)$ of the Laplacian: $\Gamma(x) = c_N \cdot |x|^{-N+2}$ as long as $N > 2$. Also Lebesgue measure is the right measure because the theory of the Laplacian is a translation invariant theory. The function spaces that arise from those balls and that measure are the right function spaces for the study of the operators that arise in the study of the Laplacian. These include fractional integral operators such as

$$\phi \mapsto \phi * \Gamma$$

and

$$\phi \mapsto \nabla(\phi * \Gamma) = (\phi * \nabla\Gamma).$$

It also includes singular integrals, which arise naturally in this context as

$$\phi \mapsto \nabla^2(\phi * \Gamma) = (\phi * \nabla^2\Gamma).$$

Likewise, for the boundary operators that arise naturally in the study of boundary value problems for the Laplacian, the harmonic analysis and Hardy spaces of isotropic balls are natural.

Now let us re-equip Ω with its complex structure. In studying holomorphic functions it is natural to consider reproducing kernels which are particularly adapted to their analysis. [Note that in one complex variable this distinction is moot; for the Poisson kernel is appropriate to real, or harmonic, function theory and the Cauchy kernel is appropriate to complex, or holomorphic function theory.

But the former is the real part of the latter.] Thus, when thinking of complex analysis in several variables, we consider the Szegö kernel and the Cauchy-Szegö kernel (see [**KRA1**] for more on these kernels).

In general we have no idea how to compute the latter two canonical kernels. However when Ω is the unit ball B then one can write down an orthonormal basis for $\mathcal{H}^2$ and actually write each of these kernels in closed form (this calculation is done in [**KRA1**]). The result is that the Szegö kernel is given by

$$S(z, \zeta) = c_n \frac{1}{(1 - z \cdot \bar{\zeta})^n}$$

and the Poisson-Szegö kernel is given by

$$\mathcal{P}(z, \zeta) = c_n \frac{(1 - |z|^2)^n}{|1 - z \cdot \bar{\zeta}|^{2n}}.$$

Here we follow tradition and use the notation $z \cdot \bar{\zeta}$ to denote $\sum_j z_j \bar{\zeta}_j$. The sublevel sets of $S(z, \zeta)$, for ζ, z in the boundary, have the form

$$\beta(\zeta, r) = \{w \in \partial B : |1 - w \cdot \bar{\zeta}| < r\}.$$

These balls are the right ones for the study of harmonic analysis of several complex variables on the boundary of the unit ball in $\mathbb{C}^n$. Notice that they are non-isotropic. See [**KRA1**], [**KRA4**] for details.

The ball $\beta(P, r)$ in ∂B has radius about r in the complex normal direction at P but it has radius about $\sqrt{r}$ in the complex tangential directions at P. The computations establishing this assertion are standard, and can be found in [**KRA1**].

In the classic book [**STE3**], Stein took the parabolic non-isotropic geometry that we have just described and applied it to *every smoothly bounded domain in any $\mathbb{C}^n$.* The result is an important theorem about the boundary behavior of holomorphic functions on domains in $\mathbb{C}^n$. Define an admissible approach region $\mathcal{A}_\alpha(P)$ for $P \in \partial\Omega$ and $\alpha > 1$ as follows:

If $z \in \Omega, P \in \partial\Omega$, then we let

$$\delta_P(z) = \min\{\text{dist}(z, \partial\Omega), \text{dist}(z, T_P(\partial\Omega))\}.$$

Here $T_P(\partial\Omega)$ is the *real* tangent hyperplane to $\partial\Omega$ at $P \in \partial\Omega$. Notice that if Ω is convex, then $\delta_P(z) = \delta_\Omega(z)$.

DEFINITION 8.1. If $P \in \partial\Omega, \alpha > 1$, let

$$\mathcal{A}_\alpha = \{z \in \Omega : |(z - P) \cdot \bar{\nu}_P| < \alpha\delta_P(z), |z - P|^2 < \alpha\delta_P(z)\}.$$

This gives an interior approach region at each boundary point of Ω with the property that it is non-tangential in complex normal directions and parabolic in shape in complex tangential directions. It is a natural artifact of complex analysis when Ω is either the unit ball or is strongly pseudoconvex (see Section 10 for definitions) because then the cross sections of $\mathcal{A}_\alpha$ are comparable to the balls $\beta(P, r)$ that we defined above. *And remember that these balls are in turn sublevel sets for the canonical kernels.* The main theorem of Stein in [**STE3**] is this:

THEOREM 8.2. *[Stein] Let* $0 < p \leq \infty$. *Let* $\alpha > 1$. *If* $\Omega \subset\subset \mathbb{C}^n$ *has* C^2 *boundary and* $f \in \mathcal{H}^p(\Omega)$, *then, for* σ-*almost every* $P \in \partial\Omega$,

$$\lim_{\mathcal{A}_\alpha(P) \ni z \to P} f(z)$$

exists.

We stress once again that, while this theorem is true all the time (that is, for any bounded domain in $\mathbb{C}^n$ with C^2 boundary), it is only optimal for strongly pseudoconvex domains. For other pseudoconvex domains (i.e. weakly pseudoconvex—see Section 10 for a discussion), the canonical kernels have a milder singularity, the shapes of the balls β in the boundary (based on examining the sublevel sets of the canonical kernels) are different, and therefore the shapes of the approach regions $\mathcal{A}_\alpha$ should be different. A consideration of how to assign approach regions to the boundary points of a domain in $\mathbb{C}^n$—using intrinsic geometry and other considerations—appears in [**NRSW**] and in [**KRA2**].

Thus the parabolic geometry inspired by the natural geometry on the boundary of the unit ball will give a viable theory on any domain, but it is the best, and it is the canonical, theory only on strongly pseudoconvex domains. By analogy, the theory coming from isotropic balls will also work on any domain, but it is only appropriate, and canonical, when studying harmonic functions. For an extreme example of why the theorem of Stein is not adequate for arbitrary domains, consider the behavior of a holomorphic function near a strictly pseudoconcave boundary point P. The function will in fact continue analytically to an entire neighborhood of P. So the correct convergence near P is *unrestricted*. The assignment of approach regions using intrinsic geometry (as in [**KRA2**]) will detect this fact, but the more traditional approaches as in [**STE3**] do not.

To summarize, each boundary of a smoothly bounded region in $\mathbb{C}^n$ will have its own natural structure as a space of homogeneous type. While there are unified theories, such as those in [**STE3**], that treat all boundaries in the same way, such treatments are not intrinsic to the given domains (unless the domains under consideration are strongly pseudoconvex). Borrowing ideas from [**KRA2**], we shall now sketch briefly what we believe to be the correct way to define a homogeneous type structure in the boundary of a domain $\Omega \subseteq \mathbb{C}^n$.

Equip Ω with the Kobayashi metric (see [**KRA1**] for definitions). Denote the integrated metric by $K(z, w)$. If $P \in \partial\Omega$ and if n_P is the inward pointing normal *segment* emanating from P and having length ϵ_0 (some $\epsilon_0 > 0$), then for $\alpha > 0$ define the approach region at P by

$$\mathcal{L}_\alpha(P) = \{z \in \Omega : K(z, n_P) < \alpha\}.$$

Calculations (which are sometimes difficult) show that for all known classes of domains that have been studied—the unit ball, strongly pseudoconvex domains in $\mathbb{C}^n$, finite type domains in $\mathbb{C}^2$ (see Section 10), finite type domains in $\mathbb{C}^n$ (Section 10), domains with strongly pseudoconcave boundary points, domains with pieces of boundary that are foliated—this definition of approach region gives the right answer. Now, with α fixed and $r > 0$, the ball with center $P \in \partial\Omega$ and radius r is defined to be the projection to the boundary of the section of $\mathcal{L}_\alpha$ at Euclidean distance r from the boundary. In all the examples that have been mentioned, and with a measure that can be defined using a Carathéodory construction but which is locally comparable to area measure, the boundary becomes a domain of homogeneous type and a Fatou theorem can be proved. See [**KRA2**] for details.

9. Hardy Spaces and Duality on Domains in $\mathbb{C}^n$

This section is the climax of the present article. We have set the stage for the questions that will be posed and answered here. Let $\Omega \subseteq \mathbb{C}^n$ be a smoothly bounded domain. Then, as already described, the holomorphic $\mathcal{H}^p$ space is well defined. By way of the Fatou theorem, we can identify any element $f \in \mathcal{H}^p$ with its boundary function f^*. Also the boundary can often be equipped with the structure of a space of homogeneous type. Therefore one can define a real variable atomic Hardy space on $\partial\Omega$ which we denote by H^p_{at}. In this section we shall use the terms "pseudoconvex", "strongly pseudoconvex", and "finite type" without having defined them. They are discussed in detail in Section 10.

THEOREM 9.1. *Let Ω be either a strongly pseudoconvex domain in any dimension or a finite type domain in $\mathbb{C}^2$. Let $0 < p \leq \infty$. Then $\mathcal{H}^p(\Omega) \subseteq H^p_{\text{at}}(\partial\Omega)$.*

In fact one can say a bit more: any $f \in \mathcal{H}^p$ can be decomposed as a sum of Szegő projections of real variable Hardy space atoms on the boundary.

An interesting feature of the analysis that goes in Theorem 9.1 is that one must modify the real variable Hardy spaces along the lines of Goldberg's theory of "local Hardy spaces" in order to make the theorem work (see [**GOL**]). This theory, in the context of a compact boundary $\partial\Omega$, necessitates having a finite dimensional space of "dummy Hardy functions". See [**KRL2**] and [**GARL**] for details.

The proof is hard work and involves estimates of various maximal functions. Details may be found in [**KRL1**] and [**KRL2**]. The techniques in these two papers follow a traditional approach, although many of the estimates are tricky and new. A powerful approach to these questions has also been developed in [**DAF**]. Indeed, [**DAF**] contains important techniques centering around the Calderón reproducing formula that should have far-reaching applications in future investigations. It should also be noted that the analogue of Theorem 9.1, and some of the other results described in this article as well, were investigated on the unit ball in [**GARL**].

THEOREM 9.2. *Let Ω be as in the last theorem or be convex and of finite type in $\mathbb{C}^n$. Then $(H^1_{\mathrm{at}}(\partial\Omega))^* = BMO(\partial\Omega)$. Moreover the dual of $\mathcal{H}^1(\Omega)$ is the space of holomorphic functions whose boundary limit function lies in $BMO(\partial\Omega)$. [This latter space is often denoted $BMOA$.]*

The duality of $\mathcal{H}^1$ and BMO was treated, when the domain Ω is the unit ball, in an unpublished version of [**CRW**] and in [**GARL**].

It should be noted that the first statement of this last theorem is, in effect, generalized nonsense. The dual of atomic H^1 is *always* a BMO space. However the second statement is difficult and involves the deep analysis of canonical kernels. That is why we do not yet know it to be true all the time. Again see [**KRL1**], [**KRL2**], and also [**DAF**] for a powerful and important approach in either the strongly pseudoconvex case or the finite type in $\mathbb{C}^2$ case.

THEOREM 9.3. *Let Ω be smoothly bounded, convex, and of finite type in $\mathbb{C}^n$, any n. Let $f \in \mathcal{H}^1(\Omega)$. Then f^* has an atomic decomposition into Szegö projections of atoms that come from the homogeneous type structure of $\partial\Omega$.*

This result follows from ideas in [**KRL3**], which in turn are based on ideas of McNeal [**MCN1**], [**MCN2**], and of McNeal/Stein [**MCS1**], [**MCS2**].

THEOREM 9.4. *Let Ω be strongly pseudoconvex in any dimension or of finite type in $\mathbb{C}^2$. Let $0 < p < \infty$. If $f \in \mathcal{H}^p(\Omega)$ then we may decompose f as*

$$f = \sum_{j=1}^{\infty} g_j h_j,$$

where $g_j, h_j \in \mathcal{H}^{2p}$.

It is well known that the canonical factorization for $\mathcal{H}^p$ functions (see [**KRA1**] or [**HOF**]) fails dramatically for holomorphic functions of several complex variables. The last theorem, first discovered in the unit ball by Coifman, Rochberg, and Weiss [**CRW**], is a useful substitute.

THEOREM 9.5. *Let $0 < p < 1$. Let Ω be either a strongly pseudoconvex domain in $\mathbb{C}^n$ or a finite type domain in $\mathbb{C}^2$. Then the dual of $\mathcal{H}^p(\Omega)$ is a non-isotropic Lipschitz space that is adapted to the homogeneous type structure of $\partial\Omega$.*

The careful formulation and proof of theorems like 9.5 require that we come to grips with formulating higher order finite differences (so that higher order Lipschitz spaces may be defined) and higher order moment conditions, that is generalizations of the mean value zero property that was used in the definition of 1-atom (so that the H^p spaces for small p may be defined). Details may be found in [**KRL1-KRL3**], [**KRA4**].

These theorems represent a proper subset of what is now known. However they give a taste of what the issues are, and what the answers look like. The analysis of a domain in $\mathbb{C}^n$ depends on the Levi geometry of the boundary. This does not happen in $\mathbb{C}^1$ because boundaries in $\mathbb{C}^1$ have no complex structure and therefore can be moved at will by holomorphic mappings. That is why we have a Riemann mapping theorem in one complex variable but none in several complex variables.

10. A Little Complex Geometry

We conclude this paper by giving a quick overview of some issues of complex Levi geometry. This will only begin to suggest why different domains in $\mathbb{C}^n$ will have boundaries with different homogeneous type structures. Further details may be found in [**KRA1**].

Let $\Omega \subseteq \mathbb{C}^n$ be a smoothly bounded domain with defining function ρ. Let $P \in \partial\Omega$. A vector $w = (w_1, \ldots, w_n) \in \mathbb{C}^n$ is defined to be in the *complex tangent space* to $\partial\Omega$ at P (written $w \in \mathcal{T}_P(\partial\Omega)$) if

$$\sum_{j=1}^n \frac{\partial\rho}{\partial z_j}(P)w_j = 0.$$

The complex tangent space is a real codimension one subspace of the ordinary real tangent hyperplane $T_P(\partial\Omega)$. The former has real dimension $(2n-2)$ while the latter has real dimension $(2n-1)$. The former is a complex linear space, while the latter obviously is not. Details on these matters may be found in [**KRA1**].

For $w \in \mathcal{T}_P(\partial\Omega)$ we define the *Levi form* at P, acting on w, to be

$$\sum_{j,k=1}^n \frac{\partial^2 \rho(P)}{\partial z_j \partial \bar{z}_k} w_j \bar{w}_k.$$

If this quadratic form is positive semi-definite on $\mathcal{T}_P(\partial\Omega)$ then we say that P is a boundary point of (Levi) pseudoconvexity. If it is positive definite then we say that P is a boundary point of *strong*, or strict, (Levi) pseudoconvexity. If all boundary points are pseudoconvex then the domain is said to be pseudoconvex. If all boundary points are strongly pseudoconvex then the domain is called strongly pseudoconvex. Any pseudoconvex domain can be exhausted by an increasing union of smoothly bounded, strongly pseudoconvex domains. It is natural to concentrate attention on pseudoconvex domains because they are precisely the domains of holomorphy—that is, the domains of existence of holomorphic functions. [A discursive treatment of all these ideas appears in [**KRA1**].]

Thirty years ago, when one was studying a pseudoconvex domain, any given boundary point was either strictly pseudoconvex or it was not (a point of the second type, for purposes of contrast, is called *weakly* pseudoconvex). Nowadays we have a stratification of pseudoconvex points. This is the point of the theory of *type*. While it is too technical to develop in detail here, we shall give the topic a few words.

In complex dimension two matters are fairly simple. A point $P \in \partial\Omega$ is said to be of finite type $m \in \mathbb{N}$ if it is possible for a non-singular one-dimensional complex curve to be tangent to $\partial\Omega$ at P to order m but there is none with tangency of order greater than m. The point is said to be of infinite type if any order of tangency can be achieved. For example, any point in the boundary of the unit ball is of finite type 2. In fact strongly pseudoconvex points are always of type 2. The boundary points of $E_m = \{(z_1, z_2) \in \mathbb{C}^2 : |z_1|^2 + |z_2|^{2k} < 1\}$ that have the form $(e^{i\theta}, 0)$ are of type $m = 2k$ (here k is a positive integer). The boundary points of $E_\infty = \{(z_1, z_2) \in \mathbb{C}^2 : |z_1|^2 + 2e^{-1/|z_2|^2} < 1\}$ that have the form $(e^{i\theta}, 0)$ are of infinite type.

At a point of finite type m as described in the last paragraph, it turns out that the shape of the canonical balls $\beta(P, r)$ in the boundary that one ought to consider when specifying a homogeneous type structure will have dimension r in the complex normal direction and dimension $r^{1/m}$ in the complex tangential directions. The balls near a point of infinite type will be much more skew. Actually fleshing out these statements requires difficult calculations, both of canonical kernels and of canonical metrics. See [**NRSW**] and [**KRA2**] for further detail.

In dimensions three and greater, the notion of type is necessarily more complicated (as follows from deep work of D'Angelo [**DAN1**] and of Catlin [**CAT1**], [**CAT2**], [**CAT3**]). Briefly, a point $P \in \partial\Omega \subseteq \mathbb{C}^n$ is of finite type if there is an *a priori* upper bound on the order of contact of (possibly singular) complex varieties of dimension 1 with $\partial\Omega$ at P. Some details of this definition may be found in [**KRA1**] and all details may be found in [**DAN1-DAN6**]. Suffice it to say that "type" measures an order of flatness of the boundary in some complex analytic sense. In dimensions three and greater, there will be more than one complex tangent direction at any given boundary point and the degree of flatness in each of these directions may be different. Thus the shapes of the canonical balls in the boundary may not be specified by two parameters (as in the $\mathbb{C}^2$ case) but rather by n parameters.

A recent development, thanks to work of McNeal [**MCN1**], [**MCN2**], is that *convex* domains of finite type—in any dimension—are much simpler than we might ever have hoped. McNeal has shown, first of all, that the type can be measured correctly using only complex lines; singular varieties need not be considered at all. Second, he has determined the local geometry of a convex finite type point; in particular, he has told us the shapes of the balls. As a result, we now know the homogeneous type structure of a convex domain of finite type in any dimension. This information is used decisively in [**MCST1**], [**MCST2**] to determine the mapping properties of the Bergman and Szegö kernels on such a domain and in [**KRL3**] to calculate the duals of various Hardy, Bloch, and Bergman spaces.

We conclude by noting that we are still a long way from understanding points of finite type (in the absence of a convexity hypothesis) in all dimensions. More troubling is that we do not know a biholomorphically invariant way to recognize convexity. What seems to be the nastiest problem of all is to study domains which are pseudoconvex at some points and pseudoconcave at others; the surfaces in the boundary where these two geometries interface are not well understood, even in dimension two.

Bibliography

[BGS]   D. Burkholder, R. Gundy, and M. Silverstein, *A maximal function characterization of the class H^p,*, Trans. Am. Math. Soc. **157** (1971), 137–153.

[CAMP] S. Campanato, *Proprieta di una famiglia di spazi funzionali*, Ann. Scuola Norm. Sup. Pisa **18** (1964), 137–160.

[CAT1] D. Catlin, *Necessary conditions for subellipticity of the $\overline{\partial}$-Neumann problem*, Ann. Math. **117** (1983), 147–172.

[CAT2] D. Catlin, *Subelliptic estimates for the $\overline{\partial}$-Neumann problem*, Ann. Math. **126** (1987), 131–192.

[CAT3] D. Catlin, *Global regularity for the $\overline{\partial}$-Neumann problem*, Proc. Symp. Pure Math. **41** (1984), Am. Math. Soc., Providence.

[CAT4] D. Catlin, *Estimates of invariant metrics on pseudoconvex domains of dimension two*, Math. Z. **200** (1989), 429–466.

[CAT5] D. Catlin, *private communication*.

[CHR]   F. M. Christ, *Lectures on Singular Integral Operators*, Conference Board of Mathematical Sciences, American Mathematical Society, Providence, 1990.

[CRW]   R. R. Coifman, R. Rochberg, and G. Weiss, *Factorization theorems for Hardy spaces in several variables*, Ann. of Math. **103** (1976), 611–635.

[COW1] R. R. Coifman and G. Weiss, *Analyse Harmonique non-commutative sur certains espaces homogenes*, Springer Lecture Notes, vol. 242, Springer Verlag, Berlin, 1971.

[COW2] R. R. Coifman and G. Weiss, *Bull. Am. Math. Soc.* **83** (1977), 569–643.

[DAF]   G. Dafni, *Hardy spaces on some pseudoconvex domains*, Jour. Geometric Analysis (to appear).

[DAN1] J. P. D'Angelo, *Several Complex Variables and Geometry*, CRC Press, Boca Raton, 1992.

[DAN2] J. P. D'Angelo Real hypersurfaces, orders of contact, and applications,, *Annals of Math.* **115** (1982), 615–637.

[DAN3] J. P. D'Angelo, *Finite type conditions for real hypersurfaces in $\mathbf{C}^n$,*, Complex Analysis Seminar, Springer Lecture Notes, vol. 1268, Springer Verlag, 1987, p. 83–102.

[DAN4] J. P. D'Angelo, *Iterated commutators and derivatives of the Levi form*, Complex Analysis Seminar, Springer Lecture Notes, vol. 1268, Springer Verlag, 1987, p. 83–102.

[DAN5] J. P. D'Angelo, *Finite type and the intersection of real and complex subvarieties*, Several Complex Variables and Complex Geometry, Proc. Symp. Pure Math., vol. 52, Part I, p. 103–118.

[DAN6] J. P. D'Angelo, *A note on the Bergman kernel*, Duke Math. Jour. **45** (1978), 259–265.

[FEF]   C. Fefferman, *Harmonic analysis and H^p spaces*, Studies in Harmonic Analysis, J. Marshall Ash, ed., Mathematical Association of America, Washington, D.C., 1976.

[FS]   C. Fefferman and E. M. Stein, *H^p spaces of several variables*, Acta Math. **129** (1972), 137–193.

[GARL]   J. Garnett and R. Latter, *The atomic decomposition for Hardy spaces in several complex variables*, Duke Math. J. **45** (1978), 815–845.

[GOL]   D. Goldberg, *A local theory of real Hardy spaces*, Duke J. Math. **46** (1979), 27–42.

[HOF]   K. Hoffman, *Banach Spaces of Holomorphic Functions*, Prentice-Hall, Englewood Cliffs, 1962.

[HOR]   L. Hörmander, *L^p estimates for pluri-subharmonic functions*, Math. Scand. **20** (1967), 65–78.

[JON]   F. John and L. Nirenberg, *Comm. Pure Appl. Math.* **14** (1961), 415-426.

[KAT]   Y. Katznelson, *An Introduction to Harmonic Analysis*, Dover, New York, 1976.

[KRA1]   S. G. Krantz, *Function Theory of Several Complex Variables*, 2nd. Ed., Wadsworth, Belmont, 1992.

[KRA2]   S. G. Krantz, *Invariant metrics and the boundary behavior of holomorphic functions on domains in $\mathbb{C}^n$*, Jour. Geometric Anal. **1** (1991), 71–98.

[KRA3]   S. G. Krantz, *Geometric Lipschitz spaces and applications to complex function theory and nilpotent groups*, J. Funct. Anal. **34** (1980), 456–471.

[KRA4]   S. G. Krantz, *Geometric Analysis and Function Spaces*, Conference Board of Mathematical Sciences, American Mathematical Society, Providence, 1993.

[KRL1]   S. G. Krantz and Song-Ying Li, *A note on Hardy spaces and functions of bounded mean oscillation on domains in $\mathbb{C}^n$*, Michigan Math. Jour., to appear.

[KRL2]   S. G. Krantz and Song-Ying Li, *On the decomposition theorems for Hardy spaces in domains in $\mathbb{C}^n$ and applications*, preprint.

[KRL3]   S. G. Krantz and Song-Ying Li, *Duality theorems for Hardy and Bergman spaces on convex domains of finite type in $\mathbb{C}^n$*, preprint.

[LAN]   R. E. Langer, *Fourier Series: The Genesis and Evolution of a Theory*, Herbert Ellsworth Slaught Memorial Paper 1, Mathematical Association of America, 1947.

[MAS]   R. Macias and C. Segovia, *Lipschitz functions on spaces of homogeneous type,*, Advances in Math. **33** (1979), 257–270.

[MCN1]   J. McNeal, *Lower bounds on the Bergman kernel near a point of finite type*, Annals of Math. **136** (1992), 339–360.

[MCN2]   J. McNeal, *Convex domains of finite type*, Jour. Funct. Anal. **108** (1992), 361–373.

[MCN3]   J. McNeal, *Estimates on the Bergman kernels of convex domains,*, Advances in Math., in press.

[MCS1]   J. McNeal and E. M. Stein, *Mapping properties of the Bergman projection on convex domains of finite type*, Duke Math. J., in press.

[MCS2]   J. McNeal and E. M. Stein, in preparation.

[NRSW]   A. Nagel, J. P. Rosay, E. M. Stein, and S. Wainger, *Estimates for the Bergman and Szegö kernels in $\mathbb{C}^2$*, Ann. Math. **129** (1989), 113–149.

[RIE]   M. Riesz, *Sur les fonctions conjuguées,*, Mat. Zeit. **27** (1927), 218–244.

[SMI]   K. T. Smith, *A generalization of an inequality of Hardy and Littlewood*, Can. J. Math. **8** (1956), 157–170.

[STE1]   E. M. Stein, *On the theory of harmonic functions of several variables*, Acta Math. **106** (1961), 137–174.

[STE2]   E. M. Stein, *Singular Integrals and Differentiability Properties of Functions*, Princeton Univ. Press, Princeton, 1970.

[STE3]   E. M. Stein, *Boundary Behavior of Holomorphic Functions of Several Complex Variables*, Princeton University Press, Princeton, 1972.

[STW1]   E. M. Stein and G. Weiss, *On the theory of harmonic functions of several variables, I: The theory of H^p spaces*, Acta Math. **103** (1960), 25–62.

[STW2]   E. M. Stein and G. Weiss, *Introduction to Fourier Analysis on Euclidean Spaces*, Princeton Univ. Press, Princeton, 1971.

Department of Mathematics
Washington University in St. Louis
St. Louis, Missouri 63130
E-mail address: SK@MATH.WUSTL.EDU

Contemporary Mathematics
Volume 185, 1995

SPECTRAL PICTURE AND INDEX INVARIANTS OF COMMUTING N-TUPLES OF OPERATORS

R.N.LEVY

Introduction. In the present paper the n-tuples of commuting or essentially commuting bounded operators in Banach space are considered from the point of view of the related topological information, including, in particular, the index of any function the operators.

It is known that if the space is Hilbert and the operators of T and their adjoints commute modulo compact operators, the relevant topological information is determined by the corresponding element $Ext(T)$ of the Brown-Douglas-Fillmore group $Ext(\sigma_e(T))$ of the joint essential spectrum $\sigma_e(T)$.

It turns out that the same topological invariant can be defined for any essentially commuting n-tuple of operators acting in Banach space. The construction uses the corresponding Koszul complex. (Recall that the Koszul complex of the commuting n-tuple of operators was introduced by J.L.Taylor in his fundamental definition of the joint spectrum and the corresponding functional calculus).

In this way for any essentially commuting n-tuple T one obtains the index class $K(T)$ as an element of the group K^0 of the complement of the essential spectrum. This element generalizes the Alexander dual of the element $Ext(T)$ mentioned above.

Our main observation is that in the case of the commuting n-tuple the topological invariant $K(T)$ is completely determined by a certain complex-analytic object. Roughly speaking, this is a family of bounded complex analytic spaces with boundaries in $\sigma_e(T)$ (forming the Fredholm spectrum $\sigma_F(T)$ of the n-tuple) with attached integers (multiplicities). In this case, the corresponding invariant in cohomology $ch(T) := ch(K(T))$ takes values in the Hodge cohomologies of the type (p,p).

The precise definition involves the Grothendieck group of the category of coherent sheaves on $\mathbb{C}^n \backslash \sigma_e(T)$ and leads to an element $\mathcal{K}(T)$ of this group. It turns out that the elements $K(T)$ and $\mathcal{K}(T)$ have good

1991 *Mathematics Subject Classification.* Primary 47A53; Secondary 47B35.

The author was partially supported by contract MM-3 with the Bulgarian Ministry of Science

functorial properties and determine explicitly the index of the functions of T under a suitable functional calculus.

If the Fredholm spectrum contains components of maximal dimension n , i.e. bounded components of $\mathbb{C}^n \backslash \sigma_e(T)$, then the corresponding multiplicities coincide with $ind(T - zI)$, z - an arbitrary point of this component; so, the invariants above generalize the notion of index of n- tuple, studied in [6] and [12].

The first part of the paper concerns the definition and the functorial properties of the invariants $K(T)$, $\mathcal{K}(T)$ and $ch(T)$. In the second part some applications are given. First we give a very short proof of the Boutet de Monvel formula for the Index of Toeplitz operators, and a generalization of this theorem to the "well situated" singular case. Another application is given to the problem of lifting, i.e. of the existence of a commuting compact perturbation of the given essentially commuting n-tuple. It turns out that the possibility of lifting could depend on the complex geometry of $\sigma_e(T)$ in $\mathbb{C}^n$.

Some of the assertions given here are proven in [10] , [9]; the proofs of the rest will be published elsewhere. In the mean time there appeared the papers [13] and [5] (see also [11]). In [5] is treated the case of 1-dimensional Fredholm spectrum and the corresponding invariants. In [13] there are considered the functorial properties and the index formulas, corresponding to the components of (maximal) dimension n of $\mathcal{K}(T)$ (see above). There are also many related results which we cannot discuss here.

1. DEFINITIONS AND PROPERTIES OF INVARIANTS.

1.1. The topological invariant. Let $T = (T_1, \ldots T_n)$ be an essentially commuting n-tuple of bounded operators acting in the Banach space X ; $[T_i, T_j] \in \mathcal{K}(X)$ - the ideal of all compact operators in X . Denote by $K(T, z)$ the Koszul complex of the n-tuple T in the point $z \in \mathbb{C}^n$. Actually, this system of spaces and differentials is not a complex; the product of two consecutive differentials is not zero but in $\mathcal{K}(X)$. Let us call such systems *essential complexes*. Denote by $\tilde{T} = (\tilde{T}_1, \ldots \tilde{T}_n)$ the n-tuple of classes of $T_1, \ldots , T_n$ in the Calkin algebra $\mathcal{A}(X)$ of the space X , and let $K(\tilde{T}, z)$ be the Koszul complex of $\tilde{T}$ acting as n-tuple of operators of multiplication in $\mathcal{A}(X)$. As usual, we shall denote by $\sigma_e(T)$ the set of all points z such that $K(\tilde{T}, z)$ is not an exact complex.

Definition 1.1.1. Let us take a continuous operator valued functions $P_i(z)$ such that $\sum_{i=1}^n (T_i - z_i I) P_i(z) = I$ modulo compact operators. The set $\left\{ \tilde{P}_i(z) \right\}$ determines a homotopy for the complex $K(\tilde{T}, z)$. Denote by X_σ and X_e the direct sum of all odd and even components of $K(T, z)$ correspondingly. Then, the collapsing procedure for this complex, described in [6] , yields an essentially invertible operator-function $D(z) : X_o \to X_e$ which depends continuously on the parameter $z \in \mathbb{C}^n \backslash \sigma_e(T)$. Then $D(z)$ determines an element of the topological K-group $K_0(\mathbb{C}^n \backslash \sigma_e(T))$ which will be denoted by $K(T)$.

An alternative definition is the following: on the domain $U = \mathbb{C}^n \backslash \sigma_e(T)$ one can take a compact perturbations of the differentials of $K_{\bullet}(T, z)$ such that the perturbed differentials form a continuous Fredholm complex. Then one can prove that there exists a continuous complex of finite-dimensional vector bundles which is quasiisomorphic to. The alternated sum of the classes of the bundles is equal to $K(T)$.

The Alexander duality determines a homomorphism of the group $K_1(\sigma_e(T))$ into $K^0(\mathbb{C}^n \backslash \sigma_e(T))$, which is an isomorphism if $\sigma_e(T)$ is homeomorphic to a finite cell complex. In this case, $K(T)$ can be considered also as an element of $K_1(\sigma_e(T))$.

Let $ch^* : K_0(U) \to H^{ev}(U) = \sum H^{2i}(U)$ be the Chern character. We shall denote by $ch^*(T) := ch^* (K(T))$ the corresponding element in cohomologies. Analogously, if $ch_* : K_1(F) \to H_{odd}(F) = \sum H_{2i+1}(F)$ is the Chern character in homology, then, applying it to the dual $DK(T)$ of $K(T)$, we obtain an element $ch_*(T) := ch_* (DK(T)) \in H_{odd} (\sigma_e(T))$.

The element $K(T)$ (or $ch^*(T)$) determines the index of the n-tuple T ; indeed, it is easy to see that the component of $ch(T)$ of degree zero has the form $\sum n_i [U_i]$, where U_i are among the bounded component of U , $[U_i]$ are the corresponding elements of $H^0(U)$, and n_i is the Euler characteristic of $K_{\bullet}(T, z)$ for arbitrary $z \in U_i$. Obviously, in many cases (e.g. when U is connected) all the indexes n_i are zero but $K(T)$ can be nontrivial.

1.2. Properties of $K(T)$. The element $K(T)$ has the natural properties of invariance and functoriality. This element is an index class for the n-tuple T , i.e. determines the index of the essentially invertible matrix functions of T . More precisely, the following properties are satisfied:

1.2.1. *Invariance.* $K(T)$ is invariant under compact or small perturbations of the n-tuple T in the class of essentially commuting n-tuples of operators.

1.2.2. *Functoriality.* Let A be a Banach algebra of functions on $\sigma_e(T)$. Suppose that there exists an A-functional calculus in the Calkin algebra $\mathcal{A}(X)$, i.e. a homomorphism $A \to \mathcal{A}(X)$ sending the coordinate functions in the corresponding operators from $\tilde{T}$. Let $f(z) = (f_1(z), \dots , f_k(z))$ be a k-tuple of functions of A and $f(T) = (f_1(T), \dots , f_k(T))$. Then

$$K\left(f(T)\right) = f_*\left(K(T)\right)$$

The action of the morphism f_* on the K-groups can be easily described in the Alexander dual groups: the dual morphism $f_* : K_1(\sigma_e(T)) \to K_1\left(f(\sigma_e(T))\right)$ is the usual morphism of direct image.

1.2.3. *Index formula.* The functoriality leads to the index property of $K(T)$; namely, let A be as above and $G(z)$ be an invertible $n \times n$ continuous matrix function with entries belonging to the function algebra A . Then $G(T)$ is a Fredholm operator in X^n and

$$ind\left(G(T)\right) = \langle [G(z)], K(T) \rangle$$

where $[G(z)]$ denotes the class of $G(z)$ in $K^1\left(\sigma_e(T)\right)$, and the pairing $\langle ., . \rangle$ is determined by the natural pairing between $K_1\left(\sigma_e(T)\right)$ and $K^1\left(\sigma_e(T)\right)$.

1.2.4. *Connection with Ext.* Suppose that X is a Hilbert space and the operators of T are essentially normal, i.e. the operators $T_1, \ldots T_n, T_1^*, \ldots T_n^*$ commute modulo compacts. Denote by $Ext(T)$ the element of the group $Ext(\sigma_e(T))$ corresponding to the extension generated by the operators of T . Then $K(T)$ is Alexander dual to $Ext(T)$. In particular, if $\sigma_e(T)$ is homeomorphic to a finite cell complex, then $K(T)$ determines T up to a Brown - Douglas - Fillmore equivalence.

1.3. The analytic invariant for a commuting n-tuple. It turns out that in the case when the operators of T have stronger commutation properties its index invariant can be calculated by the means of some analytic objects of a local nature. The most explicit case is when T is a commuting n-tuple of operators. Then the Koszul complex is a complex of Banach spaces on $\mathbb{C}^n$ with a holomorphic differentials. Taylor defined the joint spectrum $\sigma(T)$ as a set of all $z \in \mathbb{C}^n$ such that $K_\bullet(T, z)$ is not exact.

Further, by several authors (see [6] , [8] , [12]) the joint essential spectrum $\sigma_e(T)$ was defined as the set of all z such that $K_\bullet(T, z)$ has at least one infinite-dimensional homology group. Finally, by Fredholm spectrum $\sigma_F(T)$ we shall mean the set of all points of the spectrum such that $K_\bullet(T, z)$ has only finite- dimensional homology groups; $\sigma_F(T) = \sigma(T) \backslash \sigma_e(T)$.

In the case of a single operator T the situation is well-known; the Fredholm spectrum is a union of mutually disjoint open subsets $\{U_i\}$ and isolated points of the spectrum. The collection $\{n_i, U_i\}$, where $n_i = ind(T - z), z \in U_i$, is called by Brown - Douglas - Fillmore a *spectral picture* of T ; it determines the operator T up to unitary equivalence modulo compacts (BDF equivalence).

To describe the corresponding notion for several commuting operators, we start with the following simple observation which can be considered as a version of the Gleason theorem on the existence of a holomorphic structure:

Lemma 1.3.1. *Any holomorphic pointwise Fredholm complex of Banach spaces on a domain in $\mathbb{C}^n$ is locally holomorphically quasiisomorphic to a holomorphic complex of finite-dimensional spaces.*

Corollary 1.3.2. *The Fredholm spectrum of a commuting n-tuple T is a bounded complex-analytic subspace of $\mathbb{C}^n \backslash \sigma_e(T)$.*

Roughly speaking, $\sigma_F(T)$ can be assumed to be a set of bounded open domains on (singular) complex spaces with boundaries contained in $\sigma_e(T)$. The next question appearing is whether the Fredholm spectrum (possibly united with some part of the essential spectrum) is holomorphically convex:

Problem 1.3.1. Denote by $\widehat{F}$ the holomorphic hull of the set $F \subset \mathbb{C}^n$. Is there $\widehat{\sigma_F(T)}$ unifold and contained in $\sigma(T)$?

As in the case of a single operator, the geometry of the Fredholm spectrum is not sufficient to determine the index class. One needs also to know the "multiplicities" of the components of $\sigma_F(T)$. It turns out that the natural way to define it is by the use of the language of coherent sheaves and Grothedieck groups. Let U be a domain in $\mathbb{C}^n$, and let us denote by $\mathcal{O}_U$ the sheaf of germs of holomorphic functions on U . Recall that a sheaf on U is *coherent* iff it coincides with some sheaf of homologies of some finite complex of free finite-dimensional $\mathcal{O}_U$ - modules. The algebraic K-group $K_0^{alg}(U)$ is defined as the Grothendieck group of the category of all coherent sheaves on U , i.e. the group generated by the classes $[\mathcal{L}]$ of the coherent sheaves $\mathcal{L}$ on U , and with relations of the type $[\mathcal{M}] = [\mathcal{L}] + [\mathcal{N}]$ corresponding to the short exact sequences $0 \to \mathcal{L} \to \mathcal{M} \to \mathcal{N} \to 0$.

For given a Banach space X , denote by $\mathcal{O}X$ the sheaf of germs of X-valued holomorphic functions. Since the differentials of the complex $K_\bullet(T, z)$ depend holomorphically on z , one can consider the corresponding complex of sheaves of germs of holomorphic vector- functions, which will be denoted by $\mathcal{O}K_\bullet(T, z)$. Denote by $\mathcal{H}_i(T) :=$ $\mathcal{H}_i(\mathcal{O}K_\bullet(T, z))$ the i-th homology sheaf of this complex. Again from lemma 1, one sees that all the sheaves $\mathcal{H}_i(T)$ are supported on $\sigma_F(T)$ and coherent.

Definition 1.3.1. For any commutative n-tuple T of operators, we shall denote by $\mathcal{K}(T)$ the element of the Grothendieck group $K_0^{alg}(\mathbb{C}^n \backslash \sigma_e(T))$ defined by the equality

$$\mathcal{K}(T) = \sum_{i=0}^{n} (-1)^{n-i} [\mathcal{H}_i(T)]$$

We say that some element of the group $K_0^{alg}(U)$ has a *Jordan - Holder decomposition* if it can be represented as a sum of a finite number of elements $[\mathcal{O}_{M_i}]$ - the classes of the structure sheaves of irreducible complex spaces M_i . This is always possible e.g. if U is a Stein domain. At any rate, such a decomposition is possible locally.

Definition 1.3.2. Suppose that the decomposition

$$\mathcal{K}(T) = \sum n_i [\mathcal{O}_{M_i}]$$

where n_i are integers, and M_i - irreducible complex spaces, holds (globally on $\mathbb{C}^n \backslash \sigma_e(T)$ or locally near the point z^0). Then the set $\{n_i, M_i\}$ will be called a *(local) spectral picture* of the commuting n-tuple T .

1.4. Properties of $\mathcal{K}(T)$.

1.4.1. *Connection with $K(T)$.* For any complex manifold M there exists uniquely determined natural transformation of functors α_M : $K_0^{alg}(M) \rightarrow K_0(M)$ (see [1]).

Proposition 1.4.1.1. *On the domain* $U = \mathbb{C}^n \backslash \sigma_e(T)$ *we have*

$$\alpha_U \mathcal{K}(T) = K(T) .$$

Therefore, the index class of a commuting n-tuple is completely determined by its spectral picture. In particular, if the elements of T are essentially normal operators in Hilbert space, then the n-tuple T is determined by its spectral picture modulo B-D-F equivalence.

1.4.2. *Invariance.*

Proposition 1.4.2.1.   *The element $\mathcal{K}(T)$ is invariant under finite-dimensional perturbations of the operators of T in the class of commuting n- tuples.*

Problem 1.4.2.1. Is the above proposition still valid for trace-class perturbations?

1.4.3. *Functoriality.* It is difficult to describe the behavior of the spaces of homologies $H_i(T, z)$ of the Koszul complex under the holomorphic functional calculus. On the contrary, the homology sheaves of the corresponding complex of sheaves of holomorphic sections have nice functorial properties:

Proposition 1.4.3.1. *Let T be a commuting n-tuple of operators, and $f(z) = (f_1(z), \ldots, f_k(z))$ be a k-tuple of functions holomorphic in a neighborhood of $\sigma(T)$. Then*

$$\mathcal{H}_i\left(f(T)\right) = f_* \mathcal{H}_{i+n-k}(T)$$

where f_* denotes the operation of direct image of coherent sheaves.
In particular, we have $\mathcal{K}\left(f(T)\right) = f_* \mathcal{K}(T)$.

For instance, if T satisfies the Bishop β condition ($\mathcal{H}_i(T) = 0$ for $i < n$), then this condition is satisfied by $f(T)$ as well. Moreover, from the proposition above we obtain:

Corollary 1.4.3.2. *Suppose that T satisfies the condition β . Suppose also that all the terms in the local Jordan - Holder decomposition of $\mathcal{K}(T)$ (equal to $(-1)^n \mathcal{H}_n(T)$) correspond to irreducible complex spaces of a pure dimension k . Then on any connected component of $\sigma\left(f(T)\right) \backslash \sigma_e\left(f(T)\right)$ we have $H_i\left(f(T), z\right) = 0$ and $\mathcal{H}_i\left(f(T)\right) = 0$ for $i < k$, the dimension of $H_k\left(f(T), z\right)$ is locally constant, and $\mathcal{H}_k\left(f(T)\right)$ is a holomorphic vector bundle of the same dimension (perhaps different on the various components of the domain $\sigma_F(f(T))$).*
Suppose in addition that for any $z \in \sigma_F(T)$ the natural mapping $X \to \mathcal{H}_n(T)_z$ is a monomorphism. Then the adjoint $f(T)^*$ of the k-tuple $f(T)$ belongs on $\sigma_F(f(T))$ to the Cowen - Douglas class B_m.

We will see in the next point that the dimension mentioned above can be determined by methods of algebraic geometry.

1.5. Intersection number of sheaves and index formula. Suppose that $\mathcal{L}_\bullet$ and $\mathcal{M}_\bullet$ are finite complexes of coherent sheaves whose supports intersect in a unique point z_0 . One of the possible ways to define the *intersection number* $\langle \mathcal{L}_\bullet, \mathcal{M}_\bullet \rangle_{z_0}$ is the following: take a Stein neighborhood U of z_0 and finite complexes of finite-dimensional free modules $L_\bullet$, $M_\bullet$, quasiisomorphic to $\mathcal{L}_\bullet$ and $\mathcal{M}_\bullet$ correspondingly. Then $\langle \mathcal{L}_\bullet, \mathcal{M}_\bullet \rangle_{z_0}$ is equal to the Euler characteristic of the total complex, corresponding to the bicomplex of Frechet spaces $\Gamma_U\left(L_\bullet \otimes_\mathcal{O} M_\bullet\right)$, Γ_U denoting the space of sections over U.

Let the supports of the complexes $\mathcal{L}_\bullet$ and $\mathcal{M}_\bullet$ intersect in a finite set of points. Then we shall denote by $\langle \mathcal{L}_\bullet, \mathcal{M}_\bullet \rangle$ the sum of intersection numbers in all the points of intersection.

The notion above permits us to express the index of a holomorphic matrix function of T by $\mathcal{K}(T)$. We will consider a more general situation:

Proposition 1.5.1. *Suppose that $L_\bullet(z) = \{L_i, d_i(z)\}$ is a holomorphic complex of finite-dimensional spaces, defined on a neighborhood of $\sigma(T)$ and exact on $\sigma_e(T)$. Then the complex of Banach spaces $L(T) = \{L_i \otimes X, d_i(T)\}$, corresponding to $L_\bullet(z)$ by the holomorphic functional calculus, is Fredholm, and its Euler characteristic is equal to the intersection number $\langle \mathcal{K}(T), L_\bullet \rangle$*

In particular, the Euler characteristic of the complex $K_\bullet(f(T), 0)$ is equal to $\langle \mathcal{K}(T), \mathcal{N}_f \rangle$, where the sheaf $\mathcal{N}_f$ is determined on $\mathbb{C}^n \backslash \sigma_e(T)$ by the formula $\mathcal{N}_f = \mathcal{O}/\sum f_i(z)\mathcal{O}$.

Let us consider the particular case when $f(z)$ is a projection. Let $T = T' \bigcup T''$, and $z = (z', z'')$ be the corresponding decomposition of $\mathbb{C}^n$. Let $\mathcal{K}(T) = \sum n_i [\mathcal{O}_{M_i}]$ be the J. - H. decomposition. Then

$$ind\, K_\bullet(T', z'_0) = \sum n_i\, mult_{p,z'_0} M_i$$

where p is the projection $z \to z'$, and $mult_{p,z'_0} M_i$ is by definition the intersection number of M_i with the affine space $H_{z'_0} = \{z : p(z) = z'_0\}$.

1.6. Computation of $\mathcal{K}(T)$. The last formula of the preceding paragraph can be used only if $H_{z'_0}$ does not intersect the essential spectrum. In some cases it is impossible to apply this formula; for instance, suppose that the Fredholm spectrum is contained in the interior of a big sphere, belonging to $\sigma_e(T)$. Now, we shall give a local version of the formula which can be used in more general situation in order to compute the local indexes n_i.

Let $T = T' \bigcup T'' = (T'_1, \ldots, T'_k) \bigcup (T''_1, \ldots, T''_{n-k})$, z, z', z'', p be as above. Let $z_0 = (z'_0, z''_0)$ be a point of the Fredholm spectrum $\sigma_F(T)$ of T which is an isolated point of the intersection $H_{z'_0} \bigcap \sigma_F(T)$. Then the homology spaces $H_i(T', z'_0)$ decompose in the same way as if we had an isolated Fredholm point of the spectrum of a single operator:

Proposition 1.6.1. *Under the above conditions, there exist the decompositions:*

$$H_i(T', z'_0) = H'_i(T', z'_0) \bigoplus H''_i(T', z'_0), \quad \mathcal{H}_i(T')_{z'_0} = \mathcal{H}'_i(T') \bigoplus \mathcal{H}''_i(T')$$

such that

(a) *$H'_i(T', z'_0)$ is finite-dimensional and $\mathcal{H}'_i(T')$ is coherent; both are induced by the embedding of some holomorphic complex of finitely-dimensional spaces in $K_\bullet(T', z')$ near z'_0.*

(b) *The spectrum of T'' on $H_i''(T', z_0')$ does not contain z_0'', and its spectrum on $H_i'(T', z_0')$ coincides with $\{z_0''\}$ only; $H_i'(T', z_0')$ coincides with the root space of the operators $T_i'' - z_{0,i}''$, $1 \leq i \leq n - k$, in the (possibly non-separated) space $H_i(T', z_0')$.*

(c) *If U is a sufficiently small neighborhood of z_0'', then $\mathcal{H}_i'(T')_{z_0'} = p_* \mathcal{H}_{i+n-k}(T)_{z_0' \times U}$*

Denote $\chi_{z_0}(K_\bullet(T', z_0')) = \sum (-1)^i \dim H_i'(T', z_0')$. Then we have

Proposition 1.6.2. *Suppose that $\mathcal{K}(T) = \sum n_i [\mathcal{O}_{M_i}]$ is the J. - H. decomposition of $\mathcal{K}(T)$ in a small neighborhood of the point z_0. Then*

$$\chi_{z_0}(K_\bullet(T', z_0')) = \sum n_i \, mult_{p,z_0} M_i$$

In view of the above, one can compute the multiplicities n_i at least for the subspaces M_i of maximal dimension. Indeed, let M_i be a complex space of dimension k participating in the J. - H. decomposition of $\mathcal{K}(T)$, and $z_0 \in M_i$ be a point not belonging to any other M_j. One can suppose that the coordinates of z' form a local coordinate system for M_i, and therefore the assumptions of the above proposition are satisfied. Then the integer n_i can be determined from the equality

$$\chi_{z_0}(K_\bullet(T', z_0')) = n_i \, mult_{p,z_0} M_i$$

1.7. The case of an n-tuple commuting modulo operators of finite rank.

Suppose that all the commutators $[T_i, T_j]$ are finite- dimensional. The element $\mathcal{K}(T)$ can be defined in this case with a suitable modification. Namely, consider the space $\mathbb{C}^n$ as a subspace of the complex projective space $\mathbb{CP}^n$ and take the projective coordinates $Z_0, \ldots, Z_n$. Consider the Koszul complex of the operators $Z_0 T_1 - Z_1 I, \ldots, Z_0 T_n - Z_n I$ (The differentials of this complex are defined up to a constant multiple, which does not affect the next definition). Denote by $d_i(z)$ the i-th differential of the system of sheaves $\mathcal{O}K_\bullet(T, z)$. Denote $\mathcal{L}_i(T) = im d_{i-1}(z) (im d_{i-1}(z) \cap ker d_i(z))$, $\mathcal{H}_i(T) = ker d_i(z) (im d_{i-1}(z) \cap ker d_i(z))$. Then we can define

$$\mathcal{K}(T) = \sum_{i=0}^n (-1)^i ([\mathcal{H}_i(T)] - [\mathcal{L}_i(T)])$$

It can be proved that the main properties of $\mathcal{K}(T)$ stated above for commuting T, still hold in the case of an n-tuple commuting modulo finite-dimensional operators.

The main difference with the commuting case is that here the support of some sheaves can be unbounded. One can separate the bounded and unbounded parts of $\mathcal{K}(T)$:

Proposition 1.7.1. *There exists a uniquely determined decomposition* $\mathcal{K}(T) = \mathcal{K}_\infty(T) + \mathcal{K}_c(T)$, *where* $\mathcal{K}_\infty(T)$ *is a sum of classes of holomorphic vector bundles on* $\mathbb{C}\mathbb{P}^n \setminus \sigma_e(T)$ *and* $\mathcal{K}_c(T)$ *is a sum of classes of coherent sheaves, supported on a bounded complex subset of* $\mathbb{C}^n \setminus \sigma_e(T)$ *(an analog of* $\sigma_F(T)$*).*

Taking the contribution of each summand of $\mathcal{K}(T)$ in the topological invariant, we obtain, in particular, the decomposition $ch_*(T) = ch_*^\infty(T) + ch_*^c(T)$.

Denote by $\widehat{\sigma_e(T)}$ the holomorphic hull of $\sigma_e(T)$ and by $i : \sigma_e(T) \to \widehat{\sigma_e(T)}$ the natural embedding. Since all the components of the support of $\mathcal{K}_c(T)$ of dimension two and more are included in $\widehat{\sigma_e(T)}$, then the corresponding elements of $ch_*^c(T)$ are killed by the embedding i , and therefore the degree of $i_* ch_*^c(T)$ is less or equal to one. On the other hand, $i_* ch_*(T)$ can be computed via the Connes construction applied to the algebra of functions, holomorphic in a neighborhood of $\widehat{\sigma_e(T)}$. Since the corresponding operators commute modulo trace-class operators, then this element is of degree not exceeding one. Therefore, the degree of $i_* ch_*^\infty(T)$ is less than or equal to one.

Problem 1.7.1. Is the degree of $ch_*^\infty(T)$ always less or equal to one?

1.8. Calculating the Chern character. Case of operators commuting modulo trace class.

The Chern character of an element of $K_0^{alg}(U)$ is investigated in several papers, e.g. [1] , [4] , [2]. It is known that its image lies in $\oplus H^{i,i}(U)$, where $H^{p,q}(U)$ is the Hodge decomposition of $H^n(U)$. It turns out that the Chern character of a commuting n-tuple can be calculated directly from the Koszul complex of this tuple. Moreover, the same construction is valid in the case when the n-tuple commutes modulo trace class operators. The construction below is parallel to the construction of the paper [2].

1.8.1. *Trace of an endomorphism of a complex.* Let $X_\bullet(z) = \{X_i, \alpha_i(z)\}$ be a finite Fredholm complex of Banach spaces with holomorphic differentials defined on the complex manifold M . One can fix smooth homotopies for $X_\bullet(z)$, i.e. a set of operators $S_i(z) : X_i \to X_{i-1}$, smoothly depending on the parameter $z \in M$ and all the differences $\alpha_{i-1}(z) \circ S_i(z) + S_{i+1}(z) \circ \alpha_i(z) - I_{X_i}$ are finite-dimensional.

Let $F(z)$ be a holomorphically depending on z closed element of $Hom_k(X_\bullet(z), X_\bullet(z))$. More precisely, $F(z) = \{F_i(z)\}$, $F_i(z) : X_i \to X_{i+k}$ with $\Delta F(z) := (-1)^{k-1}\{\alpha_{i+k}(z) \circ F_i(z) + F_{i+1}(z) \circ \alpha_i(z)\}$ equal to zero. Then
$SF(z) := \{S_{i+k}(z) \circ F_i(z)\}$ satisfies the equation $\Delta SF(z) = F(z) + f^0(z)$, where $f^0(z)$ is a finite-dimensional operator. Denote $G^0(z) :=$

$SF(z)$, $F^1(z) := \bar{\partial}G^0(z)$. Since Δ commutes with $\bar{\partial}$, then $\Delta F^1(z) = \bar{\partial}f^0(z)$. Continuing on the same way, we put $G^j(z) := SF^j(z)$, $F^{i+1}(z) := \bar{\partial}G^j(z)$, $j = 0, \ldots, k$. Then $\Delta G^j(z) = F^j(z) + f^j(z)$, where $f^j(z) = \{f_i^j(z)\}_i$ is an element of $Hom_{k-j}(X_\bullet(z), X_\bullet(z))$ with a finite- dimensional image. In particular, $f^k(z)$ is of degree zero, and one can define

$$tr\, F(z) = tr\, f^k(z) := \sum(-1)^i tr\, f_i^k(z) \in \Omega^{0,k}(M)$$

We have $\bar{\partial}(tr F(z)) = tr\bar{\partial}f^k(z) = tr\left(\Delta\bar{\partial}G^k(z)\right) = 0$ since $\bar{\partial}G^k(z)$ is a finite-dimensional operator. In the same way one can see that $tr F(z)$ depends on the choice of the homotopies $S_i(z)$ only up to a $\bar{\partial}$-exact form. Hence we obtain $tr F(z)$ as an element of the group $H^{0,k}(M)$.

Suppose now that the product of any two consecutive differentials in $X_\bullet(z)$ is in the trace-class $\mathcal{C}^1$, $S_i(z)$ are homotopies for $X_\bullet(z)$ up to trace-class operators, and the endomorphism $F(z)$ satisfies $\Delta F(z) \in \mathcal{C}^1$. Then the operators Δ and $\bar{\partial}$ commute modulo $\mathcal{C}^1$, and the construction above can be applied. The only difference in this case is that the summands $f^j(z)$ are not finite-dimensional but in $\mathcal{C}^1$, so $tr F(z)$ can be defined as above.

Finally, let E be a holomorphic vector bundle on M , and let $F(z)$ be a closed element of $Hom_k(X_\bullet, X_\bullet \otimes E)$. Then the construction leads to a $\bar{\partial}$-closed form with coefficients sections of E.

1.8.2. *The Chern character of a complex.* Let $X_\bullet(z) = \{X_i, \alpha_i(z)\}$ be a holomorphic Fredholm complex of Banach spaces on M as above. The complex differential $d\alpha_i(z)$ of the operators $\alpha_i(z)$ can be considered as a homomorphism $d\alpha = \{d\alpha_i\}_i$ of degree one from the complex $X_\bullet(z)$ to $X_\bullet(z) \otimes \Omega^{1,0}(M)$. Differentiating the equality $\alpha_{i+1}(z) \circ \alpha_i(z) = 0$, one sees that the homomorphism $d\alpha(z)$ is closed.

Now we will consider the powers $(d\alpha)^k$ of the homomorphism $d\alpha$. Evidently $(d\alpha)^k = \{d\alpha_{i+k-1}(z) \circ \ldots \circ d\alpha_i(z)\}_i$ is a closed element of $Hom_k\left(X_\bullet, X_\bullet \otimes \Omega^{k,0}(M)\right)$ satisfying $\partial(d\alpha)^k = 0$. Hence $tr(d\alpha)^k$ is a form from $\Omega^{k,k}(M)$ which is both ∂-closed and $\bar{\partial}$- closed and therefore determines an element of $H^{k,k}(M)$.

Proposition 1.8.2.1. *The element*

$$ch^*(X_\bullet(z)) = \sum \frac{1}{k!} tr\,(d\alpha)^k \in \sum H^{k,k}(M)$$

coincides with the Chern character of the alternated sum of sheaves of homologies of the complex $\mathcal{O}X_\bullet(z)$.

The same construction holds in the case when the product of any two consecutive differentials of $X_\bullet(z)$ is in $\mathcal{C}^1$.

1.8.3. *Chern character of n-tuple.* Applying the above construction to the Koszul complex, we obtain

Proposition 1.8.3.1. *If the operators of the n-tuple T commute modulo $\mathcal{C}^1$, then the element $ch^*(T)$ belongs to $\sum H^{k,k}\left(\mathbb{C}^n\backslash\sigma_e(T)\right)$*

Take the case of commuting n-tuple T. Then one can choose the homotopies $S_i(z)$ to be zero out of arbitrary small neighborhood of $\sigma_F(T)$. So, in the commuting case $ch^*(T)$ can be considered as an element of the group $H^{k,k}_{\sigma_F(T)}\left(\mathbb{C}^n\backslash\sigma_e(T)\right)$ of cohomologies with support in $\sigma_F(T)$.

Definition 1.8.3.1. The construction above can be performed without change when T is contained in a von Neuman algebra of type II and the commutators are in the trace class with respect to the corresponding trace.

Problem 1.8.3.1. To find a similar construction for $ch^*(T)$ (or, at least, for its sufficiently high components) in the case when $[T_i, T_j] \in \mathcal{C}^p$, $p > 1$.

2. APPLICATIONS

2.1. Index formula for Toeplitz operators. The original proof of the index formula for Toeplitz operators given by Boutet de Monvel in [3] was based on the Atiyah - Singer index formula for differential operators. In fact, he proved that these theorems are in some sense equivalent. Here we will see that this theorem in a more general situation is a simple consequence of the results of part I.

Take first the regular case, considered by Boutet de Monvel. Let N be a complex submanifold of $\mathbb{C}^n$, and M be a bounded pseudoconvex domain on N. Let E be a holomorphic vector bundle defined on a some neighborhood of M in N. Denote by $H^2(M, E)$ the Hilbert space of all square-integrable holomorphic sections of E over M. Denote by $T = (T_{z_1}, \ldots, T_{z_n})$ the n-tuple of operators of multiplications by the coordinate functions in $H^2(M, E)$.

Lemma 2.1.1. *The element $\mathcal{K}(T)$ (which coincides with $\mathcal{H}_n(T)$) is equal to i_*E , where i is the embedding of M in $\mathbb{C}^n\backslash bM$.*

Suppose that the pseudoconvex domain M is pseudoregular in the sense of [15] , i.e. the C^* - algebra, generated by T (the algebra of Toeplitz operators) is commutative modulo compacts. Then T determines an element $Ext(T)$ of the group $Ext(bM)$. The connection between $\mathcal{K}(T)$ and $K(T)$, resp. $Ext(T)$, stated above, permits us to calculate explicitly this element. Then, using the Riemann-Roch-Hirtzebruch theorem for the embedding i , one obtains the Boutet de Monvell formula:

Proposition 2.1.2. *One has*

$$Ext(T) = \left(ch(E) \bigcup Td(TM)\right) \bigcap [bM]$$

where $Td(TM)$ is the Todd class of the tangential bundle of M , and $[bM]$ is the fundamental class of bM.

Corollary 2.1.3. *Let G be a continuous invertible matrix function on bM , $[G]$ be the corresponding class in $K^1(bM)$, and T_G be the corresponding Toeplitz operator in $H^2(M, E)$. Then*

$$indT_G = \langle [G], Ext(T) \rangle = \left\langle ch(G) \bigcup ch(E) \bigcup Td(TM), [bM] \right\rangle$$

The formula above can extended to some classes of singular spaces. Namely, let U be a bounded pseudoregular domain in $\mathbb{C}^n$, and $\mathcal{L}$ a coherent sheaf on U , *privileged* with respect to U in the sense of Douady. This means that for any free resolution

$$0 \to \mathcal{O}^{n_0} \to \ldots \to \mathcal{O}^{n_k} \to \mathcal{L} \to 0$$

defined in a neighborhood of $\overline{U}$, the corresponding complex of Hilbert spaces

$$0 \to H^2(U)^{n_0} \to \ldots \to H^2(U)^{n_k} \to 0$$

is exact at all terms except the last one, and the image of the last differential is closed. In this case, one can define the Hilbert space $H^2(U, \mathcal{L}) = H^2(U)^{n_k}/im\, H^2(U)^{n_{k-1}}$. Denote by T the n-tuple of operators of multiplication by the coordinate functions in this space. The C^*-algebra generated by T commutes modulo compacts. The lemma above easily extends to this case, and we have the equality

$$\mathcal{K}(T) = \mathcal{L}|_U$$

Now, a construction for $Ext(T)$ can be described in the following way: let M be the support of $\mathcal{L}$, and let F be the intersection of M with bU. The complex of finite-dimensional spaces $L_\bullet : 0 \to \mathbb{C}^{n_0} \to \ldots \to \mathbb{C}^{n_k} \to 0$, corresponding to the free resolution of $\mathcal{L}$ above, can be extended up to a complex $\tilde{L}_\bullet$ on $\mathbb{C}^n \backslash F$, exact out of M. From I.3.1 one obtains the generalization of Boutet de Monvel theorem in the singular case:

Proposition 2.1.4. *In the conditions above the complex $\tilde{L}_\bullet$ represents $K(T)$, and the element $Ext(T)$ coincides with the Alexander dual to this element in the group $K_1(F)$.*

Problem 2.1.1. It could be interesting to apply the above "local" approach to more general index problems. As a possible directions, one could consider the theory of Toeplitz operators on bounded symmetric domains, developed by Upmeier (see [17]), or the type II index problems, appearing for C^* algebras of dynamical systems or foliations (see e.g. [7]). Note that both directions meet in the case of Toeplitz operators on Reinhardt domains, considered in [16]. A possible local approach could permit us to deal with a broader class of pseudoconvex domains, namely these which could locally be modeled by symmetric or Reinhardt domains. In these cases, the corresponding sheaves are not coherent. It is not clear what category of analytic sheaves should be taken instead in order to obtain a non-trivial local and global objects containing the necessary topological information.

An alternative approach in this case could be to study the eigendistributions for the corresponding n-tuple (the corresponding approach for a single operator was developed in [14]).

2.2. Lifting problems. Let $T = (T_1, \ldots, T_n)$ be an essentially commutative n-tuple, i.e. all the commutators $[T_i, T_j]$ are compact. We shall say that *the n-tuple T can be lifted up to a commutative n-tuple* if there exists a commuting n-tuple $T' = (T'_1, \ldots, T'_n)$ such that all the differences $T_i - T'_i$ are compact. Rather unexpectedly, it turns out that some obstructions to the possibility of lifting for the topologically non-trivial essentially commuting n-tuple T lie in the complex geometry of the essential spectrum $\sigma_e(T)$ in $\mathbb{C}^n$.

We shall denote by $deg\, ch_*(T)$ the degree of the maximal non-zero component of $ch_*(T)$ (an odd number or zero). As a direct consequence of the statements in part I concerning the Fredholm spectrum of a commuting n-tuple one obtains:

Proposition 2.2.1. *Suppose that the essentially commutative n-tuple T satisfies $deg\, ch_*(T) = 2p - 1$, $p > 0$, and the n-tuple T lifts to commutative n-tuple. Then $\sigma_e(T)$ contains the boundary of some bounded complex subset of $\mathbb{C}^n \backslash \sigma_e(T)$ of complex dimension p.*

Let us mention that the condition above depends not only on the topology of $\sigma_e(T)$ but mainly on the way it is embedded in $\mathbb{C}^n$. For instance, take the integer k with $2k < n$. Then, if the sphere S^{2k-1} is embedded in $\mathbb{C}^n$ as a unit sphere of some complex subspace $\mathbb{C}^k \subset \mathbb{C}^n$, then it contains the boundary of the unit ball B^{2k} in $\mathbb{C}^k$, i.e. of a k-dimensional complex manifold. On the contrary, if S^{2k-1} is embedded as a unit sphere of some *real* subspace $\mathbb{R}^{2k} \subset \mathbb{R}^n \subset \mathbb{C}^n$, then there does not exist a bounded complex subspace (of positive dimension) of $\mathbb{C}^n \backslash S^{2k-1}$.

Definition 2.2.1. Suppose that the answer of the question 1.7.2 is positive. Then in the same way as above one could obtain an obstruction to the lifting of an essentially commuting n-tuple T up to a n-tuple commuting modulo finite- dimensional operators in the case when $deg\, ch_*(T) > 1$.

Using proposition I.8.1, one obtains

Proposition 2.2.2. *Suppose that the essentially commuting n-tuple T can be lifted up to an n-tuple commuting modulo trace-class operators. Then the element $ch^*(T)$ belongs to $\sum H^{k,k}(\mathbb{C}^n \backslash \sigma_e(T))$.*

To demonstrate an application of the proposition 1.4.1, we will consider the Hardy space $H^2(D)$ of square-integrable holomorphic functions on the unit disk $D \subset \mathbb{C}$. To any continuous function $\varphi \in C(S)$ on the unit circle S corresponds the Toeplitz operator T_φ in $H^2(D)$. The question appears when, for a given n-tuple of functions $f = (f_1, \dots, f_n) \subset C(S)$, the corresponding n-tuple of Toeplitz operators $T_f = (T_{f_1}, \dots, T_{f_n})$ lifts. Let us state some cases when this is possible. First, if all the functions f_i belong to $A(D)$, i.e. can be extended as holomorphic continuous functions on the closed unit disk, then the operators T_{f_i} commute.

Next, suppose that the functions of f are obtained from some functions belonging to $A(D)$ by a change of variable. More precisely, suppose that there exist functions $g_1, \dots, g_n \in A(D)$ and a continuous mapping $h : S \to S$ such that $f_i = g_i \circ h$, $i = 1, \dots, n$. Denote by T_h the Toeplitz operator with symbol h. Since $\|T_h\| = 1$, then T_h admits $A(D)$-functional calculus. Consider the commuting operators $g_i(T_h)$; they are compact perturbations of the operators T_{f_i} , and therefore in this case the n-tuple T_f lifts. It turns out that under some regularity conditions this is the only case when this is possible:

Proposition 2.2.3. *Suppose that $f = (f_1, \dots, f_n) : S \to \mathbb{C}^n$ is a regular $1 + \varepsilon$-smooth embedding and the n-tuple T_f lifts. Then there exist functions $g_1, \dots, g_n \in A(D)$ and a continuous mapping $h : S \to S$ such that $f_i = g_i \circ h$, $i = 1, \dots, n$.*

Problem 2.2.1. Is there a similar statement valid for the Toeplitz operators on the unit ball of $\mathbb{C}^n$, $n > 1$?

2.3. Some classification theorems. In the case of commuting n-tuples of essentially normal operators in Hilbert space one can use I.2.4 with combination with Brown - Douglas - Fillmore theory to characterize some classes of n-tuples modulo unitary equivalence. We will consider two examples.

Proposition 2.3.1. *Suppose that T is a commuting n-tuple of essentially selfadjoint operators, i.e. all $T_i - T_i^*$ are compact. If $\sigma_e(T)$ is homeomorphic to a finite cell complex, then T is a compact perturbation of a commuting n-tuple of selfadjoint operators.*

This statement is a simple consequence of the fact that there is no bounded complex subspaces of $\mathbb{C}^n$ with boundary contained in C^n.

Now consider the case of commuting essentially unitary operators, i.e. of commuting n-tuples $T = (T_1, \dots, T_n$ such that $T_i T_i^* - I$ and $T_i^* T_i - I$ are compact operators for all i , $1 \leq i \leq n$. First, let us give some examples of such an n-tuples. Let $f(z) = (f_1(z), \dots, f_n(z))$ be an n-tuple of functions holomorphic in a neighborhood of the closed unit disk in the complex plain. Suppose that all $f_i(z)$ are finite Blaschke products, i.e. a product of finitely many factors of the type $\theta(z-a)(1-z\overline{a})$ with $|\theta| = 1$, $|a| > 1$. Then for $|z| = 1$ we have $|f_i(z)| = 1$. Denote by S the operator of unilateral left shift. Then the n-tuples $f(S)$ and $f(S^*)$ are commuting and consist on essentially unitary operators.

Now, let T be an n-tuple of commuting essentially unitary operators. Then $\sigma_e(T)$ is contained in the unit torus $\mathbb{T}^n = \{z = (z_1, \dots, z_n) : |z_1| = \dots = |z_n| = 1$. The Fredholm spectrum $\sigma_F(T)$ is a bounded complex subspace of $\mathbb{C}^n \backslash \mathbb{T}^n$ and therefore its dimension does not exceed one. The irreducible components of this complex space can be easily described as images of the unit disk under suitable holomorphic mapping, and it is easy to see that the coordinate functions of this mapping are finite Blaschke products. Finally, we obtain

Proposition 2.3.2. *Suppose that T is an n-tuple of commuting essentially unitary operators and $\sigma_e(T)$ is homeomorphic to a finite cell complex. Then T is BDF- equivalent to a finite direct sum of n-tuples of the form $f(S)$ and $g(S^*)$, where all f, g are n-tuples of Blaschke products as above, plus a diagonal operator.*

The statement above suggests that from the spectral point of view the n- tuple of commuting essentially unitary operators is not very far from the case of a single operators. In this way, the following problems appear:

Problem 2.3.1. Find a finer unitary invariant discerning the non-equivalent tuples with an identical spectral picture. In particular, in the case when all the operators $T_i T_i^* - I$ and $T_i^* T_i - I$ are finite-dimensional, find a multi-dimensional generalization of the Nagy - Foias characteristic function.

Problem 2.3.2. Is there an Apostol - Brown - Pearcy - Chevreau spectral approach to the problem of invariant subspaces in this case? In particular, is it true that any n-tuple of commuting essentially unitary operators with non-empty Fredholm spectrum has a rich lattice of common invariant subspaces?

References

1. M.F.Atiyah, F.Hirzebruch, *Riemann-Roch theorem for differentiable manifold*, Bull.Amer.Math.Soc. **65** (1959), 275–281.

2. B. Angeniol, M. Lejeune-Jalabert, *Calcul differentiel and classes characteristique en geometrie algebraique*, Prepublication de l'Institute Fourier, Grenoble, **28** (1985).

3. L. Boutet de Monvel, *On the index of Toeplitz operators of several variables*, Invent. Math.,**50** (1979), 249–272.

4. N. R. O'Brian, D. Toledo, Y.L.L. Tong, *Hirzebrich-Riemann-Roch for coherent sheaves*, Amer. J. Math., **103** (1981), 253–271.

5. R. W. Carey, J. D. Pincus, *Operator theory and the boundaries of the complex curves*, preprint, 1982.

6. R. E. Curto, *Fredholm and invertible tuples of operators. The deformation problem*, Trans. Amer. Math. Soc. **226** (1981), 129– 159.

7. R. Curto, P. Muhly, J. Xia, *Toeplitz operators on flows*, J. Funct. Anal. **93**(1990), 391–450.

8. A. S. Fainstein, V. S. Schulman, *On the Fredholm complexes of Banach spaces* (Russian), Functional. Anal. i Prilozen. **14** (1984), 83- -84.

9. R. N. Levy, *Cohomological invariant of essentially commuting tuples of operators* (Russian), Functional. Anal. i Prilozen., **17** (1983), 79– 80.

10. R. N. Levy, *Algebraic and topological K-functors of commuting n-tuple of operators*, J. Operator Theory, **21** (1989), 219–253.

11. J. D. Pincus, *The principal index*, Proc. of Simp. of Pure and Appl. Math., v.**51** part I (1990) , 373–394.

12. M. Putinar, *Some invariants of semi-fredholm systems of essentially commuting operators*, J. Operator Theory, **8** (1982), 65–90.

13. M. Putinar, *Base change and Fredholm index*, Int. Eq. and Op. Th. **8** (1985), 674–692.

14. M. Putinar, *Hyponormal operators and eigendistributions*, Operator Theory: Advances and Applications, v. **17** (1986), Birkhauser v. Basel, 249–273.

15. N. Salinas, *Application of C^*-algebras and Operator Theory to proper holomorphic mappings*, Proc. of Simp. of Pure and Appl. Math., v.**51** part I (1990) ,481–492.

16. N. Salinas, A. Sheu, H. Upmeier, *Toeplitz C^*-algebras over pseudoconvex Reinhardt domains*, Ann. of Math. **130** (1989), 531- 565.

17. H. Upmeier, *Fredholm indices for Toeplitz operators on bounded symmetric domains*, Am. J. of Math. **110** (1988), 811–832.

SOFIA UNIVERSITY, FACULTY OF MATHEMATICS,, "J. BOURCHIER" STR. 5,, SOFIA 1126, BULGARIA
   E-mail address: TROYLEVBGEARN.BITNET

Contemporary Mathematics
Volume **185**, 1995

Schatten Class of Hankel and Toeplitz Operators on the Bergman Space of Strongly Pseudoconvex Domains

Huiping Li and Daniel H. Luecking

1. Introduction

Let D be a strongly pseudoconvex domain in $\mathbf{C}^n$ with the strictly plurisubharmonic

defining function ρ. (That is $D = \{z : \rho(z) < 0\}$.) Let β denote the Bergman

distance function on D. Let dV denote Lebesgue volume on D and $d\lambda(z) =$

$|\rho(z)|^{-n-1} \, dV(z)$. The Bergman space on D is the space $A^2 = L^2(D, dV) \cap \mathcal{H}$,

where $\mathcal{H}$ is the collection of holomorphic functions on D. For a function $f \in L^1(dV)$,

the Hankel operator H_f is densely defined on A^2 by the formula

$$H_f(g)(z) = (I - P)(fg)(z) = f(z)g(z) - \int_D K(z,w)f(w)g(w) \, dV(w),$$

where $K(z, w)$ is the Bergman kernel for D and P is the Bergman projection,

extended to $L^1(dV)$ via the integral above.

In [5], a characterization was obtained of the functions f for which H_f is a

bounded operator from A^2 to L^2. Here we obtain a parallel characterization of

those f for which H_f belongs to the Schatten class $\mathcal{S}_p$, $0 < p < \infty$.

In the proof, we need a corresponding characterization for multiplication

operators M_f defined from A^2 to L^2 by $M_f g = fg$. Such a characterization is

obtained by observing that $\|M_f g\|^2 = \int |g|^2 |f|^2 \, dV = \langle T_\mu g, g \rangle$. Where μ is the

measure $|f|^2 \, dV$ and T_μ is the Toeplitz operator defined by $\langle T_\mu f, g \rangle = \int_D f\bar{g} \, d\mu$.

The charcterization of μ for which T_μ is in the Schatten class was carried out

in [8] for the unit disk. In a strongly pseudoconvex domain this was done for

$p \geq 1$ by Zhimin Yan (private communication). In Section 2 we will present a

complete characterization for a strongly pseudoconvex domain. This is the bulk of

the paper, because of the technical details required when $p < 1$. These technical

details are also needed in the Hankel operator proof (Section 3) where they will

not be repeated.

One characterization of bounded Hankel operators was the boundedness of a

local measurement of the distance to analytic functions $MDA(f, z)$ (see Section 3

1991 *Mathematics Subject Classification.* Primary 47B35, 47B10; Secondary 46E20.

for definitions). The corresponding characterization for the Schatten class $\mathcal{S}_p$ is that $MDA(f, z)$ belong to $L^p(D, \lambda)$. A second characterization of boundedness, was the decomposability of f into a sum $f_1 + f_2$ where a local measurement of L^2 means, $\mathcal{A}(f_1, z)$, is bounded and the expression

$$|\rho|^{1/2} \left|\bar{\partial} f_2 \wedge \bar{\partial}\rho\right| + |\rho| \left|\bar{\partial} f_2\right|$$

is also bounded. For the Schatten classes, we find similar expressions must belong to $L^p(\lambda)$.

Another characterization of boundedness was the simple condition $(I - P)f \in$ BMO. Not surprisingly, this was the hardest to prove (see [7]) and provides the most difficulty in finding an appropriate analog for the Schatten classes. We obtain in Sections 4 and 5 two partial solutions: In Section 4 we obtain the complete picture for the case of the unit disk, but the analogous mean oscillation condition (which can be thought of as measuring the local distance to constant functions) must of necessity be replaced by a local mean distance to higher degree polynomials as p decreases. (This is analogous to the way H^p-atoms are required to have higher vanishing moments as p decreases.) In Section 5, we obtain for a strongly pseudoconvex domain the partial result that for $p > 2n$, the Hankel operator H_f belongs to $\mathcal{S}_p$ if and only if $(I - P)f \in B_p$, where $g \in B_p$ is defined by the condition that the mean oscillation $MO(g, z)$ (defined in Section 5) belongs to $L^p(\lambda)$.

2. The Toeplitz operators

Fix a number $r > 0$ and for $z \in D$ let $B(z, r)$ denote the ball about z of radius r in the Bergman metric β. Let $B(z) = B(z, r)$ for this prechosen r. $|E|$ will denote the volume of the set E.

Let $\{z_k\}$ be a sequence in D separated by $r/2$ (that is $\beta(z_m, z_k) > r/2$ when $m \neq k$) such that the collection of balls $B(z_k, r/2)$ covers D. Write B_k for $B(z_k)$. The letters C and c will denote various constants which may differ from one occurrence to the next, even in the same line.

2.1 Theorem. *Let μ be a positive finite measure on D and let the operator T_μ be densely defined on A^2 by*

$$\langle T_\mu f, g \rangle = \int f \bar{g} \, d\mu. \tag{1}$$

Then a necessary and sufficient condition on μ in order that T_μ belong to the Schatten class $\mathcal{S}_p(A^2)$ is that

$$\sum_k \left(\frac{\mu(B_k)}{|B_k|} \right)^p < \infty. \tag{2}$$

The value of the sum will depend on the sequence $\{z_k\}$ but its finiteness does not.

PROOF. The case $p \geq 1$ is so much like the proof for the unit disk in [8] that we will be somewhat sketchy. First suppose that $p = 1$ and that the sum above is finite. We will prove actually prove that the above condition implies that $T_\nu \in \mathcal{S}_p$ for any measure ν (not necessarily positive) satisfying $|\nu| \leq \mu$.

If s_m denotes the singular numbers of T_ν, then we may write

$$T_\nu(h) = \sum_m s_m \langle h, e_m \rangle f_m$$

for some orthonormal sequences $\{e_m\}$ and $\{f_m\}$. Then

$$|T_\nu|_1 = \sum_m s_m = \sum_m \langle T_\nu e_m, f_m \rangle$$

$$= \sum_m \int e_m \bar{f}_m \, d\nu$$

$$\leq \left(\sum_m \int |e_n|^2 \, d|\nu| \right)^{1/2} \left(\sum_m \int |f_n|^2 \, d|\nu| \right)^{1/2}$$

$$\leq \int K(z,z) \, d\mu(z) \leq \sum_k \int_{B_k} K(z,z) \, d\mu(z) \leq C \sum_k \frac{|\nu|(B_k)}{|B_k|},$$

which give the sufficiency of (2) for $p = 1$.

The $p = \infty$ version of (2): T_μ is bounded if and only if $\sup_k \mu(B_k)/|B_k| < \infty$, was shown in [5].

Define $E_1 = B_1$ and let $E_k = B_k \setminus \bigcup_{j<k} B_j$ and define the operator S from finite sequences to bounded operators on A^2 by $S((c_k)) = T_\nu$ where

$$\nu = \sum_k c_k \frac{|B_k|}{\mu(B_k)} \chi_{E_k} \mu$$

and the sum is taken over all k for which $\mu(B_k)$ is nonzero. It is simple to show that S is bounded from the l^1 norm to the $\mathcal{S}_1$ norm and from the l^∞ norm to the bounded operator norm. By interpolation (see [4]) S is bounded from l^p to $\mathcal{S}_p$ for all $1 \leq p < \infty$:

$$|T_\nu|_p^p \leq C_p \sum_k |c_k|^p.$$

Letting (c_k) be the sequence $(\mu(B_k)/|B_k|)$ gives us

$$|T_\mu|_p^p \le C_p \sum_k \left(\frac{\mu(B_k)}{|B_k|}\right)^p$$

That is, (2) is sufficient for all $p \ge 1$.

To obtain necessity when $p \ge 1$, we need ([4]): If $p \ge 1$ and A is any bounded operator, then for any orthonormal sequence $\{e_m\}$ and any operator T, $\sum |\langle TAe_m, Ae_m \rangle|^p \le \|A\|^2 |T|_p^p$. Let A be the operator that takes $\sum c_m e_m$ to $\sum c_m |B_m|^{1/2} K(z, z_m)$. This operator is bounded because it is the adjoint of the bounded operator which takes a function f in A^2 to $\sum |B_m|^{1/2} f(z_m) e_m$. This gives us

$$\|A\|^2 |T_\mu|_p^p \ge \sum_m |\langle A^* T_\mu A e_m, e_m \rangle|^p \ge \sum_m |\langle T_\mu A e_m, A e_m \rangle|^p$$

$$\ge \sum_m \left(|B_m| \int |K(z, z_m)|^2 \, d\mu(z)\right)^p$$

$$\ge \sum_m \left(|B_m| \int_{B_m} |K(z, z_m)|^2 \, d\mu(z)\right)^p \ge c \sum_m \left(\frac{\mu(B_m)}{|B_m|}\right)^p$$

from the C. Fefferman estimates [3], if r is small enough.

Now let $p < 1$ and assume condition (2) on the measure μ. The proof of the theorem in this case is, in general outline, also like that on the unit disk.

We use the following fact ([4]): If an operator A is both bounded and onto (that is it has a bounded right inverse) then for any operator T and orthonormal sequence $\{e_m\}$.

$$|T|_p^p \le C \sum_m \sum_k |\langle TAe_m, Ae_k \rangle|^p. \tag{3}$$

To construct an appropriate A, we use the results of Coupet [1, Théorème 1]: There is a sequence φ_m in A^2 such that the operator A sending a sequence (c_m) to $\sum_m c_m \varphi_m$ maps l^2 boundedly *onto* A^2. This alone is not enough for our purposes: We need the particular form of the functions φ_m. They have the form

$$\varphi_m(z) = \frac{|\rho(z_m)|^{(n+1)/2+\alpha-1}}{\Phi(z, z_m)^{n+\alpha}} \psi(z, z_m) \tag{4}$$

where (1) $\{z_k\}$ is a sequence as before, but with r required to be sufficiently small, (2) α is any number greater than $1/2$ (which we will choose later), (3) ψ is holomorphic in the first variable and bounded, and (4) $\Phi(z, w)$ (called $A(z, w)$ in [1]) is holomorphic in z, has positive real part, and satisfies (among other things), for $0 < \lambda < s$,

$$\sup_{z \in D} |\rho(z)|^{s-\lambda} \int_D \frac{|\rho(w)|^{\lambda-1}}{|\Phi(z, w)|^{n+s}} \, dV(w) < \infty.$$

We need a slight extension of this last property, which is proved in the same way as the above:

$$\sup_{z \in D} |\rho(z)|^{s-\lambda} \sum_{k} \frac{|\rho(z_k)|^{\lambda+n}}{|\Phi(z, z_k)|^{n+s}} < \infty. \tag{5}$$

This combined with the expression (4) above for φ_m gives the following estimate (provided α is chosen so large that $p[(n + 1)/2 + \alpha - 1] > n$):

$$\sum_{m} |\varphi_m(z)|^p \le C|\rho(z)|^{-p(n+1)/2}.$$

Use this operator A with this α (that is, α is chosen so that the above holds, r is chosen to make Coupet's theorem valid and A is defined by $Ae_m = \varphi_m$). We estimate

$$
\begin{aligned}
|T_\mu|_p^p &\le C \sum_{k} \sum_{m} |\langle T_\mu \varphi_m, \varphi_k \rangle|^p \\
&\le C \sum_{k} \sum_{m} \left(\int_D |\varphi_m(z)\varphi_k(z)| \, d\mu(z) \right)^p \\
&\le C \sum_{k} \sum_{m} \left(\sum_{j} \int_{B_j} |\varphi_m(z)\varphi_k(z)| \, d\mu(z) \right)^p \\
&\le C \sum_{k} \sum_{m} \sum_{j} \left(\int_{B_j} |\varphi_m(z)\varphi_k(z)| \, d\mu(z) \right)^p
\end{aligned}
\tag{6}
$$

Let Π_j denote a polydisk containing B_j with radius $c|\rho(z_j)|$ in the complex normal direction and radii $c\sqrt{|\rho(z_j)|}$ in the orthogonal directions. Let $\widetilde{\Pi}_j$ denote the polydisk with the same center and orientation, but with the radii expanded by a factor of 2. We can choose r small enough that $\widetilde{\Pi}_j$ is contained in $B(z_j, Cr)$ for some C. Since $|\varphi_m \varphi_k|^p$ is plurisubharmonic, we can estimate

$$\sup_{z \in B_j} |\varphi_m(z)\varphi_k(z)|^p \le C \frac{1}{|\widetilde{\Pi}_j|} \int_{\widetilde{\Pi}_j} |\varphi_m \varphi_k|^p \, dV.$$

Continuing with the estimates from (6)

$$
\begin{aligned}
|T_\mu|_p^p &\le C \sum_{k} \sum_{m} \sum_{j} \mu(B_j)^p \frac{1}{|\widetilde{\Pi}_j|} \int_{\widetilde{\Pi}_j} |\varphi_m \varphi_k|^p \, dV \\
&\le C \sum_{j} \mu(B_j)^p \frac{1}{|\widetilde{\Pi}_j|} \int_{\widetilde{\Pi}_j} \left(\sum_{m} |\varphi_m(z)|^p \right)^2 dV(z) \\
&\le C \sum_{j} \mu(B_j)^p \frac{1}{|\widetilde{\Pi}_j|} \int_{\widetilde{\Pi}_j} |\rho(z)|^{-p(n+1)} \, dV(z) \\
&\le C \sum_{j} \mu(B_j)^p |\rho(z_j)|^{-p(n+1)} \\
&\le C \sum_{j} \mu(B_j)^p |B_j|^{-p}
\end{aligned}
$$

This gives the sufficiency of (2) for the case $p < 1$.

The necessity is somewhat more difficult. We will again compose T_μ with operator A and A^*, but the goal will be to obtain a matrix form of T_μ where the diagonal is large compared to the rest of the matrix. The diagonal will be estimated from below by the expression in (2).

Thus let $|T_\mu|_p < \infty$. The first step is to observe that if $0 \le \nu \le \mu$ then $\langle T_\nu f, f \rangle \le \langle T_\mu f, f \rangle$. That is, $\|T_\nu^{1/2} f\| \le \|T_\mu^{1/2} f\|$, and so $T_\nu^{1/2}$ is the composition of $T_\mu^{1/2}$ with a contraction. Thus $|T_\nu|_p \le |T_\mu|_p$. Let $\{z_k\}$ be as above, that is, $\beta(z_k, z_m) > r/2$ and the balls $B(z_k, r/2)$ cover D. Once again the number r will have to be chosen sufficiently small to make certain later estimates valid. Given any large number $R > 0$, There is a positive integer M such that the natural numbers can be partitioned into M subsets J_i such that for each J_i the sequence $\{z_m : m \in J_i\}$ satisfies the stronger separation condition:

$$\text{dist}_\beta(B_k, B_m) \ge R, \qquad k \ne m, \ k, m \in J_i.$$

The value of R will be chosen much later, but the estimates before then will not depend on it, and therefore they will also not depend on M. Consider one of these subsequences $\{z_m : m \in J_i\}$, but call it simply $\{z_m\}$ and continue to denote the corresponding Bergman metric balls by B_m. Replace μ by the measure $\sum_m \chi_{B_m} \mu$ and continue to call it μ. If it can be shown that $\sum \mu(B_m)^p / |B_m|^p < \infty$ in this restricted case, for some choice of R, then by summing M such estimates, we will have the necessity for any μ. Finally, it does no harm to delete a finite number of the points in the sequence (altering μ so that it remains supported on the corresponding balls B_m), so that when later a $\delta > 0$ is specified, we may assume that $|\rho(z_m)|$ is always less than δ.

Let φ_m be defined exactly as in (4) for the new sequence $\{z_m\}$ and let A be defined by $A e_m = \varphi_m$ for some orthonormal basis $\{e_m\}$. For convenience we choose α so that $(n+1)/2 + \alpha - 1 = (n+1)/p$. Let X be the operator $A^* T_\mu A$ denote by Δ the operator whose matrix with respect to $\{e_n\}$ is the diagonal of the matrix of X. Then the Schatten $\mathcal{S}_p$ norm of Δ is estimated from below by

$$|\Delta|_p^p = \sum_m \langle \Delta e_m, e_m \rangle^p = \sum_m \langle T_\mu \varphi_m, \varphi_m \rangle^p$$

$$\ge \sum_m \left(\int_{B_m} \left(\frac{|\rho(z_m)|^{(n+1)/p}}{|\Phi(z, z_m)|^{(n+1)/p + (n+1)/2}} |\psi(z, z_m)| \right)^2 d\mu \right)^p.$$

At this point we need to know a little about the function $\psi(z, w)$. In [1] it is called $B(z, w)$. One property in particular that we need is that for w near enough to ∂D it satisfies

$$\inf_{z \in B(w,r)} |\psi(z, w)| > c > 0.$$

This follows from the form of $\psi(z, w)$ and the properties of the various functions involved in its definition. Essentially, $|\psi(z, w)| = |(\bar{\partial}\partial\rho(w))^n| + O(|\rho(w)| + |z - w|)$. The first term is the determinant of the matrix $(\partial^2 \rho / \partial\bar{z}_i \partial z_j)$, which is positive on $\overline{D}$. The second term is bounded on $B(w, r)$ by $C\rho(w)$, so if we take δ sufficiently small and assume $|\rho(w)| < \delta$, then we get the above lower bound. In particular, taking only z_m with $|\rho(z_m)| < \delta$ we have the following lower estimate

$$\frac{|\rho(z_m)|^{(n+1)/p}}{|\Phi(z, z_m)|^{(n+1)/p+(n+1)/2}} |\psi(z, z_m)| \geq c|\rho(z_m)|^{-(n+1)/2},$$

(because we also have $\sup_{z \in B(w,r)} |\Phi(z, w)| \leq C|\rho(w)|.$). Putting this into the estimates above for $|\Delta|_p^p$ gives

$$|\Delta|_p^p \geq c \sum_m \frac{\mu(B_m)^p}{|\rho(z_m)|^{p(n+1)}} \geq c \sum_m \frac{\mu(B_m)^p}{|B_m|^p},$$

as required. Note that in this estimate, the constant c does not depend on R.

Now we will show that for appropriate choice of R, the Schatten norm of the off diagonal operator $X - \Delta$ will be an arbitrarily small multiple of the above sum $\sum \mu(B_m)^p / |B_m|^p$. Once this is shown, we will have

$$\infty > |X|_p^p \geq |\Delta|_p^p - |X - \Delta|_p^p \geq (c - \epsilon) \sum \mu(B_m)^p / |B_m|^p.$$

We finally estimate the Schatten norm of $X - \Delta$ as follows:

$$|X - \Delta|_p^p \leq \sum_k \sum_m |\langle (X - \Delta)e_m, e_k \rangle|^p$$

$$\leq \sum_{k \neq m} |\langle T_\mu \varphi_m, \varphi_k \rangle|^p$$

$$\leq C \sum_{k \neq m} \left(\int_D \frac{|\rho(z_m)\rho(z_k)|^{(n+1)/p}}{|\Phi(z, z_m)\Phi(z, z_k)|^{(n+1)/2+(n+1)/p}} \, d\mu(z) \right)^p$$

$$\leq C \sum_{k \neq m} \sum_j \left(\int_{B_j} \frac{|\rho(z_m)\rho(z_k)|^{(n+1)/p}}{|\Phi(z, z_m)\Phi(z, z_k)|^{(n+1)/2+(n+1)/p}} \, d\mu(z) \right)^p$$

$$\leq C \sum_{k \neq m} \sum_j \frac{1}{|\widetilde{\Pi}_j|} \int_{\widetilde{\Pi}_j} \frac{|\rho(z_m)\rho(z_k)|^{(n+1)}}{|\Phi(z, z_m)\Phi(z, z_k)|^{p(n+1)/2+(n+1)}} \mu(B_j)^p$$

$$\leq C \sum_j \frac{\mu(B_j)^p}{|B_j|^p} \frac{1}{|\widetilde{\Pi}_j|} \int_{\widetilde{\Pi}_j} \sum_{k \neq m} \frac{|\rho(z)|^{(n+1)p}|\rho(z_m)\rho(z_k)|^{(n+1)}}{|\Phi(z, z_m)\Phi(z, z_k)|^{p(n+1)/2+(n+1)}} \, dV(z).$$

$$(7)$$

Where $\widetilde{\Pi}_j$ are polydisks just as before. Here again we need to make sure r is sufficiently small that $\widetilde{\Pi}_j$ is contained in some $B(z_j, Cr)$.

Our next goal is to estimate this double sum over $k \neq m$ and show that for R large enough, it is small uniformly in z. This requires us to go into some of the details behind inequality (5). Let ϵ be a positive number to be specified in more detail later. Divide the double sum into three parts: (i) The sum over all k and over only those m with $|z - z_m| > \epsilon$, (ii) a similar sum with the roles of k and m reversed, and (iii) the remaining sum over only those m and k with $\max(|z - z_k|, |z - z_m|) < \epsilon$. Because $|\Phi(z, w)| > c|z - w|$, the first sum can be estimated as follows

$$\sum_{k \neq m} \frac{|\rho(z)|^{(n+1)p}|\rho(z_m)\rho(z_k)|^{(n+1)}}{|\Phi(z, z_m)\Phi(z, z_k)|^{p(n+1)/2+(n+1)}}$$

$$\leq |\rho(z)|^{(n+1)p/2} \frac{\sum_m |\rho(z_m)|^{n+1}}{\inf_{|z-z_m|>\epsilon} |\Phi(z, z_m)|^{p(n+1)/2+n+1}} \sum_k \frac{|\rho(z)|^{(n+1)p/2}|\rho(z_k)|^{n+1}}{|\Phi(z, z_k)|^{p(n+1)/2+n+1}}$$

$$\leq C|\rho(z)|^{(n+1)p/2} \frac{|\bigcup_m B_m|}{\epsilon^{p(n+1)+2n+2}}$$

where the last inequality comes from (5). A similar estimate is valid for the second sum. Now select ϵ and δ so that on the set $\{w : |w - z| < \epsilon, |\rho(z)| < \delta\}$, the change of variables used in [2] has bounded distortion. This choice is not dependent on R and once it is made we can choose δ still smaller (relative to ϵ) so that the above is made arbitrarily small.

This change of variables takes w to $(t_1, t_2, t_3, \ldots, t_{2n}) \in \mathbf{R}^{2n}$ defined by $t_1 = t_1(w) = -\rho(w)$, $t_2 = \operatorname{Im} \Phi(z, w)$ and $(t_3 + it_4, \ldots t_{2n-1} + it_{2n})$ are coordinates for the complex tangent space to the level set $\{w : \rho(w) = \rho(z)\}$ at z. For simplicity let t_1 be denoted by ρ, t_2 by σ and $(t_3, t_4, \ldots, t_{2n})$ by t'. This transformation is uniformly continuous from $\{w \in D : |w - z| < \epsilon\}$ with the Bergman metric to the half space $\rho > 0$ in $\mathbf{R}^{2n}$ with the metric τ whose infinitesimal form is

$$\frac{d\rho^2 + d\sigma^2}{\rho^2} + \sum_{j=3}^{2n} \frac{dt_j^2}{\rho}.$$

Thus the separation properties of the sequence B_m are inherited by the images B'_m: These are contained in balls S_m in the τ metric with radius Cr which are separated by cR, with C and c independent of R.

This change of variables transforms the sum at the end of (7) into one dominated by

$$C \sum_{m \neq k} \frac{|\rho(z)|^{(n+1)p} \rho_m^{n+1} \rho_k^{n+1}}{[(|\rho(z)| + \rho_m + |\sigma_m| + |t'_m|^2)(|\rho(z)| + \rho_k + |\sigma_k| + |t'_k|^2)]^{p(n+1)/2+n+1}},$$

where (ρ_m, σ_m, t'_m) is the image of z_m under this change of variables and $(|\rho(z)|, 0, 0)$ is the image of z. The balls S_m are easily seen to have volume proportional to ρ_m^{n+1} and thus it is routine to estimate the sum above by the following integral:

$$C \iint_{\tau(x,y)>cR} \frac{|\rho(z)|^{(n+1)p} \, dV(x) \, dV(y)}{[(|\rho(z)| + x_1 + |x_2| + |x'|^2)(|\rho(z)| + y_1 + |y_2| + |y'|^2)]^{p(n+1)/2+n+1}}.$$

The metric τ is invariant under the change of variables which dilates by a factor d in the first two coordinates and by a factor of $\sqrt{d}$ in the remaining coordinates. Performing the dilation using $|\rho(z)|$ for d as a change of variables gives the following for the integral above:

$$C \iint_{\tau(x,y)>cR} \frac{dV(x) \, dV(y)}{[(1 + x_1 + |x_2| + |x'|^2)(1 + y_1 + |y_2| + |y'|^2)]^{p(n+1)/2+n+1}}.$$

The integrand is integrable over the product of the two half spaces and the domain of integration shrinks to the empty set as R tends to infinity. Thus the expression above tends to 0. Putting this finally for the double sum in the last estimate of (7) gives the result we seek: $|X - \Delta|_p$ is bounded by an arbitrarily small multiple of $\sum_m \mu(B_m)^p/|B_m|^p$ for sufficiently large R. This ends the proof. ∎

3. The Hankel operators

For a function f locally in L^2 define

$$A(f, z) = \left(\int_{B(z)} |f|^2 \, d\lambda \right)^{1/2} \sim \left(\frac{1}{|B(z)|} \int_{B(z)} |f|^2 \, dV \right)^{1/2}$$

and define the *local mean distance* from analytic functions:

$$MDA(f, z) = \inf_{h \in \mathcal{H}} A(f - h, z)$$

3.1 Theorem.   Let $0 < p < 1$ and let $f \in L^1(D)$. Then the following are equivalent:

(a) H_f belongs to the Schatten class $S_p(A^2)$.

(b) $MDA(f) \in L^p(D, d\lambda)$.

(c) $f = f_1 + f_2$ where $\mathcal{A}(f_1) \in L^p(d\lambda)$ and the function

$$z \mapsto \sup_{w \in B(z)} \left(|\rho(w)\bar{\partial}f_2(w)| + |\rho(w)|^{1/2}|\bar{\partial}\rho(w) \wedge \bar{\partial}f_2(w)| \right)$$

belongs to $L^p(d\lambda)$.

As before, let $\{z_n\}$ be a sequence in D such that $\beta(z_n, z_k) > r/2$ when $n \neq k$ and such that the balls $B(z_n, r/2)$ cover D. Then $\mathcal{A}(f) \in L^p(d\lambda)$ if and only if $\sum_n \mathcal{A}(f, z_n)^p < \infty$.

PROOF. First let $p \geq 1$. The equivalence of (a), (b) and (c) in this case was shown in [9] for the unit disk. The essence of the proof is the same for strongly pseudoconvex domains now that we have the Toeplitz operator theorem.

Assume (a): that $H_f \in \mathcal{S}_p$. This implies that for any orthonormal set $\{e_n\}$ and bounded operators A and B

$$\sum_n |\langle B^* H_f A e_n, e_n \rangle|^p \leq C |H_f|_p^p \tag{8}$$

where $|\cdot|_p$ denotes the Schatten norm on $\mathcal{S}_p$. We take for A map sending e_n to the normalized Bergman kernel k_n for the point z_n and for B we map each e_n to the function

$$g_n = \begin{cases} \dfrac{\chi_{B(z_n)} H_f k_n}{\|\chi_{B(z_n)} H_f k_n\|} & \text{when the denominator is not 0,} \\ 0 & \text{otherwise.} \end{cases}$$

Now we can estimate from below the quantity in (8):

$$\begin{aligned} |\langle H_f k_n, g_n \rangle| &= \left(\int_{B(z_n)} |f k_n - P(f k_n)|^2 \, dA \right)^{1/2} \\ &= \left(\int_{B(z_n)} |f - k_n^{-1} P(f k_n)|^2 |k_n|^2 \, dV \right)^{1/2} \\ &\geq c \inf_{h \in \mathcal{H}} \left(\frac{1}{|B(z_n)|} \int_{B(z_n)} |f - h|^2 \, dV \right)^{1/2}, \end{aligned} \tag{9}$$

where the lower estimate in the last line is from the C. Fefferman estimates ([3]) and is valid for r sufficiently small and z_n sufficiently near the boundary, i.e., for all but finitely many n. Combining (8) with (9) above proves (b).

Let (b) be assumed now. The estimates from [5] and [9] provide a method to decompose an arbitrary locally integrable function f into $f_1 + f_2$ satisfying

$(1/|B(z, r/2)|) \int_{B(z, r/2)} |f_1|^2 \, dV \leq C \cdot MDA(f, z)^2$ and $\sup_{w \in B(z, r/2)} |\rho(w) \bar{\partial} f_2(w)| + |\rho(w)| |\bar{\partial} \rho(w) \wedge \bar{\partial} f_2(w)| \leq C \cdot MDA(f, z)$. Thus (c) follows from (b).

This argument for the implication (b) $\Rightarrow$(c) is the same if $p < 1$. The following proof of (c) $\Rightarrow$(a) is also valid for all $p > 0$.

Assume (c). It is straightforward that the operator M_{f_1} of multiplication by f_1 satisfies $M_{f_1}^* M_{f_1} = T_{|f_1|^2}$, so that the conditions on f_1 are the conditions from Theorem 2.1 that are equivalent to $M_{f_1} \in \mathcal{S}_p$. Thus $H_{f_1} = (I - P)M_{f_1}$ also lies in $\mathcal{S}_p$. In addition, the conditions on f_2 imply that multiplication by $|\rho(w) \bar{\partial} f_2(w)| + |\rho(w)|^{1/2} |\bar{\partial} \rho(w) \wedge \bar{\partial} f_2(w)|$ is in $\mathcal{S}_p$ from A^2 to L^2. Moreover, it was shown in [5] that H_{f_2} can be represented as the composition of an integral operator and the projection $I - P$. The integral operator is $\bar{\partial} f_2(w) \wedge L(z, w)$ where $L(z, w)$ is a kernel with the property that $\varphi(w) \wedge L(z, w)$ defines a bounded operator whenever $|\rho \varphi| + |\rho|^{1/2} |\bar{\partial} \rho \wedge \varphi|$ is bounded on D. This integral operator can be written the composition of multiplication by $|\rho(w) \bar{\partial} f_2(w)| + |\rho(w)|^{1/2} |\bar{\partial} \rho(w) \wedge \bar{\partial} f_2(w)|$, followed by the integral operator with kernel $\varphi(w) \wedge L(z, w)$ where

$$\varphi = \frac{\bar{\partial} f_2}{|\rho \bar{\partial} f_2| + |\rho|^{1/2} |\bar{\partial} \rho \wedge \bar{\partial} f_2|}.$$

It is clear that φ satisfies the condition for the integral operator to be bounded, so $H_{f_2} \in \mathcal{S}_p$.

Now turn to $p < 1$ and assume (a): that H_f is in $\mathcal{S}_p$ with finite Schatten norm $\|H_f\|_p$. As we did for Toeplitz operators, write $\{z_n\}$ as a finite union of sequences $\{z_{kn} : n = 1, 2, 3 \ldots\}$ such that $\beta(B(z_{kn}, r), B(z_{km}, r)) > R$ when $m \neq n$, where R is a large number to be chosen later. If it can be shown that $\{MDA(f, z_{kn})\}$ has finite l^p norm then adding up a finite number of these will give us the desired finiteness of $\sum_n MDA(f, z_n)^p$. Let us consider one of these subsequences and temporarily denote it by $\{z_n\}$. Let us continue to abbreviate $B_n = B(z_n, r)$.

Let S be the operator that takes any function g to $\sum_n (g - P_n g) \chi_{B_n}$, where $P_n g$ denotes the function h which minimizes $\int_{B_n} |g - h|^2$ among all analytic functions on B_n. It is clear that S has norm at most 1 because B_n are disjoint. Now let φ_n be defined as in the proof of (a) $\Rightarrow$(b) for Toeplitz operators, $p < 1$, but replace p with $p/2$. Let A be defined in a similar way: $Ae_m = \varphi_m$ for some orthonormal sequence $\{e_n\}$ and consider the operator $W = SH_f A$. Let Δ be the diagonal of

W^*W relative to the basis $\{e_n\}$ and estimate just as for Toeplitz operators

$$|\Delta|_{p/2}^{p/2} = \sum_m \langle \Delta e_m, e_m \rangle^{p/2} = \sum_m \langle SH_f\varphi_m, SH_f\varphi_m \rangle^{p/2}$$

$$\geq \sum_m \left(\int_{B_m} |f\varphi_m - P_m f\varphi_m|^2\, dV \right)^{p/2}$$

$$\geq \sum_m \left(\frac{1}{|B_m|} \int_{B_m} |f - \varphi_m^{-1} P_m f\varphi_m|^2\, dV \right)^{p/2}$$

$$\geq \sum_m MDA(f, z_m)^p,$$

which requires the same estimate from below on the functions φ_m as in the proof for Toeplitz operators, and so is valid if we drop a finite number of points from the sequence $\{z_m\}$.

Now we estimate the $\mathcal{S}_{p/2}$ norm of the off diagonal piece $X = W^*W - \Delta$. This is bounded above by the sum

$$\sum_{m,k} |\langle Xe_k, e_m \rangle|^{p/2} = \sum_{m \neq k} \left| \sum_j \int_{B_j} (f\varphi_k - P_j f\varphi_k)\overline{(f\varphi_m - P_j f\varphi_m)} \right|^{p/2}$$

$$\leq \sum_{m \neq k} \left[\sum_j \left(\int_{B_j} |f\varphi_k - P_j f\varphi_k|^2\, dV \right)^{1/2} \left(\int_{B_j} |f\varphi_m - P_j f\varphi_m|^2\, dV \right)^{1/2} \right]^{p/2}$$

$$\leq \sum_{m \neq k} \left[\sum_j \left(\int_{B_j} |f\varphi_k - \varphi_k P_j f|^2\, dV \right)^{1/2} \left(\int_{B_j} |f\varphi_m - \varphi_m P_j f|^2\, dV \right)^{1/2} \right]^{p/2}$$

$$\leq \sum_{m \neq k} \left[\sum_j \left(\int_{B_j} |f - P_j f|^2 |\varphi_k|^2\, dV \right)^{1/2} \left(\int_{B_j} |f - P_j f|^2 |\varphi_m|^2\, dV \right)^{1/2} \right]^{p/2}$$

$$\leq C \sum_{m \neq k} \left[\sum_j MDA(f, z_j)^2 |B_j| |\varphi_m(z_{mk})\varphi_k(z_{kj})| \right]^{p/2}$$

$$\leq C \sum_j MDA(f, z_j)^p \sup_j |B_j|^{p/2} \sum_{m \neq k} |\varphi_m(z_{mj})\varphi_k(z_{kj})|^{p/2},$$

where z_{mj} is simply the point in B_j where $|\varphi_m|$ has its maximum. The constant C on the fifth line depends only on p and r, as do all the subsequent constants C.

Note that

$$\sum_{m \neq k} |\varphi_m(z_{mj})||\varphi_k(z_{kj})|^{p/2} = \sum_{k \neq m} \frac{|\rho(z_m)\rho(z_k)|^{(n+1)}}{|\Phi(z_{mj}, z_m)\Phi(z_{kj}, z_k)|^{p(n+1)/4+(n+1)}},$$

which is very much like the corresponding double sum in (7) except that p has been replaced with $p/2$ and the evaluation of both factors at $z \in B_j$ has been replaced with evaluation of one factor at z_{mj} and the other at z_{kj} (but both in B_j).

Nevertheless, the estimates proceed the same way. Again we divide the sum into 3 parts: (i) the sum over all m but only those k with $|z_k - z_j| \geq 2\epsilon$ (and hence $\min\{|z_k - z_{kj}|, |z_k - z_{mj}|\} > \epsilon$ for z_j sufficiently near the boundary). (ii) a similar sum with the rôles of k and m reversed. (iii) the sum over all m and k such that z_m and z_k are within 2ϵ of z_j. The quantity ϵ is chosen so that the change of variables has good behavior on $\{w : |w - z_j| < 2\epsilon, |\rho(w)| < \delta\}$. The resulting sums are estimated exactly as the corresponding sums for Toeplitz operators with one slight change: The change of variables used in estimating (iii) is centered at z_j (rather than at either of the points of evaluation z_{mj} or z_{kj}). This introduces an additional term of order $\max\{|z_{kj} - z_j|, |z_{mj} - z_j|\}$ in the denominator of the expression to be estimated. This will be a small multiple of $|\rho(z_j)|$ if r is small enough, and may be absorbed into that term of the denominator.

This finishes the proof that (a) implies (b) for $p < 1$ and with it the proof of the theorem. ∎

4. A BMO-like condition in the unit disk

In the case of the unit disk we can add a fourth condition equivalent to H_f being compact. It is analogous to the $(I - P)f \in$ BMO condition for the boundedness of H_f obtained in [7]. In order to state it, we introduce a "higher order mean oscillation" operator:

$$MO^{(k)}(f, z) = \inf_{p \in \mathcal{P}_k} \mathcal{A}(f - p, z)$$

where $\mathcal{P}_k$ is the collection of (analytic) polynomials of degree at most k. $MO(f)$ means the same as $MO^{(0)}(f)$.

4.1 Theorem. *Let $f \in L^1(\mathbf{D})$, $p > 0$ and choose a non-negative integer $k > 1/p - 1$. Then $H_f \in \mathcal{S}_p$ if and only if*

(d) *There exists a function g with $f - g \in \mathcal{H}$ and such that $MO^{(k)}(g) \in L^p(d\lambda)$*

PROOF. It is clear that (d) implies Theorem 3.1(b) so it remains to show that Theorem 3.1(c) implies (d). Let $f = f_1 + f_2$ as in (c). Since $MO^{(k)}(f_1, z) \leq \mathcal{A}(f_1, z)$, we need only show that $MO^{(k)}(I - P)f_2 \in L^p(\lambda)$. For simplicity, let us assume $f_1 = 0$. Fix the separated sequence z_j and let $B_j = B(z_j)$. Let $\varphi(w) = (1 - |w|^2)\bar{\partial}f_2(w)$. Then

$$g(z) = (I - P)f_2(z) = C \int_{\mathbf{D}} \frac{\varphi(w)}{(z - w)(1 - z\bar{w})} \, dA(w)$$

$$= C \int_{\mathbf{D}\backslash B_j} \frac{\varphi(w)}{(z - w)(1 - z\bar{w})} \, dA(w)$$

$$+ C \int_{B_j} \frac{\varphi(w)}{(z - w)(1 - z\bar{w})} \, dA(w).$$

The second integral above is easily estimated to be bounded by

$$\left(\sup_{w \in B_j} |\varphi(w)| \int_{B_j} |z - w|^{-1}|1 - z\bar{w}|^{-1} \, dA(w) \right) \leq C \sup_{w \in B_j} |\varphi(w)|$$

and the first integral above is analytic for z in B_j. Let $p_k(z)$ be its kth degree Taylor polynomial about z_j. It is not hard to show that

$$g(z) - p_k(z) = C \int_{D\backslash B_j} \varphi(w)G(z, w) \, dA(w) + C \int_{B_j} \frac{\varphi(w)}{(z - w)(1 - z\bar{w})} \, dA(w)$$

where $G(z, w)$ is a kernel satisfying

$$|G(z, w)| \leq C \frac{|z - z_j|^{k+1}}{|1 - z\bar{w}|^{k+3}} \leq C \frac{(1 - |z_j|^2)^{k+1}}{|1 - z_j\bar{w}|^{k+3}}$$

for $z \in B(z_j, r/2)$ and $w \notin B(z_j, r)$. Note that this makes use of the estimate $|z - w| \sim |1 - z\bar{w}|$. Now we can estimate for $z \in B(z_j, r/2)$

$$|g(z) - p_k(z)| \leq C \int_{D\backslash B_j} |\varphi(w)| \frac{(1 - |z_j|^2)^{k+1}}{|1 - z_j\bar{w}|^{k+3}} \, dA(w) + C \sup_{w \in B_j} |\varphi(w)|$$

$$\leq C \sum_i \sup_{w \in B_i} |\varphi(w)| \frac{(1 - |z_i|^2)^2(1 - |z_j|^2)^{k+1}}{|1 - z_j\bar{z}_i|^{k+3}}. \tag{10}$$

Raise this to the power p. In case $p > 1$ let $p' = p/(p - 1)$ and let $0 < \epsilon < 1/p' = 1 - 1/p$. Let $k = 0$ and use Hölder's inequality to get

$$|g(z) - g(z_j)|^p$$

$$\leq C \sum_i \sup_{w \in B_i} |\varphi(w)|^p \frac{(1 - |z_i|^2)^{2+\epsilon p}}{|1 - z_j\bar{z}_i|^{2+p}} (1 - |z_j|^2)^p \left(\sum_i \frac{(1 - |z_i|^2)^{2-\epsilon p'}}{|1 - z_j\bar{z}_i|^2} \right)^{1/p'}$$

$$\leq C \sum_i \sup_{w \in B_i} |\varphi(w)|^p \frac{(1 - |z_j|^2)^{p-\epsilon p}(1 - |z_i|^2)^{2+\epsilon p}}{|1 - z_j\bar{z}_i|^{2+p}}$$

for $z \in B(z_j, r/2)$. We have used $\sum_j (1 - |z_j|^2)^a/|1 - z_j\bar{z}_j|^M \leq C(1 - |z_j|^2)^{a-M}$ when $1 < a < M$. Now sum over j and use the same inequality again to produce the desired estimate:

$$\sum_j \sup_{z \in B(z_j, r/2)} |g(z) - g(z_j)|^p \leq C \sum_i \sup_{w \in B_i} |\varphi(w)|^p. \tag{11}$$

In case $p \leq 1$ use any k with $k + 1 > 1/p$ and again raise the left and right sides of (10) to the pth power and make use of the fact that the pth power of a sum is less than or equal to the sum of the pth powers:

$$\sup_{z \in B(z_j, r/2)} |g(z) - p_k(z)|^p \leq C \sum_i \left(\sup_{w \in B_i} |\varphi(w)| \frac{(1 - |z_i|^2)^2 (1 - |z_j|^2)^{k+1}}{|1 - z_j \bar{z}_i|^{k+3}} \right)^p.$$

Now sum on j and use the estimate mentioned above (with $M = (k + 3)p > a = (k + 1)p > 1$) to obtain the required estimate (11) again. ∎

5. A BMO-like condition in a strongly pseudoconvex domain

Define for any $f \in L^2_{\mathrm{loc}}$

$$MO(f, z) = \inf_{c \in \mathbf{C}} \left(\frac{1}{|B(z)|} \int_{B(z)} |f(u) - c|^2 \, dV(u) \right)^{1/2}$$

$$\mathrm{BMO} = \{ f \in L^2_{\mathrm{loc}} : MO(f, z) \in L^\infty \}$$

$$B_p = \{ f \in L^2_{\mathrm{loc}} : MO(f, z) \in L^p(\lambda) \}$$

Our goal here is to show that for $p > 2n$, $H_f \in \mathcal{S}_p$ if and only if $(I - P)f \in B_p$. It is clear from the second section (because $MDA(f) \leq MO(f)$) that $f \in B_p + \mathcal{H}$ implies $H_f \in \mathcal{S}_p$. For the reverse, we know that the function f satisfies the condition in part (c) of Theorem 3.1. That is $f = f_1 + f_2$ where $\mathcal{A}(f_1) \in L^p(\lambda)$ and the function defined by

$$z \mapsto \sup_{w \in B(z)} \left(|\rho(w) \bar{\partial} f_2(w)| + |\rho(w)|^{1/2} |\bar{\partial} \rho(w) \wedge \bar{\partial} f_2(w)| \right)$$

belongs to $L^p(d\lambda)$.

In [7], a kernel L was constructed which solved the $\bar{\partial}$-problem with good $L^\infty \to \mathrm{BMO}$ estimates. We will show here that it also gives good $L^p(\lambda) \to B_p$ estimates.

Recall that if φ is a $\bar{\partial}$-closed $(0, 1)$-form, then $u(z) = \int_D \varphi(w) \wedge L(z, w)$ provides a solution to the equation $\bar{\partial} u = \varphi$. Let $Q(z)$ be the function defined above:

$$Q(z) = \sup_{w \in B(z)} \left(|\rho(w) \varphi(w)| + |\rho(w)|^{1/2} |\bar{\partial} \rho(w) \wedge \varphi(w)| \right).$$

The main result of Section 6 of [7] was that if Q is bounded, then the mean oscillation of u belongs to L^∞. Here we will show that for $p > 2n$, if $Q \in L^p(\lambda)$ then the mean oscillation of u is also in $L^p(\lambda)$, that is, $u \in B_p$.

Let us denote by D_T the derivative in the tangential direction and by D the total differential. Then $|D_T f| \sim C \left(|\partial f \wedge \partial \rho| + |\bar{\partial} f \wedge \bar{\partial} \rho| \right)$.

5.1 Lemma. *Let φ denote a $\bar{\partial}$-closed $(0,1)$-form such that Q belongs to $L^p(\lambda)$, then $u(z) = \int_D \varphi(w) \wedge L(z,w)$ can be written in the form $u = g_1(z) + g_2(z)$ where both g_1 and $|D_T g_2| |\rho|^{1/2} + |D g_2| |\rho|$ belong to $L^p(\lambda)$.*

PROOF. Let L be written as a sum $L(z,w) = \sum_{k=0}^{n-1} L_k(z,w)$ as in [7], and let $u_k(z) = \int_D \varphi(w) \wedge L_k(z,w)$. We claim that for $p > 2n$, $g_1 = \sum_{k=1}^{n-2} u_k$ and $g_2 = u_{n-1}$ have the required properties.

In [7, Section 6] it was shown that for $1 \leq k \leq n-2$ we have an inequality $|\varphi(w) \wedge L_k(z,w)| \leq Q(w) W_k(z,w)$ where W_k is a positive kernel. (Actually the $Q(w)$ was missing because in that paper it was assumed to be bounded by 1.) The kernel we call W_k here was not given a name in [7], but equals the sum of the two terms

$$\frac{|\rho(z)|^{n-k-1}|\rho(w)|^{n-1/2}}{\|z-w\|_A^{2(n-k)}|\Psi(z,w)|^{n+k+1}} \left(|G(z,w)| + |\rho(w)| + |\rho(z)| + |z-w|^2 \right) |z-w|,$$

and

$$\frac{|\rho(z)|^{n-k-1}|\rho(w)|^n}{\|z-w\|_A^{2(n-k)}|\Psi(z,w)|^{n+k+1}} \left(|G(z,w)| + |\rho(z)||z-w| + |z-w|^2 \right)$$

found midway between equation (28) and inequality (29). It was also shown there that $\int W_k(z,w)\,dV(w)$ was a bounded function of z (so the kernel W_k maps L^∞ to itself). Here we need to show that the kernel W_k maps $L^p(\lambda)$ to itself. This is equivalent to the kernel $|\rho(z)|^{-(n+1)/p} W_k(z,w)|\rho(w)|^{(n+1)/p}$ mapping L^p to itself. Using the Schur method with a positive function $|\rho|^{-\epsilon/(pp')}$, it suffices to show that there exists a real number ϵ and a constant $C > 0$ such that

$$|\rho(z)|^{-(n+1)/p} \int_D W_k(z,w)|\rho(w)|^{(n+1)/p-\epsilon/p}\,dV(w) \leq C|\rho(z)|^{-\epsilon/p}$$

and

$$|\rho(w)|^{(n+1)/p} \int_D W_k(z,w)|\rho(z)|^{-(n+1)/p-\epsilon/p'}\,dV(z) \leq C|\rho(w)|^{-\epsilon/p'}$$

The argument showing the first of these inequalities is virtually the same as the corresponding argument in [7], and so is omitted. We mention only that the calculations require the following inequality:

$$\frac{1}{2} - \frac{n+1}{p} + \frac{\epsilon}{p} > 0.$$

The other half of the Schur estimates can be obtained from the estimates of [7] but using the standard change of variables (in which $t_1(z) = \rho(z) + i \operatorname{Im} G(z, w)$) rather than the one used in [7] (where $t_1(z) = G(w, z)$). This requires the inequality

$$n - k - \frac{n+1}{p} - \frac{\epsilon}{p'} > 0.$$

There exists an ϵ sattisfying both these inequalities if and only if $p > 2k + 3$ so this will be satisfied for $p > 2n - 1$.

It remains to show that u_{n-1} has the required property. Using virtually the same argument as in [7], we can show that

$$|D_T u_{n-1}(z)| \, |\rho(z)|^{1/2} + |D u_{n-1}(z)| \, |\rho(z)| \leq \int_D Q(w) W'(z, w)$$

where $W'(z, w)$ is a positive kernel similar to the W_k above. We can apply the Schur method just as above and end up with similar integrals to estimate. Again the estimates are similar to the corresponding estimates in [7] that showed the boundedness of the corresponding integrals (given the boundedness of Q). Here the estimates require the same first inequality on ϵ plus the following

$$\frac{3}{2} - \frac{n+1}{p} - \frac{\epsilon}{p'} > 0.$$

These are both satisfied by the same ϵ if and only if $p > 2n$. The result is that the left side of the inequality above belongs to $L^p(\lambda)$.   ∎

5.2 Lemma. *Let $f \in L^2_{\text{loc}}$ and $p > 2n$, then $f \in B_p$ if and only if $f = f_1 + f_2$ with $\mathcal{A}(f_1) \in L^p(\lambda)$ and $|\rho|^{1/2}|D_T f| + |\rho D f| \in L^p(\lambda)$. If f is holomorphic then f_1 may be taken to be 0.*

PROOF. Let $f = f_1 + f_2$ as given. It is a standard Sobolev estimate that $\int_\Pi \left| g - \frac{1}{|\Pi|} \int_\Pi g \right|^2 dV \leq C \int_\Pi |Dg|^2 \, dV$, where Π is the unit polydisk. Changing variables so that Π is transformed to the $\widetilde{\Pi}(z)$ of Sections 2 and 3 and letting g be f_2, we get

$$MO(f_2, z) \leq \frac{C}{|\widetilde{\Pi}(z)|} \int_{\widetilde{\Pi}(z)} |\rho| \, |D_T f_2|^2 + |\rho D f_2|^2 \, dV.$$

Holders inequality and Fubini's theorem now show that $f_2 \in B_p$. Since clearly f_1 also belongs to B_p, we have the sufficiency of the decomposition.

Conversely, the argument used in [5] and [9] to prove the decomposition analogous to (c) in Theorem 3.1 is easily adapted to provide $f = f_1 + f_2$ with

$\int \mathcal{A}(f_1)^p \, d\lambda \le \int MO(f)^p \, d\lambda$ and with

$$\sup_{B(z,r/2)} |\rho|^{1/2} \, |\mathrm{D}_T f_2| + |\rho \mathrm{D} f_2| \le MO(f,z),$$

so the decomposition is a necessary condition.

For a holomorphic function f, in addition to the above Sobolev inequality on the unit polydisk, we have also: $|\mathrm{D}f(0)|^2 \le C \int_\Pi |f - f(0)|^2$. This leads to the inequality $|\rho|^{1/2} \, |\mathrm{D}_T f| + |\rho \mathrm{D} f| \le MO(f)$ so that f itself satisfies the condition on f_2. ∎

In particular, a holomorphic function with bounded derivatives on D belongs to B_p for $p > 2n$: For then

$$\int_D \left(|\rho|^{1/2} |\mathrm{D}_T f| + |\rho \mathrm{D} f| \right)^p \, d\lambda \le C \int_D |\rho|^{p/2 - n - 1} + |\rho|^{p - n - 1} \, dV < \infty.$$

Finally, we have the following:

5.3 Lemma. *Let $p > 2n$ and $f \in L^2_{\mathrm{loc}}$. If f satisfies any one of the following conditions then $Pf \in B_p$:*

(1) *$f \in B_p$,*

(2) *$f \in L^p(\lambda)$, or*

(3) *$\mathcal{A}(f) \in L^p(\lambda)$.*

PROOF. For case (1), $f \in B_p$, write $f = f_1 + f_2$ as in the previous lemma. We claim that ρf_2 is in $L^p(\lambda)$. For this, let $F(t)$ be defined as the integral of $|f_2|^p$ over the level surface $|\rho(z)| = t$, then the condition $\rho |\mathrm{D} f_2| \in L^p(\lambda)$ implies that $F(t)$ is bounded: In local coordinates $((\zeta, t)$ with $|\rho| = t)$, estimate

$$|f_2(\zeta,t)|^p \le \left| \int_t^1 \frac{\partial f_2(\zeta,s)}{\partial s} s^{1 - (n+1)/p} s^{-1 + (n+1)/p} \, ds \right|^p$$

and use Hölder's inequality. Now integrate in ζ to obtain

$$F(t) \le C \int |\mathrm{D} f_2(\zeta,s)|^p s^{-n-1} \, ds \, d\zeta$$

In particular, if we multiply $F(t)$ by t^{p-n-1} and integrate, the result is finite so ρf_2 is in $L^p(\lambda)$, as claimed.

Since any holomorphic h with bounded derivative will belong to B_p, Pf will be in B_p for any f with compact support in D. Consequently, to show that $Pf \in B_p$

we may, without loss of generality, suppose that f_2 is supported on a neighborhood of ∂D where $|\partial \rho|$ is bounded away from 0. Let ψ be a smooth function which equals $|\partial \rho|^{-2}$ in this neighborhood of ∂D. Then

$$Pf_2(z) = \sum_{j=1}^{n} \int_D K(z,w) f_2(w) \left[\psi(w) \frac{\partial \rho}{\partial \overline{w}_j}(w) \right] \frac{\partial \rho}{\partial w_j}(w) \, dV(w).$$

Integration by parts gives

$$Pf_2(z) = \int K(z,w) \rho(w) S(f_2)(w) \, dV(w),$$

where S is a smooth first order differential operator. Since $|\rho||S(f_2)| \le C|\rho f_2| + C|\rho D f_2|$ we have $|\rho||S(f_2)| \in L^p(\lambda)$. Thus case (1) reduces to case (2) plus case (3).

Standard estimates on the Bergman kernel show that

$$|\rho(z)|^{1/2} \, |D_{T,z} K(z,w)| + |\rho(z) D_z K(z,w)| \le |\rho(z)|^{1/2} |\Psi(z,w)|^{-n-3/2}.$$

As in Lemma 5.1 it can be shown that $|\rho(z)|^{1/2-(n+1)/p} |\Psi(z,w)|^{-n-2} |\rho(w)|^{(n+1)/p}$ is bounded from L^p to L^p (using the Schur method). Differentiating under the integral sign in $Pf(z) = \int K(z,w) f(w) \, dV(w)$ gives us

$$|\rho(z)|^{1/2} \, |D_T Pf(z)| + |\rho(z) D Pf(z)| \le \int_D |f(w)| |\rho(z)|^{1/2} |\Psi(z,w)|^{-n-3/2} \, dV(w).$$

and the above argument shows that the kernel maps $L^p(\lambda)$ to itself. Thus if $f \in L^p(\lambda)$ then it satisfies the condition in Lemma 5.2 for $f \in B_p$. This proves case (2).

We reduce case (3) to case (2). Let $\Pi(w) \subset B(w)$ be a polydisc centered at w with volume greater than $c|B(w)|$. Then

$$Pf(z) = \int K(z,w) f(w) \, dV(w)$$
$$= \int_D \frac{1}{|\Pi(w)|} \int_D \chi_{\Pi(w)}(u) K(z,u) \, dV(u) f(w) \, dV(w)$$
$$= \int_D K(z,u) \left[\int_D \frac{1}{|\Pi(w)|} \chi_{\Pi(w)}(u) f(w) \, dV(w) \right] dV(u).$$

It is clear that the expression in brackets is dominated by $\mathcal{A}(f) \in L^p(\lambda)$ and the proof is complete. ∎

5.4 Theorem. Let $p > 2n$ and $f \in L^1(dV)$. Then the Hankel operator H_f belongs to $\mathcal{S}_p$ if and only if $(I - P)f \in B_p$.

PROOF. It is not hard to see from $(I - P)f \in B_p$ or from the conditions in Theorem 3.1 that $f \in L^2_{\text{loc}}$ and we will use this fact without further mention. We have seen that the condition $(I-P)f \in B_p$ is sufficient. Now suppose that $H_f \in \mathcal{S}_p$. By Theorem 3.1(c), $f = f_1 + f_2$ with $\mathcal{A}(f_1) \in L^p(\lambda)$ and the usual condition on f_2. Clearly then $f_1 \in B_p$ and by Lemma 5.3 so is Pf_1. Thus $(I - P)f_1 \in B_p$. According to Lemma 5.1 we can write $f_2 = g + h$ where h is holomorphic and g satisfies the condition of Lemma 5.2. Thus $g \in B_p$ by that lemma. By the previous lemma we have $Pg \in B_p$ and so $(I - P)f_2 = (I - P)g$ is also in B_p. ∎

The following simple argument shows that Theorem 5.4 cannot be valid as it stands for $p \leq 2n$. Let f be a continuous function with compact support in D. Then $H_f \in \mathcal{S}_p$ for all $p > 0$. If we had $(I-P)f \in B_p$ then we would have $Pf \in B_p$. By Lemma 5.2 we could then conclude $|D_T Pf| \, |\rho|^{1/2} \in L^p(\lambda)$. But it is shown in [10] and [6] that when this occurs, Pf must be a constant. Since the projections of functions with compact support are dense in space of holomorphic function that extend continuously to the boundary [11], we have a contradiction.

References

[1]   B. Coupet, Décomposition atomique des espace be Bergman, *Indiana Univ. Math. J.*, **38** (1989), 917–941.

[2]   Sh. A. Dautov and G. M. Henkin, Zeros of holomorphic functions of finite order and weighted estimates for solutions of the $\bar{\partial}$-equation, *Math. USSR Sbornik*, **35** (1979), 449–459. Translated from *Mat. Sbornik*, **107** (1978).

[3]   C. Fefferman, The Bergman kernel and biholomorphic mappings of pseudo-convex domains, *Invent. Math.*, **26** (1974), 1–65.

[4]   I. C. Gohberg and M. G. Kreĭn, "Introduction to the Theory of Non-selfadjoint Operators." Nauka, Moskow, 1965; translation in *Translations of Mathematical Monographs*, Vol. 18, Amer. Math. Soc., Providence, RI, 1969.

[5]   Huiping Li, Hankel operators on the Bergman space of strongly pseudoconvex domains, *Intgeral Equations and Operator Theory*, to appear.

[6]   Huiping Li, Schatten classes Hankel operators on strongly pseudoconvex domains, Proc. Amer. Math. Soc., **119** (1993), 1211–1221

[7]   Huiping Li and D. H. Luecking, BMO on strongly pseudoconvex domains: Hankel operators, duality and BMO estimates, *Trans. Amer. Math. Soc.*, to appear.

[8] D. H. Luecking Trace ideal criteria for Toeplitz operators, *J. Functional Anal.* **73** (1987), 345–368.

[9] D. H. Luecking, Characterizations of certain classes of Hankel operators on the Bergman spaces of the unit disk, *J. Functional Anal.*, **110** (1992), 247–271.

[10] M. Peloso, Hankel operators on weighted Bergman spaces on strongly pseudo-convex domains, *Ill. J. Math.*, **38** (1994), 223–249.

[11] R. M. Range, *Holomorphic Functions and Integral Representations in Several Complex Variables*, Springer-Verlag, New York, 1986.

Huiping Li
SUNY at Buffalo
Buffalo NY 14214
hli@newton.math.buffalo.edu

Daniel H. Luecking
University of Arkansas
Fayetteville, AR 72701
luecking@comp.uark.edu

Contemporary Mathematics
Volume **185**, 1995

Operator Equations with Elementary Operators

Martin Mathieu

Abstract. We study a number of situations where operator equations with elementary operators on C*-algebras arise, and discuss tools for their solution.

The purpose of the present paper is to overview some of the work on elementary operators done by various authors during the past few years from the point of view of operator equations. Though some of the results can be established in a wider setting, we have chosen that of C*-algebras in order to obtain a unified framework.

An elementary operator S on a C*-algebra A is defined by

$$(1) \qquad Sx = \sum_{j=1}^{n} a_j x b_j \qquad (x \in A)$$

where $a = (a_1, \ldots, a_n)$, $b = (b_1, \ldots, b_n) \in M(A)^n$ and $M(A)$ is the multiplier algebra of A. A system of operator equations involving elementary operators is thus of the form

$$(2) \qquad S^{(\alpha)} x = c^{(\alpha)}$$

where $c^{(\alpha)} \in A$ and the parameter α runs through a possibly infinite index set. We can distinguish three different types of the equation

$$(3) \qquad \sum_{j=1}^{n} a_j x b_j = c.$$

1991 *Mathematics Subject Classification.* Primary 47 B 47; Secondary 46 L 99, 47 A 62, 47 B 48.
Participation of the author in the AMS Summer Research Conference was gratefully supported by the NSF.
This paper is in final form, and no version of it will be submitted for publication elsewhere.

Let the element $c \in A$ and the n-tuples $a, b \in M(A)^n$ be given. Then we may either look for (at least) one solution of (3), or we may investigate the conditions on a and b imposed if *all* $x \in A$ solve (3). Or, if x and c are specified, we can look for $a, b \in M(A)^n$ fulfilling (3). We will now discuss these three questions separately as well as some of their possible applications.

1. c, a, and b are given, x is to be found.

This question either belongs to the *spectral theory* of elementary operators, or to *factorization problems*. Suppose that $c = 0$. Then, non-zero solutions to the equation $(\rho - S)x = 0$ exist if and only if ρ is an eigenvalue of S, which in turn depends on the relative position of the joint point spectra $P\sigma(a)$ and $P\sigma(b^*)$ in $\mathbf{C}^n$. A sample result in this direction is the following.

PROPOSITION 1.1 [**33**, Theorem 3.8] *Let A be a prime C^*-algebra and $a, b \in M(A)^n$. If $\lambda = (\lambda_1, \ldots, \lambda_n) \in P\sigma(a)$ and $\mu = (\mu_1, \ldots, \mu_n) \in P\sigma(b^*)$, then $\sum_{j=1}^n \lambda_j \bar{\mu}_j \in P\sigma(S)$, the point spectrum of S.*

Similarly, describing the range of $\rho - S$ is equivalent to asking for a solution of $(\rho - S)x = c$ for *any* $c \in A$. For example, the following holds.

PROPOSITION 1.2 [**33**, Theorem 3.9] *Let A be a prime C^*-algebra and $a, b \in M(A)^n$ be commuting n-tuples (i.e., all the a_j's commute and all the b_j's commute). For every $\rho \in \mathbf{C}$, the elementary operator $\rho - S$ is not surjective if and only if $\rho = \sum_{j=1}^n \bar{\lambda}_j \mu_j$ for some $\lambda = (\lambda_1, \ldots, \lambda_n) \in AP\sigma(a^*)$ and $\mu = (\mu_1, \ldots, \mu_n) \in AP\sigma(b)$ (where $AP\sigma(\cdot)$ stands for the joint approximate point spectrum).*

A comprehensive survey of the spectral theory of elementary operators is provided in [**13**]. However, spectral methods generally give no further information on the solution x nor a way to construct it. Moreover, if merely one non-zero $c \in A$ is specified, global information on the range of S may be useless. Such factorization problems, i.e., finding a solution of (3) or (2) with only one c specified, for example arise in connection with positive completions of matrices [**39**], [**40**] [1]. In [**40**], a C^*-algebra A is defined to be QF (a *quasifactorization algebra*) if for all $a \in A_+$, $c = (c_1, \ldots, c_n) \in A^n$, and $b = (b_{jk}) \in M_n(A)_+$, the positive $n \times n$-matrices over A, such that $\begin{pmatrix} a^2 & c \\ c^* & b^2 \end{pmatrix} \in M_{n+1}(A)_+$ there exists $x = (x_1, \ldots, x_n) \in M(A)^n$ with $\|x\| \leq 1$ satisfying $axb = c$. We may rewrite this last equation to obtain a system of operator equations similar to (2):

$$\sum_{j=1}^n ax_j b_{jk} = c_k \qquad \text{for all } 1 \leq k \leq n.$$

For example, every von Neumann algebra and every corona algebra $M(A)/A$ for a σ-unital C^*-algebra A is QF [**40**, Theorem 4.3]. QF algebras enjoy the property that

---

[1] Factorization of operators, and in particular matrices, into finite products of operators with prescribed properties have attracted a great deal of attention over the past years, see e.g. [**44**].

certain partially defined matrices over them have positive completions. Putting $n = 1$ in the definition we see that the equation $axb = c$ always has a solution $x \in M(A)$ with $\|x\| \leq 1$ provided that $\begin{pmatrix} a^2 & c \\ c^* & b^2 \end{pmatrix} \geq 0$. Suppose that A is unital and $b = 1$. It then follows that $ax = c$ has a solution $x \in A$ with $\|x\| \leq 1$ if and only if $cc^* \leq a^2$. An elaboration of this argument (which we now reproduce from [40]) shows that every unital QF algebra is *UMF* (a *uniform majorization-factorization algebra*), i.e., whenever $cc^* \leq aa^*$ there is $x \in A$ with $\|x\| \leq 1$ such that $ax = c$.

The condition $cc^* \leq aa^*$ is equivalent to $\begin{pmatrix} aa^* & c \\ c^* & 1 \end{pmatrix} \geq 0$; this follows from the identity

$$\begin{pmatrix} (aa^* + \varepsilon)^{-1/2} & 0 \\ 0 & 1 \end{pmatrix} \begin{pmatrix} aa^* + \varepsilon & c \\ c^* & 1 \end{pmatrix} \begin{pmatrix} (aa^* + \varepsilon)^{-1/2} & 0 \\ 0 & 1 \end{pmatrix} =$$

$$\begin{pmatrix} 1 & (aa^* + \varepsilon)^{-1/2}c \\ c^*(aa^* + \varepsilon)^{-1/2} & 1 \end{pmatrix}$$

and the fact that the matrix on the right hand side is positive if and only if $\|(aa^* + \varepsilon)^{-1/2}c\| \leq 1$ for all $\varepsilon > 0$. Thus, by the QF property, there is $x \in A$, $\|x\| \leq 1$ such that $c = (aa^*)^{1/2}x$. Specializing to $c = a$ this yields a weak polar decomposition $a = (aa^*)^{1/2}v^*$ for some $v \in A$, $\|v\| \leq 1$. As a result, for general c, we have that $c = ay$ with $y = vx$. This last claim follows from the fact that

$$\begin{pmatrix} 1 & a^* & vx \\ a & aa^* & c \\ x^*v^* & c^* & 1 \end{pmatrix} \geq 0$$

and therefore

$$0 \leq \begin{pmatrix} 1 & a^* & vx \\ a & aa^* & c \\ x^*v^* & c^* & 1 \end{pmatrix} - \begin{pmatrix} 1 \\ a \\ x^*v^* \end{pmatrix} \begin{pmatrix} 1 & a & vx \end{pmatrix} = \begin{pmatrix} 0 & 0 & 0 \\ 0 & 0 & c - avx \\ 0 & c^* - x^*v^*a^* & 1 - x^*v^*vx \end{pmatrix}$$

which implies that $c - avx = 0$.

Specializing this argument to $c = (aa^*)^{1/2}$ shows that $(aa^*)^{1/2} = av$, so that a similar reasoning yields the following result illustrating the intimate relation between positivity and factorization.

PROPOSITION 1.3.   *Let A be a unital QF algebra. For all $a, b, c \in A$, the following conditions are equivalent.*

(a) $\begin{pmatrix} aa^* & c \\ c^* & b^*b \end{pmatrix} \geq 0$;

(b) *There is $x \in A$, $\|x\| \leq 1$ such that $axb = c$.*

UMF algebras were studied in [17], [18], and [19] with some preliminary work in [24] and [37], e.g., and operator factorization is applied to similarity problems in [14]. The following characterization is a combination of results in [3] and [19].

THEOREM 1.4 [**3**], [**19**] *Suppose that the C*-algebra A does not contain an un-countable orthogonal family of self-adjoint elements. Then the following conditions are equivalent.*

(a) *A is* UMF*;*
(b) $M_n(A)$ *is* UMF *for all* $n \in \mathbf{N}$*;*
(c) *A is* AW*;*
(d) *A is a Rickart C*-algebra;*
(e) *A is* SAW*.*

On the other hand, little seems to be known on the solubility of the more general equation (3) in these algebras. Also, the following question is open (for the terminology, see Section 3).

PROBLEM 1.5.   Which boundedly centrally closed C^*-algebras are QF?

2. c and x are given, a and b are to be found.

Let $\mathscr{El}(A)$ denote the algebra of all elementary operators on the C^*-algebra A. If the elements x and c are fixed in (3), then it is a question of the transitivity of the action of $\mathscr{El}(A)$ on A whether or not (3) can be fulfilled. The following immediate consequence of the definition shows that it is inevitable to impose some additional conditions on x and c.

PROPOSITION 2.1.   *In the unital case, the operator equation* (3) *can be satisfied universally, i.e., for each pair* (x, c)*, if and only if A is a simple C*-algebra.*

Finite systems of equations of the form (3) have been studied by Magajna in [**27**], [**29**], and [**30**]. Here the problem is the following: Let $x = (x_1, \ldots, x_m)$ and $c = (c_1, \ldots, c_m)$ in A^m be given. When does there exist $S \in \mathscr{El}(A)$ such that

$$(4) \qquad\qquad Sx_i = c_i, \ 1 \leq i \leq m$$

holds? An answer which can be considered as next to the simple case, and which is essentially a consequence of the Jacobson density theorem, is the following. Recall that an algebra is said to be *primary* if it is unital and contains exactly one maximal ideal.

THEOREM 2.2 [**27**] *Let A be a primary C*-algebra with unique maximal ideal K. For each pair* $(x, c) \in A^{2m}$ *such that* $\{c_1, \ldots, c_m\}$ *is linearly independent* mod K*, there exists* $S \in \mathscr{El}(A)$ *such that* $Sx_i = c_i$ *for all* $1 \leq i \leq m$*.*

Note that every von Neumann factor is primary so that Theorem 2.2 in particular applies to $A = B(H)$. As a consequence, Magajna obtained an affirmative answer to the Fong-Sourour conjecture on the nonexistence of non-zero compact elementary operators on the Calkin algebra. (This conjecture was independently affirmed in a wider context in [**32**] using methods similar to those discussed in Section 3 below.) In the special case $A = B(H)$, the conclusion can even be strengthened to the existence of elements a, $b \in A$ such that $ax_i b = c_i$ for all $1 \leq i \leq m$ [**27**, Theorem 1].

Theorem 2.2 also has applications to the range inclusion problem for elementary operators, cf. [**17**].

PROBLEM 2.3. Give necessary and sufficient conditions for the solubility of (4).

A necessary condition close at hand is the following: for every $z = (z_1, \ldots, z_m) \in Z(A)^m$, where $Z(A)$ denotes the center of A, $\sum_{i=1}^{m} z_i c_i$ is contained in the ideal generated by $\sum_{i=1}^{m} z_i x_i$. A stronger version of this turns out to be sufficient too, and a variant invoking closed ideals yields an approximate solution of (4).

THEOREM 2.4 [**29**, Proposition 1.1] *Let A be a unital C*-algebra. If $x \in A^m$ has the property that the ideal generated by $\sum_{i=1}^{m} \lambda_i x_i$ is dense in A for every $\lambda = (\lambda_1, \ldots, \lambda_m) \in \mathbf{C}^m \setminus \{0\}$, then (4) is soluble for all right hand sides $c \in A^m$.*

THEOREM 2.5 [**29**, Theorem 2.1] *Let A be a C*-algebra and $(x, c) \in A^{2m}$ be such that $\sum_{i=1}^{m} \lambda_i c_i$ is contained in the closed ideal generated by $\sum_{i=1}^{m} \lambda_i x_i$ for all $\lambda = (\lambda_1, \ldots, \lambda_m) \in \mathbf{C}^m$. Then, there exists a sequence $S_n \in \mathcal{El}(A)$ such that*

$$\lim_{n \to \infty} S_n x_i = c_i \qquad \text{for all } 1 \leq i \leq m.$$

As a consequence, Magajna obtains, among others, that a bounded linear operator on a C*-algebra A fixes every closed ideal of A if and only if it lies in the strong closure of $\mathcal{El}(A)$. However, Problem 2.3 in general seems to be open although some progress in the case $A = B(H)$ is made in [**30**]. There, it is also shown that for any pair $(x, c) \in A^{2m}$, A an arbitrary C*-algebra, there is $a \in A$ satisfying $ax_i = c_i$ for all $1 \leq i \leq m$ if and only if for each $y \in A^m$ the element $\sum_{i=1}^{m} c_i y_i \in A \sum_{i=1}^{m} x_i y_i$, bringing us back to the factorization problems of the previous section.

3. c is given, a and b are to be found such that (3) holds for all x.

We now turn our attention to a problem somewhat different to those discussed in Sections 1 and 2. So far, we were looking for solutions of (2), (3), or related operator equations for *single* x (given or to be determined). Now, the solutions have to suffice for *all* $x \in A$, thus turning the problem into an operator equation *on* A, the solution depending on coefficients in A. There are two cases. Either $c = \phi(x)$ depends functionally on x, or $c = 0$. Among the elaborated and prominent examples of the first kind is the question of innerness of derivations, that is, given a derivation δ on A, when does there exist $a \in M(A)$ such that

$$xa - ax = \delta(x) \qquad (x \in A)\,?$$

(Among the prominent answers is Sakai's for simple C*-algebras.) In the sequel, we will not enlarge any further on this case, but confine ourselves with the following two sample results.

THEOREM 3.1 [28] *Let A be a simple C*-algebra and $\phi\colon A \to A$ be a bounded linear map. For each finite dimensional subspace X of A, there is an elementary operator $S_X \in \mathcal{E}\ell(A)$ such that $\phi_{|X} = S_{X|X}$. If A is a σ-finite type III factor, then S_X may be chosen of the form $S_X(x) = axb$ (a and b depending on X, of course).*

THEOREM 3.2 [11, Theorem 4.6] *A complete contraction on a von Neumann algebra $A \subseteq B(H)$ is in the point-weak closure of $\{S \in \mathcal{E}\ell(A) \mid \|S\| \leq 1\}$ if and only if it has a completely contractive A'-bimodule extension to $B(H)$.*

The reader will recognize the connection between Theorem 3.1 and the previous section, while Theorem 3.2 is one of the many results on approximation by elementary operators (taking us somewhat away from strict equations).

The problem of which $a, b \in M(A)^n$ solve

$$(5) \qquad\qquad \sum_{j=1}^{n} a_j x b_j = 0 \qquad (x \in A)$$

looks rather particular, but has a number of interesting applications two of which will be discussed in Section 4. Apparently, (5) was first studied by Fong and Sourour in [20] for $A = B(H)$ and A the Calkin algebra. Their answer was put into a comprehensive context in [33] where we proved that $a_j = 0$ for all $1 \leq j \leq n$ provided that $\{b_1, \ldots, b_n\}$ is linearly independent (the general case can be reduced to this one), whenever A is a *prime* C*-algebra. The true reason for this result is revealed by the fact that the extended centroid of a prime C*-algebra coincides with $\mathbf{C}$ [33, Proposition 2.5]; thus, to present the full picture some more terminology is needed.

DEFINITION 3.3. Let A be a C*-algebra and let $\mathcal{I}_{ce}$ denote the collection of all closed essential ideals of A. The *bounded symmetric algebra of quotients of A*, $Q_b(A)$, is the complex unital *-superalgebra of A which is maximal with respect to the following properties

(i) $\forall q \in Q_b(A) \; \exists I \in \mathcal{I}_{ce} : Iq + qI \subseteq A$

(ii) $\forall q \in Q_b(A), \; I \in \mathcal{I}_{ce} : qI = 0$ or $Iq = 0 \Rightarrow q = 0$.

The center of $Q_b(A)$ is the *bounded extended centroid of A*, denoted by C_b.

This concept, due to Ara and the author, is the C*-analogue of the symmetric ring of quotients $Q_s(R)$ which is available for every semiprime ring R, cf. e.g. [38] [2]. Indeed, many features of $Q_s(R)$ and its center C, the extended centroid of R, take over to $Q_b(A)$ and C_b. For instance, C_b is a field if and only if A is prime. There is a way to construct $Q_b(A)$ via *local double centralizers* ([35], [36]), but it can also be recovered within $Q_s(A)$ as its bounded part in the sense of Handelman and Vidav:

$$Q_b(A) = \{q \in Q_s(A) \mid q^*q \leq n1 \text{ for some } n \in \mathbf{N}\}.$$

---

[2] Just delete "closed" in front of "essential ideals" in Definition 3.3.

Define a C^*-norm $\|\cdot\|$ on $Q_b(A)$ by

$$\|q\|^2 = \inf\{\lambda > 0 \mid q^*q \le \lambda 1\} \qquad (q \in Q_b(A))$$

to turn $Q_b(A)$ into a pre-C^*-algebra [1]. The completion of $Q_b(A)$ with respect to this norm is called *the local multiplier algebra of* A and will be denoted by $M_{\mathrm{loc}}(A)$. The name stems from an alternative description which appeared earlier in work by Elliott [16] and Pedersen [41], but it turned out to be very fruitful to exploit the connection with the algebraic counterpart. That it is not sufficient to use $Q_s(A)$ alone is shown by the following result of Ara.

THEOREM 3.4 [2] *For every C^*-algebra A, $Q_s(A)$ is a pre-C^*-algebra if and only if* $\dim C < \infty$.

The alternative construction of $M_{\mathrm{loc}}(A)$ is based on the universal property of the multiplier algebra $M(A)$: whenever A can be embedded as a closed essential ideal into a C^*-algebra B, the canonical embedding of A into $M(A)$ extends uniquely to an embedding of B into $M(A)$, thus making the following diagram commutative

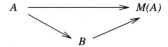

Applying this to the downwards directed set $\mathcal{I}_{ce}$, we obtain a directed system of C^*-algebras and embeddings as in the following commutative diagram

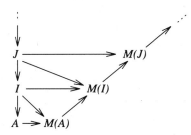

The C^*-inductive limit $\varinjlim M(I)$ of this system is $M_{\mathrm{loc}}(A)$.

Since the embeddings are compatible with the centers, we immediately obtain another commutative diagram

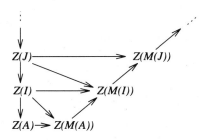

and an inductive limit $\varinjlim Z(M(I))$. One of the fundamental results relates this one to the center of $M_{\mathrm{loc}}(A)$ and can be regarded as a local version of the Dauns-Hofmann theorem.

THEOREM 3.5 [4] *For every C*-algebra A, the center Z of $M_{\mathrm{loc}}(A)$ is an AW*-algebra and coincides with $\varinjlim Z(M(I))$ (and thus with $\overline{C_b}$), and its structure space is the projective limit $\varprojlim \beta\check{I}$ over the Stone-Čech compactifications $\beta\check{I}$ of the open dense subsets $\check{I}$ in the primitive spectrum $\check{A}$ of A.*

DEFINITION 3.6.   For every C*-algebra A, *the bounded central closure of A is the C*-subalgebra $\overline{AC_b}$ of $M_{\mathrm{loc}}(A)$ and will be denoted by ^cA. A C*-algebra A is called* boundedly centrally closed *if $^cA = A$.*

As an important consequence of Theorem 3.5 we have

COROLLARY 3.7 [6] *For every C*-subalgebra B of $M_{\mathrm{loc}}(A)$ that contains both A and C_b, we have that $Z(B) = Z$. In particular, $Z(^cM(A)) = Z$.*

It turns out that the class of boundedly centrally closed C*-algebras, which includes all AW*-algebras, $M_{\mathrm{loc}}(A)$ and ^cA itself, is the correct generalization of prime C*-algebras from the viewpoint of solubility of the operator equation (5).

THEOREM 3.8 [6] *Let A be a boundedly centrally closed C*-algebra and $a, b \in M(A)^n$. Suppose that $\{b_1, \ldots, b_n\}$ is linearly independent over $Z = Z(M(A))$. Then, the only solution to (5) is $a = 0$.*

This theorem itself is a consequence of a more general result yielding an isometric embedding of the central Haagerup tensor product $M(A) \otimes_{Zh} M(A)$ into the space of completely bounded operators on a boundedly centrally closed C*-algebra A [6]. Partial results of this type for $B(H)$ or (certain) von Neumann algebras had been obtained in [10], [11], [23], and [42]. Related results for von Neumann subalgebras can be found in [31].

4. Applications.

In this final section, we briefly discuss two of the applications of the results in Section 3. Both belong to the following circle of problems. Suppose a (bounded) linear mapping ϕ on a C*-algebra has some additional, e.g. algebraic properties. Is it possible to give a concrete description of ϕ in terms of 'basic' maps? Often, these turn out to be combinations of elementary operators with other fundamental classes of operators such as Jordan homomorphisms, for instance. An example of such a result is the fact that every bijective linear map ϕ on $B(H)$ (or on $K(H)[1]$) that preserves invertibility essentially is of the form $\phi(x) = axb$ for some invertible elements $a, b \in B(H)$ [43]. Seemingly, this is more related to the first part of Section 3, but we shall see below that this is not necessarily the case.

Our first example is motivated from mathematical physics. The irreversible evolution of an open quantum system modeled by a C^*-algebra A is described by a semigroup $(T_t)_{t \geq 0}$ of completely positive operators on A. It was first observed by Lindblad that the infinitesimal generator L of such a semigroup has to fulfill the following set of inequalities

$$(6) \qquad L_n(x^*x) + x^*L_n(1)x - x^*L_n(x) - L_n(x^*)x \geq 0 \qquad (x \in M_n(A), \ n \in \mathbf{N})$$

where L_n is the canonical extension of L on $M_n(A)$. A bounded linear self-adjoint operator satisfying (6) is called *completely dissipative* or a *Lindblad generator*. There are two basic examples of such operators:

(i) L is completely positive itself;

(ii) L is a generalized inner *-derivation, i.e., $L(x) = xk + k^*x =: \delta_{k,k^*}(x)$ for some $k \in M(A)$ and all $x \in A$. The name comes from the fact that δ_{k,k^*} is *-preserving and satisfies a generalized Leibniz rule

$$\delta_{k,k^*}(xyz) = \delta_{k,k^*}(xy)z - x\delta_{k,k^*}(y)z + x\delta_{k,k^*}(yz) \qquad (x, y, z \in A)$$

from which (6) easily follows.

In many concrete physical situations, it has been observed that every completely dissipative operator allows a canonical decomposition of the form

$$(7) \qquad\qquad\qquad L = L^0 + \delta_{k,k^*}$$

where L^0 is completely positive. In [12], Christensen and Evans proved that (7) is in fact always possible if $A \subseteq B(H)$, $L^0 : A \to A''$, and $k \in A''$. This was built on preliminary work by Gorini, Kossakowski and Sudarshan [22] and by Lindblad [25], [26]. In general, a decomposition of the form (7) is not unique, but the different possibilities have different physical interpretations (of what kind the influence of the surrounding heat bath is). Therefore, suppose that

$$L = L_1^0 + \delta_{k_1,k_1^*} = L_2^0 + \delta_{k_2,k_2^*}$$

are two decompositions of the above type. Putting $a = k_1 - k_2$ we thus arrive at the following question.

PROBLEM 4.1. Given two completely positive operators L_1^0, L_2^0, which $a \in M(A)$ satisfy the operator equation

$$(8) \qquad\qquad\qquad L_1^0 - L_2^0 + \delta_{a,a^*} = 0?$$

Uniqueness, of course, occurs in (7) if and only if a is such that $\delta_{a,a^*} = 0$. For the case $A = B(H)$, dim $H = \infty$, Davies [15] proved that under the assumption that $L_j^0(1)$ is compact, all solutions of (8) lie in $i\mathbf{R}$ and thus uniqueness is guaranteed. This was extended to general infinite dimensional prime unital C^*-algebras in [21] as a consequence of the following result, the proof of which uses techniques related to the ones explained in Section 3.

THEOREM 4.2 [21] *Let A be a C*-algebra and P a proper closed prime ideal of $M(A)$ not containing A. Under the hypothesis $L^0 A \subseteq P$, each two decompositions of a completely dissipative operator L on A of the form $L = L^0 + \delta_{k,k^*}$ with L^0 completely positive and $k \in M(A)$ only differ by an addition by δ_{c,c^*} with $c \in P$.*

The second example comes from ring theory. Amongst the well-studied classes of additive mappings on non-commutative rings are those which are *centralizing*, i.e., every commutator $[x, F(x)]$ is central for all x in the ring R on which F is defined. One of the first results on centralizing mappings is Posner's theorem from 1957 which states that, on a prime ring R, the existence of a non-zero centralizing derivation forces R to be commutative. For C*-algebras, this has the consequence that there exist no non-zero centralizing derivations. In a first step of deducing this from Posner's theorem as well as for the further development, we note the following property. Denote by δ_a the inner derivation $\delta_a(x) = xa - ax$ $(x \in A)$.

PROPOSITION 4.3. *Every centralizing additive mapping F on a C*-algebra A satisfies*

$$(9) \qquad \delta_{F(y)} - \delta_y F = 0 \qquad (y \in A).$$

For illustration, we include a proof of Proposition 4.3 for the case $F = \delta$ of a derivation on A. From $[x, \delta(x)] \in Z(A)$ for all $x \in A$ we obtain

$$[x + y, \delta(x + y)] \in Z(A) \qquad (x, y \in A)$$

and hence

$$(10) \qquad [x, \delta(y)] + [y, \delta(x)] \in Z(A) \qquad (x, y \in A).$$

We compute that

$$\begin{aligned}
[x, \delta(x^2)] &= [x, x\delta(x)] + [x, \delta(x)x] \\
&= x[x, \delta(x)] + [x, \delta(x)]x \\
&= 2\,x[x, \delta(x)] \\
&= [x^2, \delta(x)] \qquad (x \in A)
\end{aligned}$$

whence specializing (10) to $y = x^2$ yields that $x[x, \delta(x)] \in Z(A)$ for all x. As a result,

$$[x, \delta(x)]^2 = [x[x, \delta(x)], \delta(x)] = 0 \qquad (x \in A)$$

wherefore $[x, \delta(x)] = 0$. Hence, (10) reduces to

$$[x, \delta(y)] + [y, \delta(x)] = 0 \qquad (x, y \in A)$$

which is nothing but the assertion.

Note that (9) implies for a centralizing derivation δ that the product $\delta_y \delta$ is a derivation for each $y \in A$ whence δ has to be zero [34].

There are two basic examples of centralizing mappings:

(i) $F = L_c$, the (left) multiplication by some $c \in Z(M(A))$;

(ii) any F such that $FA \subseteq Z(A)$.

Suppose that $F = L_c + \zeta$ where $c \in Z(M(A))$ and ζ is an additive mapping into the center of A. Then,

$$[x, F(y)] = [x, cy] + [x, \zeta(y)] = [x, cy] \qquad (x, y \in A),$$

or equivalently,

(11) $$\delta_{F(y)} - \delta_{cy} = 0 \qquad (y \in A).$$

Conversely, if there exists $c \in Z(M(A))$ such that (11) holds, then $\zeta = F - L_c$ defines an additive mapping from A into $Z(A)$. Thus, the system of operator equations (11) has a central solution c if and only if there exists a canonical decomposition

$$F = L_c + \zeta \quad \text{with } c \text{ central, } \zeta \text{ has central values only.}$$

Observe that (11) is a system of the form (2) parametrized by all elements in A.

The connection between centralizing mappings and the results of Section 3, in particular Theorem 3.8, is provided by the following result.

PROPOSITION 4.4 [5] *If F is a mapping on a C*-algebra A satisfying*

$$\delta_{F(y)} - \delta_y F = 0 \qquad (y \in A)$$

then

(12) $$M_{\delta_{F(y)}(x),\delta_u(v)} - M_{\delta_y(x),\delta_{F(u)}(v)} = 0 \qquad (x, y, u, v \in A).$$

Solving these operator equations first yields a family $\{c_x^y \mid x, y \in A\} \subseteq C$ as well as a family of projections $\{e_x^y \mid x, y \in A\} \subseteq C_b$ such that

$$e_x^y \delta_{F(y)}(x) - c_x^y \delta_y(x) = 0 \qquad (x, y \in A)$$

from which an element $c \in C$ is constructed using the self-injectivity of C; this satisfies $ce_x^y = c_x^y$ for all x, y and solves (11). An additional argument then establishes the following canonical decomposition.

THEOREM 4.5 [5, Theorem 3.2] *Let A be a C*-algebra with commutator ideal K. Every centralizing additive mapping F on A has a unique decomposition $F = L_c + \zeta$, where $c \in Z(M(K))$ and $\zeta: A \to C_b$ is additive.*

As can be seen by examples, in general, $c \notin Z(M(A))$ and $\zeta A \not\subseteq Z(A)$. If, however, A is boundedly centrally closed, then the decomposition is valid within

$M(A)$ and hence Theorem 4.5 in particular includes the case of a von Neumann algebra previously treated by Brešar [7] [3].

Similar questions for other types of structure-preserving mappings are close at hand. We wish to conclude with the following sample problem.

PROBLEM 4.6. Is there a general description of commutativity-preserving mappings on C^*-algebras?

Brešar proved that, on a prime C^*-algebra $A \neq M_2(\mathbf{C})$, every bijective linear mapping F such that both F and F^{-1} preserve commutativity is of the form $F = \lambda\phi + f$ where $\lambda \in \mathbf{C} \setminus \{0\}$, ϕ is a Jordan isomorphism, and f is a linear functional on A. The case of von Neumann algebras is treated in [9], but at the time of this writing, an answer to the general question was out of reach.

References

1. P. Ara, *The extended centroid of C*-algebras*, Arch. Math. (Basel) **54** (1990), 358–364.

2. ———, *On the symmetric algebra of quotients of a C*-algebra*, Glasgow Math. J. **32** (1990), 377–379.

3. P. Ara and D. Goldstein, *A solution of the matrix problem for Rickart C*-algebras*, Math. Nachr. **164** (1993), 259–270.

4. P. Ara and M. Mathieu, *A local version of the Dauns-Hofmann theorem*, Math. Z. **208** (1991), 349–353.

5. ———, *An application of local multipliers to centralizing mappings on C*-algebras*, Quart. J. Math. Oxford (2) **44** (1993), 129–138.

6. ———, *On the central Haagerup tensor product*, Proc. Edinburgh Math. Soc. **37** (1994), 161–174.

7. M. Brešar, *Centralizing mappings on von Neumann algebras*, Proc. Amer. Math. Soc. **111** (1991), 501–510.

8. ———, *Centralizing mappings and derivations in prime rings*, J. Algebra **156** (1993), 385–394.

9. M. Brešar and C. R. Miers, *Commutativity preserving mappings of von Neumann algebras*, Canad. J. Math. **45** (1993), 695–708.

10. A. Chatterjee and A. M. Sinclair, *An isometry from the Haagerup tensor product into completey bounded operators*, J. Operator Theory, in press.

11. A. Chatterjee and R. R. Smith, *The central Haagerup tensor product and maps between von Neumann algebras*, J. Funct. Anal. **112** (1993), 97–120.

12. E. Christensen and D. E. Evans, *Cohomology of operator algebras and quantum dynamical semigroups*, J. London Math. Soc. (2) **20** (1979), 358–368.

13. R. E. Curto, *Spectral properties of elementary operators*, in: "Elementary operators and applications", Proc. Int. Workshop, Blaubeuren, June 1991; World Scientific, Singapore 1992, 3–52.

14. R. E. Curto and L. A. Fialkow, *Similarity, quasisimilarity, and operator factorizations*, Trans. Amer. Math. Soc. **314** (1989), 225–254.

15. E. B. Davies, *Uniqueness of the standard form of the generator of a quantum dynamical semigroup*, Rep. Math. Phys. **17** (1980), 249–255.

---

[3] Our method in fact also generalizes the purely algebraic situation of a prime ring [8] to the semiprime case [5].

16. G. A. Elliott, *Automorphisms determined by multipliers on ideals of a C*-algebra*, J. Funct. Anal. **23** (1976), 1–10.

17. L. A. Fialkow, *Structural properties of elementary operators*, in: "Elementary operators and applications", Proc. Int. Workshop, Blaubeuren, June 1991; World Scientific, Singapore 1992, 55–113.

18. _____, *Invertible factorization and operator equations*, Acta Sci. Math. (Szeged), to appear.

19. L. A. Fialkow and H. Salas, *Majorization, factorization and systems of linear operator equations*, Math. Balkanica **4** (1990), 22–34.

20. C. K. Fong and A. R. Sourour, *On the operator identity* $\Sigma A_k X B_k \equiv 0$, Canad. J. Math. **31** (1979), 845–857.

21. Th. S. Freiberger and M. Mathieu, *Uniqueness of a Lindblad decomposition*, in: "Elementary operators and applications", Proc. Int. Workshop, Blaubeuren, June 1991; World Scientific, Singapore 1992, 179–187.

22. V. Gorini, A. Kossakowski, and E. C. G. Sudarshan, *Completely positive dynamical semigroups of N-level systems*, J. Math. Phys. **17** (1976), 821–825.

23. U. Haagerup, *The α-tensor product of C*-algebras*, 1981, unpublished.

24. D. Handelman, *Homomorphisms of C*-algebras to finite AW*-algebras*, Mich. Math. J. **28** (1981), 229–240.

25. G. Lindblad, *On the generators of quantum dynamical semigroups*, Commun. Math. Phys. **48** (1976), 119–130.

26. _____, *Dissipative operators and cohomology of operator algebras*, Lett. Math. Phys. **1** (1976), 219–224.

27. B. Magajna, *A system of operator equations*, Canad. Bull. Math. **30** (1987), 200–209.

28. _____, *An application of reflexivity of finite dimensional subspaces of factors*, in: "Elementary operators and applications", Proc. Int. Workshop, Blaubeuren, June 1991; World Scientific, Singapore 1992, 211–221.

29. _____, *A transitivity theorem for algebras of elementary operators*, Proc. Amer. Math. Soc. **118** (1993), 119–127.

30. _____, *Interpolation by elementary operators*, Studia Math. **105** (1993), 77–92.

31. _____, *The Haagerup norm on the tensor product of operator modules*, preprint.

32. M. Mathieu, *Elementary operators on prime C*-algebras*, II, Glasgow Math. J. **30** (1988), 275–284.

33. _____, *Elementary operators on prime C*-algebras*, I, Math. Ann. **284** (1989), 223–244.

34. _____, *Properties of the product of two derivations of a C*-algebra*, Canad. Math. Bull. **32** (1989), 490–497.

35. _____, *On C*-algebras of quotients*, Sem.ber. Funkt.anal. **16**, Tübingen 1989, 107–120.

36. _____, *The symmetric algebra of quotients of an ultraprime Banach algebra*, J. Austral. Math. Soc. (Series A) **50** (1991), 75–87.

37. C. Olsen and G. K. Pedersen, *Corona C*-algebras and their applications to lifting problems*, Math. Scand. **64** (1989), 63–86.

38. D. S. Passman, *Computing the symmetric ring of quotients*, J. Algebra **105** (1987), 207–235.

39. V. I. Paulsen, S. C. Power and R. R. Smith, *Schur products and matrix completions*, J. Funct. Anal. **85** (1989), 151–178.

40. V. I. Paulsen and L. Rodman, *Positive completions of matrices over C*-algebras*, J. Operator Theory **25** (1991), 237–253.

41. G. K. Pedersen, *Approximating derivations on ideals of C*-algebras*, Invent. Math. **45** (1978), 299–305.

42. R. R. Smith, *Completely bounded module maps and the Haagerup tensor product*, J. Funct. Anal. **102** (1991), 156–175.

43. A. R. Sourour, *Invertibility preserving linear maps on $L(X)$*, preprint.

44. P. Y. Wu, *The operator factorization problems*, Linear Algebra Appl. **117** (1989), 35–63.

Mathematisches Institut, Universität Tübingen, Auf der Morgenstelle 10, D-72076 Tübingen, Germany

e-mail: mmima01@mailserv.zdv.uni-tuebingen.de

Contemporary Mathematics
Volume **185**, 1995

MEMBERSHIP IN THE CLASS $\mathbb{A}^{(2)}_{\aleph_0}(\mathcal{H})$

ALFREDO OCTAVIO

ABSTRACT. In this work we develop some of the theory of doubly-generated dual algebras. We study those pairs of contractions that belong to a special class, where certain systems of equations can be solved. We give some necessary and sufficient conditions for pairs to belong in this class. We use some recently developed techniques to study this problem.

1. INTRODUCTION

Recently some attention has been paid to the theory of doubly-generated dual algebras, the attention, so far, has been given to functional calculi (cf., [12] and [15]) and to the reflexivity of algebras (cf., [5], [13], and [14]). We attempt in this paper to push forward using the techniques developed in [13] in order to study those pairs of contractions that belong to a certain class of operators. This classes are analogous to those study in the singly-generated case (cf., [3]). This classes and some other general preliminaries are discussed in Section 2.

In this work we study the classes $\mathbb{A}^{(2)}_{\aleph_0}(\mathcal{H})$ and $\mathbb{A}^{(2)}(\mathcal{H})$ (to be defined later). We describe some conditions under which a pair belongs to $\mathbb{A}^{(2)}(\mathcal{H})$ and some consequences of belonging to it. This issues are treated in Section 3. In Section 4 we discuss the relation between membership in $\mathbb{A}^{(2)}_{\aleph_0}(\mathcal{H})$ universal compressions, and reflexivity there our main result can be found.

As usual, some unsolved and, perhaps, interesting problems are sparse around the paper. This study is far from over and, we think, very much necessary, in order to put doubly–generated dual algebras in the same footing as singly–generated ones.

2. PRELIMINARIES

Let $\mathcal{H}$ be a (complex, separable, infinite dimensional) Hilbert space. The algebra $\mathcal{L}(\mathcal{H})$ of continuous linear operators from $\mathcal{H}$ into itself, is endowed with a weak* topology, since it is the dual of the Banach ideal composed of compact operators with finite trace. A dual algebra $\mathcal{A}$ is a weak* closed subalgebra of $\mathcal{L}(\mathcal{H})$ containing the identity. A dual algebra is the dual of a quotient which we denote by $\mathcal{Q}_\mathcal{A}$, its

1991 *Mathematics Subject Classification.* Primary: 47A60; Secondary: 47A15.
Key words and phrases. Contractions, Dual algebras, Functional Calculus

This paper is in final form and no version of it will be submitted for publication elsewhere.

elements are equivalence classes and we shall enclosed them in square brackets (cf., [3]). Given $x, y \in \mathcal{H}$ we define a rank–one operator by

$$(x \otimes y)(v) = <v, y> x, \qquad v \in \mathcal{H}.$$

Clearly $(x \otimes y)$ is of finite trace. Furthermore, any finite trace operator is an infinite sum of rank–one operators.

An n-tuple of commuting contractions $(T_1, \ldots, T_n)$ in $\mathcal{L}(\mathcal{H})$ generates the minimal dual algebra, $\mathcal{A}_{T_1, \ldots, T_n}$, that contains every T_j, $j = 1, \ldots, n$ plus the identity. In this case we denote by $\mathcal{Q}_{T_1, \ldots, T_n}$ the predual of $\mathcal{A}_{T_1, \ldots, T_n}$. Given a pair of commuting contractions (T_1, T_2) in $\mathcal{L}(\mathcal{H})$, there may exists a functional calculus generated by the pair. This means there is an algebra homomorphism from the algebra of bounded analytic functions on the bidisk $(\mathbb{D}^2)$, denoted by $H^\infty(\mathbb{D}^2)$, into $\mathcal{A}_{T_1, T_2}$. We shall assume the reader is familiar with the theory of dual algebras (as developed in [3]) and with the functional calculus for pairs of commuting contractions and its properties, the later is studied in several papers (cf.,[5], [13], [14], [15]). This functional calculus can be constructed provided certain elementary measures related to the pair (of commuting contractions) (T_1, T_2) have an absolute continuity property (cf., [13]), we shall denote the class of all such pairs by $ACC^{(2)}(\mathcal{H})$.

We denote by $\sigma_T(T_1, \ldots, T_n)$ the (joint) Taylor spectrum of an n-tuple of commuting operators in $\mathcal{L}(\mathcal{H})$ and by $\sigma_H(T_1, \ldots, T_n)$ the (joint) Harte spectrum of an n-tuple of commuting operators in $\mathcal{L}(\mathcal{H})$. For the definitions of the different joint spectra see [11].

Let $\mathcal{A}$ be a dual algebra. We remark that if $[L] \in \mathcal{Q}_\mathcal{A}$, then $[L] = \sum_{j=1}^\infty [x_j \otimes y_j]$ (where convergence is in the norm of $\mathcal{Q}_\mathcal{A}$), for some square summable sequences $\{x_j\}$ and $\{y_j\}$ in $\mathcal{H}$. One of the central techniques in the theory of dual algebras consists in finding conditions under which systems of equations of the form $[L] = [x \otimes y]$ have solutions. This is the key ingredient of the pioneering paper [6]. Since then several similar properties for dual algebras have been studied (cf.,[3]) in one and several variables (cf., [13]). We recall the definitions of these properties.

For $n \in \mathbb{N}$, a dual algebra $\mathcal{A}$ is said to have property $(\mathbb{A}_n)$ if for every doubly indexed family $[L_{j,k}] \in \mathcal{Q}_\mathcal{A}$, $0 \le j, k < n$, there exist vectors x_j, y_k in $\mathcal{H}$, $0 \le j, k < n$, such that

$$[L_{j,k}] = [x_j \otimes y_k], \quad 0 \le j, k < n.$$

We now define the classes $\mathbb{A}^{(2)}(\mathcal{H})$ and $\mathbb{A}_n^{(2)}(\mathcal{H})$. The analogous classes $\mathbb{A}(\mathcal{H})$ and $\mathbb{A}_n(\mathcal{H})$ have been a central topic of study in the theory of dual algebras (cf., [3]). Let $(T_1, T_2) \in ACC^{(2)}(\mathcal{H})$. We say that $(T_1, T_2) \in \mathbb{A}^{(2)}(\mathcal{H})$ if its functional calculus is an isometry. Furthermore, for n a cardinal number with $1 \le n \le \aleph_0$, we say that $(T_1, T_2) \in \mathbb{A}_n^{(2)}(\mathcal{H})$ if $(T_1, T_2) \in \mathbb{A}^{(2)}(\mathcal{H})$ and $\mathcal{A}_{T_1, T_2}$ has property $(\mathbb{A}_n)$.

One immediate consequence of the definition of the classes $\mathbb{A}_n^{(2)}(\mathcal{H})$ is

$$\mathbb{A}^{(2)}(\mathcal{H}) \supset \mathbb{A}_1^{(2)}(\mathcal{H}) \supset \cdots \supset \mathbb{A}_n^{(2)}(\mathcal{H}) \supset \cdots \supset \mathbb{A}_{\aleph_0}^{(2)}(\mathcal{H}).$$

From this we can conclude that

$$\bigcap_{n=1}^\infty \mathbb{A}_n^{(2)}(\mathcal{H}) \supseteq \mathbb{A}_{\aleph_0}^{(2)}(\mathcal{H}).$$

In the case of dual algebras generated by a single contraction we have (cf. [3], Theorem 6.3) that

$$\bigcap_{n=1}^{\infty} \mathbb{A}_n(\mathcal{H}) = \mathbb{A}_{\aleph_0}(\mathcal{H}).$$

Whether the analogous equality holds for pairs of contractions is an open problem.

Problem 2.1. *Let* $(T_1, T_2) \in \mathbb{A}^{(2)}(\mathcal{H})$, *does*

$$\bigcap_{n=1}^{\infty} \mathbb{A}^{(2)}_n(\mathcal{H}) = \mathbb{A}^{(2)}_{\aleph_0}(\mathcal{H})?$$

A word about notation: We have try to denote a pair of operators by (T_1, T_2) when the result (or problem, or definition) can be easily extended to n–tuples and by (S, T) when we don't know how to extend it to n–tuples.

3. MEMBERSHIP IN $\mathbb{A}^{(2)}(\mathcal{H})$

Two of the most beautiful and interesting theorems in the theory of dual algebras generated by a single contraction are the following:

Theorem 3.1. (cf., [7]) *If* $T \in \mathcal{L}(\mathcal{H})$ *is a contraction with* $\sigma(T) \supseteq \mathbb{T}$, *then* T *has a nontrivial invariant subspace.*

Theorem 3.2. $\mathbb{A}(\mathcal{H}) = \mathbb{A}_1(\mathcal{H})$.

The second theorem is an improvement of the first proved, independently of each other, in [2] and [8]. One important step in the proof of these results is the following theorem of Apostol (cf., [1]).

Theorem 3.3. *If* $T \in \mathcal{L}(\mathcal{H})$ *is an absolutely continuous contraction with* $\sigma(T) \supseteq \mathbb{T}$, *then either* $T \in \mathbb{A}(\mathcal{H})$ *or* T *has a nontrivial hyperinvariant subspace.*

In order to generalize 3.1 and 3.2, it seems reasonable to expect a generalization of 3.3. We pose the following:

Problem 3.4. *Let* $(T_1, T_2) \in ACC^{(2)}(\mathcal{H})$ *with* $\sigma_T(T_1, T_2) \supseteq \mathbb{T}^2$. *Show that either* $(T_1, T_2) \in \mathbb{A}^{(2)}(\mathcal{H})$ *or* (T_1, T_2) *has a common invariant subspace.*

Note that by a well known result of Chō and Takaguchi (cf., [10]), it is equivalent to state the above problem with the condition $\sigma_H(T_1, T_2) \supseteq \mathbb{T}^2$ instead of $\sigma_T(T_1, T_2) \supseteq \mathbb{T}^2$. This is no longer true in higher dimensions.

We now present a partial result towards a complete solution of 3.4. First, we prove a "weak" analytic mapping property for the Harte spectrum.

Lemma 3.5. *If* $T_1, T_2 \in ACC^{(2)}(\mathcal{H})$ *and* $\lambda = (\lambda_1, \lambda_2) \in \sigma_H(T_1, T_2) \cap \mathbb{D}^2$, *then* $h(\lambda) \in \sigma(h(T_1, T_2))$, *for any* $h \in H^{\infty}(\mathbb{D}^2)$.

Proof. For $w = (w_1, w_2) \in \mathbb{D}^2$, write

$$h(w) - h(\lambda) = (w_1 - \lambda_1)g_1(w_1, w_2) + (w_2 - \lambda_2)g_2(w_1, w_2).$$

Thus,

$$h(T_1, T_2) - h(\lambda) = (T_1 - \lambda_1)g_1(T_1, T_2) + (T_2 - \lambda_2)g_2(T_1, T_2).$$

If $h(\lambda) \notin \sigma(h(T_1, T_2))$, we let $A = (h(T_1, T_2) - h(\lambda))^{-1}$, and multiply the above equation by A to get

$$I = (T_1 - \lambda_1)g_1(T_1, T_2)A + (T_2 - \lambda_2)g_2(T_1, T_2)A.$$

Since A commutes with any element of $\mathcal{A}_{T_1, T_2}$, we have $\lambda \notin \sigma_H(T_1, T_2)$, which is a contradiction, and the lemma is proved. $\blacksquare$

For a given number $0 < \theta \leq 1$, we shall define the following set

$$\zeta_\theta(T_1, T_2) = \mathbb{D}^2 \cap \sigma_H(T_1, T_2) \cup \{(\lambda_1, \lambda_2) \in \mathbb{D}^2 \setminus \sigma_H(T_1, T_2) \ : \ \text{There is a}$$
$$u \in \mathcal{H}, \|u\| = 1 \text{ such that there are numbers } 0 < \theta_1, \theta_2 < 1$$
$$\text{with } \theta_1 + \theta_2 = \theta \text{ and } \|(T_j - \lambda_j)u\| \leq \theta_j(1 - |\lambda_j|)\}.$$

We have the following sufficient condition for membership in $\mathbb{A}^{(2)}(\mathcal{H})$.

Theorem 3.6. *Let $(T_1, T_2) \in ACC^{(2)}(\mathcal{H})$, and assume that for some θ satisfying $0 < \theta < 1/2$ the set $\zeta_\theta(T_1, T_2)$ is dominating for $\mathbb{T}^2$. Then $(T_1, T_2) \in \mathbb{A}^{(2)}(\mathcal{H})$.*

Proof. Since the functional calculus is contractive it suffices to prove that, for every $h \in H^\infty(\mathbb{D}^2)$ we have

$$(1) \qquad\qquad\qquad\qquad \|h(T_1, T_2)\| \geq \|h\|_\infty.$$

To show this we fix $h \in H^\infty(\mathbb{D}^2)$, and choose θ with $0 < \theta < 1/2$ such that $\zeta_\theta(T_1, T_2)$ is dominating for $\mathbb{T}^2$. We shall prove that

$$(2) \qquad\qquad\qquad\qquad \|h(T_1, T_2)\| \geq \|h\|_\infty(1 - 2\theta),$$

which implies (1). Indeed, it follows from (2) that

$$(3) \quad \|h(T_1, T_2)\|^n \geq \|h^n(T_1, T_2)\| \geq \|h^n\|_\infty(1 - 2\theta) = \|h\|_\infty^n(1 - 2\theta), \quad n \in \mathbb{N}.$$

Taking nth roots and limits (as $n \to \infty$) we obtain (1). Since $\zeta_\theta(T_1, T_2)$ is dominating, it is enough to show that for every $\lambda = (\lambda_1, \lambda_2) \in \zeta_\theta(T_1, T_2)$,

$$(4) \qquad\qquad\qquad\qquad \|h(T_1, T_2)\| \geq |h(\lambda)| - 2\theta\|h\|_\infty.$$

If $\lambda \in \sigma_H(T_1, T_2) \cap \mathbb{D}^2$, by 3.5, we have that $h(\lambda) \in \sigma(h(T_1, T_2))$, so (4) holds (since $\|h(T_1, T_2)\| \geq r(h(T_1, T_2)) \geq |h(\lambda)|$, where r denotes the spectral radius). Thus we may assume that $\lambda \in \zeta_\theta(T_1, T_2) \setminus \sigma_H(T_1, T_2)$, and from the definition of the set $\zeta_\theta(T_1, T_2)$ we get the existence of numbers $0 < \theta_1, \theta_2 < 1$ with $\theta_1 + \theta_2 = \theta$ and a vector $u \in \mathcal{H}$ such that

$$\|(T_j - \lambda_j)u\| \leq \theta_j(1 - |\lambda_j|), \quad j = 1, 2.$$

Write, for $w = (w_1, w_2) \in \mathbb{D}^2$,

$$h(w) = h(\lambda) + g_1(w)(w_1 - \lambda_1) + g_2(w)(w_2 - \lambda_2),$$

where $g_j \in H^\infty(\mathbb{D}^2)$ and satisfies

$$\|g_j\|_\infty \leq 2\|h\|_\infty(1 - |\lambda|_j)^{-1}, \quad j = 1, 2.$$

Then

$$\begin{aligned}
\|h(T_1, T_2)\| &\geq \|h(T_1, T_2)u\| = \|h(\lambda)u + g_1(T_1, T_2)(T_1 - \lambda_1)u + \\
&\quad + g_2(T_1, T_2)(T_2 - \lambda_2)u\| \geq \\
&\geq |h(\lambda)| - \|g_1(T_1, T_2)(T_1 - \lambda_1)u\| - \\
&\quad - \|g_2(T_1, T_2)(T_2 - \lambda_2)u\| \geq |h(\lambda)| - \\
&\quad - \|g_1\|_\infty \theta_1(1 - |\lambda_1|) - \|g_2\|_\infty \theta_2(1 - |\lambda_2|) \geq \\
&\geq |h(\lambda)| - 2\|h\|_\infty \theta_1 - 2\|h\|_\infty \theta_2 = \\
&\geq |h(\lambda)| - 2\theta\|h\|_\infty,
\end{aligned}$$

and the proof is complete. $\blacksquare$

The following is an immediate consequence of the previous theorem.

Corollary 3.7. *If* $(T_1, T_2) \in ACC^{(2)}(\mathcal{H})$ *is such that* $\sigma_H(T_1, T_2) \cap \mathbb{D}^2$ *is dominating for* $\mathbb{T}^2$, *then* $(T_1, T_2) \in \mathbb{A}^{(2)}(\mathcal{H})$.

A solution to the following problem would, therefore, provide a solution for 3.4.

Problem 3.8. *If* $(T_1, T_2) \in ACC^{(2)}(\mathcal{H})$ *and there is a* $0 < \theta \leq 1$ *such that* $\zeta_\theta(T_1, T_2)$ *is not dominating for* $\mathbb{T}^2$, *does* (T_1, T_2) *have a nontrivial common invariant subspace?*

We end this section by showing the following consequence of membership in $\mathbb{A}^{(2)}(\mathcal{H})$.

Theorem 3.9. *If* $(T_1, T_2) \in \mathbb{A}^{(2)}(\mathcal{H})$, *then* $\sigma_H(T_1, T_2) \supset \mathbb{T}^2$.

Proof. Let $(\lambda_1, \lambda_2) \in \mathbb{T}^2$. It is enough to show that given $\epsilon > 0$ there is a unit vector $x \in \mathcal{H}$ such that

$$\|(T_1 - \lambda_1)x\| < \epsilon \text{ and } \|(T_2 - \lambda_2)x\| < \epsilon.$$

Let $\delta = \min\{\dfrac{\epsilon}{2\|T_1 - \lambda_1\|}, \dfrac{\epsilon}{2\|T_2 - \lambda_2\|}\}$ and let U be the open neighborhood of (λ_1, λ_2) in $\mathbb{T}^2$ defined by

$$U = \{(w_1, w_2) \in \mathbb{T}^2 \; : \; |w_1 - \lambda_1| + |w_2 - \lambda_2| < \epsilon/2\}.$$

By [16], Theorem 3.5.3, there is a function $h \in H^\infty(\mathbb{D}^2)$ such that $|h| = \epsilon$ on $\mathbb{T}^2 - U$ and $|h| = 1$ on U. Since $\|h(T_1, T_2)\| = \|h\|_\infty = 1$, there exists a unit vector $v \in \mathcal{H}$ such that

$$1 - \delta \leq \|h(T_1, T_2)v\| \leq 1.$$

Let $y = h(T_1, T_2)v$ and $x = y/\|y\|$. Then

$$\begin{aligned}
\|(T_1 - \lambda_1)x\| &= \|(T_1 - \lambda_1)x - (T_1 - \lambda_1)y + (T_1 - \lambda_1)y\| \leq \\
&\leq \|(T_1 - \lambda_1)\|\|x - y\| + \|(T_1 - \lambda_1)y\| \leq \\
&\leq \|(T_1 - \lambda_1)\|\delta + \|(T_1 - \lambda_1)h(T_1, T_2)v\| \leq \\
&\leq \epsilon/2 + \|(w_1 - \lambda_1)h(w_1, w_2)\|_\infty < \epsilon.
\end{aligned}$$

Similarly we show that $\|(T_2 - \lambda_2)x\| < \epsilon$. ∎

4. Membership in $\mathbb{A}_{\aleph_0}^{(2)}(\mathcal{H})$ and universal compressions

Let $T \in \mathbb{A}^{(2)}(\mathcal{H})$ and assume $\mathcal{A}_T$ has property $(\mathbb{A}_{\aleph_0})$. It is known (cf., [3], Theorem 5.11) that given a strict contraction $A \in \mathcal{L}(\mathcal{H})$ (i.e., $\|A\| < 1$), there exist subspaces $\mathcal{M}$ and $\mathcal{N}$ in $\mathrm{Lat}(T)$ with $\mathcal{M} \supset \mathcal{N}$ such that, with respect to the decomposition

$$\mathcal{H} = \mathcal{N} \oplus (\mathcal{M} \ominus \mathcal{N}) \oplus (\mathcal{H} \ominus \mathcal{M}),$$

the operator T can be written as a matrix of the form

$$T = \begin{pmatrix} X & Y & Z \\ 0 & A' & R \\ 0 & 0 & S \end{pmatrix},$$

where A' is unitarily equivalent to A.

A weaker statement holds for pairs of contractions and it is a restatement of [4], Theorem 6.6.

Theorem 4.1. *Let $(T_1, T_2) \in \mathbb{A}_{\aleph_0}^{(2)}(\mathcal{H})$, and let (A_1, A_2) be a pair of commuting contractions from $\mathcal{L}(\mathcal{H})$ with $\sigma_T(A_1, A_2) \subset \mathbb{D}^2$. Then there are common invariant subspaces $\mathcal{M}$ and $\mathcal{N}$ of (T_1, T_2) with $\mathcal{M} \supset \mathcal{N}$ such that, with respect to the decomposition*

$$\mathcal{H} = \mathcal{N} \oplus (\mathcal{M} \ominus \mathcal{N}) \oplus (\mathcal{H} \ominus \mathcal{M}),$$

the operators T_1 and T_2 have matrices of the form

$$T_1 = \begin{pmatrix} * & * & * \\ 0 & A_1' & * \\ 0 & 0 & * \end{pmatrix}, \quad T_2 = \begin{pmatrix} * & * & * \\ 0 & A_2' & * \\ 0 & 0 & * \end{pmatrix},$$

where A_1' and A_2' are such that there is an invertible operator $X : \mathcal{H} \to \mathcal{M} \ominus \mathcal{N}$ with $A_j' = X A_j X^{-1}$ for $j = 1, 2$.

If $(T_1, T_2) \in \mathbb{A}^{(2)}(\mathcal{H})$, then for each $\lambda \in \mathbb{D}^2$, the map $[C_\lambda] : h(T_1, T_2) \mapsto h(\lambda)$ is a (well-defined) weak∗-continuous linear functional on $\mathcal{A}_{T_1, T_2}$, and is, hence, an element of $\mathcal{Q}_{T_1, T_2}$.

Let $\mathcal{A}$ be a dual algebra. For $0 \leq \theta \leq 1$, we denote by $\mathcal{X}_\theta(\mathcal{A})$ the set of all $[L] \in \mathcal{Q}_\mathcal{A}$ such that there exist sequences $\{x_i\}$ and $\{y_i\}$ in $\mathcal{H}$ satisfying the following conditions:

(5) $$\limsup_{i \to \infty} \|[x_i \otimes y_i] - [L]\| \leq \theta,$$

(6) $$\|x_i\| \leq 1, \qquad \|y_i\| \leq 1, \qquad 0 \leq i < \infty,$$

and

(7) $$\|[x_i \otimes z]\| + \|[z \otimes y_i]\| \to 0, \qquad z \in \mathcal{H}.$$

Let $\mathcal{A} \subset \mathcal{L}(\mathcal{H})$ be a dual algebra and $0 \leq \theta < \gamma$. We follow [3] by saying that $\mathcal{A}$ has property $X_{\theta,\gamma}$ if the set $\mathcal{X}_\theta(\mathcal{A})$ contains the closed ball $B_{0,\gamma}$ of radius γ centered at the origin in $\mathcal{Q}_\mathcal{A}$.

The definitions above are equivalent to Definitions 2.7 and 2.8 of [3] (cf., [9], see also [5]). In general, it is known that every dual algebra with property $X_{\theta,\gamma}$, for some $0 \leq \theta < \gamma \leq 1$, also has property $(\mathbb{A}_{\aleph_0})$ (cf., [3], Theorem 3.7). Thus, we have the following

Theorem 4.2. *If $(T_1, T_2) \in \mathbb{A}^{(2)}(\mathcal{H})$ and $\mathcal{A}_{T_1,T_2}$ has property $X_{\theta,\gamma}$, for some $0 \leq \theta < \gamma \leq 1$, then $(T_1, T_2) \in \mathbb{A}_{\aleph_0}^{(2)}(\mathcal{H})$.*

Let $T \in \mathbb{A}(\mathcal{H})$ be a contraction and assume $\mathcal{A}_T$ has property $(\mathbb{A}_{\aleph_0})$. Then $\mathcal{A}_T$ also has property $X_{\theta,\gamma}$, for some $0 \leq \theta < \gamma \leq 1$ (cf., [3], Theorem 6.3). Furthermore, $\mathcal{A}_T$ has property $X_{\theta,\gamma}$, for some $0 \leq \theta < \gamma \leq 1$, if and only if it has property $X_{\theta,\gamma}$, for every $0 \leq \theta < \gamma \leq 1$, and hence, if and only if it has property $X_{0,1}$. Thus, we are interested in finding a converse for 4.2, but, so far, our methods have produced only partial results. We pose the following problem.

Problem 4.3. *Let $(T_1, T_2) \in \mathbb{A}_{\aleph_0}^{(2)}(\mathcal{H})$. Does $\mathcal{A}_{T_1,T_2}$ have property $X_{\theta,\gamma}$, for some $0 \leq \theta < \gamma \leq 1$? Is property $X_{0,1}$ equivalent to property $X_{\theta,\gamma}$, for every $0 \leq \theta < \gamma \leq 1$?*

We now present a partial solution of 4.3. We first present a collection of lemmas which are, basically, from [13].

Lemma 4.4. *Let $(S, T) \in \mathbb{A}_{\aleph_0}^{(2)}(\mathcal{H})$, assume that either $\|S^n h\| \to 0$ or $\|T^n h\| \to 0$ (as $n \to \infty$), for all $h \in \mathcal{H}$, and let $\{x_n\}$ be a sequence of vectors in $\mathcal{H}$ converging weakly to zero. Then for any vector $y \in \mathcal{H}$, $\|[y \otimes x_n]\| \to 0$.*

Proof. The proof is exactly the same as the proof of Lemma 3.3 of [13]. With the notation of that proof, the key turning point is to show that $\|V_1^* y_1 \otimes x_n\|_{Q_{W_2^*}} \to 0$ as $n \to \infty$. It is enough to show that W_2^* is in the class $\mathbb{A}_{\aleph_0}(\mathcal{K})$ and then use Theorem 6.3 of [3] instead of Lemma 3.2 of [13]. But $W_2^* \in \mathbb{A}_{\aleph_0}(\mathcal{K})$ since it is an extension of $T \in \mathbb{A}_{\aleph_0}(\mathcal{H})$ (cf., [3], Proposition 4.11). ∎

Lemma 4.5. *Let $(S, T) \in \mathbb{A}_{\aleph_0}^{(2)}(\mathcal{H})$, assume that either $\|S^{*n} h\| \to 0$ or $\|T^{*n} h\| \to 0$ (as $n \to \infty$), for all $h \in \mathcal{H}$, and let $\{x_n\}$ be a sequence of vectors in $\mathcal{H}$ converging weakly to zero. Then for any vector $y \in \mathcal{H}$, $\|[x_n \otimes y]\| \to 0$.*

Proof. For any $x, y \in \mathcal{H}$, we have

$$\|[x \otimes y]\|_{Q_{S,T}} = \sup_{A \in \mathcal{A}_{S,T}, \|A\|=1} |<Ax, y>| = \sup_{A \in \mathcal{A}_{S,T}, \|A\|=1} |<A^* y, x>|$$
$$= \sup_{A \in \mathcal{A}_{S^*,T^*}, \|A\|=1} |<Ay, x>| = \|[y \otimes x]\|_{Q_{S^*,T^*}}.$$

Since $(S^*, T^*) \in \mathbb{A}^{(2)}(\mathcal{H})$ and either $\|S^{*n} h\| \to 0$ or $\|T^{*n} h\| \to 0$ (as $n \to \infty$), for all $h \in \mathcal{H}$ the lemma follows from 4.4. ∎

Main Theorem. *If $(S, T) \in \mathbb{A}_{\aleph_0}^{(2)}(\mathcal{H})$, $\|S^n h\| \to 0$, and $\|T^{*n} h\| \to 0$ (as $n \to \infty$), for all $h \in \mathcal{H}$, then $\mathcal{A}_{S,T}$ has property $X_{0,1}$.*

Proof. Let $\lambda = (\lambda_1, \lambda_2) \in \mathbb{D}^2$ and let $A_j \in \mathcal{L}(\mathcal{H})$ be a scalar operator with respect to some orthonormal basis $\{e_n\}$ with $A_j e_n = \lambda_j e_n$, $j = 1, 2$. By 4.1 there are common invariant subspaces $\mathcal{M}, \mathcal{N}$ of (S, T) with $\mathcal{M} \supset \mathcal{N}$ such that the compression of T_j to $\mathcal{M} \ominus \mathcal{N}$ is equal

$$h(S, T) = \begin{pmatrix} * & * & * \\ 0 & D & * \\ 0 & 0 & * \end{pmatrix},$$

with respect to the decomposition

$$\mathcal{H} = \mathcal{N} \oplus (\mathcal{M} \ominus \mathcal{N}) \oplus (\mathcal{H} \ominus \mathcal{M}),$$

where D is the scalar operator (with respect to the basis $\{e_n\}$) defined by $D e_n = h(\lambda) e_n$. Therefore,

$$< h(S, T) e_n, e_n >= h(\lambda),$$

and by the definition of $[C_\lambda]$ we have that $[e_n \otimes e_n] = [C_\lambda]$. Since $e_n \to 0$ in the weak topology, we can conclude from 4.4 and 4.5 that for every $y \in \mathcal{H}$,

$$\|[e_n \otimes y]\| + \|[y \otimes e_n]\| \to 0.$$

Thus, $[C_\lambda] \in \mathcal{X}_0(\mathcal{A}_{S,T})$ and $\mathcal{A}_{S,T}$ has property $X_{0,1}$. ∎

We note here that in [13] it is shown that a pair S, T of commuting contractions whose (joint) left essential spectrum is dominating and that satisfies some other condition (either $\|S^n h\| \to 0$ (as $n \to \infty$), for all $h \in \mathcal{H}$ or the pair is "diagonally extendable" (cf., [15])), has property $X_{0,1}$ and, thus, by 4.2, $(S, T) \in \mathbb{A}_{\aleph_0}^{(2)}(\mathcal{H})$.

Some open problems remain. They concern a weakening of the hypothesis in S and T. Relating this again with the results in [13]. We could pose the following

Problem 4.6. *With the notation above. If $\|S^n x\| \to 0$ for all $x \in \mathcal{H}$ and the right joint essential spectrum of (S, T) is dominating for T. Is it true that $(S, T) \in \mathbb{A}_{\aleph_0}^{(2)}(\mathcal{H})$?*

REFERENCES

1. C. Apostol, *Ultraweakly closed operator algebras*, J. Operator Theory **2** (1979), 49–61.
2. H. Bercovici, *Factorization theorems and the structure of operators in Hilbert space*, Ann. Math. **128** (1988), 399–404.
3. H. Bercovici, C. Foias, and C. Pearcy, *Dual algebras with applications to invariant subspaces and dilation theory*, CBMS Regional Conf. Ser. in Math., vol. 56, Amer. Math. Soc., Providence, R.I., 1985.
4. H. Bercovici, C. Hernandez-Garciadiego, and V. Paulsen, *Universal compressions of representations of $h^\infty(g)$*, Math. Ann. **281** (1988), 177–191.
5. H. Bercovici and W. S. Li, *Isometric functional calculus on the bidisk and invariant subspaces*, Bull. London Math. Soc. **25** (1993), 582–590.
6. S. Brown, *Some invariant subspaces of subnormal operators*, Integral Equations Operator Theory **1** (1978), 310–330.
7. S. Brown, B. Chevreau, and C. Pearcy, *On the structure of contraction operators ii*, J. Funct. Anal. **76** (1986?), 30–55.
8. B. Chevreau, *Sur les contractions a calcul fonctionnel isometrique ii*, J. Operator Theory **20** (1988), 269–293.
9. B. Chevreau and C. Exner, G. nad Pearcy, *On the structure of contractions operators iii*, Mich. Math. J. **36** (1989), 29–62.
10. M Chō and M. Takaguchi, *Boundary of taylor's joint spectrum for two commuting contractions*, Sc. Rep. Hirosaki Univ. **28** (1981), 1–4.

11. R. Curto, Applications of several complex variables to multiparameter spectral theory, 25–90, Surveys of recent results in operator theory, II, Longman, London, 1988.

12. M. Kosiek, *Representation generated by a finite number of hilbert space operators*, Ann. Polon. Math. **44** (1984), 309–315.

13. M. Kosiek, A. Octavio, and M. Ptak, *On the reflexivity of pairs of contractions*, Proc. Amer. Math. Soc., to appear, 1993.

14. M. Kosiek and M. Ptak, *Reflexivity of n-tuples of contractions with rich joint left essential spectrum*, Integral Equations Operator Theory **13** (1990), 395–420.

15. A. Octavio, *Coisometric extension and functional calculus for pairs of contractions*, J. of Operator Theory, to appear, 1992.

16. W. Rudin, *Function theory on polydisks*, Benjamin, New York, 1969.

IVIC, M–581, P.O. BOX 20010, MIAMI, FL 33102–0010
E-mail address: aoctavio@ivic.ivic.ve

Contemporary Mathematics
Volume **185**, 1995

HIGHER ORDER HANKEL FORMS

Jaak Peetre[1] and Richard Rochberg[1,2]

I. Introduction and Summary

Suppose H is the Hardy space or the Bergman space associated to the unit disk, D, or the upper half plane, U. The (small) Hankel operator on these spaces has been studied extensively. ([Z] is a good background for this.) However, sometimes it is more convenient to look instead at the associated Hankel bilinear form, B. This is the viewpoint, for instance, in [JPR]. B extends from a functional on $H \times H$ to an element of $(H \otimes H)^*$, the space of linear functionals on $H \otimes H$. (We are intentionally vague for now about which tensor product we are using.) If there is a group acting on H, for instance the Möbius group acting on the Bergman space of the disk, there will be an induced action on $(H \otimes H)^*$. In the case of classical spaces this action is not irreducible and, in fact, the irreducible component of lowest weight gives the Hankel forms. In [JP1] the bilinear forms corresponding to the other irreducible components, called Hankel forms of higher weight, were introduced and studied in detail.

Here we consider these higher weight forms from several viewpoints not based on group actions. This allows for new insights onto the general results and specific formulas of [JP1] and it shows how to extend the ideas of [JP1] to contexts in which there is no relevant group action. Because we are moving away from representation theory we will change terminology and refer to the forms we introduce as Hankel forms of "higher order" or "higher type". (A caveat: our forms of order k or type k are the forms of weight $2k + 2$ in [JP1].) We now give an informal preview of the contents. Definitions and precise formulations are given the later sections.

If B is a bilinear form on the Hilbert space H then we can regard B as a functional on the (for now algebraic) tensor product $H \otimes H$ by setting $B(f \otimes g) = B(f, g)$ and extending by linearity. If H consists of functions on a space X then $H \otimes H$ can be

[1] Much of this work developed while the authors were guests at Institut Mittag-Leffler. They wish to take this opportunity to thank the institute for its hospitality.

[2] Professor Rochberg's work was partly supported by NSF Grant DMS 9007491.

1991 *Mathematics Subject Classification* 47B35.

regarded as a space of functions on $X \times X$ by setting $(f \otimes g)(x_1, x_2) = f(x_1)g(x_2)$ and extending by linearity. From this point of view Hankel forms are (or, rather, extend to) linear functionals on a space of functions of two variables. The characteristic property is that their value at $f(x, y)$ depends only on $f(x, x)$, the restriction of f the diagonal $\Delta = \{(x, x) : x \in X\}$. The Hankel forms of higher order depend on the value of f at Δ as well as the higher order Taylor behavior of f near Δ. There are various ways to make this precise and this will lead to several alternative definitions of Hankel forms of higher type or higher order. Of course when H is one of the spaces considered in [JP1] we want to recapture the classes of forms defined there. (We verify that in some, but not all, cases.) In particular Hankel forms of order zero are classical (small) Hankel forms in the sense of [JPR].

Our first interpretation of "higher order behavior" will use derivatives. Hankel forms are sensitive to the value of $f(x, y)$ on Δ. Hence such forms also respond to the change of $f(x, y)$ as one moves along Δ; however, such forms do not note the rate of change traverse to Δ. A form will be called a Hankel form of type 1 or order 1 if its values are allowed to also depend on this traverse rate of changes and thus on the full first order Taylor expansion of f on Δ. Similarly higher order forms will involve higher order Taylor data. We present this approach in Section 2 using a vector field, the Hirota field, that differentiates in a direction traverse to Δ. We also investigate which forms of order one are orthogonal (in the Hilbert-Schmidt inner product) to forms of order 0. In fact we also look a bit at the analogous question for higher orders.

In Section 3 we revisit the same themes from a different viewpoint without explicit differential operators. We start with a Hilbert space H which is Hilbert space of functions on X and is a reproducing kernel Hilbert space, a HSRK. A Hilbert-Schmidt bilinear form on H is an element of $(H \otimes H)^*$ where now we are using the *Hilbert space* tensor product. $H \otimes H$ is a Hilbert space and hence we may identify $H \otimes H$ with $(H \otimes H)^*$. With this identification, we are regarding bilinear forms as elements of $H \otimes H$. The Hilbert space tensor product $H \otimes H$ is a HSRK on $X \times X$. Let V_Δ be the subspace of $H \otimes H$ consisting of functions that vanish on Δ. We have

$$(1.1) \qquad\qquad H \otimes H = V_\Delta^\perp \oplus V_\Delta.$$

The Hankel forms are exactly the forms in $V_\Delta^\perp$.

If we were doing algebraic geometry and wanted to study a class of functions on Δ we could start by looking at a class of functions on $X \times X$ and then taking the quotient

$$[(\text{Appropriate class of) functions on } X \times X] /$$

[the ideal of those which vanish on the diagonal].

We have done the Hilbert space analog of this and used the isometric identification

$$(H \otimes H)/V_\Delta = V_\Delta^\perp.$$

If we follow this algebraic lead then we recall that one way to get more refined local information is to look at

(the ideal)/(the square of the ideal)

etc. In the Hilbert space setting we can't square elements of V_Δ. However, in many cases we have a dense class which can be used to define V_Δ^2. We can then refine (1.1) to

$$(1.2) \qquad H \otimes H = V_\Delta^\perp \oplus (V_\Delta \ominus V_\Delta^2) \oplus V_\Delta^2.$$

The sum of the first two summands would be the Hankel forms of order 1. In cases such as the Bergman space of D this definition agrees with the previous one. The agreement of the two definitions in the most general contexts is unclear.

In Section 4 we again change viewpoints. The Bergman space of the disk is a Hilbert module over the space H^∞ of bounded analytic functions on D. Given $a \in H^\infty$ we can define a map δ_a from bilinear forms on the Bergman space to bilinear forms on the Bergman space by

$$(\delta_a B)(f, g) = B(af, g) - B(f, ag).$$

Using this notation we recast an observation in Section 6 of [JP1] as a definition: B is a Hankel form if $\delta_a B \equiv 0 \ \forall a \in H^\infty$. We then continue by saying that B is a Hankel form of type 1 if $(\delta_a)^2 B = 0 \ \forall a \in H^\infty$. This viewpoint, which is elaborated in Section 4, lets establish a connection with the theory of Hilbert modules.

Many of the known results about classical Hankel forms suggest questions for forms of higher order. Although we don't consider such questions systematically we do, in Section 5, give the analog of Kronecker's theorem for higher order Hankel forms on the Fock space.

In the final section we comment on relations between this work and other topics and mention some questions.

Other work that considers this class of forms, or closely related ones, includes [BP], [GP], [P1], [P2], [P3], [P4], [W], and [Zh].

II. HIGHER ORDER FORMS DEFINED USING DERIVATIVES.

A. The general set up.

Suppose Ω is a domain in $\mathbb{C}$ whose boundary consists of finitely many smooth curves and μ is a measure on Ω of the form $d\mu = h(z)dxdy$ for some smooth h. Let $H = A^2(\Omega, \mu)$ be the associated Bergman space, $L^2(\Omega, \mu) \cap \mathrm{Hol}(\Omega)$, and for $z \in \Omega$ let k_z be the associated Bergman kernel function. Thus for $f \in H$,

$$(2.1) \qquad f(z) = \langle f, k_z \rangle.$$

A Hankel form on H is a bilinear map from $H \times H$ to $\mathbb{C}$ with the characteristic property that $B(f, g)$ is actually a linear functional of the analytic function fg. Thus, formally at least, B is specified by a unique analytic symbol function b and

$$B(f, g) = \langle fg, b \rangle.$$

(The word "formally" reminds us that if $f, g \in H$ then we have $fg \in L^1(\Omega)$ and hence the pairing only makes sense for $b \in L^\infty(\Omega)$. However in many cases we want to allow more general b. There are various standard density and limiting considerations to deal with this issue and we leave such details to the reader.)

The simplest Hankel form is the rank one form given by setting $b = k_{\zeta_0}$ for some $\zeta_0 \in \Omega$. In that case, noting (2.1)

$$(2.2) \qquad B(f,g) = \langle fg, k_{\zeta_0} \rangle = f(\zeta_0)g(\zeta_0).$$

Another interesting example is the rank two form obtained by setting $b = \frac{\partial}{\partial \zeta}\big|_{\zeta=\zeta_0} k_\zeta$. In this case

$$
\begin{aligned}
B(f,g) &= \langle fg, \frac{\partial}{\partial \bar{\zeta}}\Big|_{\zeta=\zeta_0} k_\zeta \rangle \\
&= \frac{\partial}{\partial \zeta}\Big|_{\zeta=\zeta_0} \langle fg, k_\zeta \rangle \\
(2.3) \qquad &= \frac{\partial}{\partial \zeta}\Big|_{\zeta=\zeta_0} f(\zeta)g(\zeta) = f'(\zeta_0)g(\zeta_0) + f(\zeta_0)g'(\zeta_0)
\end{aligned}
$$

Thus, if we think of these forms as linear functionals of the function of two variables $f(x)g(y)$ they can involve values of $f(x)g(y)$ on $\Delta = \{(\zeta,\zeta) : \zeta \in \Omega\}$, as in (2.2), as well as derivatives along Δ, as in (2.3). They do not however, involve values of derivatives in the direction given by the vector field $D = (1,-1)$. The higher order Hankel forms are those which do involve such derivatives. To describe these derivatives we introduce the Hirota differential operator, D, which takes pairs of functions of one variable to functions of one variable

$$D^p(f \bullet g) = \left(\frac{\partial}{\partial y}\right)^p \Big|_{y=0} f(x+y)g(x-y).$$

That is $D^p(f \bullet g) = \sum_{j=0}^{p} \binom{n}{k}(-1)^{p-j} f^{(j)}(x)g^{(p-j)}(x)$.

The Hirota operator is quite useful in the study of certain PDE's, see [AS]. Here we find it a useful notation but don't know if there is any substantial relation to the work in [AS]. We record some properties of D,

$$D^p(f \bullet g) = (-1)^p D(g \bullet f).$$

in particular

$$D^p(f \bullet f) = 0 \quad \text{if } p \text{ is odd.}$$

Also,

$$D^p(e^{ix\xi} \bullet e^{ix\eta}) = 2^p(\xi - \eta)^p e^{ix(\xi+\eta)}.$$

Thus D^p, considered as a bi-differential operator (see [JP2], [CM]), has symbol $2^p(\xi - \eta)^p$.

We say that a bilinear form on H is a Hankel form of type s or order s if it is a linear functional of the $s+1$ functions $D^0(f \bullet g), D^1(f \bullet g), \ldots, D^s(f \bullet g)$. Symbolically,

$$
\begin{aligned}
(2.4) \qquad T_s(f,g) &= L(D^0(f \bullet g), \ldots, D^s(f \bullet g)) \\
&= \sum_{j=0}^{s} L_j(D^j(f \bullet g))
\end{aligned}
$$

where L_j are linear functionals on $\mathrm{Hol}(\Omega)$. Thus Hankel forms of type 0 are Hankel forms. A simple Hankel form of type 1 is the rank 2 form given by, for $\zeta_0 \in \Omega$,

$$(2.5) \qquad \begin{aligned} T(f,g) &= \langle D(f \bullet g), k_{\zeta_0} \rangle \\ &= \langle f'g - fg', k_{\zeta_0} \rangle \\ &= f'(\zeta_0)g(\zeta_0) - f(\zeta_0)g'(\zeta_0). \end{aligned}$$

Notice that the forms (2.2), (2.3), and (2.5) together give the full Taylor data up to order 1 at $(\zeta_0, \zeta_0) \in \Delta$. That is, for $F(x,y) = f(x)g(y)$ we can use those forms to find $F(\zeta, \zeta), F_x(\zeta, \zeta)$ and $F_y(\zeta, \zeta)$. Similarly we can extract Taylor data up to order s using the Hankel forms of type at most s.

Let us concentrate for a moment on the individual terms of the sum in (2.4). $D^j(f \bullet g)$ is a function in $\mathrm{Hol}(\Omega)$ and a rich class of functionals on $\mathrm{Hol}(\Omega)$ can be realized by a formulas such as

$$L_j(D^j(f \bullet g)) = \int_\Omega D^j(f \bullet g)\bar{b} \, d\mu.$$

for some b in $\mathrm{Hol}(\Omega)$. However we could use a different measure, $d\nu$, and find that the *same* functional is also given by

$$L_j(D^j(f \bullet g)) = \int_\Omega D^j(f \bullet g)\bar{\tilde{b}} \, d\nu$$

for a different $\tilde{b}$ in $\mathrm{Hol}(\Omega)$. Although at first glance it would seem natural to stay with μ, the flexibility to choose different ν sometimes helps in studying the invariance properties of the operators (see the discussion of "different gauges" in [JPR]). In particular in [JP1] in was natural to use a different ν for each index j. Thus we realize the form in (2.4) as

$$(2.6) \qquad T_s(f,g) = \sum_{p=0}^{s} \int_\Omega D^p(f \bullet g)\bar{A}_p \, d\nu_p$$

where $A_p \in \mathrm{Hol}(\Omega)$.

B. Orthogonality.

Suppose now that all the bilinear forms being considered are in the Hilbert-Schmidt class $\mathcal{S}_2$. Hence, using the Hilbert-Schmidt inner product, which we recall in a moment, there is a notion at orthogonality. We will say a Hankel form of type s is a *special form of type s* if it is orthogonal to all Hankel forms of type $j, j = 0, \ldots, s-1$.

Our basic examples here are the Bergman and Fock spaces.

Fix $\alpha > -1$ and let H be $A^2(D, (1 - |z|^2)^\alpha \, dxdy)$ or $A^2(U, y^\alpha \, dxdy)$. For convenience we set

$$\lambda = \alpha + 2$$

and recall the Pochhammer symbols

$$(\lambda)_p = \lambda(\lambda + 1) \cdots (\lambda + p - 1).$$

It is shown in [JP1] that the special Hankel forms of type s are exactly those which are linear functionals of

$$(2.7) \qquad J_s(f,g) = \sum_{p=0}^{s} (-1)^p \binom{s}{p} \frac{f^{(p)}}{(\lambda)_p} \frac{g^{(s-p)}}{(\lambda)_{s-p}}.$$

In particular we have, for $s = 1$

$$(2.8) \qquad J_1(f,g) = f'g - fg'$$

and for $s = 2$

$$(2.9) \qquad J_2(f,g) = f''g - 2\frac{\lambda+1}{\lambda}f'g' + fg''.$$

On U, for $\alpha > -1$ set

$$d\mu_\alpha = \frac{\alpha+1}{\pi}(2y)^\alpha \, dxdy.$$

When working on $H = A^2(U, d\mu_\alpha)$ it turns out that a natural choice for $d\nu_p$ is

$$d\nu_p = d\mu_{2\alpha+2p+2}.$$

Thus the special Hankel forms of type s are realized as

$$(2.10) \qquad T_s(f,g) = \int_U J_s(f,g)\bar{b} \, d\nu_s$$

where $b \in \mathrm{Hol}(\Omega)$ is the symbol function of the form (with respect to the selected gauge). We return to this formula later.

If B_i, $i = 1, 2$, are Hilbert-Schmidt bilinear forms on H then there are unique A_i in $L^2(\Omega \times \Omega, \mu \times \mu) \cap \mathrm{Hol}(\Omega \times \Omega)$ such that

$$(2.11) \qquad B_i(f,g) = \iint_{\Omega \times \Omega} f(x)g(y)A_i(x,y) \, d\mu(x)d\mu(y).$$

A_i is called the (integral) kernel of the form.

The Hilbert-Schmidt inner product is given by

$$\langle B_1, B_2 \rangle_{H.S.} = \langle A_1, A_2 \rangle_{L^2(\Omega \times \Omega, \mu \times \mu)}.$$

To go further we note that we can pass from B_i to A_i using the reproducing kernels. Using the fact that for $a, b \in \Omega$ and for nice $H \in \mathrm{Hol}(\Omega \times \Omega)$ we have

$$\langle H(x,y), k_a(x)k_b(y) \rangle_{L^2(\Omega \times \Omega, \mu \times \mu)} = H(a,b).$$

we find that if A and B are related as in (2.8) then

$$(2.12) \qquad\qquad A_i(x,y) = B_i(k_x, k_y).$$

We will now use (2.12) to try to understand the special Hankel forms.

Suppose T_s is given by (2.6). By (2.12) it has integral kernel

$$F(x,y) = \sum_p \int D^p(k_x \bullet k_y) \overline{A_p} \, d\nu_p.$$

Fix $q < s$ and $r \in \Omega$. If T_s is special then it is orthogonal to the form

$$R(f,g) = D^q(f \bullet g)(r).$$

By (2.12) that form has kernel

$$G(x,y) = D^q(k_x \bullet k_y)(r).$$

To have $T_s \perp R$ we must have $F \perp G$, thus,
$$(2.13)$$
$$0 = \langle G, F \rangle = \sum_p \int_\Omega \overline{A_p(t)} \left(\iint_{\Omega \times \Omega} D^p(k_x \bullet k_y)(t) \, \overline{D^q(k_x \bullet k_y)(r)} \, d\mu(x) d\mu(y) \right) d\nu_p(t).$$

Writing $k_a(b) = k(b,a)$ the inner integrand is

$$\left(\frac{\partial}{\partial u}\right)^p \bigg|_{u=0} k(t+u,x) k(t-u,y) \left(\frac{\partial}{\partial v}\right)^q \bigg|_{v=0} \overline{k(r+v,x) k(r-v,y)}$$

$$= \frac{\partial^{p+q}}{\partial u^p \partial \bar{v}^q} \bigg|_{u=v=0} k(t+v,x) \overline{k(r+v,x)} k(t-u,y) \overline{k(r-v,y)}.$$

Now note that

$$\int_\Omega k(a,x) \overline{k(b,x)} \, d\mu(x) = \int k(x,b) \overline{k(x,a)} \, d\mu(x) = \langle k_b, k_a \rangle = k_b(a) = k(a,b).$$

Hence the inner integral is

$$\frac{\partial^{p+q}}{\partial u^p \partial \bar{v}^q} \bigg|_{u=v=0} k(t+u, r+v) k(t-u, r-v) = D^p \bar{D}^q(k \bullet k)$$

where the expression on the left defines the "double Hirota operator" $D^p \bar{D}^q$. Explicitly

$$D^p \bar{D}^q(k \bullet k) = \sum_{j=0}^{p} \sum_{k=0}^{s} (-1)^{p+q-j-k} \binom{p}{j} \binom{q}{k} \frac{\partial^{j+k}}{\partial t^j \partial \bar{r}^k} k(t,r) \frac{\partial^{p+q-j-k}}{\partial t^{p-j} \partial \bar{r}^{q-k}} k(t,r).$$

Thus, from (2.13) we have, for all $q < s$, all $r \in \Omega$

$$\sum_p \int_\Omega \bar{A}_p(t) D^p D^q(k \bullet k)(t, r) \, d\nu_p(t) = 0.$$

We now examine this for small s.

When $s = 1$ we have

$$0 = \int \bar{A}_0(t) k(t, r)^2 \, d\nu_0(t) + \int \bar{A}_1(t) \Big(\frac{\partial k(t, r)}{\partial \bar{r}} k(t, r) - k(t, r) \frac{\partial k(t, r)}{\partial \bar{r}} \Big) \, d\nu_1(t).$$

The second integral vanishes identically and we conclude

$$0 = \int \bar{A}_0(t) k_r^2(t) \, d\nu_0(t).$$

In reasonable cases this forces A_0 to vanish thus what remains, writing $A_1 = A$, $d\nu_1 = d\nu$, is

$$T(f, g) = \int (f'g - fg') \bar{A} \, d\nu.$$

Thus the special Hankel forms of type 1 are given by superposition of the basic rank 2 forms given by (2.5).

We now go on to $s = 2$. In the previous computation we saw $D^1 \bar{D}^0(k \bullet k)$ vanishes identically. Using this we find, for $q = 0$

$$\int D^0 \bar{D}^0(k \bullet k) \bar{A}_0 + \int D^2 \bar{D}^0(k \bullet k) \bar{A}_2 = 0$$

and, for $q = 1$, using $D^0 \bar{D}^1(k \bullet k) \equiv 0$ and $D^2 \bar{D}^1(k \bullet k) \equiv 0$ we have

$$\int D^1 \bar{D}^1(k \bullet k) \bar{A}_1 = 0.$$

More explicitly,

$$(2.14) \qquad \int k_r^2 \bar{A}_0 \, d\nu_0 + 2 \int (k_r' k_r - k_r'^2) \bar{A}_2 \, d\nu_2 = 0$$

and

$$(2.15) \qquad \int \Big(\frac{\partial^2}{\partial \bar{r} \partial t} k_r(t) \cdot k_r(t) - \frac{\partial}{\partial t} k_r(t) \frac{\partial}{\partial \bar{r}} k_r(t) \Big) \overline{A_1(t)} \, d\nu_1(t) = 0.$$

Thus in the common cases, and presumably in much greater generallity, we have $A_1 \equiv 0$. Also, from (2.14), we see that A_0 is determined by A_2 which we may think of as the "symbol function".

We now want to check that (2.14) and (2.15) give (2.9). Recall that on the space $L^2(U, d\mu_\alpha) \cap \text{Hol}(U)$ the reproducing kernel is given by

$$(2.16) \qquad\qquad k_r(z) = (z - \bar{r})^{-2-\alpha}.$$

Direct substitution of (2.16) in (2.15) gives

$$0 = c \int_U (z - \bar{r})^{-6-2\alpha} \overline{A_1(z)} \, d\mu_{2\alpha+4}.$$

Using (2.16) for $2\alpha + 4$ instead of α, we can evaluate the integral and find

(2.17) $$0 = c \, A_1(r).$$

r is arbitrary thus $A_1 \equiv 0$.

When we substitute (2.16) into (2.14) and do the differentiation and arithmetic we find

$$0 = \int \overline{A_0(z)}(z - \bar{r})^{-2\alpha-4} \, d\mu_{2\alpha+2} + 2(\alpha + 2) \int \overline{A_2(z)}(z - \bar{r})^{-2\alpha-6} \, d\mu_{2\alpha+4}.$$

Using (2.6) for $2\alpha + 2$ and $2\alpha + 4$ we find

(2.18) $$0 = A_0 + 2(\alpha + 2)A_2.$$

For a special Hankel form of type 2 we have (2.17) and (2.18). Hence, writing $A = A_0$ we have the form

$$B(f, g) = \int \bar{A} f g \, d\mu_{2\alpha+2} - \frac{1}{2(\alpha + 2)} \int \bar{A} D^2(f \bullet g) \, d\mu_{2\alpha+4}$$

Writing $\frac{\partial^2}{\partial z^2}(2 \operatorname{Im} z)^{2\alpha+4} = (2\alpha + 3)(2\alpha + 2)(2 \operatorname{Im} z)^{2\alpha+2}$ we find the first integral equals

$$\int \bar{A}(fg)'' \frac{1}{2\alpha + 5} \frac{1}{2\alpha + 4} \, d\mu_{2\alpha+4}.$$

Combining the two integrals, doing the arithmetic, and remembering that $\lambda = \alpha + 2$, we have

$$B(f, g) = c \int \bar{A}(f''g - 2\frac{\lambda + 1}{\lambda} f'g' + fg'') d\mu_{2\alpha+2}$$

(with $c = -(2\alpha + 5)^{-1}$). In short, B is a linear functional of $J_2(f, g)$.

We know of no reason why these computations can't be continued past $s = 2$. On the other hand we know of no effective way to organize the numerics. The formula in [JP1] was obtained by exploiting the group action.

There is one case in which we can identify the special Hankel forms of order s for every s; that is the Fock space. Let H be the entire functions in $L^2(\mathbb{C}, \exp(-|z|^2))$. This is a HSRK and $k_w(z) = e^{z\bar{w}}$. We claim that for positive integers n, k with $n \neq k$ and any ζ_1 and ζ_2 in $\mathbb{C}$ the finite rank (and hence certainly Hilbert-Schmidt) functionals

$$B_1(f, g) = D^n(f \bullet g)(\zeta_1)$$

and

$$B_2(f, g) = D^k(f \bullet g)(\zeta_2)$$

are orthogonal.

First note that

(2.19) $D^{(j)}(k_x \bullet k_y) = (\bar{x} - \bar{y})^j e^{z(\bar{x}+\bar{y})}.$

By this and (2.12) we need to check the orthogonality of

$$(x-y)^n e^{\bar{\zeta}_1(x+y)}$$

and

$$(x-y)^k e^{\bar{\zeta}_2(x+y)}$$

in $H \otimes H$. We must evaluate

$$\iint_{\mathbb{C}\times\mathbb{C}} (x-y)^n (\bar{x}-\bar{y})^k e^{\bar{\zeta}_1(x+y)} e^{\zeta_2(\overline{x+y})} e^{(|x|^2+|y|^2)} \, dA(x)\, dA(y).$$

(Here $dA(x)$ is area measure in $\mathbb{C}$.) Introduce new integration variables $u = x - y$, $v = x+y$. The integrand factors and the u integral is the integral over $\mathbb{C}$ of $u^n \bar{u}^k$ times a Gaussian density. This vanishes identically unless $n = k$. This is enough to conclude that for any positive m and ζ in $\mathbb{C}$

$$B(f,g) = D^m(f \bullet g)(\zeta)$$

is a special Hankel form of type m. It follows that linear combinations of such are dense in the space of special Hankel forms of type n and that the most general is an integral

$$B(f,g) = \int_{\mathbb{C}} \overline{A(s)} D^m(f \bullet g)(\zeta) e^{-|\zeta|^2} \, dA(\zeta).$$

The integral kernel of this form is, by (2.12)

$$K(x,y) = \int \overline{A(\zeta)} D^m(k_x \bullet k_y)(\zeta) e^{-|\zeta|^2} \, dA(\zeta).$$

Using (2.19) we have

$$K(x,y) = \int A(\zeta)(x-y)^m e^{\bar{\zeta}(x+y)} e^{-|\zeta|^2} \, dA(\zeta)$$
$$= (x-y)^m \langle A, k_{x+y} \rangle$$
$$= (x-y)^m A(x+y).$$

In [JP1] it was noted that bilinear forms on the Fock space which had kernels of this form, $\overline{(x-y)^m A(x+y)}$, had nice transformation properties under the natural Heisenberg group action.

III. Tensor Products

A. Type 0.

Suppose now that H is a HSRK. Thus the elements of H are functions on some set X and for each $x \in X$ there is $k_x \in H$ so that for all $h \in H$

$$h(x) = \langle h, k_x \rangle_H.$$

Suppose B is a bilinear form on H. B maps $H \times H$ to $\mathbb{C}$ and extends by linearity to the algebraic tensor product of H with itself. If B is a Hilbert-Schmidt form then it extends to a map of $H \otimes H$, the Hilbert space tensor product, to $\mathbb{C}$. For the rest of this section we shall consider only Hilbert-Schmidt forms. B is a linear functional on $H \otimes H$ and hence an element of $(H \otimes H)^*$ which, because it is a Hilbert space, we identify with $H \otimes H$.

Pick $\zeta \in X$. The rank one bilinear form

$$B_\zeta(f, g) = f(\zeta)g(\zeta)$$

only depends on the product of f and g and hence is a Hankel form in the classical cases and can also be called a Hankel form in this general context. In the classical cases we can realize B_ζ by

$$B_\zeta(f, g) = \langle f, P(k_\zeta \bar{g}) \rangle$$

and hence B_ζ is the Hankel form associated with the Hankel operator with symbol k_ζ. (If we leave the classical picture we need not have H as a subspace of a natural L^2 space so we don't necessarily have P.) We now want to understand B_ζ as an element of $H \otimes H$. First recall that by identifying $f \otimes g$ with the function on $X \times X$ whose value at (x_1, x_2) is $f(x_1)g(x_2)$ we can realize $H \otimes H$ as a space of functions on $X \times X$. In fact $H \otimes H$ is itself a HSRK and the kernel function for $(\zeta_1, \zeta_2) \in X \times X$ is given by

$$k_{(\zeta_1, \zeta_2)}(x_1, x_2) = k_{\zeta_1}(x_1) k_{\zeta_2}(x_2).$$

Thus, tracking through the definitions, we see that $k_{(\zeta, \zeta)}$ is the element of $H \otimes H$ corresponding to B_ζ. We can define S mapping the symbol function of a Hankel form to the element of $H \otimes H$ associated to the form by setting

$$S(k_\zeta)(x_1, x_2) = k_\zeta(x_1)k_\zeta(x_2) = k_{(\zeta, \zeta)}(x_1, x_2)$$

and extending by linearity. We also note that $S^* : H \otimes H \to H$ is given by

$$S^*(f(\cdot))(x) = f(x, x).$$

That is S^* evaluates functions on $X \times X$ on the diagonal $\Delta\{(x, x); x \in X\}$.

In the language of the previous section S maps from the symbol function to the integral kernel.

We will say a Hilbert-Schmidt form on H is a Hankel form if it is in the closed linear span of $\{S(k_\zeta) : \zeta \in X\}$. Let

$$V_\Delta = \{f \in H \otimes H; f(\zeta, \zeta) = 0 \ \forall \zeta \in X\}.$$

Proposition 3.1: *The Hankel forms in $H \otimes H$ are exactly $V_\Delta^\perp$.*

Note: This shows that we didn't lose any of what we wanted when we just take limits of linear combinations of rank 1 Hankel forms.

Proof. If $f \in V_\Delta$ then $\langle f, S(k_\zeta) \rangle = \langle f, k_{(\zeta, \zeta)} \rangle = f(\zeta, \zeta) = 0$. Thus S maps into $V_\Delta^\perp$. On the other hand if the span of $\{S(k_\zeta)\}$ is not dense in $V_\Delta^\perp$ then we can find h in $V_\Delta^\perp$ which is orthogonal to all the $S(k_\zeta)$. By the computation we just did $h \in V_\Delta$. Thus $h \equiv 0$ and we are done.

Thus to understand the (Hilbert-Schmidt) normed theory of Hankel forms we need to understand the normed space $V_\Delta^\perp$. By identifying $V_\Delta^\perp$ with $(H \otimes H)/V_\Delta$ we can realize $V_\Delta^\perp$ as a space of functions on X. However, description of the norm on this function space is not trivial even for the standard examples. The identification of the norm when H is the Hardy space on D (and for other examples also) is done in [DP] in the context of module tensor products (about which more in the next section). (However the computation for the Hardy space goes back at least to [Ru].)

Suppose now H is the Hardy space of the disk. Thus $X = D$. For $\zeta \in X$, $k_\zeta(w) = (1 - \bar\zeta w)^{-1}$. In [DP] we find that for this H, $V_\Delta^\perp$, as a space of functions on D, is, isometrically, the Bergman space. Thus if we regard $V_\Delta^\perp$ as the natural space of symbols of Hilbert-Schmidt Hankel forms we conclude that a Hankel form on the Hardy space is in the Hilbert-Schmidt class if and only if its symbol is in the Bergman space.

Of course this is not the standard answer. If we start with a classical Hankel operator with symbol function k_ζ then we associated it with $S(k_\zeta) \in H \otimes H$. We noted that $S(k_\zeta) \in V_\Delta^\perp$ and we identify that space with the Bergman space of Δ by restrictions

$$S(k_\zeta)|_\Delta = k_\zeta(x_1) k_\zeta(x_2)|_\Delta = k_\zeta{}^2.$$

Thus it is $k_\zeta{}^2$ which must be in the Bergman space. To see what is going on in general we must see how to extend the map

$$(3.1) \qquad\qquad\qquad\qquad k_\zeta \to k_\zeta{}^2$$

to general holomorphic functions as a *linear* map. Although it is not clear what to do for general HSRK, we can find an explicit formula in the case of the Hardy space. The map in (3.1) is the map

$$(3.2) \qquad\qquad\qquad\qquad f(z) \to (zf(z))'.$$

Thus the Hankel operator with classical symbol f will induce a Hilbert-Schmidt form exactly if $(zf)'$ is in the Bergman space. This is true exactly if f is in the Dirichlet space – which is the traditional answer.

For the Bergman space and the Fock space it is also easy to find the *linear* form of the map in (3.1) and recapture the known results in those cases also.

Before going to Hankel forms of higher type we would like to mention a possible direction for further work. Except for the explicit description of $V_\Delta^\perp$ and the description of $S(f)|_\Delta$ in (3.2) we didn't use any specific facts about the Hardy space. In fact we didn't need the description of $V_\Delta^\perp$. If we start with H *any* HSRK and f any finite linear combination of kernel functions $f = \sum a_i k_i$ then

$$S(f) = \sum a_i S(k_i)$$

will be in $V_\Delta^\perp$ and its norm will be

(3.3) $\|(S^*S)^{1/2}(f)\|_H$

where

$$S^*S(f) = \sum a_i S^*S(k_i) = \sum a_i k_i^2.$$

Of course to go further we still need to understand S^*S as a linear map. However we have removed any reference to finding the norm on $V_\Delta^\perp$ (or, rather we have noted that $\|S(f)\|_{V_\Delta^\perp} = \|(S^*S)^{1/2}(f)\|_H$ and now don't know, in general, how to find $(S^*S)^{1/2}$). There are many spaces where Hankel forms have not traditionally been studied where it might be interesting to pursue these ideas; for example, $H = \{\text{functions } f \text{ on } [0,1] : \int_0^1 |f'|^2 < \infty\}$.

B. Higher type.

We saw that (one candidate to be called) the symbol of a Hankel form lived in the space $(H \otimes H)/V_\Delta$. That quotient construction is analogous to the one in algebraic geometry in which we look at functions on a large set (in this case $X \times X$) modulo those that vanish on a subset (here Δ) as a way to study the placement of that subset in the larger set. Algebraic geometers often carry the analysis further by considering not only the ideal of functions which vanish on the small set but also higher powers of that ideal. That is the plan we follow here.

For $n = 1, 2, 3, \ldots$ let V_Δ^n be the closure of the span of products of n elements of V_Δ. Actually, of course, although such products are defined pointwise thay may fail to be in $H \otimes H$. We must restrict consideration to those products which are in $H \otimes H$. For the standard examples involving Hilbert spaces of holomorphic functions it is easy to find convenient dense classes in $H \otimes H$ for which all powers are in $H \otimes H$. However for examples not involving holomorphy or involving complicated geometry it is not clear what happens. For instance it is easy to give examples of H's which are HSRK's (but not of holomorphic functions) for which $V_\Delta^2 = V_\Delta$ and hence for which this further analysis gives nothing new.

The analysis in part A of this section led us to split $H \otimes H$ as

$$H \otimes H = V_\Delta^\perp \otimes V_\Delta$$

and to identify the first summand as Hankel forms, rather Hankel forms of type 0. We now refine the decomposition to

$$H \otimes H = V_\Delta^\perp \oplus (V_\Delta \ominus V_\Delta^2) \oplus (V_\Delta^2 \ominus V_\Delta^3) \oplus \cdots$$

We will regard the summands after the first as Hankel forms of higher type – more specifically the nth summand contains the special Hankel forms of type $n-1$. We would like to describe these elements of $H \otimes H$ as bilinear forms, to identify appropriate symbol functions (presumably functions of a single variable) and see how to pull back the norm in $H \otimes H$ to the space of symbol functions. Although we won't get very far with the general program we will make a start.

Note: This analysis is analogous to a construction abstract algebra. If we let F_n, $n = 0, 1, \ldots$ be the space of Hankel forms of type at most n then we have a *filtration* of $H \otimes H$, an increasing family of subspaces that, in some sense, exhausts $H \otimes H$. Associated with a filtration is a *graded* structure

$$H \otimes H = G_0 \oplus G_1 \oplus \cdots$$

where $G_i = F_i / F_{i-1}$. In the presence of Hilbert space structure, in this case the Hilbert-Schmidt inner product, we have a cannonical realization of this quotient as the orthocomplement $F_i \ominus F_{i-1}$.

We assume for the rest of this section that H consists of holomorphic functions of one complex variable. (Clearly some of the discussion extends – even to spaces not defined by holomorphy – but the full pattern isn't clear.)

We start by looking at $V_\Delta \ominus V_\Delta^2$. In (2.5) we saw a rank two Hankel form of type one given by $B_\zeta(f, g) = f'(\zeta)g(\zeta) - f(\zeta)g'(\zeta)$. The corresponding element of $H \otimes H$ is

$$L_\zeta = \frac{\partial}{\partial \bar{\zeta}} k_\zeta \otimes k_\zeta - k_\zeta \otimes \frac{\partial}{\partial \bar{\zeta}} k_\zeta.$$

That is,

$$B_\zeta(f, g) = \langle L_\zeta, f \otimes g \rangle_{H \otimes H}.$$

Let $D = D_{(1,-1)}$ be the differential operator on functions on $X \times X$ given by

$$Df = \left(\frac{\partial}{\partial x_1} - \frac{\partial}{\partial x_2} \right) f.$$

We then have

(3.4) $$L_\zeta = D^* S k_\zeta$$

where D^* is the adjoint of D in the Hilbert space $H \otimes H$. Recall that the special type one Hankel forms we saw earlier were the superposition of the B_ζ. The next proposition shows our current definition is compatible with that one, at least for the classical spaces.

Proposition 3.2:

A) *For each ζ in x, $L_\zeta \in V_\Delta \ominus V_\Delta^2$.*

B) *If H is one of the classical spaces (i.e. the Hardy space, Bergman space, or standard weighted Bergman space of D or U, or the Fock space on $\mathbb{C}$) then the finite linear combinations of L_ζ's are dense in $V_\Delta \ominus V_\Delta^2$.*

Proof. For Part A, $L_\zeta(x,x) = 0$ so $L_\zeta \in V_\Delta$. On the other hand if $g \in V_\Delta^2$ then $g = \lim g_n = \lim h_n f_n$ with h_n, f_n in V_Δ. Thus

$$\begin{aligned}
\langle L_\zeta, g \rangle &= \lim \langle L_\zeta, h_n f_n \rangle \\
&= \lim D(h_n f_n)(\zeta, \zeta) = 0.
\end{aligned}$$

For Part B, suppose we are in the Hardy space of the disk (the other cases are similar). Suppose $f \in V_\Delta \ominus V_\Delta^2$ and $f \perp L_\zeta$ for all ζ. We will show $f \in V_\Delta^2$ and hence conclude f vanishes identically and will be done. $f \in V_\Delta$ insures $f(x,x) \equiv 0$. Thus $f(x,y) = (x-y)g(x,y)$ for some holomorphic g. Thus,

$$\begin{aligned}
0 &= \langle L_\zeta, f \rangle \\
&= D((x-y)g(x,y))(\zeta, \zeta) \\
&= g_x(\zeta, \zeta) - g_y(\zeta, \zeta).
\end{aligned}$$

Thus $g(x,y) = (x-y)h(x,y)$ for some holomorphic h. Hence

(3.5) $$f(x,y) = (x-y)^2 h(x,y).$$

Now note that f is the norm limit of the functions $f_r(x,y) = f(rx, ry)$. By (3.5) the f_r are clearly of the form $g_r \cdot h_r$ with $g_r, h_r \in V_\Delta$. Thus $f_r \in V_\Delta^2$ and hence so is f. We are done.

Next we would like to identify the one variable symbol functions associated with these type one forms and compute the (Hilbert-Schmidt) norms of the forms. We have set things up so that there is an association between k_ζ and L_ζ. Hence we will say that k_ζ is the (one variable) symbol function of the form L_ζ. More generally finite sums $h = \sum^N \alpha_i k_{z_i}$ will be symbol functions for a dense class of forms

$$D^* S(h) = \sum \alpha_i L_{\zeta_i}.$$

We wish to compute $\|D^* S(h)\|_{H \otimes H}$ or, equivalently,

(3.6) $$\|(S^* DD^* S)^{1/2} h\|_H.$$

We don't know how to do this in general but it is not hard for the weighted Bergman spaces of U. In that case $S^* DD^* S$ has a simple form and we can extract the square root "with bare hands". For the Bergman space on U with weight $y^{-\alpha} dx dy$ the kernel function is $k_\zeta(z) = c_\alpha(z - \bar{\zeta})^{-2-\alpha}$. For the rest of this computation we denote inessential multiplicative constants by c. Pick $\zeta, \zeta' \in u$

$$\begin{aligned}
\langle S^* DD^* S k_\zeta, k_{\zeta'} \rangle_H &= \langle D^* S k_\zeta, D^* S k_{\zeta'} \rangle_{H \otimes H} \\
&= \langle L_\zeta, L_{\zeta'} \rangle_{H \otimes H} \\
&= \langle \frac{\partial}{\partial \bar{\zeta}} k_\zeta \otimes k_\zeta - k_\zeta \otimes \frac{\partial}{\partial \bar{\zeta}} k_\zeta, \frac{\partial}{\partial \bar{\zeta}} k_{\zeta'} \otimes k_{\zeta'} - k_{\zeta'} \otimes \frac{\partial}{\partial \bar{\zeta}} k_{\zeta'} \rangle_{H \otimes H}.
\end{aligned}$$

Recalling $\langle a \otimes b, c \otimes d \rangle_{H \otimes H} = \langle a, c \rangle_H \langle b, d \rangle_H$ and evaluating everything we find

$$\langle S^* DD^* Sk_\zeta, k'_\zeta \rangle = 2 \Big(\frac{\partial}{\partial \zeta'} \frac{\partial}{\partial \bar{\zeta}} k_\zeta(\zeta') \Big) k_\zeta(\zeta') - \Big(\frac{\partial}{\partial \bar{\zeta}} k_\zeta(\zeta') \Big) \Big(\frac{\partial}{\partial \zeta'} k_\zeta(\zeta') \Big).$$

This formula is completely general. However, for the weighted Bergman space we can compute explicitly

$$\langle \cdot, \cdot \rangle = c \ (\zeta' - \bar{\zeta})^{-(6+2\alpha)}$$
$$= c \ (\partial^{4+\alpha} k_\zeta)(\zeta').$$

Here $\partial^{4+2\alpha}$ is a (possibly fractional) power of $\frac{\partial}{\partial z}$ which can be regarded as defined via Fourier transform in the x variable. Hence,

$$S^* DD^* S = c \ \partial^{4+\alpha}$$

and

$$(S^* DD^* S)^{1/2} = c \ \partial^{2+\alpha/2}.$$

Thus

$$\| (S^* DD^* S)^{1/2} h \|_H = c \ \| \partial^{2+\alpha/2} h \|_H.$$

But,

$$\| \partial^{2+\alpha/2} h \|_H \sim \| h' \|_{B2}$$

where B_2 is the Dirichlet space. This is the result from [JP1] which we had expected to recapture.

It is only partially clear how to continue. Certainly the map from k_ζ to

$$L_\zeta^{(n)} = (D^*)^n Sk_\zeta$$

which extends by linearity to a dense subset of H will generate a large class of Hankel forms of type n. However, in the absence of special formulas it is not clear how to build special forms of type n, even for $n = 2$. Also, we have an abstract description of the norm of forms – however to make it explicit we need to study the operator $(S^* D^n (D^*)^n S)^{1/2}$ acting on H. Although we could go further on the weighted Bergman space, even the case $n = 1$ is mysterious for general spaces.

There is one case where we can deal with all n without much effort, that is the Fock space. Suppose H is the Fock space. Ignoring inessential multiplicative constants, for f in $H \otimes H$

$$(D^* f)(x_1, x_2) = (x_1 - x_2) f(x_1, x_2).$$

(In fact the Fock space is much studied in part because of the fact that on it differentiation and multiplication by the coordinate function are adjoints.) Thus, the basic linear functionals are given by

$$L_\zeta^{(n)} = (D^*)^n Sk_\zeta = (x_1 - x_2)^n k_\zeta(x_1) k_\zeta(x_2).$$

The computation that $L_\zeta^{(n)} \perp L_{\zeta'}^{(m)}$ if $\zeta, \zeta' \in \mathbb{C}, n \neq m$ is the same as the one done at the end of section II. Finally

$$S^* D^n D^{*n} S k_\zeta = (\partial_{x_1} - \partial_{x_2})^n (x_1 - x_2)^n e^{i\bar\zeta(x_1 + x_2)}\Big|_{x_1 = x_2}.$$

Most of the terms resulting from expanding using Leibnitz's rule will vanish upon evaluation on the diagonal. We find

$$(S^* D^n D^{*n} S k_\zeta)(x) = c \, k_\zeta(2x).$$

Thus, more generally

$$(S^* D^n D^{*n} S h)(x) = c \, h(2x)$$

and hence, after looking at power series

$$(S^* D^n D^{*n} S)^{1/2} h(x) = c \, h(\sqrt{2} x).$$

We have established the following.

Proposition 3.3. Let H be the Fock space and $G \in H$. Let $L_G^{(n)}$ be the special Hankel form of type n with integration kernel

$$(D^{*n} S) G(x_1, x_2) = (x_1 - x_2)^n G(x_1 + x_2).$$

The Hilbert-Schmidt norm of $L_G^{(n)}$ is

$$\|L_G^{(n)}\|_{H.S.} = c_n \|G(\sqrt{2} x)\|_H.$$

IV. BILINEAR FORMS ON HILBERT MODULES

We have seen that Hankel forms and their higher order generalizations are closely related to multiplicative structure. One context in which there is multiplicative structure is when the Hilbert space being considered is a module over a function algebra. Such Hilbert modules are studied systematically in [DP] (see also [S]). We refer to those sources for definition, background, and more examples. For our purposes the Hardy and Bergman spaces of a domain Ω viewed as a module over $H^\infty(\Omega)$, the algebra of bounded analytic functions on Ω, carry all the features of interest. The module action $H^\infty(\Omega) \times H \to H$ is, of course, the pointwise product $a(x) \times h(x) \to a(x)h(x)$. It is interesting to note that this viewpoint excludes some natural HSRK's. For instance the Fock space can't be studied in this context (at least not without modification).

Suppose H is a Hilbert space which is a Hilbert module over the function algebra A. For $a \in A, h \in H$ we denote the module action by juxtaposition $a(h) = ah$. Suppose $B(\cdot, \cdot)$ is a bilinear functional on H. We can ask how B interacts with the action of A. That is we could compare $B(af, g)$ with $B(f, ag)$. To do this we define, for any a in A, and any bilinear form $B, \delta_a B$ given by

$$(\delta_a B)(f, g) = B(af, g) - B(f, ag).$$

It is immediate that $\delta_a B$ will be bounded, Hilbert-Schmidt, etc. if B is.

For the Hardy space and the Bergman space, both regarded as modules over H^∞, B is a Hankel form exactly if

$$(4.1) \qquad\qquad \delta_a B = 0 \quad \forall a \in A.$$

With this fact as motivation we call a bilinear form acting on a Hilbert module H over a function algebra A a Hankel forms if (4.1) holds.

As before, let $H \otimes H$ be the Hilbert space tensor product of H with itself. A can act on either factor. Those actions agree on the quotient

$$(H \otimes H)/Z$$

where

$$Z = \text{ closed linear span of } \{af \otimes g - f \otimes ag : f, g \in H, a \in A\}.$$

This quotient is called the module tensor product. A chase through the definitions shows that the Hankel forms are exactly the bilinear forms which extend to the module tensor product. The construction of the module tensor product occurs naturally in other contexts and its structure is computed in [DP] for several examples including the Hardy and Bergman spaces on D. For instance, if H is the Hardy space then the module tensor product of H with itself is the Bergman space. (What is at issue is the normed structure *as Hilbert modules*.) We will see in a moment that fact is equivalent to the identification of $V_D^\perp$ for the Hardy space with Z.

If H is also a HSRK then we can regard $H \otimes H$ as a space of functions on $X \times X$. In that realization Z is the closure of the span of functions of the form

$$(a(x_1) - a(x_2))f(x_1)g(x_2)$$

with $a \in A; f, g \in H$. Thus it is certainly true that $Z \subseteq V_\Delta$. In nice cases such as the Hardy and Bergman spaces we have $Z = V_\Delta$. The requirement (4.1) forces the integral kernel, $h(x_1, x_2)$, of a Hankel bilinear form

$$h(x_1, x_2) \perp Z.$$

If $Z = V_\Delta$ we have

$$h \in V_\Delta^\perp$$

and we have the situation in the previous section.

We can proceed to higher order Hankel forms by saying that the forms just described are of type 0 and that a bilinear form is a Hankel form of type n if

$$(4.2) \qquad\qquad \forall a \in A, \quad \delta_a B \text{ is a Hankel form of type } n - 1.$$

Polarization shows that this is equivalent to

$$\forall a \in A, \quad (\delta_a)^n B = 0.$$

It had been noted in Section 6 of [JP1] that if B was a Hankel forms of type n then for any bounded holomorphic function a, $\delta_a B$ was a form of type $n-1$. Here we have turned that observation around to use it as a definition.

If H is a HSRK this is equivalent to saying that the integral kernel $h(x_1, x_2)$ satisfies

$$(4.3) \qquad h(x_1, x_2) \perp \prod_{j=1}^{n} (a_j(x_1) - a_j(x_2)) f(x_1) g(x_2)$$

for every $f, g \in H$ and $a_1, \ldots, a_n \in A$. Of course in the favorable cases the functions on the right hand side of (4.3) span a dense subset of V_Δ^n and we are back in the situation described in the previous section.

We noted that the norm on module tensor product of H with H is related to the Hilbert-Schmidt norm of Hankel operators of type 0. We don't know what the analogous statement is for higher weight Hankel forms.

There are many examples of Hilbert modules different from the ones we have been discussing. It might be interesting to know what class of bilinear forms are singled out by (4.1) and (4.2) for such examples. A simple example is to note that the Hardy space of the disk is a module over the subalgebra of H^∞ consisting a for which $a(0) = a(1/2)$.

V. KRONECKER'S THEOREM AND FUNCTIONAL EQUATIONS

Kronecker's theorem for classical Hankel operators (or forms) is the explicit characterization of the finite rank forms. The basic conclusion is that any finite rank Hankel form on the Hardy space is a finite linear combination of the rank one forms given by (2.2) (or possibly limiting cases such as $(fg)'(\zeta)$ etc.). An analogous result holds for Hankel forms on many other spaces of holomorphic functions, see [JPR] [Po].

Here we wish to raise the question of how to characterize the finite rank Hankel forms of higher order. One suspects that the answer is that the finite rank Hankel forms of order s will be finite linear combinations of forms

$$(5.1) \qquad \left(\left(\frac{d}{dz} \right)^j D^k (f \bullet g) \right)(\zeta)$$

with $k \leq s$ and j arbitrary. (Here we stay with holomorphic functions of a single variable.)

There may be an algebraic proof of this fact in the style of [JPR] but we have not been able to find one. An alternative approach is to work with the integral kernel functions.

Suppose, for instance, that we want to work with the Fock space. If $B(\cdot, \cdot)$ is a special Hankel form of type s then its integral kernel $K(z, w)$ is of the form $K(z, w) = (z - w)^s b(z + w)$ for an entire function b. If this operator is to be of rank N then there must be $2N$ entire functions $f_1, \ldots, f_N, g_1, \ldots, g_N$ so that

$$(z - w)^s b(z + w) = \sum_{i=1}^{N} f_i(z) g_i(w).$$

We will now find the functions b which satisfy this functional equation. Changing variables this equation becomes

$$\zeta^s b(\omega) = \sum_{i=1}^{N} \tilde{f}_i(\zeta + \omega)\tilde{g}_i(\zeta - \omega).$$

Here $\tilde{f}_i$ and $\tilde{g}_i$ are the obvious modifications of f_i and g_i. We now differentiate s times with respect to ζ. For new entire functions h_i, k_i we have

$$b(\omega) = \sum_{i=1}^{M} h_i(\zeta + \omega)k_i(\zeta - \omega)$$

or

$$(5.2) \qquad\qquad b(z + w) = \sum_{i=1}^{M} \tilde{h}_i(z)\tilde{k}_i(w).$$

Thus we have reduced to the case $s = 0$. We can now appeal to the Kronecker theorem of [JPR] and conclude

$$(5.3) \qquad\qquad b(z) = \sum_{i=1}^{n} z^{n_i} \exp(\alpha_i z)$$

for some $\alpha_i \in \mathbb{C}$, and integers n_i. Thus the original kernel gives the bilinear form

$$\sum \left(\left(\frac{\partial}{\partial z}\right)^{n_i} D^s(f \bullet g) \right)(\bar{\alpha}_i)$$

as expected. This is for special forms of order s. The general ones will include lower order terms as in (5.1).

(We should note that we don't need the algebraic approach of [JPR] to solve (5.2). The $M + 1$ functions

$$\left(\frac{\partial}{\partial w}\right)^j b(z + w)\Big|_{w=0} \qquad j = 0, \ldots, M$$

lie in the M dimensional vector space spanned by $\tilde{h}_i(z), i = 1 \ldots, M$. Hence b must satisfy a constant coefficient ODE and (5.3) follows.)

The fact that (5.3) holds for any s led the second author to conjecture that if

$$(5.4) \qquad\qquad h(z - w)b(z + w) = \sum_{i=1}^{N} f_i(z)g_i(w)$$

holds for entire h, b, f_i, g_i then b must be of the form (5.3). Although that is correct (and is easy to see) if $N = 1$, if $N = 2$, there are solutions of (5.4) in which h and

b are elliptic functions. The full analysis of (5.4) for $N = 2$ and 3 is in [RR]. The analysis for larger N is open.

What happens in the Hardy space of U. For $s = 0$ the integral kernel of the Hankel form is a function $k(x, y)$ on $\mathbb{R} \times \mathbb{R}$. From that viewpoint Kronecker's theorem characterizes the functions f which satisfy a functional equation

$$\frac{f(x) - f(y)}{x - y} = \sum_{i=1}^{n} h_i(x) g_i(y).$$

The conclusion is that

$$f(x) = \sum_{i=1}^{n} \alpha_i \frac{1}{(x - \zeta_i)^{n_i}(y - \zeta_i)^{n_i}}$$

for complex α_i, ζ_i and integer n_i (as well as limiting degenerate forms). There is an analogous question for the operators higher type. The integral kernels are described on p.317 of [JP1]. When $s = 1$ we want describe g which satisfy

$$\frac{g(x) - g(y) - (x - y)(g'(x) + g'(y))/2}{(x - y)^2} = \sum_{i=1}^{n} h_i(x) k_i(y).$$

This, and the higher order cases, are interesting functional equations but we don't know how to solve them.

VI. COMMENTS AND QUESTIONS

1. What is the theory of higher order Hankel forms for Hardy and Bergman spaces associated with general domains in $\mathbb{C}$? The definitions here give an algebraic framework but analytical questions (boundedness, S_p properties, etc.) as well as structural questions (the form of the integral kernel for special Hankel forms of type s) are open. What about domains in $\mathbb{C}^n$? Notions such as $H \otimes H = V_\Delta^\perp \oplus V_\Delta$ give a natural starting point. Many interesting analytic questions seem to arise. What if H is a Hilbert space or Hilbert module which is not a space of holomorphic functions? What about bilinear forms on $H_1 \otimes H_2$ for $H_1 \neq H_2$? Some results bearing on these questions are in [W].

2. Why haven't we discussed higher order Toeplitz operators?

The classical definition of Hankel operators on the Hardy space of the line is that the Hankel operator R_b with symbol b maps f to $R_b(f) = (I - P)(bf)$. Here P is the Szegö projection of L^2 onto H^2. Thus R maps H^2 to $\overline{H^2}$ which is the linear dual of H^2. Thus to get a bilinear form we should pair with a function in H^2, the linear dual of $\overline{H^2}$. Thus

$$B(f, g) = \int g R_b f$$

gives a Hankel bilinear form mapping $H \otimes H$ to $\mathbb{C}$.

If we start with the Toeplitz operator T_b mapping H^2 to H^2 by $T_b f = P(bf)$ we can try to follow the same path. We can pair $T_b f$ with g in $\overline{H^2}$, the linear dual of H^2. Thus we have a Toeplitz bilinear form

$$B(f, g) = \int g T_b f$$

mapping $H \otimes \overline{H}$ to $\mathbb{C}$. If we proceed as before, we identify B with a kernel function inside $H \otimes \overline{H}$. Let us shift viewpoint for a moment and suppose H is the unweighted Bergman space of U. In that case the symbols of Toeplitz operators are functions on U but we can pass to limits including the case in which $b = \delta_{\zeta_0}$, a point mass at ζ_0. This generates the rank one Toeplitz operator

$$f \to f(\zeta_0) k_{\zeta_0}$$

where k_{ζ_0} is the Bergman kernel function. The bilinear form would be

$$(6.1) \qquad\qquad (f, g) \to f(\zeta_0) g(\zeta_0)$$

for f in H, g in $\overline{H}$. (The reason we left the Hardy space is that Toeplitz theory there, as traditionally formulated, doesn't include compact Toeplitz operators; and hence certainly not the basic rank one example we just described. However, that example is part of the more general theory of Leucking in [L].)

The integral kernel of the form (6.1) is

$$(6.2) \qquad\qquad k_{\zeta_0} \otimes \overline{k_{\zeta_0}}$$

where k_{ζ_0} is the reproducing kernel for H and, hence, $\overline{k_{\zeta_0}}$ is the reproducing kernel for $\overline{H}$. If we follow our earlier analysis then we would consider generalized Toeplitz forms to be the closure of the linear span of the functionals with kernels (6.2). The theory of higher order Toeplitz forms would look at functionals orthogonal to those given by (6.2). If $K(x, y) \in H \otimes \overline{H}$ and $K(x, y) \perp k_\zeta \otimes \overline{k_\zeta} \ \forall \zeta \in D$ then $K(\zeta, \zeta) = 0$. However, K is a function on $D \times D$ which is holomorphic in the first variable and conjugate holomorphic in the second. Thus $L(z, \zeta) = K(z, \bar\zeta)$ is holomorphic in the bidisk $D \times D$. But $L(z, \bar z) = 0$. Hence $L \equiv 0$. Thus there are no non-zero higher order Toeplitz operators in this context.

Actually the discussion we just had only precludes Hilbert-Schmidt higher order Toeplitz operators. However a systematic study of linear functionals of expressions such as

$$f' \overline{g} \quad \text{or} \quad f \overline{g'} \qquad f, g \in H$$

would be interesting. Such operators can be found, among other places, in [Zh].

3. Let H be a HSRK on x and let θ be the linear map defined on kernel functions by $\theta(k_\zeta) = k_\zeta \otimes k_\zeta$. This extends by linearity to a map of H to $H \otimes H$. We can map $H \otimes H$ to $(H \otimes H) \otimes H$ by mapping $e \otimes f$ to $(\theta \otimes \mathrm{Id})(e \otimes f) = \theta(e) \otimes f$ and can map $H \otimes H$ to $H \otimes (H \otimes H)$ by $(\mathrm{Id} \otimes \theta)(e \otimes f) = e \otimes \theta(f)$. If we have a map

θ of H into $H \otimes H$ so that under the natural identification of $(H \otimes H) \otimes H$ with $H \otimes (H \otimes H)$ it is true that

$$(6.3) \qquad\qquad (\mathrm{Id} \otimes \theta)\theta = (\theta \otimes \mathrm{Id})\theta$$

then θ is called a coassociative coproduct and H, with this coproduct, is called a coalgebra. Certainly (6.3) holds for $\theta(k_\zeta) = k_\zeta \otimes k_\zeta$, thus the map from symbol function to integral kernel is a comultiplication which makes H a coalgebra. Although (6.3) is obvious at the level of kernel functions it is less obvious when written out explicitly for general symbols in the Hardy or Bergman spaces.

Coalgebras and the related Hopf algebras are currently active areas of research. We don't know if the fact that the map from symbol to kernel is a comultiplication is a signpost to further relations between Hankel theory and coalgebra theory or if it is just a curiosity. Certainly there are relations between some of the formulas of [JP1] and those of Section 12 of [JR].

4. The theory of the boundedness of higher order Hankel forms is related to understanding linear functionals of expressions such as $D^n(f \bullet g)(\zeta)$. We now rewrite this Hirota form in slightly different notation. We start with f and g, both functions of one variable and form

$$(6.4) \qquad\qquad \left(\frac{\partial}{\partial x} - \frac{\partial}{\partial \xi} \right)^n f(x)g(\xi) \Big|_{x=\xi} .$$

This is just $D^n(f \bullet g)(x)$.

Suppose we start with $u(x,y)$ and $v(x,y)$ a pair of functions of two real variables. For each positive integer n we can then form

$$(6.5) \qquad\qquad \left(\frac{\partial^2}{\partial x \partial \eta} - \frac{\partial^2}{\partial \xi \partial y} \right)^n u(x,y)v(\xi,\eta) \Big|_{\substack{x=\eta \\ y=\eta}} .$$

Thus, when $n = 1$, we get $u_x v_y - u_y v_x$; the determinant of u and v. (The expression we get for general n is a scalar multiple of the nth transvectant of u and v.) In [G] and [CG] Coifman and Grafakos develop a number of interesting mapping properties of the operator in (6.5). Some of their results amount to studying linear functionals of the expression in (6.5). There is certainly a visual similarity between (6.4) and (6.5). At an admittedly vague level there also seems to be some analogies between the results of [JP1] and those of [CG]. It seems natural to ask if there is a more explicit and systematic relation between the two sets of ideas.

5. We mentioned various open problems and possible directions for further research as we went along. The problem of describing the finite rank higher order Hankel forms for the classical domains seems like a particularly nice question. Although it is not central to the Hankel theory, it seems as if it ought to be a nice problem in commutative algebra (as was the case $s = 0$ in [JPR], [Po]) or in functional equations as was suggested in the previous section.

References

[AS]   M.J. Ablowitz and H. Segur, *Solitons and the inverse scattering transform*, SIAM (1981).

[BP]   J. Boman and J. Peetre, *Big Hankel operators of higher weight*, Rend. Circ. Mat. Palermo **38** (1989), 65-78.

[CG]   R. R. Coifman and L. Grafakos, *Hardy space estimates for multilinear operators I*, Rev. Mat. Iberoamericana **8** (1992), 45-67.

[CM]   R. Coifman and Y. Meyer, *Au-delà des opèrateurs pseudo-différentils*, Astérisque **57** (1978).

[DP]   R. Douglas, and V. Paulsen, *Hilbert modules over Function Algebras*, Pitman Research Notes in Mathematics, series, 217 (1989), Longman Scientific and Technical, Essex.

[G]   L. Grafakos, *Hardy space estimates for multilinear operators II*, Rev. Mat. Iberoamericana **8** (1992), 69-92.

[GP]   B. Gustafsson and J. Peetre, *Hankel forms on multiply connected domains. Part Two. The case of higher connectivity.*, Complex Variables **13** (1990), 239–250.

[JP1]   S. Janson and J. Peetre, *A new generalization of Hankel operators (the case of higher weights)*, Math. Nachr. **132** (1987), 318–328.

[JP2]   ———, *Paracommutators-boundedness and Schatten-von Neumann properties*, Trans. Amer. Math. Soc. **305** (1988), 467-504.

[JPR]   S. Janson, J. Peetre, and R. Rochberg, *Hankel forms and the Fock space*, Rev. Mat. Iberoamericana **3** (1987), 61–138.

[JR]   S.A. Juni and G.C. Rota, *Coalgebras and Bialgebras in combinatories* in Umbral Calculus and Hopf Algebras, Contemp. Math. **6**, AMS, Providence.

[L]   D. Luecking, *Trace ideal criteria for Toeplitz operators*, J. Funt. Anal. **73** (1987), 345–368.

[P1]   J. Peetre, *Hankel forms of Arbitrary weight over a symmetric domain via the transvectant*, Rocky Mountain Math. J. (to appear).

[P2]   ———, *Orthogonal polynomials arising in connection with Hankel forms of higher weight*, Bull. Sci. Math..

[P3]   ———, *Hankel Kernels of Higher Weight for the Ball*, Nagoya Math. J. **130** (1993), 183–192.

[P4]   ———, *Comparison of two kinds of Hankel type operators*, Rend. Cric. Mat. Palermo **62** (1993), 181–194.

[Po]   S. C. Power, *Finite rank multivariable Hankel forms*, Linear Alg. and its Appl. **48** (1982), 237-244.

[RR]   R. Rochberg and L. Rubel, *A functional equation*, Indiana U. Math. J. **41** (1992), 363–376.

[R]   W. Rudin, *Function Theory on the Polydisk*, Benjamin, New York, 1969.

[S]   N. Salinas, *Products of kernel functions and module tensor products,* Operator Theory: Advances and Applications, Birkhäuser, Verlag-Basel, (1988), 219–241.

[W]   R. Wallsten, Ph. D. Thesis, Uppsala, 1991.

[Zh]   G. Zhang, Ph. D. Thesis, Lund, 1991.

[Z]   K. Zhu, *Operator Theory in Function Spaces*, Marcel Dekker Inc., New York, 1990.

Jaak Peetre, Department of Mathematics, Lund University, Box 118, S-22100, Lund, Sweden

E-mail: jaak@maths.lth.se

Richard Rochberg, Department of Mathematics, Box 1146, Washington University, St. Louis, MO 63130-4899, USA

E-mail: rr@math.wustl.edu

Contemporary Mathematics
Volume **185**, 1995

Real Valued Spectral Flow

VICUMPRIYA S. PERERA

ABSTRACT. The notion of spectral flow as introduced by [**APS1**], provides an isomorphism between the fundamental group of the non trivial path component $\mathcal{F}_*^{sa}$, of self adjoint Fredholm operators, $\mathcal{F}^{sa}$ and the additive group of integers, $\mathbb{Z}$. In this paper, we generalize these ideas to type II_∞ von Neumann algebra, N and obtain a real valued spectral flow, sf as an isomorphism between $\pi_1(\mathcal{F}_{II,*}^{sa})$ and $\mathbb{R}$, where $\mathcal{F}_{II,*}^{sa}$ is the corresponding non trivial path component of the self adjoint Breuer-Fredholm elements, $\mathcal{F}_{II}^{sa}$ in N.

1. Introduction

In this short expository article, we describe the notions of spectral flow and essential codimension in the case of the bounded operators, $\mathcal{L}(H)$ and their generalizations to type II_∞ von Neumann algebras. We begin with the descriptions of spectral flow, sf and essential codimension, ec in Section **2** and extend these definitions to the type II_∞ case in Section **3**. Then we obtain the following theorems.

THEOREM 1. $\pi_1(\mathcal{F}_{II,*}^{sa}) \cong \mathbb{R}$. $\square$

Let P be an infinite coinfinite projection in a type II_∞ von Neumann algebra, N and let U be a unitary that commutes with P modulo the Breuer ideal in N. Then

THEOREM 2. *Let* $S = 2P - 1$, $T = UPU^*$ *and* $\delta(t) = tS + (1-t)T$, $t \in [0,1]$. *Then*

$$sf_{II}\{\delta(t)|_{t \in [0,1]}\} = ec_{II}(UPU^*, P). \quad \square$$

1991 *Mathematics Subject Classification*. Primary 46L10, 46L80; Secondary 58G10.
This paper is in the final form and will not be submitted for publication elsewhere.

1.1 Notation. We denote a type II_∞ von Neumann algebra factor acting on $\mathcal{H}$ with faithful, normal, semifinite trace, Tr by N. Let $\mathcal{K}_N$ denote the Breuer ideal, that is the norm closed 2-sided ideal in N generated by the projections of finite trace. Let $\mathcal{F}_{II}$ and $\mathcal{F}_{II}^{sa}$ denote the Breuer-Fredholm elements in N, and the self adjoint Breuer- Fredholm elements in N respectively. Let $\mathcal{G}(N)$ and $\mathcal{U}(N)$ denote the invertibles and the unitaries in N. The set of unitaries in N of the form $1+k$ for $k \in \mathcal{K}_N$ is denoted by $U_{\mathcal{K}_N}$. Let $\pi : N \longrightarrow N/\mathcal{K}_N$ be the natural projection. The type II_∞ Calkin algebra is denoted by $\mathcal{Q}_{II}(N) = N/\mathcal{K}_N$. Let $\sigma(x)$ be the spectrum of an element x $\in$ N. An element x in N is essentially positive if $\pi(x)$ is positive. Essentially negative elements are defined in an analogous manner.

2. Spectral Flow in $\mathcal{L}(H)$

Let $\{B_t\}_{t\in[0,1]}$ be a family of self adjoint elliptic differential operators of order ≥ 1 on a closed manifold. Then one has

i)   Discrete spectrum of finite multiplicity

ii) No essential spectrum and

iii) Eigenvectors span whole of $L^2(E)$.

The Spectral Flow of the family $\{B_t\}_{t\in[0,1]}$ is the number of eigenvalues with multiplicity passing from negative to positive minus the number passing from positive to negative.

One can give a definition of Spectral Flow using bounded operators. Let $\mathcal{F}^{sa}$ be self adjoint Fredholm operators. Then $\mathcal{F}^{sa}$ has three path components $\mathcal{F}_+^{sa}, \mathcal{F}_-^{sa}$ and $\mathcal{F}_*^{sa}$. First two are contractible, but the third $\mathcal{F}_*^{sa}$ satisfies $\pi_1(\mathcal{F}_*^{sa}) \cong \mathbb{Z}$ [**AS**]. Thus, loops of self adjoint Fredholm operators has an integer associated to it. This integer is related to spectral flow in two ways. First, there is a deformation retract $F^{sa} \subsetneq \mathcal{F}_*^{sa}$, [**AS**] where

$$F^{sa} = \{T \in \mathcal{F}_*^{sa} | \sigma(T) \text{ is a finite subset of} [-1,1] \text{ and } \sigma_{ess}(T) = \{-1,+1\}\}.$$

Given a loop of operators in F^{sa}, there will be a finite number of eigenvalues crossing 0 and the net crossing is the same as the integer associated to the loop using

$$\pi_1(F^{sa}) \cong \pi_1(\mathcal{F}_*^{sa}) \cong \mathbb{Z}.$$

Second, given a family of elliptic self adjoint differential operators, $\{B_t\}_{t\in[0,1]}$, one can associate a loop of bounded self adjoint Fredholm operators, by

$$\{f(B_t)\}_{t\in[0,1]}$$

where $f(x) = \frac{x}{\sqrt{1+x^2}}$. Then one has

PROPOSITION [**AS**].

$$sf\{B_t\} = [f(B_t)] \in \pi_1(\mathcal{F}_*^{sa}) \cong \mathbb{Z}. \quad \square$$

2.1 Essential Codimension in $\mathcal{L}(H)$.

Let $\pi : \mathcal{L}(H) \longrightarrow \mathcal{Q}(H)$ be the natural projection. Let P, Q be infinite and coinfinite projections in $\mathcal{L}(H)$ such that $\pi(P) = \pi(Q) \neq 0$. If $Q \leq P$ then P–Q is a finite projection and Codimension(Q,P) = Dim $((P-Q)\mathcal{H})$. If $Q \not\leq P$ then one has a way to generalize codimension. For this, we obtain isometries U, V such that $V^*V = U^*U = 1$ and $UU^* = P$ $VV^* = Q$. Then V^*U is a Fredholm operator and

DEFINITION [**BDF**]. $ec(Q, P) = Index(V^*U)$

PROPERTIES.
i) ec is independent of the choices of U and V.
ii) ec generalizes usual codimension.
iii) $ec(Q, P) = 0 \iff \exists W \in U\mathcal{H} \ni Q = W^*PW$ and $\pi(W) = 1$
iv) $\|P - Q\| < 1 \Longrightarrow ec(Q, P) = 0.$ $\square$

We define the restricted unitary group to be $U_{res}(\mathcal{H}) = \{U \in U\mathcal{H} \mid UP - PU \in \mathcal{K}\}$ and the Grassmannian space to be $Gr_P(\mathcal{H}) = \{Q \mid Q \text{ is a projection and } P - Q \in \mathcal{K}\}$. Hence an element $U \in U_{res}(\mathcal{H})$ can be written as

$$U = \begin{pmatrix} X & b \\ c & Y \end{pmatrix}$$

with respect to the decomposition $P\mathcal{H} \oplus (1 - P)\mathcal{H} = \mathcal{H}$. Furthermore, it can be shown that $ec(UPU^*, P) = Index(X)$.

THEOREM.
$$ec : \pi_0(Gr_P(\mathcal{H})) \xrightarrow{\cong} \mathbb{Z}$$
$$\hat{ec} : \pi_0(U_{res}(\mathcal{H})) \xrightarrow{\cong} \mathbb{Z}$$
via $\hat{ec}([U]) = ec(UPU^*) = Index(X).$ $\square$

Now suppose $P \sim 1 \sim 1 - P$ be given. Let $B_0 = 2P - 1$, $Q = UPU^*$, $\pi(Q) = \pi(P)$ and $B_1 = 2UPU^* - 1$. We define B_t to be the linear path $(1 - t)B_0 + tB_1 \ \forall \ t \in [0, 1]$. Then B_t is a path in $\mathcal{F}_*^{sa}$ and can be completed to a loop in $\mathcal{F}_*^{sa}$ via the contractibility of $U(\mathcal{H})$. Then we have the following:

THEOREM [**BW**].

$$sf(\{B_t\}_{t\in[0,1]}) = ec(UPU^*, P) \square$$

3. Type II_∞ Von Neumann algebras

We now proceed to generalize these ideas to a type II_∞ von Neumann algebra factor, N.

DEFINITION [**B3-4**]. $T \in N$ is Breuer-Fredholm if
i) There exist finite projection P in N such that range(1–P) $\subseteq$ range(T) and
ii) Null projection of T, N_T is finite.

DEFINITION [**B3-4**]. There exist a real valued index, Ind_{II}, defined on $\mathcal{F}_{II}$ by

$$Ind_{II}(T) = Dim(N_T) - Dim(N_{T^*})$$

where Dim is the restriction of a faithful normal semifinite trace, Tr to projections of finite trace.

THEOREM [**B3-4**]. *Let N be a type II_∞ von Neumann algebra factor. Then*

$$Ind_{II} : \pi_0(\mathcal{F}_{II}) \xrightarrow{\cong} \mathbb{R}. \quad \square$$

3.1 Definition of ec in II_∞ case.

We now define ec map for type II_∞ von Neumann algebras. Let P be an infinite coinfinite projection in N and $\varepsilon = 2P - 1$. Let [a,b] denote the commutator ab $-$ ba of two elements a and b in N. Analogous to the case of bounded operators define $\mathcal{U}(N, \mathcal{K}_N)$ by,

$$\mathcal{U}(N, \mathcal{K}_N) = \{a \in \mathcal{U}(N) \mid [a, \varepsilon] \in \mathcal{K}_N\}.$$

An element A in $\mathcal{U}(N, \mathcal{K}_N)$ can be written in the matrix form

$$A = \begin{pmatrix} X & b \\ c & Y \end{pmatrix}$$

with respect to the decomposition $P \oplus (1 - P)$ of the identity, 1 of N. Then b, c are in $\mathcal{K}_N$ and X, Y are Breuer-Fredholm elements. Let $Fred(N_+)$ be the space of Breuer-Fredholm elements in N_+, for $N_+ = PNP$. Since P is an infinite projection N_+ is also a Type II_∞ factor. Hence

THEOREM [**CP**]. *The map $\psi : \mathcal{U}(N, \mathcal{K}_N) \to Fred(N_+)$ defined by*

$$A = \begin{pmatrix} X & b \\ c & Y \end{pmatrix} \longmapsto X$$

is a homotopy equivalence. $\square$

DEFINITION. We define the restricted unitary group in N, denoted $U_{res}(N)$, to be

$$U_{res}(N) = \{U \in \mathcal{U}(N) \mid UP - PU \in \mathcal{K}_N\}.$$

Thus $U_{res}(N) = \mathcal{U}(N, \mathcal{K}_N)$.

Now using the Bott periodicity theorem together with the above Theorem we have the following:

THEOREM [**CP**].

$$\pi_n(U_{res}(N)) = \mathbb{R}, \ n \ even$$
$$= 0, \ n \ odd. \quad \square$$

DEFINITION. The Grassmannian relative to a projection P in a von Neumann algebra N, denoted $Gr_P(N)$, is defined to be

$$Gr_P(N) = \{UPU^* \mid U \in U_{res}(N)\}.$$

We now define the map, ec for von Neumann algebras in analogy with bounded operators, $\mathcal{L}(H)$.

DEFINITION. Let $U = \begin{pmatrix} X & b \\ c & Y \end{pmatrix}$ be an element in $U_{res}(N)$. Then we define the real valued essential codimension map $ec : Gr_P(N) \to \mathbb{R}$ by

$$ec(UPU^*) = Index(X).$$

Note that ec infact compares two projections UPU^* and P. Hence it is a generalization of the notion from Chapter 1. Therefore to be more precise one should denote the above map by $ec(UPU^*, P)$. But for the simplicity of notation, we have avoided mentionning P explicitly.

THEOREM. ec induces a 1-1 correspondence

$$\pi_0(Gr_P(N)) \longrightarrow \mathbb{R}. \quad \square$$

In view of the induced group operation on $\pi_0(Gr_P(N))$, we get an isomorphism between $\pi_0(Gr_P(N))$ and the additive group of real numbers, $\mathbb{R}$. We will call this isomorphism also the essential codimension, ec.

Let

$$U = \begin{pmatrix} X & b \\ c & Y \end{pmatrix}$$

be an element in $U_{res}(N)$. We will denote its path component in $U_{res}(N)$ by [U]. Then

THEOREM.

$$\widehat{ec} : \pi_0(U_{res}(N)) \longrightarrow \mathbb{R} \quad defined \ by \quad \widehat{ec}([U]) = Index(X)$$

is a group isomorphism. $\quad \square$

3.2 Homotopy Groups of $\mathcal{F}_{II}^{sa}$.

We now compute all the homotopy groups of the space of self adjoint Breuer-Fredholm elements in N. In particular, we will also show that the loops on the self adjoint Breuer-Fredholm elements are a classifying space for $K^1( ) \otimes \mathbb{R}$ which is a von Neumann algebra analogue of a theorem from [AS].

Let $\overline{\mathcal{G}}_{II}$, $\overline{\mathcal{U}}_{II}$, $\overline{\mathcal{G}}_{II}^{sa}$, and $\overline{\mathcal{U}}_{II}^{sa}$ denote invertibles, unitaries, self adjoint invertibles and self adjoint unitaries in $\mathcal{Q}_{II}(N)$ respectively. In analogy with bounded operators, we call an element x in N essentially positive if $\pi(x)$ is positive. Essentially negative elements in N are defined in an analogous way. We will denote the spectrum of an element x in N by $\sigma(x)$.

DEFINITION. Let $\mathcal{F}_{II,+}^{sa}$ and $\mathcal{F}_{II,-}^{sa}$ denote essentially positive and essentially negative self adjoint Breuer-Fredholm elements in Type II_∞ factor N respectively. Let $\mathcal{F}_{II,*}^{sa}$ be the complement of $\mathcal{F}_{II,+}^{sa} \bigcup \mathcal{F}_{II,-}^{sa}$ in $\mathcal{F}_{II}^{sa}$.

Define

$$Gr(\mathcal{Q}_{II}(N)) = \{p \in \mathcal{Q}_{II}(N) \mid p^2 = p^* = p \neq 0, 1\}$$

PROPOSITION 1. *Let r be the standard unitary contaction map and ϕ : $\overline{\mathcal{U}}_{II,*}^{sa} \longrightarrow Gr(\mathcal{Q}_{II}(N))$ be defined by $\phi(u) = \frac{u+1}{2}$. Then*

$$\mathcal{F}_{II,*}^{sa} \overset{\pi}{\simeq} \overline{\mathcal{G}}_{II,*}^{sa} \overset{r}{\simeq} \overline{\mathcal{U}}_{II,*}^{sa} \overset{\phi}{\cong} Gr(\mathcal{Q}_{II}(N)). \quad \square$$

THEOREM. *$\mathcal{F}_{II,+}^{sa}, \mathcal{F}_{II,-}^{sa}$ and $\mathcal{F}_{II,*}^{sa}$ are the path components of $\mathcal{F}_{II}^{sa}$. Furthermore, $\mathcal{F}_{II,+}^{sa}$ and $\mathcal{F}_{II,-}^{sa}$ are contractible.* $\quad \square$

SKETCH OF PROOF. Contractibility of $\mathcal{F}_{II,+}^{sa}$ is implied by the linear contaction map $T_t = (1-t)T + t(1)$, $0 \leq t \leq 1$ from $\mathcal{F}_{II,-}^{sa}$ to 1 and the path connectedness of $\mathcal{F}_{II,*}^{sa}$ follows by Proposition 1 together with the path connectedness of $Gr(N)$, the set of proper projections in N. $\quad \square$

We now proceed to calculate the homotopy groups of $\mathcal{F}_{II,*}^{sa}$. For this we obtain the following principal fiber bundle and then obtain its long exact sequence of homotopy groups. We then calculate each homotopy group in this sequence and deduce our next theorem.

Let $j_0 \in \overline{\mathcal{U}}_{II,*}^{sa}$. Choose $P_0 \in Gr(N)$ such that $\pi(2P_0 - 1) = j_0$. Define $r_{j_0} : \overline{\mathcal{U}}_{II} \longrightarrow \overline{\mathcal{U}}_{II,*}^{sa}$ by $r_{j_0}(u) = u j_0 u^*$. Let $p_0 = \pi(P_0)$.

PROPOSITION 2. *$r_{j_0} : \overline{\mathcal{U}}_{II} \longrightarrow \overline{\mathcal{U}}_{II,*}^{sa}$ defines a principal fiber bundle with fiber $\overline{\mathcal{U}}_{II,0} = r_{j_0}^{-1}(j_0) = \{ u \in \overline{\mathcal{U}}_{II} \mid j_0 = u j_0 u^* \}$.* $\quad \square$

Since $\overline{\mathcal{U}}_{II,*}^{sa}$ is paracompact by a result of Dold (For example, see p.56, [**Swi**]), r_{j_0} is a fibration. Therefore,

PROPOSITION 3. *We get the following long exact sequence of homotopy groups:*

$$\ldots\ldots \to \pi_{k+1}(\overline{\mathcal{U}}_{II,*}^{sa}) \to \pi_k(\overline{\mathcal{U}}_{II,0}) \to \pi_k(\overline{\mathcal{U}}_{II}) \to \pi_k(\overline{\mathcal{U}}_{II,*}^{sa}) \to \ldots..$$

$$\ldots.. \to \pi_1(\overline{\mathcal{U}}_{II,*}^{sa}) \to \pi_0(\overline{\mathcal{U}}_{II,0}) \to \pi_0(\overline{\mathcal{U}}_{II}) \to \pi_0(\overline{\mathcal{U}}_{II,*}^{sa}) = 0. \quad \square$$

The following proposition calculates all homotopy groups of $\overline{\mathcal{U}}_{II,0}$ and $\overline{\mathcal{U}}_{II}$.

PROPOSITION 4.

$$\pi_k(\overline{\mathcal{U}}_{II,0}) \cong \pi_k(\overline{\mathcal{U}}_{II}) \oplus \pi_k(\overline{\mathcal{U}}_{II})$$

$$\pi_k(\overline{\mathcal{U}}_{II}) = \mathbb{R}, \ k \ even$$

$$= 0, \ k \ odd. \quad \square$$

Hence the above long exact sequence reduces to a short exact sequence. Furthermore, one can show that $\overline{\mathcal{U}}_{II,*}^{sa}$ is path conntected and the middle map in that sequence is the addition. Hence

$$0 \to \pi_{2k+1}(\overline{\mathcal{U}}_{II,*}^{sa}) \overset{1-1}{\to} \pi_{2k}(\overline{\mathcal{U}}_{II,0}) \overset{+}{\to} \pi_{2k}(\overline{\mathcal{U}}_{II}) \overset{onto}{\to} \pi_{2k}(\overline{\mathcal{U}}_{II,*}^{sa}) \to 0$$

$$0 \to \pi_1(\overline{\mathcal{U}}_{II,*}^{sa}) \overset{1-1}{\to} \pi_0(\overline{\mathcal{U}}_{II,0}) \overset{+}{\to} \pi_0(\overline{\mathcal{U}}_{II}) \overset{onto}{\to} \pi_0(\overline{\mathcal{U}}_{II,*}^{sa}) = 0.$$

THEOREM. *For $k \geq 0$,*

$$\pi_k(\mathcal{F}_{II,*}^{sa}) = \pi_k(\overline{\mathcal{U}}_{II,*}^{sa}) = \mathbb{R}, \ k \ odd$$
$$= 0, \ k \ even. \quad \square$$

In particular we have

THEOREM 1 [**Per1**].

$$\pi_1(\mathcal{F}_{II,*}^{sa}) \cong \mathbb{R}. \quad \square$$

3.3 Real Valued Spectral flow.

In this section we define the real valued spectral flow on the space of self adjoint Breuer-Fredholm elements in a type II_∞ von Neumann algebra factor using the map between $\pi_1(\mathcal{F}_{II,*}^{sa})$ and $\mathbb{R}$ and obtain this map explicitly. We conclude this section by proving a relation between real-valued sf and real-valued ec. Let X be a topological space. We use ΩX to denote the loop space of X.

THEOREM. *The map*

$$\alpha_P : \mathcal{U}(N) \longrightarrow Gr(\mathcal{Q}_{II}(N))$$

defined by $\alpha_P(U) = \pi(UPU^*)$ is a fibration with fiber

$$\alpha_P^{-1}(\pi(P)) = U_{res}(N). \quad \square$$

REMARK. One can obtain the corresponding long exact sequence of homotopy groups of the above fibration.

$$\ldots \longrightarrow \pi_{k+1}(\mathcal{U}(N)) \longrightarrow \pi_{k+1}(Gr(\mathcal{Q}_{II}(N)) \longrightarrow \pi_k(U_{res}(N))$$

$$\longrightarrow \pi_k(\mathcal{U}(N)) \longrightarrow \pi_k(Gr(\mathcal{Q}_{II}(N)) \longrightarrow \ldots$$

Since $\mathcal{U}(N)$ is contractible, $\pi_k(\mathcal{U}(N)) = 0$ *for all* $k \geq 0$. Hence this long exact sequence reduces to,

$$\pi_{k+1}(Gr(\mathcal{Q}_{II}(N)) \cong \pi_k(U_{res}(N)) \ \forall \ k \geq 0.$$

Let ρ be a lifting map for α_P. Then one can construct a homotopy equivalence map $\bar{\rho} : \Omega(Gr(\mathcal{Q}_{II}(N))) \longrightarrow U_{res}(N)$ giving this isomorphism. Therefore $\bar{\rho}$ will map loops in $Gr(\mathcal{Q}_{II}(N))$ to the end point of the lifted path which starts at the identity in $U_{res}(N)$.

But from Proposition **1**,

$$\mathcal{F}^{sa}_{II,*} \overset{\pi}{\underset{\simeq}{\longrightarrow}} \overline{\mathcal{G}}^{sa}_{II,*} \overset{r}{\underset{\simeq}{\longrightarrow}} \overline{\mathcal{U}}^{sa}_{II,*} \overset{\phi}{\underset{\simeq}{\longrightarrow}} Gr(\mathcal{Q}_{II}(N)),$$

Let

$$f : \Omega(\mathcal{F}^{sa}_{II,*}) \longrightarrow U_{res}(N)$$

be defined by $f = \bar{\rho} \circ \Omega(\phi \circ r \circ \pi)$. Then f is a homotopy equivalence map and

$$\pi_1(\mathcal{F}^{sa}_{II,*}) \cong \pi_0(\Omega\mathcal{F}^{sa}_{II,*}) \overset{\pi_0(f)}{\cong} \pi_0(U_{res}(N)) \overset{\widehat{ec}}{\cong} \mathbb{R}.$$

DEFINITION. The real valued spectral flow map, denoted sf_{II}, is defined to be

$$sf_{II} = \widehat{ec} \circ \pi_0(f)$$

Hence

THEOREM.

$$sf_{II} : \pi_1(\mathcal{F}^{sa}_{II,*}) \overset{\cong}{\longrightarrow} \mathbb{R}. \quad \square$$

Furthermore, we also obtain the following homotopy equivalences, $\Omega\mathcal{F}^{sa}_{II,*} \simeq U_{res}(N) \simeq \Omega U_{\mathcal{K}_N} \simeq \mathcal{F}_{II}$ which proves that the space $\Omega\mathcal{F}^{sa}_{II,*}$ is a classifying space for the functor $K^0(\ ) \otimes \mathbb{R}$. Now consider Let $S, T \in \mathcal{F}^{sa}_{II,*}$ such that $S = U^*TU$ for some $U \in \mathcal{U}(N)$. Let $\delta(t)$ be a path from T to S. Complete it to a loop $\bar{\delta}$ at T via $U_t^*TU_t$, provided $\{U_t\}_{t\in[0,1]}$ is the path in $\mathcal{U}(N)$ connecting U and 1. Then we define the sf_{II} of the loop $\delta(t)$ to be

$$sf_{II}\{\delta(t)|_{t\in[0,1]}\} = sf_{II}(\bar{\delta}).$$

It can be shown that the definition of sf_{II} for a path in $\mathcal{F}^{sa}_{II,*}$ is independent of the choice of the unitary path $\{U_t\}_{t\in[0,1]}$. Furthermore,

THEOREM 2 [**Per1**]. *Let* $S = 2P-1$, $T = UPU^*$ *and* $\delta(t) = tS+(1-t)T$, $t \in [0,1]$. *Then*

$$sf_{II}\{\delta(t)|_{t\in[0,1]}\} = ec(UPU^*, P). \quad \square$$

3.4 Generalizations to Hilbert C^*-modules.

We extend the definitions of essential codimension and spectral flow to Hilbert C^*-modules and derive results similar to the Type II case. Our definition of spectral flow will utilize the property that

$$\pi_1(\mathcal{F}^{sa}_{A,*}, T_0) \cong K_0(A).$$

Hence the spectral flow in this setting will take values in K-theory of A. The choice of the notion sf to denote this isomorphism is justified by analogous result in type II_∞ case. A more detailed version is presented in [**Per2**].

Let A be a unital C^*-algebra. The Hilbert space over A , $\mathcal{H}_A$, is defined to be the space of all l^2-sequences of elements in A. If $A = \mathbb{C}$ then $\mathcal{H}_A$ reduces to the standard separable Hilbert space l^2. Let $\mathcal{L}(\mathcal{H}_A)$, the Hilbert C^*-module over

A, be the C^*-algebra of all bounded A-module operators on $\mathcal{H}_A$ whose adjoints exist. Let $\mathcal{K}(\mathcal{H}_A)$ denote the generalized compact operators in $\mathcal{L}(\mathcal{H}_A)$.

DEFINITION. An operator T in $\mathcal{L}(\mathcal{H}_A)$ is a generalized Fredholm operator if $\pi(T)$ is invertible in $\mathcal{Q}(\mathcal{H}_A)$. The set of all generalized Fredholm operators in $\mathcal{L}(\mathcal{H}_A)$ is denoted by $\mathcal{F}_A$.

Any generalized Fredholm operator $T \in \mathcal{F}_A$ has a compact perturbation g that can be polar decomposed into $g = v|g|$, where v is a partial isometry such that $1 - v^*v$, $1 - vv^* \in \mathcal{K}(\mathcal{H}_A)$ (16.4 of [**Weg**]). Since $\pi(T) = \pi(g)$ and $v - (gg^*)^{-1/2}g \in \mathcal{K}(\mathcal{H}_A)$, we have $v - (TT^*)^{-1/2}T \in \mathcal{K}(\mathcal{H}_A)$. Therefore the following definition makes sense.

DEFINITION. An element T in $\mathcal{F}_A$ has a K_0-valued index defined by

$$Index(T) = [1 - v^*v] - [1 - vv^*] \in K_0(A),$$

where $v - (TT^*)^{-1/2}T \in \mathcal{K}(\mathcal{H}_A)$.

This Index induces a map on path components of $\mathcal{F}_A$. Furthermore

$$Index : \pi_0(\mathcal{F}_A) \longrightarrow K_0(A)$$

is a group isomorphism [**Min**]. Let A be a C^* -algebra with strictly positive element. Then the unitary group of $\mathcal{L}(\mathcal{H}_A)$ is contractible [**CH**]. Examples of C^*-algebras with strictly positive element include unital C^*-algebras. By Prop. 3.10.5 [**Ped**], it is equivalent to the existence of a countable approximate identity for A.

Let $P \sim 1 \sim 1 - P$ be a proper projection in $\mathcal{L}(\mathcal{H}_A)$.

DEFINITION. We define the generalized restricted unitary group $U_{res}(\mathcal{H}_A)$ to be

$$U_{res}(\mathcal{H}_A) = \{U \in \mathcal{U}(A) \,|UP - PU \in A \otimes \mathcal{K}\}$$

and generalized Grassmannian space $Gr_P(\mathcal{H}_A)$ to be

$$Gr_P(\mathcal{H}_A) = \{Q \in \mathcal{L}(\mathcal{H}_A) \,|Q = UPU^* \; for \; U \; \in U_{res}(\mathcal{H}_A)\}.$$

Define essential codimension

$$ec : Gr_P(\mathcal{H}_A) \longrightarrow K_0(A)$$

by $ec(Q,P) = Index(a)$, where $Q = UPU^*$ and $U = \begin{pmatrix} a & b \\ c & d \end{pmatrix}$ with respect to $P \oplus (1 - P) = 1$.

We now define $\widehat{ec}$ on $U_{res}(\mathcal{H}_A)$. Let [U] denote the path component of U in $U_{res}(\mathcal{H}_A)$. Then

DEFINITION. We define $\widehat{ec} : \pi_0(U_{res}(\mathcal{H}_A)) \longrightarrow K_0(A)$ by

$$\widehat{ec}([U]) = ec(UPU^*, P).$$

The next result follows using results in Zhang, [**Z1**].

THEOREM. *The map $\widehat{ec} : \pi_0(U_{res}(\mathcal{H}_A)) \longrightarrow K_0(A)$ is a group isomorphism.*
□

We now, generalize this to the self-adjoint case./

DEFINITION. Let $\mathcal{F}^{sa}_{A,+}$, $\mathcal{F}^{sa}_{A,-}$ be the essentially positive and essentially negative self adjoint generalized Fredholm operators and $\mathcal{F}^{sa}_{A,*}$ be the complement of $\mathcal{F}^{sa}_{A,+} \cup \mathcal{F}^{sa}_{A,-}$ in $\mathcal{F}^{sa}_A$.

LEMMA. $\mathcal{F}^{sa}_{A,+}$ *and* $\mathcal{F}^{sa}_{A,-}$ *are contractible path components of* $\mathcal{F}^{sa}_A$. *In general,* $\mathcal{F}^{sa}_{A,*}$ *is not path connected.* □

PROPOSITION. *If A is purely infinite simple C^*-algebra then*

$$\pi_0(\mathcal{F}^{sa}_{A,*}) \cong K_1(A). \quad \square$$

THEOREM [**Per2**]. *Let $T_0 \in \mathcal{F}^{sa}_{A,*}$ be such that there exists a $P_0 \in Gr(\mathcal{L}(\mathcal{H}_A))$ with $T_0 = 2P_0 - 1$. Let $j_0 \in \overline{\mathcal{U}}^{sa}_{A,*}$ be a fixed element with $\pi(T_0) = j_0$, then*

$$\pi_k(\mathcal{F}^{sa}_{A,*}, T_0) \cong \pi_k(\overline{\mathcal{U}}^{sa}_{A,*}, j_0) \cong K_0(A), \ k \text{ odd}, \ k > 0$$
$$\cong K_1(A), \ k \text{ even}, \ k > 0. \quad \square$$

DEFINITION. Let the notations be as above. Then

$$\pi_1(\mathcal{F}^{sa}_{A,*}, T_0) \cong K_0(A)$$

via the above theorem. Let

$$sf_A : \pi_1(\mathcal{F}^{sa}_{A,*}, T_0) \longrightarrow K_0(A)$$

denote the isomorphism. Then sf_A is defined to be the generalized spectral flow on Hilbert C^*-modules.

4. Acknowledgement

The author wishes to express his sincere thanks to Prof. Jerome Kaminker for his valuable advice and support during this research. Author also wishes to thank the refree for his/her valuable comments.

REFERENCES

[A]   M.F. Atiyah, *Elliptic operators, discrete groups and von Neumann algebras*, Ast'erisq -ue **32/33** (1976), 43-72.

[APS1] M.F. Atiyah, V.K. Patodi and I.M. Singer, *Spectral Assymetry and Riemannian Geometry III*, Math. Proc. Cambridge Phil. Soc. **79** (1970), 71-99.

[APS2] _____, *Spectral Assymetry and Riemannian Geometry*, Bull. Lond. Math. Soc. **5** (1973), 229-234.

[APS3] _____, *Index Theory for Skew adjoint Fredholm operators*, Publ. Math. IHES **37** (1969), 305-326.

[B]   B. Blackadar, *K-Theory for Operator Algebras*, Springer - Verlag, New York, 1986.

[BW]   B. Booss and K.P. Wojciechowski, *Elliptic Boundary Value Problems for Operators of Dirac Type* (1993) (to appear).

[B1]   M. Breuer, *On the Homotopy type of the group of Regular elements of Semifinite von Neumann Algebras*, Math. Ann. **185** (1970), 61-74.

[B2]   _____, *Theory of Fredholm Operators and Vector Bundles relative to a von Neumann Algebra*, Rocky Mountain Math. J. **3** [3] (1973), 383-429.

[B3]   _____, *Fredholm theories in von Neumann Algebras, I*, Math. Ann. **178** (1968), 243-254.

[B4]   _____, *Fredholm theories in von Neumann Algebras, II*, Math. Ann. **180** (1969), 313-325.

[Bro]   L.G. Brown, *Ext of certain free product C^*-algebras*, J. Operator Theory **6** (1981), 135-141.

[BDF]   L.G. Brown, R.G. Douglas and P.A. Fillmore, *Unitary equivalence modulo the compact operators and extensions of C^*-algebras*, Proc. Conf. On Operator Theory, Lecture Notes in Math., vol. 345, Springer-Verlag, Berlin and New York, 1973, pp. 58-128.

[CP]   A. Carey and J. Phillips, *Algebras almost commuting with Clifford Algebras in a type II_∞ factor*, K Theory **4** [5] (1991), 445-478.

[C]   J. Cuntz, *K Theory for certain C^*-algebras*, Ann. of Math. **113** (1981), 181-197.

[CH]   J. Cuntz and N. Higson, *Kuiper's Theorem for Hilbert C^*-Modules*, Contemp. Math. **62** (1987), 429-435.

[Dix]   J. Dixmier, *Von Neumann Algebras*, North Holland, New York, 1981.

[Do]   A. Dold, *Partition of unity in the theory of fibrations*, Ann. of Math. **78** (1963), 223-255.

[Dou]   R.G. Douglas, *Banach algebra techniques in operator theory*, Academic Press, New York, 1972.

[DHK1]   R.G. Douglas, S. Hurder and J. Kaminker, *Eta invariants and von Neumann algebras*, Bull. AMS **21** (1988), 83-87.

[DHK2]   _____, *Cyclic cocycles, renormalization and eta-invariants*, Inventiones Mathematicae **103** (1991), 101-179.

[DHK3]   _____, *Toeplitz operators and the eta-invariant : The case of S^1*, Contemp. Math. **70** (1988), 11-41.

[FO1]   K. Furutani and N. Otsuki, *A space of generalized Atiyah-Patodi-Singer boundary conditions and a formula for a spectral flow* (1993) (to appear).

[FO2]   _____, *Spectral flow and intersection number*, J. Math. Kyoto Uni. **33-1** (1993), 261-283.

[KR]   R. V. Kadison and J.R. Ringrose, *Fundamentals of the theory of Operator Algebras*, vol. II, Academic Press, New York, 1986.

[Kam]   J. Kaminker, *Operator algebraic invariants for elliptic operators*, Proc. Symp. in Pure Math. **51 -I** (1990), 307-314.

[Min]   J. A. Mingo, *K-Theory and Multipilers of stable C^*-algebras*, Trans. AMS **299-[1]** (1987), 397-411.

[Ped]   G.K. Pedersen, *C^*-algebras and Their Automorphism Groups LMS 14*, Academic Press, Newyork, 1979.

[Per1]   V.S. Perera, *Real Valued Spectral Flow in a Type II_∞ Factor* (1993) (to appear).

[Per2]   V.S. Perera, *Homotopy groups of self adjoint Fredholm operators in a Hilbert C^*-module* (1993) (to appear).

[PS]   A. Pressely and R.G. Segal, *Loop Groups*, Oxford University Press, New York, 1986.

[Swi]   R.M. Switzer, *Algebraic Topology : Homotopy and Homology*, Springer-Verlag, Berlin, 1975.

[Weg]   N.E. Wegge-Olsen, *Masters Thesis at University of Copenhagen*, University of Copenhagen, Copenhagen, Denmark, 1989.

[Woj]   K.P. Wojciechowski, *A note on the space of pseudodifferential projections with the same principal symbol*, J. Operator Theory **15** (1986), 207-216.

[Z1]   S. Zhang, *K-Theory and Bi-variable Index$(x, [p]_e)$: Properties , Invariants and Applications, Parts I,II & III* (to appear).

[Z2]   _____ , *K-Theory , K-skeleton factorizations and Bi-variable Index (x,p) : Parts I,II & III* (to appear).

[Z3]   _____ , *Certain C^*-algebras with real rank zero and their Corona & Multiplier Algebras*, K Theory (to appear).

[Z4]   _____ , *K-theory and Homotopy of Certain groups and Infinite Grassmannian spaces associated with C^*-algebras*, K Theory (to appear).

[Zsi]   L. Zsido, *The Weyl-Von Neumann Theorem in Semifinite Factors*, J. of Func. Analysis **18** (1975), 60-72.

DEPARTMENT OF MATHEMATICS, INDIANA UNIVERSITY-PURDUE UNIVERSITY AT INDIANAPLOIS, 402N BLACKFORD STREET, INDIANAPLOIS, INDIANA 46202

Current address: Department of Mathematics, The Ohio State University at Newark, 1179 University Drive, Newark OH 43055

E-mail address: vicum@math.ohio-state.edu

Contemporary Mathematics
Volume **185**, 1995

Abstract $\overline{\partial}$-resolutions for several commuting operators

Mihai Putinar

May 23, 1994

ABSTRACT. We relate several spectral decomposition and spectral localization properties of commuting systems of Banach space operators to some distinguished resolutions of the corresponding analytic modules. A universal dilation for commuting tuples of operators and the relationship between spectral sets and generalized scalar dilations are also included in this framework. A few applications to function theory and complex geometry are derived from the main abstract results.

Introduction.

The generic examples in operator theory are provided by multiplication operators on topological spaces of analytic functions. Thus a series of specific constructions and results from the theory of analytic functions have shaped the present form of operator theory. These remarks are especially relevant for the relatively young multivariable spectral theory. The present paper is devoted to a very specific aspect of the implementation of several variables complex analysis methods into the field of abstract spectral theory. More specifically, we report below a series of results related to commutative systems of Banach space operators, results which were obtained in the last decade by techniques inspired from the abstract de Rham theorem in sheaf theory. Roughly speaking, the conclusion of the following pages is that the familiar $\overline{\partial}$-complex represents the core of both extension and dilation theories for several commuting operators.

1991 *Mathematics Subject Classification.* Primary 47A11,46M20; Secondary 32C35.
Paper supported by the National Science Foundation Grant DMS-9201729.
This paper is in final form and no version of it will be submitted for publication elsewhere.

Let T be a commutative n-tuple of linear bounded operators acting on the Banach space X. It is well known that we can interpret T as a structure of a Banach module on X, over the algebra of entire functions $\mathcal{O}(\mathbf{C}^n)$. Various topologically free resolutions of the module X are known as they appear for instance in the study of the analytic functional calculus for T or in the homological approach to the classical dilation theory, see Taylor [26] and Douglas and Paulsen [3]. These resolutions correspond to the projective resolutions in algebra. However we will focus below on another important class of resolutions, namely some topological analogues of the injective resolutions. More precisely, beginning with the Dolbeault complex as a model we will study those analytic modules which admit similar Fréchet soft resolutions. As proved in [20] these modules are the multidimensional analogue of the Banach space operators with Bishop's property (β). The analytic modules which admit a resolution to the right with modules over the algebra of smooth functions generalize the class of sub- scalar operators. The mere existence of good soft resolutions for a module facilitates various homological computations, in particular the computation of joint spectra for large classes of tuples of analytic Toeplitz operators, see for instance [20] and [8].

By combining a topologically free resolution of a module with the Dolbeault complex on each factor one obtains a complex of fine modules which is quasi- isomorphic to the original module. This provides in the terminology of operator theory a canonical functional model for the initial module, with some additional rich symmetries. A deep application of this construction was recently obtained by Eschmeier [5] in a series of positive invariant subspace results for tuples of operators with Bishop's property (β).

Speaking locally, at the level of analytic sheaves, the existence of Fréchet soft resolutions is equivalent to the quasi-coherence of a sheaf. Thus the localization of the modules with abstract $\bar{\partial}$-resolutions has much in common with the well known theory of coherent analytic sheaves. The forthcoming monograph [9] is mainly concerned with these topics.

The first three sections of this note are devoted to different classes of resolutions, first in local algebra and then over a topological algebra. Section 4 contains the construction of a canonical complex of fine topologically free sheaves which is quasi-isomorphic to a given analytic Banach module. Interpreted in terms of operators this construction produces the pattern for several dilation theorems. In Section 5 we compute explicitly some of this dilation spaces and we relate them to the possible spectral sets of a tuple of commuting operators. Section 6 contains a quasi-coherence criterion for

Fréchet analytic sheaves on complex manifolds (see [23]).

1 Modules and their resolutions

One of the principal problems of multivariable operator theory is to classify commutative n-tuples of operators up to unitary equivalence or up to similarity. Having as model the Jordan form of a finite matrix we discuss first the joint similarity of a system of commuting matrices. Even this apparently simple classification problem does not have a simple solution. To be more specific, let V denote a finite dimensional complex vector space and let $T = (T_1, \ldots, T_n)$ be a commuting system of linear transformations of V. This is equivalent to endowing V with a structure of module over the algebra of polynomials given by:

$$p.v = p(T)v, v \in V, p \in \mathbf{C}[z_1, \ldots, z_n].$$

For n=1 the structure of the module V is fairly simple, namely it decomposes into a direct sum over the spectrum of T (or equivalently the support of V):

$$V = \bigoplus_{w \in \sigma(T)} V_w,$$

where each V_w is a finite dimensional module over the localization $\mathbf{C}[z]_w$ of the ring of polynomials at the point w. The latter module decomposes in turn into factors:

$$V_w = \bigoplus_{i \in I} \mathbf{C}[z]_w / m_w^{k_i},$$

where m_w is the maximal ideal of the respective local ring. Each k_i represents the size of a Jordan block associated to the eigenvalue w of the operator T.

In the case $n \geq 2$ the second decomposition is missing. Namely,keeping the same notation the finite dimensional $\mathbf{C}[z]_w$-module V_w cannot in general be decomposed into simpler factors. It is well known in algebra that these (Artinian) modules over a local ring have continuous moduli. A suggestive example in this direction was proposed by Gelfand and Ponomarev[12]. More precisely, they prove that the classification of a pair of nilpotent matrices contains in particular the classification of all non necessarily commuting pairs of matrices.

Thus in any classification problem involving several commuting operators we cannot expect to go further than the above algebraic picture. A standard way to study the structure of the module V_w is to consider a free resolution of it:

$$0 \longrightarrow \mathbf{C}[z]^{p_n} \longrightarrow \ldots \longrightarrow \mathbf{C}[z]^{p_0} \longrightarrow V_w \longrightarrow 0.$$

Such a resolution of length n exists by the Syzygy Theorem of Hilbert. (See for instance [16]).

Of a particular interest are some canonical resolutions with additional symmetry, such as the Koszul resolution of the quotient field of a maximal ideal:

$$K.((z-w), \mathbf{C}[z]) \longrightarrow \mathbf{C}[z]/m_w \longrightarrow 0.$$

See [16] for details.

A second classical, this time coordinateless, resolution of a module M over a $\mathbf{C}$-algebra A is the Bar resolution:

$$B.^A(A, M) \longrightarrow M \longrightarrow 0.$$

The typical term in this resolution is $A \otimes A \otimes \ldots \otimes A \otimes M$,where the tensor product is taken over the complex field. A completion of this tensor product in an adequate norm gives the topological Bar complex which computes for instances the topological Tor's over a Fréchet algebra. (See [16] and [26] for details).

In passing from modules to sheaves of modules, we encounter a series of canonical resolutions to the right which are used in the computation of the Čech cohomology. The typical example is the de Rham complex and its correspondent in complex geometry, the Dolbeault complex:

$$0 \longrightarrow \mathcal{O} \longrightarrow \mathcal{E}^{(0,0)} \xrightarrow{\overline{\partial}} \mathcal{E}^{(0,1)} \longrightarrow \ldots \longrightarrow \mathcal{E}^{(0,n)} \longrightarrow 0.$$

It is well known for instance that the Dolbeault complex is exact at the level of global sections of an open subset U of $\mathbf{C}^n$ if and only if U is a domain of holomorphy. (See Grauert and Remmert [13] for details).

The principal object of study for us is a Banach module over the algebra of entire functions $\mathcal{O}(\mathbf{C}^n)$ (or equivalently a commutative n-tuple of linear bounded operators on X). Thus the completion of the elements of the Dolbeault complex in different norms will provide canonical examples for X (i.e. the completion of $\mathcal{O}(U)$) and resolutions with "better localizable" spaces.

2 Decomposable resolutions

A linear bounded operator T acting on a Banach space X is called *decomposable* if for any finite open covering $(U_i)_{i \in I}$ of the complex plane there are closed invariant subspaces $X_i, (i \in I)$ of T, such that:

$$X = \sum_{i \in I} X_i, \quad \sigma(T|X_i) \subset U_i, (i \in I).$$

This concept is due to C.Foiaş [10] and it encodes one of the most general and at the same time flexible spectral decompositon properties of a linear operator. The theory of decomposable operators is extensively treated in the survey [24] and in the monograph [27].

It was E.Bishop [2] who has forseen a non-pointwise behaviour of the resolvent of the operator T which would control decomposability like properties of T. Namely we say after Bishop that T satisfies property (β) if, for any open set U of the complex plane, the map

$$T - z : \mathcal{O}(U, X) \longrightarrow \mathcal{O}(U, X)$$

is one to one with closed range in the Fréchet topology of the space of X-valued analytic functions defined on U. The next theorem clarifies the realtion between (β) and the class of decomposable operators.

Theorem 1. *Let T be a linear bounded operator on a Banach space.*
  a). T is decomposable if and only if T and T' have property (β);
  b). T is sub-decomposable if and only if T has property (β).

For a proof of a) see Eschmeier and Putinar[6] and for b) see Albrecht and Eschmeier[1]. By a sub-decomposable operator we mean an operator which is similar to the restriction of a decomposable operator to a closed invariant subspace.

Let T be an operator with property (β) and let us denote by $\mathcal{O}^X$ the sheaf of X-valued analytic functions on **C**. Then the exact sequence of analytic sheaves:

$$0 \longrightarrow \mathcal{O}^X \overset{T-z}{\longrightarrow} \mathcal{O}^X \longrightarrow \mathcal{F} \longrightarrow 0, \tag{1}$$

defines a sheaf of Fréchet analytic modules which localizes the module X.

Indeed, a Taylor series argument shows immediately that $X \cong \mathcal{F}(\mathbf{C})$, as topological $\mathcal{O}(\mathbf{C})$-modules. Moreover, one remarks that the sheaf $\mathcal{F}$ is supported by the spectrum of T and that in general one can transfer the properties of T into sheaf theoretical properties of $\mathcal{F}$. The analytic sheaf $\mathcal{F}$ has another important property derived from the exact sequence (1). Namely, since the sheaves $\mathcal{O}^X$ are topologically free, then $\mathcal{F}$ has a (finite) global topological free resolution. Such a Fréchet analytic module is called ,in analytic geometry ,*quasi-coherent* and it has most of the properties of coherent analytic sheaves. It is easy to see that $\mathcal{F}$ is the unique quasi-coherent module which localizes the global module X. For that reason it was called *the sheaf model* of the operator T in [18].

From now on we will switch freely the terminology between operators and the corresponding modules over the algebra of entire functions. The next result complements Theorem 1.

Theorem 2.   *Let X be a Banach $\mathcal{O}(\mathbf{C})$-module.*

a). X is decomposable if and only if there is a Fréchet analytic soft sheaf $\mathcal{F}$ on $\mathbf{C}$ with the property that $X \cong \mathcal{F}(\mathbf{C})$ as topological $\mathcal{O}(\mathbf{C})$-modules;

b). X is sub-decomposable if and only if there is a finite resolution of analytic Banach modules:

$$0 \longrightarrow X \longrightarrow D^0 \longrightarrow \ldots \longrightarrow D^p \longrightarrow 0, \qquad (2)$$

with $D^0, \ldots, D^p$ decomposable Banach $\mathcal{O}(\mathbf{C})$-modules.

In fact one can prove that p can be taken equal to 1 in b). The sheaf $\mathcal{F}$ with the properties stated in a) is unique and hence it is the sheaf model of the module X.

Thus (2) is an abstract analogue of the Dolbeault resolution for the module X. The classes of sub-scalar, or in particular hyponormal, operators provide examples of modules with such decomposable resolutions. In almost all particular cases these resolutions are canonically related to the module X via the ususal $\bar{\partial}$-operator. Let us consider for instance a bounded open set U of the complex plane and the Bergman space $A^2(U)$ consisting of the square integrable analytic functions in U with respect to the planar Lebesgue measure. We have then an exact sequence:

$$0 \longrightarrow A^2(U) \longrightarrow D \xrightarrow{\bar{\partial}} L^2(U) \longrightarrow 0,$$

where D is the domain of $\bar{\partial}$. From this sequence one immediately reads the sheaf model of the Bergamn space $A^2(U)$:

$$\mathcal{F}(V) = \{f \in \mathcal{O}(V \cap U); f \in A^2(K \cap U), K \subset\subset V\},$$

for an arbitrary open set V of **C**.

For several commuting operators the same picture holds. To be more precise, let X be a Banach $\mathcal{O}(\mathbf{C}^n)$-module. Following Frunză, we say that X has property (β) if, for every open polydisk D of $\mathbf{C}^n$ the Koszul complex $K.(z-w, \mathcal{O}(D, X))$ is exact in positive degree and it has Hausdorff homology in degree zero. Thus at the level of analytic sheaves we still have an exact sequence:

$$K.(z - w, \mathcal{O}^X) \longrightarrow \mathcal{F} \longrightarrow 0,$$

with $\mathcal{F}$ a Fréchet quasi-coherent sheaf. As a matter of fact the definition of property (β) does not depend on the Koszul resolution. Its invariant meaning can be expressed in terms of the topological Tor's introduced by Taylor [26] (see also [19] for details). In the multivariable theory we adopt as definition of decomposability the existence of a Fréchet soft sheaf model. The original definition is slightly weaker than this property (cf. [19]). Exactly as in the single variable case we have the following result, reproduced below from [20].

Theorem 3. *A Banach $\mathcal{O}(\mathbf{C}^n)$-module X has property (β) if and only if there is a resolution with Banach $\mathcal{O}(\mathbf{C}^n)$-modules:*

$$0 \longrightarrow X \longrightarrow D^0 \longrightarrow \ldots \longrightarrow D^p \longrightarrow 0,$$

with $D^i, i \in \{0, 1, \ldots, p\}$, decomposable modules.

In the above statement one can take p to be less or equal than n.

The Bergman space of a bounded domain of holomorphy in $\mathbf{C}^n$ is again a typical example of a module with such a resolution and hence with property (β). The sheaf model of a module with property (β) can be defined as before.

This localization theory has a series of applications to operator theory or to function theory. We list below only two examples of such consequences. The monograph [9] contains various other applications of the theory of the sheaf model and in general of the homological algebra which is behind this concept.

Proposition 1. *Two commutative tuples of Banach space operators which have property (β) and are quasi-similar have equal joint spectra and equal essential joint spectra.*

Proposition 2. *Let Ω be a bounded strictly pseudoconvex domain of $\mathbf{C}^n$ with smooth boundary. Then any analytically invariant subspace of the Bergman space $A^2(\Omega)$ of finite codimension is of the form $(P_1, \ldots, P_k)A^2(\Omega)$, where $P_1, \ldots, P_k$ are polynomials with finitely many common zeroes.*

Proofs of these facts appear in [21],[22] or [9].

3   Smooth resolutions

Looking at the typical resolution of the Bergman space of a bounded domain of holomorphy $\Omega \subset \mathbf{C}^n$:

$$0 \longrightarrow A^2(\Omega) \longrightarrow Dom(\overline{\partial}^{(0,0)}) \longrightarrow \ldots \longrightarrow Dom(\overline{\partial}^{(0,n-1)}) \longrightarrow L^2(\Omega) \longrightarrow 0,$$

we remark first that the sequence is exact (by the well known Hörmander's Theorem [15]) and secondly that the domains of the $\overline{\partial}$-operators are not only soft modules but even Banach modules over the algebra of smooth functions $\mathcal{E}(\mathbf{C}^n)$. Indeed, for a fixed $i, (0 \le i \le n-1)$ we have by definition:

$$Dom(\overline{\partial}^{(0,i)}) = \{\xi \in L^2{}_{(0,i)}(\Omega); \overline{\partial}^{(0,i)}\xi \in L^2{}_{(0,i+1)}(\Omega)\},$$

and obviously such a form ξ can be multiplied by a smooth function. The norm of the differential form ξ is by definition:

$$\|\xi\|^2 = \|\xi\|_2{}^2 + \|\overline{\partial}^{(0,i)}\|_2{}^2.$$

A property resembling (β) characterizes the modules with a more special smooth resolution as in the above example. Namely we say that a Banach module X over the algebra of entire functions $\mathcal{O}(\mathbf{C}^n)$ satisfies property $(\beta)_{\mathcal{E}}$ if the Koszul complex $K.(z - w, \mathcal{E}(\mathbf{C}^n, X))$ is exact in positive degree and it has Hausdorff homology in degree zero. In order to state the main result concerning smooth resolutions we need the following definition. A Fréchet module E over the algebra $\mathcal{E}(\mathbf{C}^n)$ is said to possess the Mittag-Leffler property (in short the (ML) property) if it is the inverse limit of a countable linearly ordered system of Banach $\mathcal{E}(\mathbf{C}^n)$- modules having all the structural maps with dense range. Most of the standard function spaces on $\mathbf{C}^n$ have this property.

Theorem 4. *A Banach $\mathcal{O}(\mathbf{C}^n)$-module has property $(\beta)_{\mathcal{E}}$ if and only if there is a resolution with Fréchet $\mathcal{O}(\mathbf{C}^n)$-modules:*

$$0 \longrightarrow X \longrightarrow E^0 \longrightarrow \ldots \longrightarrow E^p \longrightarrow 0,$$

where each $E^i, 0 \leq i \leq p$ is a (ML)-module over $\mathcal{E}(\mathbf{C}^n)$.

One can show that p can be chosen at most equal to n. A proof of this theorem appears in [20]. Even for $n = 1$ the last result brings a novel characterization of the class of operators that are similar to the restriction to an invariant subspace of a generalized scalar operator. For simplicity these operators will be called *subscalar*.

Corollary 1. *A Banach space operator is subscalar if and only if it has property $(\beta)_\mathcal{E}$.*

This result was originally obtained in [7] prior to Theorem 4. The same paper contains a few application of Corollary 1 to the division of vector valued distributions by analytic functions.

Another consequence of Theorem 4 is the following result, originally proved by Malgrange[17] in the more general case of real analytic functions.

Theorem (Malgrange). *Let U be an open subset of $\mathbf{C}^n$ and let $f_1, \ldots, f_k$ be complex analytic functions defined on U. Then the ideal $(f_1, \ldots, f_k)\mathcal{E}(U)$ is closed in the Fréchet topology of $\mathcal{E}(U)$.*

The idea of the proof is the following. Let I denote the closed ideal generated in by the system of functions $f_1, \ldots, f_k$ in $\mathcal{E}(U)$. First one proves locally in a neighbourhood of a fixed point of U that the complex derived from the Dolbeault complex:

$$0 \longrightarrow \mathcal{O}(U)/(f_1, \ldots, f_k) \longrightarrow \Lambda^{(0,0)}(\mathcal{E}(U)/I) \xrightarrow{\bar{\partial}} \ldots \xrightarrow{\bar{\partial}} \Lambda^{(0,n)}(\mathcal{E}(U)/I) \longrightarrow 0,$$

is exact. This proves in view of Theorem 4 that the module $\mathcal{O}(U)/(f_1, \ldots, f_k)$ has property $(\beta)_\mathcal{E}$, therefore its tensor product over $\mathcal{O}(U)$ with $\mathcal{E}(U)$ is Hausdorff.

In the next sections we discuss some other examples of canonical $\bar{\partial}$-resolutions for analytic modules.

4   Dilations and resolutions.

Returning to the dictionary between linear operators and analytic modules we remark that in the previous two sections the possibility of extending an operator to a decomposable or scalar operator was characterized by the existence of a resolution to the right of the corresponding module with soft, respectively smooth modules. For operators which are not extendable to better operators there is no hope of obtaining such resolutions. However, we know from operator theory that in general we expect that an operator possesses a good dilation rather that extension. This conclusion can easily be adapted to our framework based on resolutions. The only difference is that this time the corresponding good resolution is not on a single side of the original module. The pioneering work of Douglas and Foiaş [3] contains the germ of this idea (see for further developments [4]). The construction of a universal dilation contained in this section applies to any operator and it has a variety of applications, in general less accurate than those obtained from the classical unitary dilation of a contraction (cf. [25]).

First we present an algebraic construction which is common to all examples below.

Let A be a **C**-algebra with unit and let B be an algebra over A, both commutative. Let M be an A-module together with a free resolution:

$$\ldots \longrightarrow F_1 \longrightarrow F_0 \longrightarrow M \longrightarrow 0.$$

Consider, in the category of A-modules, a finite resolution to the right of A with B-modules:

$$0 \longrightarrow A \longrightarrow E^0 \longrightarrow E^1 \longrightarrow \ldots \longrightarrow E^N \longrightarrow 0.$$

Denote by C. the total complex associated to the double complex with exact columns $F.\otimes_A E\cdot$. Endowed with the standard degree, the complex C. is exact in any degree except zero and its homology at zero is isomorphic to the module M. (One can prove this assertion by a diagram chasing or by a spectral sequence argument).

Thus the original A-module M is quasi-isomorphic (i.e. isomorphic to the single non-zero homology space) to the complex:

$$\ldots \longrightarrow C_1 \longrightarrow C_0 \longrightarrow C_{-1} \longrightarrow \ldots \longrightarrow C_{-N} \longrightarrow 0. \tag{3}$$

First we follow the previous algebraic pattern in a general construction which applies to any commutative tuple of operators. Let T be a commutative n-tuple of linear bounded operators acting on the Banach space X. Let Ω be a bounded domain of holomorphy in $\mathbf{C}^n$ which contains the joint spectrum of T. Then for the analytic module X canonically associated to T there is a finite topologically free resolution given by the Koszul complex:

$$K.(T - z, \mathcal{O}(\Omega, X)) \longrightarrow X \longrightarrow 0.$$

On the other hand the space of analytic functions $\mathcal{O}(\Omega)$ can be resolved by the Dolbeault complex:

$$0 \longrightarrow \mathcal{O}(\Omega) \longrightarrow \mathcal{E}^{(0,0)}(\Omega) \longrightarrow \ldots \longrightarrow \mathcal{E}^{(0,n)}(\Omega) \longrightarrow 0.$$

Putting together these two resolutions we obtain a finite complex of Fréchet $\mathcal{E}(\Omega)$-modules:

$$0 \longrightarrow E_n \longrightarrow E_{n-1} \longrightarrow \ldots \longrightarrow E_0 \longrightarrow \ldots \longrightarrow E_{-n} \longrightarrow 0, \qquad (4)$$

which is exact except in zero degree where its homology is isomorphic to X.

As explained in [20] or [9] one can replace in the preceding complexes the space of analytic functions in Ω by the Bergman space and the Dolbeault resolution with smooth coefficients by the same resolution with L^2 coefficients. By taking appropriate completions of the respective tensor products one finds the following result.

Proposition 3. *Let X be a Banach $\mathcal{O}(\Omega)$-module, where Ω is a bounded domain of holomorphy in C^n. Then there is a finite complex of Banach $\mathcal{E}(\overline{\Omega})$-modules (4) which is exact everywhere except in degree zero, where its homology is topologically isomorphic to X.*

Needless to say that the boundary morphisms in the complex (4) are linear over the algebra of analytic functions (in $\overline{\Omega}$). It is important to observe that the support of the module X (i.e. the joint spectrum of T) is contained in Ω, while the support of each term of the complex (4) is supported by $\overline{\Omega}$. This quite serious weakness of the above universal resolution will be corrected in some particular cases considered in the next section.

By replacing the module E_0 in (4) with the quotient module $D_+ = E_0/Im(E_1 \to E_0)$ one obtains a resolution of X:

$$0 \longrightarrow X \longrightarrow D_+ \longrightarrow E_{-1} \longrightarrow \ldots \longrightarrow E_{-n} \longrightarrow 0. \qquad (5)$$

Similarly, defining $D_- = Ker(E_0 \to E_{-1})$ one finds a resolution to the left of X:

$$0 \longrightarrow E_n \longrightarrow \ldots \longrightarrow E_1 \longrightarrow D_- \longrightarrow X \longrightarrow 0. \qquad (6)$$

Then simple homological arguments prove the following result. (See [9] for details).

Corollary 2. *Let X be a Banach $\mathcal{O}(\Omega)$-module with the canonical Banach module resolution (4) and its truncations (5) and (6). Then:*

a). X has property (β) if and only if D_+ is a decomposable module;

b). The dual module X' has (β) if and only if D_- is a decomposable module;

c). X is decomposable if and only if D_+ and D_- are decomposable modules.

Notice that all other elements E_k in the resolutions (5) and (6) are decomposable. Thus D_+ is a universal decomposable extension of the module X (whenever such an extension exists) and analogously D_- is a universal decomposable co-extension of X (that is X is a quotient module of it).

The analysis of these universal (virtually decomposable) modules associated to X has recently led to the following remarkable invariant subspace result.

Theorem(Eschmeier). *Let T be a commutative n-tuple of linear bounded Banach space operators. If the Taylor joint spectrum of T is thick, then T has a non-trivial common invariant subspace.*

Above by thick we mean a compact set in $\mathbf{C}^n$ which is dominating in a non-empty open set. In particular a set with interior points is thick. This theorem culminates a long series of results derived from Scott Brown's technique of producing invariant subspaces for operators that possess extensions with good spectral decomposition properties. (See for details [5]).

5 Spectral sets and scalar dilations

The present section illustrates at the level of multivariable spectral theory the following general principle of , or rather result to aim at in , dilation theory : (uniform) estimates of the analytic functional calculus imply the existence of "good" dilations with spectrum contained in the support of the estimate. The classical example in this direction is the interplay between von Neumann's inequality and the unitary dialtion (both for Hilbert space contractions). Both in the single variable case as well as for commuting systems of operators there are serious difficulties along these lines, to mention only the long standing open question whether an analogue of von Neumann's inequality holds for tuples of operators. The compromise we are making in what follows consists in replacing the unitary dilations by generalized scalar ones and secondly by working modulo similarity rather than unitary equivalence. The same algebraic construction presented in the prevoius section produces under suitable estimates of the functional calculus scalar resolutions and in particular the needed dilations.

First we recall some notation. Let Ω be a bounded domain of $\mathbf{C}^n$. The space of bounded holomorphic functions in Ω is denoted by $H^\infty(\Omega)$. The space of Lipschitz continuous functions of positive order α which are holomorphic in Ω will be denoted by $A^\alpha(\overline{\Omega})$. For an integer α, this is the space of continuously differentiable functions of order α on the closure of Ω which are holomorphic inside Ω. Similarly one denotes by C^α the Lipschitz spaces without any analyticity assumption. Finally by $C^\alpha{}_b(\Omega)$ one denotes the space of functions which are uniformly bounded in the Lipschitz α-norm on Ω. By $C^n_{\overline{b}}$ we denote the space of those functions in Ω which have uniformly bounded $\overline{\partial}$-derivatives up to order n.

Theorem 5. *Let $\Omega \subset \mathbf{C}^n$ be a bounded strictly pseudoconvex domain and let X be a Banach $H^\infty(\Omega)$-module. Then there is a complex of Banach $C^1_{\overline{b}}(\Omega)$-modules which is exact except in degree zero where its homology is topologically isomorphic to X:*

$$\ldots \longrightarrow K_1 \longrightarrow K_0 \longrightarrow K_{-1} \longrightarrow \ldots \longrightarrow K_{-n} \longrightarrow 0. \qquad (7)$$

Proof. Let us denote the algebra $H^\infty(\Omega)$ by A for simplicity. Then there is a topologically **C**-split resolution of the A-module X given by the Bar complex:

$$B.^A(A, X) \longrightarrow X \longrightarrow 0.$$

See for details [26].

The algebra A has a resolution with smooth modules given by the following $\overline{\partial}$-complex:

$$0 \longrightarrow A \longrightarrow (\Lambda^{(0,.)}(C^0{}_b(\Omega)), \overline{\partial}^{(0,.)}).$$

The exactness of this complex of closed operators is a non-trivial fact obtained as a consequence of the Koppelman-Leray-Henkin integral formula for analytic functions in a strictly pseudoconvex domain. Moreover, one proves with the aid of some singular integral operators derived from the same integral formula that the latter $\overline{\partial}$-complex is topologically **C**-split. (See [14] Theorem 3.1.3 for details). Therefore, denoting by D^q the domain of the operator $\overline{\partial}^{(0,q)}$ between the corresponding spaces of forms with coefficients in

$C^0{}_b(\Omega)$ one obtains a **C**-split complex of Banach A-modules with bounded boundary operators:

$$0 \longrightarrow A \longrightarrow D^0 \longrightarrow \ldots \longrightarrow D^n \longrightarrow 0. \qquad (8)$$

The typical term in the Bar resolution is of the form $A \overline{\otimes} A \overline{\otimes} \ldots \overline{\otimes} A \overline{\otimes} X$, where we choose $\overline{\otimes}$ to be the completed projective topological tensor product of two locally convex spaces. Following the algebraic scheme we replace each term of the Bar resolution by the complex $D^{\cdot} \overline{\otimes} A \overline{\otimes} \ldots \overline{\otimes} A \overline{\otimes} X$ and we remark that this is still a topologically split **C**-complex. By putting together these pieces into a double complex and passing to the associated simple complex we complete the proof of Theorem 5.

Remark that the complex (7) is unbounded to the left and the set Ω is strictly pseudoconvex, but not necessarily with smooth boundary.

Since the multiplication map

$$A \overline{\otimes} X \longrightarrow X$$

has obviously a continuous inverse to the right, and because the algebra A is topologically complemented in D^0, the above proof has the following consequence.

Corollary 3. *Let T be a commutative n-tuple of linear bounded operators acting on a Banach space X and let Ω be a bounded strictly pseudoconvex open subset of $\mathbf{C}^n$. If T admits a continuous $H^\infty(\Omega)$ functional calculus, then there is a commutative n-tuple S of linear bounded operators which acts in the larger Banach space $K \supset X$ with the following properties:*

a). There is a linear bounded projection $P : K \to X$;

b). S admits a continuous functional calculus with elements of $C^1_{\overline{b}}(\Omega)$;

c). For any function $f \in H^\infty(\Omega)$ the following relation:

$$Pf(S)|X = f(T), \qquad (9)$$

holds.

Moreover, the preceding dilation is canonical, in the sense that to tuples T intertwined by some operators correspond scalar tuples S intertwined by some dilated operators and formula (9) is functorial. Let us also remark that the inclusion $X \subset K$ was chosen to be isometric.

A more involved analysis of the Koppelman-Leray-Henkin integrals reveals that the above results hold for the algebra $A^0(\overline{\Omega})$ instead of $H^\infty(\Omega)$ and all the other function spaces that extended to the boundary of Ω, (cf. [14] Theorem 3.3.4). We recall that $A^0(\overline{\Omega})$ denotes the Banach algebra of analytic functions in Ω which are continuous on $\overline{\Omega}$. Moreover, in that case it is known that the algebra of analytic functions in neighbourhoods of $\overline{\Omega}$ is dense in $A^0(\overline{\Omega})$, (see [14] Theorem 3.5.1). Consequently we obtain the following result.

Theorem 6. *Let T be a commutative n-tuple of linear bounded operators acting on the Banach space X. Let $\Omega \subset \mathbf{C}^n$ be a bounded strictly pseudoconvex open set with the property that, for any analytic function $f \in \mathcal{O}(\overline{\Omega})$ one has:*

$$\|f(T)\| \leq C\|f\|_{\infty,\overline{\Omega}},$$

with C a positive constant.

Then there is a larger Banach space $K \supset X$ and a commutative n-tuple S of linear bounded operators on K with the following proerties:

a). There is a linear continuous projection $P : K \to X$;

b). S admits a continuous functional calculus with elements of $C^1_{\bar{b}}(\overline{\Omega})$;

c). For any function $f \in \mathcal{O}(\overline{\Omega})$ one has:

$$Pf(S)|X = f(T).$$

As before, the construction of the dilation S is functorial, in the sense that a morphism of analytic modules $X \longrightarrow X'$ extends to a morphism between their dilations $K \longrightarrow K'$. In the familiar terminology of dilation theory we are in this case in the presence of the property of lifting the commutants.

The proof of Theorem 6 repeats the same arguments with the only new observation that, whenever the tuple T admits a continuous functional calculus with elements of $\mathcal{O}(\overline{\Omega})$, then the Taylor joint spectrum of T is contained in $\overline{\Omega}$,(see for instance [9] Chapter 5). Thus the notation f(T) in the statement makes perfect sense.

In the more restrictive case of strictly pseudoconvex domains with smooth boundary the same results can be formulated for any Lipschitz space. The main ingredient is in this case the Henkin integral formula which splits again topologically the corresponding $\bar{\partial}$-complex. (See [14] Theorem 2.6.1).

6   Quasi-coherent analytic sheaves

The Banach modules over the algebra of entire functions in n complex variables have property (β) if and only if they can be localized by a Fréchet
analytic quasi-coherent sheaf , cf. Section 2 above. The construction of
different resolutions for Banach modules discussed in the previous sections
has the following application to the theory of analytic sheaves.

Theorem 7.   *Let $\mathcal{F}$ be a Fréchet analytic sheaf on a domain of holomorphy
$\Omega \subset \mathbf{C}^n$. The following conditions are equivalent:*

*a). There is a Fréchet topologically free resolution (to the left) of $\mathcal{F}$ on
Ω;*

*b). $\mathcal{F}$ is acyclic on Ω and it has a resolution to the right with Fréchet
soft analytic sheaves;*

*c). $\mathcal{F}$ is acyclic on Ω and it coincides with the only non-trivial homology
space of a complex of Fréchet $\mathcal{E}$-modules on Ω.*

By an acyclic sheaf on Ω we mean a sheaf with trivial Čech cohomology in positive degree. Let us remark that in the above statements we do
not assume that the respective resolutions are finite. However, we know a
posteriori that finite resolutions exist in all three cases.

Proof. $b) \Rightarrow a)$. Let

$$0 \longrightarrow \mathcal{F} \longrightarrow \mathcal{S}^0 \longrightarrow \ldots \longrightarrow \mathcal{S}^{n-1} \longrightarrow \mathcal{G} \longrightarrow 0$$

be a resolution of $\mathcal{F}$ with Fréchet analytic sheaves and suppose that $\mathcal{S}^q, 0 \leq q \leq n-1$ are soft sheaves. A repeated application of the long exact sequence
of cohomology shows that:

$$H^1{}_c(V, \mathcal{G}) \cong H^{n+1}{}_c(V, \mathcal{F})$$

for any open subset V of Ω. By dimension reasons the second cohomology
group vanishes, whence $\mathcal{G}$ is a soft sheaf. Since a Fréchet soft analytic module
is quasi-coherent (by a variant of Theorem 3 above, cf. [19]) one proves by
descending induction on the previous resolution that $\mathcal{F}$ is a quasi- coherent
analytic module, that is assertion a).

The implication $a) \Rightarrow b)$ follows from Theorem 3, applied to Fréchet
modules instead of Banach modules, (see also [20]).

The implication $a) \Rightarrow c)$ was proved before as a preparation for Proposition 3, in the process of constructing the universal scalar dilation of an analytic module.

Finally, the implication $c) \Rightarrow a)$ has the following proof. Let $\mathcal{L}$. be a complex of Fréchet $\mathcal{E}$-modules on Ω which is exact except in zero degree, where its homology coincides with $\mathcal{F}$. Then we have a short exact sequence;

$$0 \longrightarrow \mathcal{G} \longrightarrow \mathcal{F} \longrightarrow \mathcal{H} \longrightarrow 0,$$

where $\mathcal{G}$ is a Fréchet analytic module with a reslution to the left with Fréchet $\mathcal{E}$-modules and $\mathcal{H}$ is a Fréchet analytic module with a resolution to the right with Fréchet $\mathcal{E}$-modules. We will use several times the observation that, in a sequence of acyclic Fréchet analytic sheaves if two are quasi-coherent, then the third has the same property. First we infer from the given resolution and by dimenson reasons that the sheaf $\mathcal{G}$ is acyclic on Ω. Then it follows from the same resolution that it is quasi-coherent. Then we prove from the above short exact sequence that $\mathcal{H}$ is an acyclic sheaf. From point b) and the given resolution we know that $\mathcal{H}$ is in fact quasi-coherent. Then we conclude that $\mathcal{F}$ is quasi-coherent.

The preceding proof shows that any Fréchet quasi-coherent analytic module on a Stein manifold of complex dimension n is quasi-isomorphic to a finite complex of length at most 2n of topologically free Fréchet $\mathcal{E}$-modules having $\mathcal{O}$-linear differential operators as boundaries.

Returning to abstract $\bar{\partial}$-resolutions we reproduce from [9] the following pseudoconvexity criterion.

Proposition 4. *Let $\Omega \subset \mathbf{C}^n$ be a bounded domain and let $A \subset \mathcal{O}(\Omega)$ be a Fréchet $\mathcal{O}(\overline{\Omega})$-submodule which contains the function identical equal to one. Then each of the following conditions implies the pseudoconvexity of Ω:*

a). A has property (β) and it is supported by $\overline{\Omega}$;

b). There is a Fréchet quasi-coherent module $\mathcal{F}$ supported by $\overline{\Omega}$ and with $\mathcal{F}(\overline{\Omega}) \cong A$;

c) There is an abstract $\bar{\partial}$-resolution of A with Fréchet soft $\mathcal{O}(\overline{\Omega})$-modules;

d) A is the single non-trivial homology space of a complex of Fréchet $\mathcal{E}(\overline{\Omega})$-modules.

This result extends the well known pseudoconvexity criteria based on the vanishing of various groups of cohomology with bounds. (See [9] Chapter 7 for examples and details).

To draw a conclusion of the whole discussion we remark that any commutative n-tuple T of linear bounded Banach space operators is equivalent , via a canonical construction, to a finite complex of length 2n of Banach-free analytic vector bundles with $\mathcal{O}$-linear first order differential operators as boundary maps. Moreover, the tuple T is spectrally localizable (i.e. it has property (β)) if and only if this equivalence persists at the level of sheaves.

Thus the classical $\bar{\partial}$-complex is not an isolated object of study in multivariable spectral theory.

References

1. Albrecht,E. and Eschmeier,J., *Functional models and local spectral theory*, preprint 1987.

2. Bishop,E., *A duality theorem for an arbitrary operator*, Pacific J. Math. 9(1959), 379-394.

3. Douglas,R.G. and Foiaş,C., *A homological view in dilation theory*, preprint, 1976.

4. Douglas,R.G. and Paulsen,V. *Hilbert modules over function algebras* , Pitman Res.Notes Mat. vol. 219, Harlow, 1989.   5. Eschmeier Inv spaces

6. Eschmeier,J. and Putinar,M., *Spectral theory and sheaf theory.III* , J.reine angew.Math. 354(1984), 150-163.

7. Eschmeier,J. and Putinar,M . *On quotients and restrictions of generalized scalar operators*, J.Funct.Analysis 84(1989), 115-134.

8. Eschmeier,J. and Putinar,M., *Spectra of analytic Toeplitz tuples on Bergman spaces*, Acta Math.Sci.(Szeged), to appear.

9. Eschmeier,J. and Putinar,M. *Spectral decompositions and analytic sheaves*, Oxford University Press, to appear.

10. Foiaş,C., *Spectral maximal spaces and decomposable operators*, Arch.Math. 14(1963), 341-349.

11. Frunză,S., *The Taylor spectrum and sepctral decompositions*, J.Funct. Analysis 19(1975), 390-421.

12. Gelfand,I.M. and Ponomarev,V.A., Remarks on the classification of a pair of commuting linear transformations in a finite dimensional space (Russian), Funct. Anal. Appl. 3(1969), 81-82.

13. Grauert,H. and Remmert,R., *Coherent analytic sheaves*, Springer Verlag, Berlin, 1984.

14. Henkin,G. and Leiterer,J., *Function theory on complex manifolds*, Akademie Verlag, Berlin, 1984

15. Hormander,L., L^2-*estimates and existence theorems for the* $\bar{\partial}$-*operator*, Acta Math. 113(1965), 89-152.

16. Mac Lane,S., *Homology*, Springer Verlag, Berlin, 1963.

17. Malgrange,B., *Ideals of differentiable functions*, Oxford Univ. Press, Oxford, 1966.

18. Putinar,M., *Spectral theory and sheaf theory.I*, in vol. "Dilation theory,Toeplitz operators and other topics", Birkhäuser Verlag, Basel et al., 1983, pp.283-298.

19. Putinar,M., *Spectral theory and sheaf theory.II*, Math.Z. 192(1986), 473-490.

20. Putinar,M., *Spectral theory and sheaf theory.IV*, Proc. Symp.Pure Math. 51(1990), 273-293.

21. Putinar,M., *Invariant subspaces of several variable Bergman spaces*, Pacific J. Math. 147(1990),355-364.

22. Putinar,M., *Quasi-similarity of tuples with Bishop's property* (β), Integral Eq. Operator Theory 15(1992), 1047-1052.

23. Putinar,M. and Salinas,N., *Analytic transversality and Nullstellensatz in Bergman spaces*, Contemp.Math. 137(1992), 367-381.

24. Radjabalipour,M., *Decomposable operators*, Bull.Iran Math, Soc. 9(1978), 1-49.

25. Sz.Nagy,B. and Foiaş,C., *Harmonic analysis of operators on Hilbert space*, American Elsevier, New York, 1970.

26. Taylor,J.L., *Homology and cohomology for topological algebras* , Adv.Math. 9(1972), 147-182.

27. Vasilescu,F.H., *Analytic functional calculus and spectral decompositions*, D.Reidel Publ.Co., Dordrecht, 1982.

DEPARTMENT OF MATHEMATICS, UNIVERSITY OF CALIFORNIA, RIVERSIDE, CA 92521. *E-mail address*: mputinar@ucrmath.ucr.edu

Contemporary Mathematics
Volume 185, 1995

TOEPLITZ C^*-ALGEBRAS AND
SEVERAL COMPLEX VARIABLES

By Norberto Salinas

In this note, we present a brief survey of the C^*-algebraic properties of Toeplitz operators on domains $\Omega \subseteq \mathbf{C}^n$. The main object of study is the C^*-algebra generated by Toeplitz operators on a canonical Hilbert space of holomorphic functions on Ω, with symbols in a natural function algebra over Ω.

The classical situation is $\Omega = \mathbf{D}$, the open unit disk ($n = 1$). There is a natural (orthonormal) basis of $L^2(\partial \mathbf{D})$ given by $\{\epsilon_n(z) = z^n/\sqrt{2\pi}, \quad n \in \mathbf{Z}\}$. Every function $\phi \in L^\infty(\partial \mathbf{D})$ gives rise to the (normal) operator M_ϕ on $L^2(\partial \mathbf{D})$ (via pointwise multiplication). The matrix of M_ϕ with respect to the above basis is invariant under a shift translation in the direction of the principal diagonals. The Toeplitz operator associated with ϕ is the compression via the (Szego) projection $S: L^2(\partial \mathbf{D}) \to H^2(\partial \mathbf{D})$ to the Hardy space $H^2 = H^2(\partial \mathbf{D})$, i.e., $T_\phi = SM_\phi|_{H^2}$. Thus, T_ϕ has a Toeplitz matrix representation on the basis $\{\epsilon_n: n \in \mathbf{Z}_+\}$ of H^2 (see [10, Chap 7]).

Notice that the (standard) Toeplitz C^*-algebra, i.e., the C^*-algebra generated by $\{T_\phi: \phi \in C(\partial \mathbf{D})\}$, is actually generated by Toeplitz operators with symbols in the *disk algebra*.

There are many ways one may go in generalizing this set up. One may use the fact that H^2 can be represented as a Sobolev space on Ω, and hence use a different measure to define square integrability (see [3]). A typical example of this approach is the study of Bergman Toeplitz operators (here the functional Hilbert space is the Bergman space over $\mathbf{D}$, see [1]). One may also change the domain and consider a simply connected domain Ω with smooth boundary (using, for instance, the Riemann mapping from $\mathbf{D}$ onto Ω), and work with the Hilbert space of holomorphic functions on Ω whose boundary values are square integrable with respect to the linear Lebesgue measure of the boundary. Or, more generally, one may take any bounded domain $\Omega \subseteq \mathbf{C}$ and considered the "Hardy" space over $\partial\Omega$ (in which case, one needs to impose some smoothness restrictions to $\partial\Omega$ so that we may define the linear measure on $\partial\Omega$) or one may consider the Bergman space over Ω (see [1]).

Of course, one may also generalize the classical Toeplitz operators by considering domains in $\mathbf{C}^n$, $n > 1$. It is in this context that the interplay between the complex geometry of $\partial\Omega$ and Toeplitz operators on Ω becomes more relevant. The absence of a Riemann mapping theorem in several complex variables, makes this interplay an interesting tool. In the present article, we shall outline some of the developments in this direction.

---

[1] AMS SubjectClassification: 46-05, 47-33, 32-37

Let's begin with a bounded domain Ω in $\mathbf{C}^n$. The first examples were considered in [6] where $\Omega = B_n$ the unit ball in $\mathbf{C}^n$, and in [11] where $\Omega = \mathbf{D}^n$ the unit polydisk in $\mathbf{C}^n$, generated by symbols in the algebra $A(\Omega)$ of continuous functions on $\overline{\Omega}$ and holomorphic on Ω. Let $\widetilde{\partial}\Omega$ be the distinguished boundary of Ω (i.e., the Sylov boundary with respect to $A(\Omega)$). Thus, if $\Omega = B_n$, the unit ball in $\mathbf{C}^n$, then $\widetilde{\partial}\Omega = \partial B_n$, and if $\Omega = \mathbf{D}^n$, the unit polydisk in $\mathbf{C}^n$, then $\widetilde{\partial}\Omega = (\partial \mathbf{D})^n$. In general, we must impose a smoothness condition on $\widetilde{\partial}\Omega$ to allow the existence of a natural measure λ supported on $\widetilde{\partial}\Omega$. Let $L^2(\widetilde{\partial}\Omega)$ be the space of square integrable functions with respect to the natural measure λ on $\widetilde{\partial}\Omega$ and let $H^2(\widetilde{\partial}\Omega)$ be the subspace of functions that can be extended to Ω. In order to see that $H^2(\widetilde{\partial}\Omega)$ is closed in $L^2(\widetilde{\partial}\Omega)$, one uses the reasonable working assumption (usually easy to verify) that every point of Ω yields a bounded evaluation on $H^2(\widetilde{\partial}\Omega)$. The (orthogonal) projection S from $L^2(\widetilde{\partial}\Omega)$ onto $H^2(\widetilde{\partial}\Omega)$ has a kernel function $s_w(z)$, $w \in \Omega$, $z \in \widetilde{\partial}\Omega$, usually called the Szego kernel. Given $\phi \in L^\infty(\widetilde{\partial}\Omega)$, the Toeplitz operator on $H^2(\widetilde{\partial}\Omega)$ with symbol ϕ is the compression to $H^2(\widetilde{\partial}\Omega)$ of the multiplication operator induced by ϕ on $L^2(\widetilde{\partial}\Omega)$. Thus, $T_\phi f = S\phi f$, for every $f \in H^2(\widetilde{\partial}\Omega)$.

Definition 1 *The (Hardy) Toeplitz C^*-algebra $\mathcal{T}(\partial\Omega)$ over Ω is the C^*-algebra generated by $\{T_\phi : \phi \in A(\Omega)\}$.*

Another reasonable assumption we make (and also easy to check) is that $A(\Omega)$ is dense in $H^2(\widetilde{\partial}\Omega)$. We use it in the following (standard) lemma.

Lemma 2 *$\mathcal{T}(\partial\Omega)$ is an irreducible C^*-algebra.*

Proof: Assume that P is a self-adjoint idempotent that commutes with T_ϕ for every $\phi \in A(\Omega)$. Let $\theta = P1$. Then, $\theta \in H^2(\widetilde{\partial}\Omega)$, and

$$\theta\phi = \phi P1 = T_\phi P1 = PT_\phi 1 = P\phi,$$

for every $\phi \in A(\Omega)$. Now,

$$\left| \int \theta\phi\overline{\psi}\, d\lambda \right| = |\langle \theta\phi, \psi \rangle| = |\langle P\phi, \psi \rangle| \leq 1,$$

for every $\phi, \psi \in A(\Omega)$ with $\|\phi\|_{H^2} = \|\psi\|_{H^2} = 1$, so $|\theta| \leq 1$ a.e. λ. Since $A(\Omega)$ is dense in $H^2(\widetilde{\partial}\Omega)$, we deduce that $Pf = T_\theta f$ for every $f \in H^2(\widetilde{\partial}\Omega)$. Thus, given $w \in \Omega$, we have:

$$\theta(w) = (P1)(w) = \langle P1, s_w \rangle = \langle P^2 1, s_w \rangle =$$
$$\langle P1, T_\theta^* s_w \rangle = \theta(w)\langle P1, s_w \rangle = \theta^2(w).$$

It follows that either $\theta = 1$ or $\theta = 0$, as desired. $\qquad\qquad\square$

Remark 3 (a) In order to insure that $\mathcal{T}(\partial\Omega)$ contains a non-zero compact operator, so that $\mathcal{K} \subseteq \mathcal{T}(\partial\Omega)$ (because of the previous lemma), we shall assume that the Szego kernel s of Ω is a generalized Bergmann kernel function. In particular, this means (see [8, Definition 4.11]) that the operator D_{Z-w}, defined on $H^2(\widetilde{\partial\Omega})$ by

$$D_{Z-w} = \sum_{j=1}^{n} (T_{Z_j} - w_j)(T_{Z_j}^* - \overline{w_j}),$$

has closed range and one-dimensional null-space, for every $w \in \Omega$. Here $Z_j : \Omega \to \mathbf{C}$ denotes the j-th coordinate function $1 \leq j \leq n$, and Z is the n-tuple $Z = (Z_1, \ldots, Z_n)$.

(b) Let $\mathcal{C}$ be the commutator ideal of $\mathcal{T}(\partial\Omega)$, so that $\mathcal{K} \subseteq \mathcal{C}$. We claim that $\mathcal{C}$ coincides with the two-sided ideal $\mathcal{J}$ generated by the self-commutators of $T_\phi : \phi \in A(\Omega)$, i.e., $[T_\phi^*, T_\phi] = T_\phi^* T_\phi - T_\phi T_\phi^*$. Indeed, it is clear that $\mathcal{J} \subseteq \mathcal{C}$. On the other hand, since the image of T_ϕ is a normal element in $\mathcal{T}(\partial\Omega)/\mathcal{J}$ and T_ϕ commutes with T_ψ for every $\phi, \psi \in A(\Omega)$, it follows that $\mathcal{T}(\partial\Omega)/\mathcal{J}$ is abelian. This implies that $\mathcal{C} \subseteq \mathcal{J}$, as required. Then, it is not hard to show (using for example [18, Lemma 2.2]) that the map that sends $f \in C(\widetilde{\partial\Omega})$ to the image of T_f in $\mathcal{T}(\partial\Omega)/\mathcal{C}$ is a *-homomorphism. Therefore, $\mathcal{T}(\partial\Omega)/\mathcal{C} \simeq C(X)$, where X can be identified with a closed subset of $\widetilde{\partial\Omega}$.

The main questions are:

(I) When is $X = \widetilde{\partial\Omega}$?

(II) What is the structure of $\mathcal{C}/\mathcal{K}$? What is its relationship with $\partial\Omega$?

Remark 4 (a) If Ω is strongly pseudoconvex then $X = \widetilde{\partial\Omega} = \partial\Omega$ and $\mathcal{C} = \mathcal{K}$ (see [13] or [17]).

(b) If $\Omega = \Omega_1 \times \Omega_2$, where Ω_1 and Ω_2 are strongly pseudoconvex, then $X = \widetilde{\partial\Omega} = (\partial\Omega_1) \times (\partial\Omega_2)$, and $\mathcal{C}/\mathcal{K} \simeq (C(\partial\Omega1) \otimes \mathcal{K}) \oplus (\mathcal{K} \otimes C(\partial\Omega_2))$ (this follows from part (a) above and the Kunet formula). One observes, in this example, the C^*-algebraic implications of the presence of corners and flat faces in $\partial\Omega$. This phenomenon was studied more intensively for the case of Reinhardt domains and Toeplitz operators on Bergmann spaces (see [22] and [20]). It is easy to see that facts similar to those pointed out in the above lemma and remarks are also valid when one substitutes the Hardy space $H^2(\widetilde{\partial\Omega})$ by the Bergman space $H^2(\Omega)$ (i.e., the space of holomorphic functions on Ω which are square integrable with respect to the volume Lebesgue measure). Before explaining the resulting interplay between the structure of $\mathcal{T}(\Omega)$ (the Bergman-Toeplitz C^*-algebra on $H^2(\Omega)$) and $\partial\Omega$ for Reinhardt domains, we proceed to present a general class of domains for which $X = \partial\Omega$ and $\mathcal{C} = \mathcal{K}$, for the case of Bergman-Toeplitz C^*-algebras.

Definition 5 *A domain $\Omega \subseteq \mathbf{C}^n$ is called pseudoregular if $\partial\Omega$ is piecewise smooth, pseudoconvex (i.e., a domain of holomorphy), and the $\overline{\partial}$-Neumann problem satisfies the compactness property (see [19, Definition 1.10]).*

Remark 6 (a) We review briefly here the compactness property of the $\overline{\partial}$-Neumann problem (see, for example, [19, Section 2]). Let $\overline{\partial}$ be the maximal (closed) extension of the standard $\overline{\partial}$ operator on the Grassmann algebra $\Lambda(\mathbf{C}^n) \otimes C_0^\infty(\Omega)$ of differential forms with coefficients in $C_0^\infty(\Omega)$ to $\Lambda(\mathbf{C}^n) \otimes L^2(\Omega)$. Then $\overline{\partial}^*$ is the minimal (closed) extension (i.e., the closure) of the formal adjoint of the standard $\overline{\partial}$ operator on $\Lambda(\mathbf{C}^n) \otimes C_0^\infty(\Omega)$. If $\overline{\Omega}$ is bounded and pseudoconvex, then $\mathrm{Ran}(\overline{\partial})$ is closed, and it follows that $\overline{\partial} + \overline{\partial}^*$ is self-adjoint and has closed range, and that $\mathrm{Ker}(\overline{\partial}) \cap \mathrm{Ker}(\overline{\partial}^*) = H^2(\Omega)$. Then the operator K defined on $\Lambda(\mathbf{C}^n) \otimes L^2(\Omega)$ to be zero on $H^2(\Omega)$ and to be the inverse of $\overline{\partial} + \overline{\partial}^*$ on $[\Lambda(\mathbf{C}^n) \otimes L^2(\Omega)] \ominus H^2(\Omega)$ is bounded and self-adjoint. The compactness property for the $\overline{\partial}$-Neumann problem is equivalent to stating that the restriction of K^2 to the subspace of $(0,1)$ forms with coefficients in $L^2(\Omega)$ (which is reducing for K^2) is compact.

(b) Pseudoregular domains include strongly pseudoconvex domains, domains of finite type (and hence pseudoconvex domains with real analytic boundary), and more generally, domains whose boundary is weakly regular (see [5]).

(c) When $\partial\Omega$ is C^∞-smooth, one can define the Cauchy-Riemann tangential operator $\overline{\partial}_b$ on smooth differential forms. Even when Ω is pseudoconvex, the maximal extension $\overline{\partial}_b$ to $\Lambda(\mathbf{C}^n) \otimes L^2(\partial\Omega)$ does not necessarily have closed range. Thus, in the assumption of the second part of the following proposition, this property is tacitly assumed.

Proposition 7 *Assume that Ω is pseudoregular. Then $\mathcal{C} = \mathcal{K}$, and $\mathcal{T}(\Omega)/\mathcal{K} \simeq C(\partial\Omega)$. If in addition, we assume that the $\overline{\partial}_b$-Neumann problem satisfies the compactness property on $\partial\Omega$, then $\mathcal{T}(\partial\Omega)/\mathcal{K} \simeq C(\partial\Omega)$, and $\mathcal{T}(\Omega)$ and $\mathcal{T}(\partial\Omega)$ are unitarily equivalent.*

Proof: The first assertion follows as in [19, Theorem 2.3]. For the second statement, let $\overline{T} = (T_1, \ldots, T_n)$ and $\overline{S} = (S_1, \ldots, S_n)$ be the n-tuples of multiplication by the coordinate functions on $H^2(\Omega)$ and on $H^2(\partial\Omega)$, respectively. A similar argument to one used in [19, Theorem 2.3] shows that $\mathcal{C} = \mathcal{K}$, and that $\mathcal{T}(\partial\Omega)/\mathcal{K} \simeq C(\partial\Omega)$. This implies that $\overline{T}$ and $\overline{S}$ are essentially normal n-tuples and the essential spectra of $\overline{T}$ and $\overline{S}$ coincide with $\partial\Omega$. Further, it easily follows that $\mathrm{ind}(\overline{T} - \lambda) = \mathrm{ind}(\overline{S} - \lambda)$ for every $\lambda \notin \partial\Omega$. Thus, by the results in [4] there exists a unitary transformation $U \colon H^2(\Omega) \to H^2(\partial\Omega)$ such that $UT_jU^* - S_j \in \mathcal{K}$, $1 \le j \le n$. Since $C^*(\overline{T}) \simeq \mathcal{T}(\Omega)$ and $C^*(\overline{S}) \simeq \mathcal{T}(\partial\Omega)$ contain the ideal of compact operators, it follows that $U^*\mathcal{T}(\partial\Omega)U = \mathcal{T}(\Omega)$, as desired. $\square$

Conjecture 8 *If Ω is pseudoconvex with "regular" boundary, then $\mathcal{T}(\Omega) \simeq \mathcal{T}(\partial\Omega)$.*

In all known examples, the above conjecture is valid. Thus, for the next stage in our discussion, we shall work with the Toeplitz C^*-algebra $\mathcal{T}(\Omega)$ on the Begman space $H^2(\Omega)$.

Definition 9 *A domain $\Omega \subseteq \mathbf{C}^n$ is called Reinhardt if it is invariant under the action of the n-torus $\mathbf{T}^n$ (under componentwise multiplication).*

Remark 10 It follows that a Reinhardt domain Ω is uniquely determined by its intersection with the positive octant, i.e., by $|\Omega| = \{\overline{x} = (x_1, \ldots, x_n) \in \Omega : x_j > 0, 1 \leq j \leq n\}$. It also follows that Ω is pseudoconvex if and only if $\log(|\Omega|) = \{\overline{y} \in \mathbf{R}^n : \exp(\overline{y}) = (\exp(y_1), \ldots, \exp(y_n)) \in |\Omega|\}$ is convex (see [17]).

Definition 11 *Let Ω be a domain in $\mathbf{C}^n$. Two points z, w in $\overline{\Omega}$ are said to be holomorphically equivalent if there exists a chain of overlapping analytic disks in $\overline{\Omega}$ (i.e., holomorphic mappings $f_j : \mathbf{D} \to \overline{\Omega}$, $1 \leq j \leq m$) such that z is the center of the first disk and w is the center of the last disk (i.e., $z = f_1(0)$, $w = f_m(0)$, and $f_j(\mathbf{D}) \cap f_{j+1}(\mathbf{D}) \neq \emptyset$, $1 \leq j < n$). The holomorphic equivalence classes form a foliation called the holomorphic foliation of Ω.*

Theorem 12 *Let Ω be a (bounded) pseudoconvex Reinhardt domain in $\mathbf{C}^n$ such that $C = \log(|\Omega|)$ is a tamed (convex) set (i.e., the map $F : [0,1] \times \overline{C} \times \overline{C} \to \overline{C}$ given by $F(\lambda, x, y) = \lambda x + (1 - \lambda)y$ is open, [16] and [20]). Then there exists a decreasing sequence of ideals of length $n+1$ in $T(\Omega)$, i.e., $\mathcal{J}_0 = T(\Omega) \supseteq \mathcal{J}_1 \supseteq \cdots \supseteq \mathcal{J}_{n+1} = \{0\}$, so that $\mathcal{J}_{k+1}/\mathcal{J}_k \simeq C^*(\mathcal{F}_k)$, where $\mathcal{F}_k$ is the holomorphic foliation of dimension k, $0 \leq k \leq n$. Here, $\mathcal{J}_1 = C$ is the commutator ideal of $T(\Omega)$ and $\mathcal{J}_n = \mathcal{K}$ is the ideal of compact operators.*

Corollary 13 *Let Ω be a pseudoconvex Reinhardt domain in $\mathbf{C}^n$. Then, $C = \mathcal{K}$ if and only if $\partial\Omega$ contains no analytic disks.*

Conjecture 14 *Theorem 12 is valid for every (bounded) pseudoconvex domain Ω with "regular" boundary.*

Remark 15 (a) Theorem 12 can be proved using the groupoid approach of [7] (see [22] and [20]; see also [15], [9] and [16] for similar applications of groupoids). When the group of (transverse) symmetries that acts on Ω is not abelian (such as when Ω is a generalized Reinhardt domain in the sense of [14]), more sophisticated tools are needed to classify $T(\Omega)$ (or $T(\partial\Omega)$). We survey briefly some aspects of this development in the rest of the paper.

(b) Let K be a real compact Lie group, and let L be a closed subgroup, so that the quotient space $S = K/L$ has the structure of a compact Riemannian symmetric space (see [12, page 174]). This means that K has an involutive automorphism θ whose fixed point subgroup K_θ satisfies $K_\theta^\circ \subseteq L \subseteq K_\theta$, where K_θ° is the connected component of K_θ containing the identity e of K. Then the Lie algebras $\mathbf{l}$ and $\mathbf{k}$ of L and K, respectively, give rise to the Cartan decomposition

$$\mathbf{k} = \mathbf{l} \oplus i\mathbf{m}, \tag{1}$$

via θ. Here, $i\mathbf{m} \sim T_\theta(S)$ is identified with the tangent space of S at the coset $[e]$. Via the standard complexification, we obtain the complex Lie groups $L^{\mathbf{C}}$ and $K^{\mathbf{C}}$, and we let $S^{\mathbf{C}} = K^{\mathbf{C}}/L^{\mathbf{C}}$. It follows that $S^{\mathbf{C}}$ has a natural complex structure, and we have: $\mathbf{k}^{\mathbf{C}} = \mathbf{l}^{\mathbf{C}} \oplus \mathbf{m}^{\mathbf{C}}$, with $\mathbf{m} = \mathbf{m}^{\mathbf{C}} \cap i\mathbf{k}$.

(c) With the above notation, let $\mathbf{r}$ be a maximal abelian subspace of $\mathbf{m}$, and let $\mathbf{r}_\leq \subseteq \mathbf{r}$ be a closed Weyl subchamber for the action of the Weyl group on $\mathbf{r}$. It follows that $S^{\mathbf{C}}$ has a polar decomposition $S^{\mathbf{C}} = K \exp(\mathbf{r}_\leq)[e]$. Thus, every point

$z \in S^{\mathbf{C}}$ can be written as $z = k \exp(a)[e]$, with $k \in K$ and $a \in \mathbf{r}_{\leq}$. Moreover, a is uniquely determined by z. Formally, we write $a = \log|z|$.

Definition 16 *Following [14], an open (connected) domain $\Omega \subseteq S^{\mathbf{C}}$ is called a K-circular (or K-Reinhardt) domain if it is K-invariant, i.e., $z \in \Omega \implies kz \in \Omega, \forall k \in K$.*

Remark 17 (a) For any K-circular domain $\Omega \subseteq S^{\mathbf{C}}$, there exists a polar decomposition $\Omega = K \exp(\Lambda)$, where $\Lambda = \log(|\Omega|) \subseteq \mathbf{r}_{\leq}$ is called the logarithmic base of Ω. It follows that Ω is a pseudoconvex domain if and only if it is logarithmically convex, i.e., $\log(|\Omega|)$ is convex. Furthermore, if Ω is logarithmically conical, i.e., if $\log(|\Omega|)$ is a sharp convex cone, then $S = \widetilde{\partial}\Omega$ is the Sylov's boundary of $A(\Omega)$ (the uniform subalgebra of $C(S)$ consisting of those functions that can be extended holomorphically to Ω).

(b) Of course, Reinhardt domains satisfy the above definition. Just take $K = \mathbf{T}^n$, and $L = \{(1, \ldots, 1)\}$. We now specialize the above definition to the case when $K = \mathcal{U}_n \times \mathcal{U}_n$ where $\mathcal{U}_n$ is the unitary group of the algebra $\mathcal{M}_n$ of $n \times n$ complex matrices. In this case, we let $L = \{(u, v): u = v\}$, so that $S = K/L \simeq \mathcal{U}_n$. Thus, $S^{\mathbf{C}} = \mathrm{Gl}_n(\mathbf{C})$ is the group of invertible matrices in $\mathcal{M}_n$. The action of K on $S^{\mathbf{C}}$ is given by $(U, V)A = UAV^*$. Also, $\mathbf{k}$ is the Lie algebra of all pairs of hermitian matrices in $\mathcal{H}_n = \{A \in \mathbf{M}_n: A = A^*\}$, $\mathbf{l} = \{(A, A): A \in \mathcal{H}_n\}$, and $\mathbf{m} = \{(A, -A): A \in \mathcal{H}_n\}$. Thus, we may identify $\mathbf{m}$ with $\mathcal{H}_n$. A natural maximal abelian subspace of $\mathcal{H}_n$ is the set $\mathcal{D}\mathcal{H}_n$ of diagonal matrices. We let $\mathbf{r} = \mathcal{D}\mathcal{H}_n$, and we let $\mathbf{r}_<$ be the set of diagonal matrices with decreasing non-zero diagonal elements. Then, $\mathbf{r}_{\leq}$ is an appropriate Weyl chamber. It follows that a K-circular domain is a matricial Reinhardt domain in the sense of [2].

(c) The "matricial" Hartogs's Wedge is perhaps the simplest logarithmically conical K-circular domain, where $S = K/L = \mathcal{U}_n$, as in part (b) above. It can be defined as follows: Let η be an $n-1$-tuple of real numbers greater than 1. Then we let

$$\Omega_\eta = \{A \in \mathcal{M}_n: \sigma_j^{\eta_j}(A) \leq \sigma_{j+1}(A) < 1, \qquad 1 \leq j < n\},$$

where $\sigma_j(A)$ denotes the j-th singular number of the matrix A. Notice that the "polar decomposition" of the matrix A coincides with the singular value decomposition of A.

Definition 18 *Let Ω be a logarithmically conical K-circular domain in $S^{\mathbf{C}}$, where $S = K/L$, as above. Let $L^2(S)$ be the Hilbert space of square integrable functions with respect to the natural measure on S induced by the Haar measure of K. Further, let $H^2(\widetilde{\partial}\Omega)$ be the subspace of $L^2(S)$ consisting of functions which have a holomorphic extension to Ω. Given a function $\phi \in A(\Omega)$ (i.e., holomorphic on Ω and continuous on S), we let T_ϕ the operator multiplication by ϕ on $H^2(\widetilde{\partial}\Omega)$. As in Definition 1, we define $\mathcal{T}(\partial\Omega)$ to be the (Hardy) Toeplitz C^*-algebra which is generated by $\{T_\phi: \phi \in A(\Omega)\}$.*

The following result will appear in [23].

Theorem 19 *Let Ω_η be the matricial Hartogs's wedge defined above. Then, the conclusion of Theorem 12 is valid when $T(\Omega)$ is replaced by $T(\partial\Omega_\eta)$. In particular, $T(\widetilde{\partial}\Omega_\eta)$ is type I if and only if all components of η are rational.*

Conjecture 14 was also verified for logarithmically conical K-Reinhardt domains which are (bounded) symmetric domains in [24] using similar techniques to those needed to prove 19.

References

[1] S. Axler, J. Conway, and G. McDonald, *Toeplitz Operators on Bergman Spaces*, Can. J. of Math., 34 (1982), 466-483.

[2] E. Bedford, and J. Dadok, *Generalized Reinhardt Domains*, J. Geom. Anal., 1 (1991), 1-17.

[3] H. Boas,*Holomorphic Reproducing Kernels in Reinhardt Domains*, Pacific J. Math., 112 (1982), 273-292.

[4] L. G. Brawn, R. G. Douglas, and P. A. Fillmore, *Extensions of C^*-Algebras and K-homology*, Ann. of Math., 105 (1977), 265-324.

[5] D. Catlin, *Global Regularity of the $\overline{\partial}$-Neumann Problem*, Proc. Sympos. Pure Math., 41 (1984), A.M.S. Providence, Rhode Island.

[6] L. Coburn, *Singular Integral Operators and Toeplitz Operators on Odd Spheres*, Indiana Univ. Math. J., 43 (1973), 433-439.

[7] R. Curto, and P. Muhly, *C^*-Algebras of Multiplication Operators on Bergman Spaces*, J. Func. Anal. 61 (1985), 315-329.

[8] R. Curto, and N. Salinas, *Generalized Bergman Kernels and the Cowen-Douglas Theory*, Amer. J. Math., 106 (1984), 447-488.

[9] R. Curto, and K. Yan, *The Taylor Spectrum of Infinite Direct Sums*, Contemp. Math., 120 (1991), 19-27.

[10] R. G. Douglas, *Banach Algebra Techniques in Operator Theory*, Academic Press, New York, (1972).

[11] R. G. Douglas, and R. Howe, *On the C^*-Algebra of Toeplitz Operators on the Quarter Plane*, Trans. Amer. Math. Soc., 158 (1971), 203-217.

[12] S. Helgason *Differential Geometry and Symmetric Spaces*, Academic Press, New York, (1962).

[13] S. Krantz, *Function Theory of Several Complex Variables*, John Wily and Sons, New York, (1992).

[14] M. Lassalle, *L'epace de Hardy d'un Domaine de Reinhardt Generalise*, J. Funct. Anal., 60 (1985), 309-340.

[15] P. Muhly P, and J. Renault, *C^*-Algebras of Multivariable Weiener-Hopf Operators*, Trans. of Amer. Math. Soc. 274 (1982), 1-44.

[16] A. Nica, *Some Remarks on the Groupoid Approach to Wiener-Hopf Operators*, J. of Operator Theory, 18 (1987), no. 1, 163-198.

[17] M. Range, *Holomorphic Functions and Integral Representations in Several Complex Variables*, Springer Verlag, New York, (1986).

[18] N. Salinas, *Smooth Extensions and Smooth Joint Quasitriangularity*, Operator Theory: Advances and Applications, 11 (1983), 303-332.

[19] N. Salinas, *The $\overline{\partial}$-Formalism and the C^*-Algebra of the Bergman n-Tuple*, J. of Operator Theory, 22 (1989), 325-343.

[20] N. Salinas, *Toeplitz Operators and Weighted Wiener-Hopf Operators, Pseudoconvex Reinhardt and Tube Domains*, Trans. of Amer. Math. Soc., 336 (1993), 675-699.

[21] N. Salinas, *Toeplitz C^*-Algebras on Pseudoconvex Domains with Transverse Symmetries*, Operator Algebras, Mathematical Physics and Low Dimensional Topology, edited by R. Herman and B. Tanbay, Research Notes in Math., A. K. Peters, 1993, 271-286.

[22] N. Salinas, A. Sheu, and H. Upmeier, *Toeplitz Operators on Pseudoconvex Domains, and Foliation C^*-Algebras*, Ann. of Math., 130 (1989), 531-565.

[23] N. Salinas and H. Upmeier, *Holomorphic Foliations and Toeplitz C^*-Algebras*, in preparation.

[24] H. Upmeier, *Toeplitz C^*-Algebras and Non-commutative Duality*, 26 no. 2 (1991), 407-432.

Contemporary Mathematics
Volume **185**, 1995

POSITIVITY CONDITIONS AND STANDARD MODELS
FOR COMMUTING MULTIOPERATORS

F.-H. VASILESCU

19 January 1994

ABSTRACT. Positivity plays an important role in the Hilbert space operator theory. The simplest positivity condition is, perhaps, that characterizing a Hilbert space contraction C, i.e., $1 - C^*C \geq 0$, where 1 denotes the identity of the space. The aim of this work is to present natural extensions of this positivity condition, valid for one or several (commuting) operators, and to derive some of their consequences. More precisely, we exhibit examples of operators that satisfy such conditions, which are called standard models. Then we partially describe the structure of arbitrary families of operators satisfying a given positivity condition, showing that such a family is essentially the restriction of the standard model to an invariant subspace, modulo an additional term of a related type. We obtain statements which extend and unify the corresponding ones from [Vas 2], [MuVa], [CuVa 2], using similar techniques.

1. POSITIVITY CONDITIONS

Let $n \geq 1$, $m \geq 1$ be fixed integers. Let also $p = (p_1, \ldots, p_m)$ be a family of complex polynomials

$$p_j(z) := 1 - c_{j1}z_1 - \cdots - c_{jn}z_n (j = 1, \ldots, m) , \qquad (1.1)$$

where $z = (z_1, \ldots, z_n) \in \mathbb{C}^n$, with the following properties:

   i. $c_{jk} \geq 0$ for all indices j, k;
  ii. for every $k \in \{1, \ldots, n\}$ there is a $j \in \{1, \ldots, m\}$ such that $c_{jk} \neq 0$;
 iii. $p_j \not\equiv 1$ for all indices j.

If $\mathbb{Z}_+^m$ is the set of all m-tuples of nonnegative integers (i.e., multi-indices of length m), then for every $\gamma = (\gamma_1, \ldots, \gamma_m) \in \mathbb{Z}_+^m$ we set

$$p^\gamma(z) := p_1(z)^{\gamma_1} \cdots p_m(z)^{\gamma_m} (z \in \mathbb{C}^n) . \qquad (1.2)$$

Let $\mathcal{H}$ be a complex Hilbert space and let $\mathcal{L}(\mathcal{H})$ be the algebra of all bounded linear operators on $\mathcal{H}$. An n-tuple $T = (T_1, \ldots, T_n) \in \mathcal{L}(\mathcal{H})^n$ will be designated as a *multioperator*. If $T_1, \ldots, T_n$ mutually commute, then T is said to be a *commuting multioperator* (briefly, a c.m.).

Let $T \in \mathcal{L}(\mathcal{H})^n$ be a c.m. We define the operators M_{T_k} on $\mathcal{L}(\mathcal{H})$ by the formula

$$M_{T_k}(X) := T_k^* X T_k (X \in \mathcal{L}(\mathcal{H}), k = 1, \ldots, n) . \qquad (1.3)$$

1991 *Mathematics Subject Classification.* 47A13, 47A45, 47B37.

We also set $M_T := (M_{T_1}, \dots, M_{T_n})$, which is, in fact, a c.m. on $\mathcal{L}(\mathcal{H})$.

Let p be a family of polynomials as in (1.1). We define

$$R_{T,j} := \sum_{k=1}^{n} c_{jk} M_{T_k} \qquad (j = 1, \dots, m) . \tag{1.4}$$

Then $R_T := (R_{T,1}, \dots, R_{T,m})$ is also a c.m. on $\mathcal{L}(\mathcal{H})$.

For a given $\gamma \in \mathbb{Z}_+^m$ we set

$$\Delta_T^{p,\gamma} := p^\gamma(M_T)(1) , \tag{1.5}$$

where $1 = 1_\mathcal{H}$ is the identity on $\mathcal{H}$. Note that

$$\Delta_T^{p,\gamma} = (I - R_{T,1})^{\gamma_1} \cdots (I - R_{T,m})^{\gamma_m}(1) ,$$

where I is the identity on $\mathcal{L}(\mathcal{H})$.

Definition 1.1.   The c.m. $T \in \mathcal{L}(\mathcal{H})^n$ satisfies the *positivity condition* (p, γ) (briefly, the p.c. (p, γ)) if

$$\Delta_T^{p,\beta} \geq 0 \quad (0 \leq \beta \leq \gamma) . \tag{1.6}$$

The main purpose of the present paper is to derive as much information as possible about the structure of those c.m. that satisfy the p.c. (p, γ).

If $T \in \mathcal{L}(\mathcal{H})^n$ is a c.m., then $T^* := (T_1^*, \dots, T_n^*)$, and $T^\alpha := T_1^{\alpha_1} \cdots T_n^{\alpha_n}$ for each $\alpha \in \mathbb{Z}_+^n$.

Lemma 1.2.   *Let $\mathcal{G}$ be another Hilbert space, let $T \in \mathcal{L}(\mathcal{H})^n$, $R \in \mathcal{L}(\mathcal{G})^n$ be c.m., and let $V : \mathcal{H} \to \mathcal{G}$ be an isometry such that $VT_j = R_j V (j = 1, \dots, n)$. Then we have*

$$\Delta_T^{p,\gamma} = V^* \Delta_R^{p,\gamma} V \quad (\gamma \in \mathbb{Z}_+^m) . \tag{1.7}$$

In particular, if R satisfies the p.c. (p, γ), then T has the same property.

Proof.   Note that

$$M_T^\alpha(1) = T^{*\alpha} T^\alpha = T^{*\alpha} V^* V T^\alpha = V^* R^{*\alpha} R^\alpha V = V^* M_R^\alpha(1) V$$

for all $\alpha \in \mathbb{Z}_+^n$.

If we write

$$p^\gamma(z) = \sum_{\alpha \in \mathbb{Z}_+^n} d_{p,\gamma}(\alpha) z^\alpha , \tag{1.8}$$

where only a finite number of coefficients $d_{p,\gamma}(\alpha)$ are nonnull, then we have:

$$\Delta_T^{p,\gamma} = \sum_{\alpha \in \mathbb{Z}_+^n} d_{p,\gamma}(\alpha) M_T^\alpha(1)$$

$$= V^* (\sum_{\alpha \in \mathbb{Z}_+^n} d_{p,\gamma}(\alpha) M_R^\alpha(1)) V = V^* \Delta_T^{p,\gamma} V ,$$

i.e., (1.7) holds.

Remark 1.3.   If the c.m. T satisfies the p.c. (p, γ), then any unitary transformation of T and any restriction of T to an invariant subspace also satisfy the p.c. (p, γ), as a consequence of Lemma 1.2.

From now on we denote by $e^{(m)}$, $e_k^{(m)}$ the multi-indices $(1, \dots, 1)$, $(0, \dots, 0, \overset{k}{1}, 0, \dots, 0)$ $(k = 1, \dots, m)$ from $\mathbb{Z}_+^m$, respectively. When no confusion is possible, we designate these symbols by e, e_k, respectively.

Definition 1.4.   A c.m. $T \in \mathcal{L}(\mathcal{H})^n$ is said to be *p-contractive* (briefly, T is a p-c.c.m.) if T satisfies the p.c. (p, e).

Note that if T is a p-c.c.m., then we have, in particular, $R_{T,j}(1) \leq 1$ $(j = 1, \dots, m)$.

The next result is inspired by [MüVa, Lemma 2] (see also [CuVa 2]).

Proposition 1.5.   *Let $T \in \mathcal{L}(\mathcal{H})^n$ be a c.m., let $\gamma \in \mathbb{Z}^m$, and let $\gamma^s \in \mathbb{Z}_+^m$ be given by $\gamma_j^s = 1$ if $\gamma_j \geq 1$, $\gamma_j^s = 0$ if $\gamma_j = 0$ $(j = 1, \dots, m)$. Suppose $\Delta_T^{p, \beta} \geq 0$ for all $\beta \leq \gamma^s$, and $\Delta_T^{p, \gamma} \geq 0$. Then $\Delta_T^{p, \beta} \geq 0$ for all $\beta \leq \gamma$, $\beta \geq \gamma^s$.*

Proof.   Assume first $\gamma^s = e$. We prove the assertion by induction with respect to $|\gamma| := \gamma_1 + \cdots + \gamma_m$.

If $|\gamma| = m$, then $\gamma = e$, and the property is true by the hypothesis. Suppose the property holds for all $\gamma \geq e$ with $|\gamma| = q$ for some $q \geq m$, and let us prove it for a $\gamma' \geq e$ with $|\gamma'| = q + 1$. We may assume with no loss of generality that $\gamma' = \gamma + e_1$, where $\gamma \geq e$. Fix a vector $h \in \mathcal{H}$ and set

$$a_k := \langle (R_{T,1})^k (\Delta_T^{p, \gamma}) h, h \rangle \qquad (k = 0, 1, 2, \dots) .$$

We have

$$a_k - a_{k+1} = \langle (R_{R,1})^k (\Delta_T^{p, \gamma'}) h, h \rangle \geq 0 , \qquad (1.9)$$

since $\Delta_T^{p, \gamma'} \geq 0$ by the induction hypothesis, and the operators $(R_{T,1})^k$ are positive on $\mathcal{L}(\mathcal{H})$. Note also that

$$\left| \sum_{k=0}^r a_k \right| = \left| \left\langle \sum_{k=0}^r (R_{T,1})^k (\Delta_T^{p, \gamma}) h, h \right\rangle \right|$$

$$= \left| \left\langle \Delta_T^{p, \beta} h, h \right\rangle - \left\langle (R_{T,1})^{r+1} (\Delta_T^{p, \beta}) h, h \right\rangle \right|$$

$$\leq \| \Delta_T^{p, \beta} \| [1 + \| (R_{T,1})^{r+1} \|] \| h \|^2 , \qquad (1.10)$$

for each integer $r \geq 0$, where $\beta = \gamma - e_1$.

We prove now the estimate

$$\| R_{T,j} \| \leq 1 \qquad (j = 1, \dots, m) . \qquad (1.11)$$

Indeed, we have

$$R_{T,j}(1) = \sum_{k=1}^{n} c_{jk}T_k^*T_k \leq 1 \qquad (j = 1, \ldots, m) , \qquad (1.12)$$

by the hypothesis. Then for all $h_1, h_2 \in \mathcal{H}$ and $X \in \mathcal{L}(\mathcal{H})$ we can write

$$\left|\langle R_{T,j}(X)h_1, h_2\rangle\right| \leq \|X\| \sum_{k=1}^{n} c_{jk}\|T_kh_1\|\|T_kh_2\|$$

$$\leq \|X\| \left(\sum_{k=1}^{n} c_{jk}\|T_kh_1\|^2\right)\left(\sum_{k=1}^{n} c_{jk}\|T_kh_2\|^2\right) \leq \|X\|\|h_1\|^2\|h_2\|^2,$$

via (1.12), whence we infer (1.11).

Using (1.11), we can assert that the left hand side of (1.10) is uniformly bounded for $r \geq 0$. This property and (1.9) show that a_k must be nonnegative for all $k \geq 0$. In particular $a_0 \geq 0$, and so $\Delta_T^{p,\gamma} \geq 0$.

¿From the previous argument we derive easily that the property holds when T is a p-c.c.m. and $\gamma \geq e$. The condition $\gamma \geq e$ is not essential. Indeed, by diminishing m if necessary, we apply the same argument to the family $\{R_{T,j}; \gamma_j \geq 1\}$.

Remark 1.6.   1. If we consider the polynomial

$$p_1(z) := 1 - z_1 - \cdots - z_n \ (z \in \mathbb{C}^n) ,$$

we obtain the case studied in [MüVa], related to the geometry of the unit ball in $\mathbb{C}^n$. With this choice for $p = (p_1)$, a c.m. T is a p-c.c.m. if and only if

$$1 - T_1^*T_1 - \cdots - T_n^*T_n \geq 0 .$$

2. If we consider the polynomials

$$p_j(z) := 1 - z_j(z \in \mathbb{C}^n; j = 1, \ldots, n) ,$$

we obtain the case studied in [CuVa2], related to the geometry of the unit polydisc in $\mathbb{C}^n$. With this choice for $p = (p_1, \ldots, p_n)$, a c.m. T is a p-c.c.m. if and only if

$$\sum_{\alpha \leq \gamma}(-1)^{|\alpha|}T^{*\alpha}T^\alpha \geq 0 \qquad (\gamma \in \mathbb{Z}_+^n, \gamma \leq e) .$$

3.   The previous proof shows that (1.12) implies (1.11). The converse is also true. Indeed, if $\|R_{T,j}\| \leq 1$ for some j, then $\|R_{T,j}(1)\| \leq 1$, and so $R_{T,j}(1) \leq 1$, since $R_{T,j}(1)$ is positive.

2. STANDARD MODELS

Let $p = (p_1, \ldots, p_m)$ be a fixed family of complex polynomials as in (1.1). For each $\gamma \in \mathbb{Z}_+^m$ we define the coefficients $d_{p,\gamma}(\alpha)$ via formula (1.8). For z in a neighborhood of the origin of $\mathbb{C}^n$ we write the expansion

$$[p^\gamma(z)]^{-1} = \sum_{\alpha \in \mathbb{Z}_+^n} b_{p,\gamma}(\alpha) z^\alpha . \tag{2.1}$$

Lemma 2.1.   *We have the following:*

i. *The coefficients $b_{p,\gamma}(\alpha)$ are nonnegative for all $\alpha \in \mathbb{Z}_+^n$. If $\gamma \geq e^{(m)}$, then $b_{p,\gamma}(\alpha) \neq 0$ for every α.*

ii. *For each $\beta \in \mathbb{Z}_+^m$, $\beta \leq \gamma$, one has*

$$\sum_{\delta \leq \alpha} d_{p,\beta}(\delta) b_{p,\gamma}(\alpha - \delta) \geq 0 .$$

iii. *The equality*

$$\sum_{\delta \leq \alpha} d_{p,\gamma}(\delta) b_{p,\gamma}(\alpha - \delta) = 1 \quad if \quad \alpha = 0 ,$$
$$= 0 \quad if \quad \alpha \neq 0 ,$$

also holds.

Proof.   i. If $w = (w_1, \ldots, w_m) \in \mathbb{C}^m$ satisfies $|w_1| < 1, \ldots, |w_m| < 1$, then we have the series expansion

$$(1 - w_1)^{-\gamma_1} \cdots (1 - w_m)^{-\gamma_m} = \sum_{\beta \in \mathbb{Z}_+^m} \rho_\gamma(\beta) w^\beta \tag{2.2}$$

for every $\gamma \in \mathbb{Z}_+^m$, $\gamma \geq e$, where $\rho_\gamma(\beta) = (\gamma + \beta - e)! [\beta!(\gamma - e)!]^{-1}$ (with the usual notation for multi-indices: $w^\beta = w_1^{\beta_1} \cdots w_m^{\beta_m}$, $\beta! = \beta_1! \cdots \beta_m!$ etc.). If $z = (z_1, \ldots, z_n) \in \mathbb{C}^n$ satisfies the conditions

$$\left| \sum_{k=1}^n c_{jk} z_k \right| < 1 \quad (j = 1, \ldots, m) ,$$

then we can perform in (2.2) the substitution

$$w_j = \sum_{k=1}^n c_{jk} z_k \quad (j = 1, \ldots, m) ,$$

and obtain the representation (2.1). Since all coefficients c_{jk} are nonnegative, we clearly have $b_{p,\gamma}(\alpha) \geq 0$ for all α.

The above argument has used the fact that $\gamma \geq e$. That $b_{p,\gamma}(\alpha) \geq 0$ for all α follows even if this condition is not fulfilled. Indeed, if $\gamma \neq 0$ is arbitrary, by diminishing m if necessary, we apply the previous argument to the product of those expressions $p_j(z)^{-\gamma_j}$ with $\gamma_j \geq 1$. The case $\gamma = 0$ is trivial.

We have to show that $b_{p,\gamma}(\alpha) \neq 0$ for all $\alpha \in \mathbb{Z}_+^n$ if $\gamma \geq e$. To see this, fix an $\alpha \in \mathbb{Z}_+^n$, put $A := \{1, \ldots, n\}$, and set

$$A_1 := \{k \in A; c_{1k} \neq 0\} ,$$
$$A_2 := \{k \in A \backslash A_1 : c_{2k} \neq 0\} ,$$
$$\ldots\ldots\ldots\ldots\ldots$$
$$A_{m-1} := \{k \in A \backslash (A_1 \cup \cdots \cup A_{m-2}); c_{m-1,k} \neq 0\}$$
$$A_m := A \backslash (A_1 \cup \cdots \cup A_{m-1}) ,$$

Note that if $k \in A_m$, then $c_{mk} \neq 0$.
We define the integers

$$\beta_j := \sum_{k \in A_j} \alpha_k \quad (j = 1, \ldots, m) ,$$

where $\beta_j := 0$ if $A_j = \emptyset$. Then the polynomial

$$\prod_{j=1}^{m} \left(\sum_{k=1}^{n} c_{jk} z_k \right)^{\beta_j}$$

contains a monomial of the form

$$(c_{j_1 1} z_1)^{\alpha_1} \cdots (c_{j_n n} z_n)^{\alpha_n}$$

with $c_{j_k k} \neq 0$ for all $k \in A$. It follows from the first part of the proof that

$$b_{p,\gamma}(\alpha) \geq c(\alpha)(c_{j_1 1})^{\alpha_1} \cdots (c_{j_n n})^{\alpha_n} \neq 0 ,$$

for some constant $c(\alpha) > 0$. Hence (i) is established.

ii. If $\beta \leq \gamma$, we have for z in a neighborhood of the origin of $\mathbb{C}^n$ that

$$p^\beta(z)[p^\gamma(z)]^{-1} = [p^{\gamma-\beta}(z)]^{-1}$$
$$= \sum_{\delta \in \mathbb{Z}_+^n} d_{p,\beta}(\delta) z^\delta \sum_{\alpha \in \mathbb{Z}_+^n} b_{p,\gamma}(\alpha) z^\alpha$$
$$= \sum_{\alpha \in \mathbb{Z}_+^n} \left[\sum_{\delta \leq \alpha} d_{p,\beta}(\delta) b_{p,\gamma}(\alpha - \delta) \right] z^\alpha = \sum_{\alpha \in \mathbb{Z}_+^n} b_{p,\gamma-\beta}(\alpha) z^\alpha ,$$

in virtue of (1.8) and (2.1). Thus

$$\sum_{\delta \leq \alpha} d_{p,\beta}(\delta) b_{p,\gamma}(\alpha - \delta) = b_{p,\gamma-\beta}(\alpha) \geq 0 \qquad (2.3)$$

for all $\alpha \in \mathbb{Z}_+^n$ by (i), which establishes (ii).

iii. If $\beta = \gamma$, as we clearly have $b_{p,0}(0) = 1$ and $b_{p,0}(\alpha) = 0$ if $\alpha \neq 0$, the assertion follows from (2.3), and so (iii) is also established.

Let $\mathcal{H}$ be a fixed Hilbert space. We denote by $\mathcal{K}$ the Hilbert space $\ell^2(\mathbb{Z}_+^n, \mathcal{H})$, i.e., the space of those functions $f : \mathbb{Z}_+^n \to \mathcal{H}$ such that

$$\|f\|^2 := \sum_{\alpha \in \mathbb{Z}_+^n} \|f(\alpha)\|^2 . \tag{2.4}$$

Let $\mathcal{K}_0 \subset \mathcal{K}$ be the subspace of those functions that have finite support. Clearly, $\mathcal{K}_0$ is dense in $\mathcal{K}$.

Let $\gamma \in \mathbb{Z}_+^m$, $\gamma \geq e$ be fixed. We define the operators $S_k^{(p,\gamma)}$ on $\mathcal{K}_0$ by the formula

$$(S_k^{(p,\gamma)} f)(\alpha) := \left[\frac{b_{p,\gamma}(\alpha)}{b_{p,\gamma}(\alpha + e_k)} \right]^{\frac{1}{2}} f(\alpha + e_k) \quad (f \in \mathcal{K}_0) , \tag{2.5}$$

for all $\alpha \in \mathbb{Z}_+^n$ and $k = 1, \ldots, n$. The definition is correct, since $b_{p,\gamma}(\alpha) \neq 0$ for all α, by Lemma 2.1. We shall show that each $S_k^{(p,\gamma)}$ extends to a bounded operator on $\mathcal{K}$, also denoted by $S_k^{(p,\gamma)}$ and that $S^{(p,\gamma)} := (S_1^{(p,\gamma)}, \ldots, S_n^{(p,\gamma)})$ is a c.m. that satisfies the p.c. (p,γ). When no confusion is possible, we denote $S_k^{(p,\gamma)}$, $S^{(p,\gamma)}$ simply by S_k, S, respectively.

Next, we consider the operators

$$(S_k^* f)(\alpha) : = \left[\frac{b_{p,\gamma}(\alpha - e_k)}{b_{p,\gamma}(\alpha)} \right]^{\frac{1}{2}} f(\alpha - e_k) \quad \text{if} \quad \alpha_k \geq 1 ,$$
$$: = 0 \qquad\qquad \text{if} \quad \alpha_k = 0 , \tag{2.6}$$

for all $f \in \mathbb{Z}_+^n$ and $k = 1, \ldots, n$. Note that S_k^* is the "formal adjoint" of S_k, i.e.,

$$\left\langle S_k^* f, g \right\rangle = \left\langle f, S_k g \right\rangle \quad (f, g \in \mathcal{K}_0) , \tag{2.7}$$

where the inner product on $\mathcal{K}$ corresponds to the norm (2.4). After proving that S_k extends to a bounded operator on $\mathcal{K}$, it will follow from (2.7) that S_k^* has a unique extension to $\mathcal{K}$, which is precisely the adjoint of the extension of S_k.

An inductive argument shows that

$$(S^\delta f)(\alpha) = \left[\frac{b_{p,\gamma}(\alpha)}{b_{p,\gamma}(\alpha + \delta)} \right]^{\frac{1}{2}} f(\alpha + \delta) \tag{2.8}$$

for all $\alpha, \delta \in \mathbb{Z}_+^n$, $f \in \mathcal{K}_0$, and that

$$(S^{*\delta} f)(\alpha) = \left[\frac{b_{p,\gamma}(\alpha - \delta)}{b_{p,\gamma}(\alpha)} \right]^{\frac{1}{2}} f(\alpha - \delta) \quad \text{if} \quad \delta \leq \alpha ,$$
$$= 0 \qquad\qquad \text{otherwise} , \tag{2.9}$$

for all α, $\delta \in \mathbb{Z}_+^n$, $f \in \mathcal{K}_0$. Therefore

$$(S^{*\delta}S^\delta f)(\alpha) = \frac{b_{p,\gamma}(\alpha - \delta)}{b_{p,\gamma}(\alpha)} f(\alpha) \qquad \text{if } \delta \leq \alpha \,,$$
$$= 0 \qquad\qquad\qquad \text{otherwise} \,, \qquad (2.10)$$

for all α, $\delta \in \mathbb{Z}_+^n$, $f \in \mathcal{K}_0$.

Using (2.10) we can give a sense to the right hand side of (1.5) for $S :=$ $(S_1, \dots, S_n)$, by setting

$$(p^\beta(M_S)(1)f)(\alpha) := \sum_{\delta \in \mathbb{Z}_+^n} d_{p,\beta}(\delta)(S^{*\delta}S^\delta f)(\alpha)$$
$$= [b_{p,\gamma}(\alpha)]^{-1} \sum_{\delta \leq \alpha} d_{p,\beta}(\delta) b_{p,\gamma}(\alpha - \delta) f(\alpha) \qquad (2.11)$$

for all $\alpha \in \mathbb{Z}_+^n$, $\beta \in \mathbb{Z}_+^m$, $f \in \mathcal{K}_0$.

We can state now a result proving the existence of a special c.m. that satisfies a given positivity condition.

Theorem 2.2. *Let $p = (p_1, \dots, p_m)$ be a family of polynomials as in (1.1), and let $\gamma \in \mathbb{Z}_+^m$, $\gamma \geq e$ be given. Then there exists a c.m. $S = S^{(p,\gamma)} \in \mathcal{L}(\mathcal{K})^n$ satisfying the p.c. (p, γ). Moreover,*

$$\lim_{k \to \infty} (R_{S,j})^k(1)f = 0 \quad (f \in \mathcal{K}, j = 1, \dots, m) \,, \qquad (2.12)$$

where $R_{S,j}$ is given by (1.4).

Proof. As previously mentioned, we show that the operators $S_1, \dots, S_n$ given by (2.5) extend to $\mathcal{K}$, and they will provide the desired c.m.

¿From (2.11) we obtain

$$\langle p^\beta(M_S)(1), f, f \rangle$$
$$= \sum_{\alpha \in \mathbb{Z}_+^n} [b_{p,\gamma}(\alpha)]^{-1} \sum_{\delta \leq \alpha} d_{p,\beta}(\delta) b_{p,\gamma}(\alpha - \delta) \|f(\alpha)\|^2 \geq 0 \qquad (2.13)$$

for all $\beta \in \mathbb{Z}_+^m$, $\beta \leq \gamma$, $f \in \mathcal{K}_0$, in virtue of Lemma 2.1.

In particular,

$$\langle p_j(M_S)(1), f, f \rangle = \langle (1 - \sum_{k=1}^n c_{jk} S_k^* S_k)f, f \rangle \geq 0$$

for all $j = 1, \dots, m$ and $f \in \mathcal{K}_0$, whence

$$\|S_k f\|^2 \leq (c_{j_k k})^{-1} \|f\|^2 \qquad (2.14)$$

for all $k = 1, \ldots, n$ and $f \in \mathcal{K}_0$, with $c_{j_k k} \neq 0$ (whose existence is insured by the hypothesis). Estimate (2.14) clearly implies that each S_k has a bounded extension to the space $\mathcal{K}$, also denoted by S_k. It is easily seen that $S_1, \ldots, S_n$ mutually commute. Hence $S := (S_1, \ldots, S_n) \in \mathcal{L}(\mathcal{K})^n$ is a c.m. Moreover, from (2.13) we deduce that $\Delta_S^{p,\beta} \geq 0$ for all $\beta \leq \gamma$, i.e., S satisfies the p.c. (p, γ).

To prove (2.12) take first an $f \in \mathcal{K}_0$ and notice that

$$[(R_{S,j})^k(1)f](\alpha)$$

$$= \sum_{\substack{\delta \in \mathbb{Z}_+^n \\ |\delta|=k}} \left(\frac{k!}{\delta!} c_j^\delta S^{*\delta} S^\delta f \right)(\alpha) \right)$$

$$= [b_{p,\gamma}(\alpha)]^{-1} \sum_{\substack{\delta \leq \alpha \\ |\delta|=k}} \frac{k!}{\delta!} c_j^\delta b_{p,\gamma}(\alpha - \delta) f(\alpha) \qquad (2.15)$$

for all $\alpha \in \mathbb{Z}_+^n$, $j = 1, \ldots, m$, and $k \geq 1$ an integer, via (2.10), where $c_j = (c_{j1}, \ldots, c_{jn})$. Since f has finite support, there exists an $\alpha^0 \in \mathbb{Z}_+^n$ such that $f(\alpha) = 0$ if α is not $\leq \alpha^0$. Then it follows from (2.10), (2.15) that $[(R_{S,j}^k(1)f](\alpha) = 0$ for all α if $k > |\alpha^0|$. Hence (2.12) holds for $f \in \mathcal{K}_0$. Since $\mathcal{K}_0$ is dense in $\mathcal{K}$, and $\|R_{S,j}\| \leq 1$ for all j by (1.11), we derive easily that (2.12) holds for all $f \in \mathcal{K}$ and $j = 1, \ldots, m$. This completes the proof of the theorem.

Definition 2.3. Then c.m. $S = S^{(p,\gamma)} \in \mathcal{L}(\mathcal{K})^n$ given by Theorem 2.2 will be called the *backwards multishift* of type (p, γ).

Definition 2.3 contains, in particular the concept of backwards multishift of type (n, m) defined in [MüVa] (obtained for $p_1(z) = \cdots = p_m(z) = 1 - z_1 - \cdots - z_n$), as well as that of backwards multishift of type (n, γ) defined in [CuVa2] (obtained for $p_j(z) = 1 - z_j$, $j = 1, \ldots, n$, and $\gamma \in \mathbb{Z}_+^n$, $\gamma \geq e$).

Remark 2.4. If $S = S^{(p,\gamma)}$ and if $\beta \in \mathbb{Z}_+^m$ is not $\leq \gamma$, then $\Delta_S^{p,\beta}$ is not positive. Indeed, the polynomial $p^\delta(z)$ has some negative coefficients for every $\delta \in \mathbb{Z}_+^m$, $\delta \neq 0$. For this reason the function $p^\beta(z)[p^\gamma(z)]^{-1}$ also has some negative coefficients. Then the computation (2.13) shows that $p^\beta(M_S)(1)$ cannot be positive.

3. STRUCTURE THEOREMS

In this section we show that a c.m. $T \in \mathcal{L}(\mathcal{H})^n$ satisfying the p.c. (p, γ) is essentially equal to the restriction of the backwards multishift of type (p, γ) plus a term of a related form to an invariant subspace. This is the reason for which the concept introduced by Definition 2.3 can be also designated as a *standard model* (for the p.c. (p, γ)).

Let $p = (p_1, \ldots, p_m)$ be a fixed family of complex polynomials as in (1.1). The family p will be associated with the domain

$$\Omega_p := \{z \in \mathbb{C}^n; \sum_{k=1}^n c_{jk} |z_k|^2 < 1, j = 1, \ldots, m\} . \qquad (3.1)$$

For every c.m. T (acting in a Hilbert or Banach space) we denote by $\sigma(T)$ its (Taylor) joint spectrum.

Lemma 3.1. *Let $T \in \mathcal{L}(\mathcal{H})^n$ be a p-c.c.m. Then $\sigma(T) \subset \bar{\Omega}_p$.*

Proof. Since T is a p-c.c.m., the estimate (1.11) holds. In particular, $\sigma(R_{T,j}) \subset \bar{\mathbb{D}}$, where $\mathbb{D} \subset \mathbb{C}$ is the open unit disc ($j = 1, \ldots, m$). ¿From the spectral mapping theorem for the joint spectrum (see, for instance, [Vas 1]) we derive

$$\sigma(R_{T,j}) = (\sigma(\sum_{k=1}^{n} c_{jk} M_{T_k})$$

$$= \{\sum_{k=1}^{n} c_{jk} u_k; u = (u_1, \ldots, u_n) \in \sigma(M_T)\}$$

$$= \{\sum_{k=1}^{n} c_{jk} z_k \bar{w}_k; z, w \in \sigma(T)\} \subset \bar{\mathbb{D}}, \tag{3.2}$$

where we use the equality

$$\sigma(M_T) = \{(z_1 \bar{w}_1, \ldots, z_n \bar{w}_n); z, w \in \sigma(T)\}, \tag{3.3}$$

which follows from [CuFi].

Therefore, if $z \in \sigma(T)$, it follows from (3.2) that

$$\sum_{k=1}^{n} c_{jk} |z_k|^2 \leq 1 \quad (j = 1, \ldots, m), \tag{3.4}$$

and so $z \in \bar{\Omega}_p$.

The next result extends [CuVa2, Prop. 4.6].

Proposition 3.2. *Let $T \in \mathcal{L}(\mathcal{H})^n$ be a c.m. such that $p^\gamma(M_T)(1) \geq 0$ for some $\gamma \in \mathbb{Z}_+^m$. If $\sigma(T) \subset \Omega_p$, then $p^\beta(M_T)(1) \geq 0$ for all $\beta \leq \gamma$.*

Proof. We assume first that $\gamma_1 \geq 1$ and $\beta := \gamma - e_1$. Since $\sigma(T) \subset \Omega_p$, it follows from (3.3) and (3.2) that we actually have $\sigma(R_{T,j}) \subset \mathbb{D}$ for all $j = 1, \ldots, m$. In particular, $I - R_{T,1}$ is invertible and $(I - R_{T,1})^{-1}$ has a power series expansion. Hence

$$p^\beta(M_T)(1) = (I - R_{T,1})^{-1} p^\gamma(M_T)(1) \geq 0,$$

since the powers of $R_{T,1}$ preserve positivity.

Using this idea, an inductive argument proves easily the general assertion.

Lemma 3.3.   *Let $T \in \mathcal{L}(\mathcal{H})^n$ be a c.m. such that $R_{T,1}, \ldots, R_{T,m}$ are contractions. Let also $A \in \mathcal{L}(\mathcal{H})$ be positive, and let $\gamma \in \mathbb{Z}_+^m$ have the property $p^\beta(M_T)(A) \geq 0$ for all $\beta \leq \gamma$. Then for every $r \in (0,1)$ the operator $p^\gamma(rM_T)$ is invertible, and*

$$0 \leq [p^\gamma(rM_T)]^{-1} p^\gamma(M_T)(A) \leq A .$$

Proof.   We follow the lines of [CuVa2, Lemma 3.6]. Note first that $\sigma(rR_{T,j}) \subset r\bar{\mathbb{D}}$ for all j, by the hypothesis. Hence

$$p^\gamma(rM_T) = (I - rR_{T,1})^{\gamma_1} \cdots (I - rR_{T,m})^{\gamma_m}$$

is invertible. Since all monomials $R_T^\eta (\eta \in \mathbb{Z}_+^m)$ are positive on $\mathcal{L}(\mathcal{H})$, using the power series expansion of $[p^\gamma(rM_T)]^{-1}$ and that $p^\gamma(M_T)(A) \geq 0$, we obtain

$$[p^\gamma(rM_T)]^{-1} p^\gamma(M_T)(A) \geq 0 .$$

To prove the remaining estimate, we apply an inductive argument on $|\gamma|$. If $|\gamma| = 0$, the assertion is obvious. Suppose that the property holds for all γ's with $|\gamma| = q$, where $q \geq 0$ is an integer, and let us prove it for some γ' with $|\gamma'| = q + 1$. We write $\gamma' = \gamma + e_j$, where $|\gamma| = q$. Note that

$$\begin{aligned}
&[p^{\gamma'}(rM_T)]^{-1} p^{\gamma'}(M_T)(A) \\
&= [p^\gamma(rM_T)]^{-1} p^\gamma(M_T)(A) \\
&\quad + (r-1)R_{T,j}[p_j(rM)]^{-1}[p^\gamma(rM_T)]^{-1} \; p^\gamma(M_T)(A) .
\end{aligned} \quad (3.5)$$

Since

$$0 \leq [p^\gamma(rM_T)]^{-1} \; p^\gamma(M_T)(A) \leq A$$

by the first part of the proof and the induction hypothesis, and because $R_{T,j}[p_j(rM_T)]^{-1}$ preserves positivity, from (3.5) we derive the desired conclusion.

Corollary 3.4.   *Let $T \in \mathcal{L}(\mathcal{H})^n$ be a c.m. satisfying the p.c. (p, γ) for some $\gamma \geq e$. Then we have*

$$\sum_{\alpha \in \mathbb{Z}_+^n} b_{p,\gamma}(\alpha) \|(\Delta_T^{p,\gamma})^{\frac{1}{2}} T^\alpha h\|^2 \leq \|h\|^2 \quad (h \in \mathcal{H}) . \quad (3.6)$$

Proof.   In virtue of (2.1), (2.2), and Lemma 3.2, we can write

$$\begin{aligned}
[p^\gamma(rM_T)]^{-1}(\Delta_T^{p,\gamma}) &= \sum_{\beta \in \mathbb{Z}_+^m} \rho_\gamma(\beta) r^{|\beta|} R_T^\beta(\Delta_T^{p,\gamma}) \\
&= \sum_{\alpha \in \mathbb{Z}_+^n} b_{p,\gamma}(\alpha) r^{|\alpha|} T^{*\alpha} \Delta_T^{p,\gamma} T^\alpha \leq 1
\end{aligned}$$

for all $r \in (0,1)$. Letting $r \to 1$, we obtain

$$\sum_{\beta \in \mathbb{Z}_+^m} \rho_\gamma(\beta) R_T^\beta(\Delta_T^{p,\gamma}) = \sum_{\alpha \in \mathbb{Z}_+^n} b_{p,\gamma}(\alpha) T^{*\alpha} \Delta_T^{p,\gamma} T^\alpha \leq 1 , \quad (3.7)$$

where the series are convergent in the strong operator topology. From (3.7) we obtain, in particular, (3.6).

Lemma 3.5. *Let* $(a_\delta)_{\delta \in \mathbb{Z}_+^m}$ *be a non-increasing sequence of nonnegative numbers, and let* $\alpha \in \mathbb{Z}_+^m$ *be fixed. Assume that*

$$\sum_{\delta \in \mathbb{Z}_+^m} \delta^\alpha a_\delta < \infty \ .$$

Then

$$\lim_\delta \delta^{\alpha + e} a_\delta = 0 \ .$$

This is a multivariable version of [Mül, Lemma 3.8], whose proof can be found in [CuVa2]. We omit the details.

We shall denote in the following the limit in the strong operator topology of $\mathcal{L}(\mathcal{H})$ by s-lim.

If $\delta, \epsilon \in \mathbb{Z}_+^m$, we set $\delta \circ \epsilon = (\delta_1 \epsilon_1, \ldots, \delta_m \epsilon_m)$.

Lemma 3.6. *Let* $T \in \mathcal{L}(\mathcal{H})^n$ *be a c.m. satisfying the p.c.* (p, γ) *for some* $\gamma \geq e$. *Then*

$$s - \lim_\delta R_T^{\epsilon \circ \delta}(1) \tag{3.8}$$

exists for all $\epsilon \in \mathbb{Z}_+^m$, $\epsilon \leq e$. *In addition,*

$$s - \lim_\delta \delta^{\beta \circ \eta} R_T^{\delta \circ \eta}(\Delta_T^{p, \beta \circ \eta}) = 0 \tag{3.9}$$

for all $\beta \leq \gamma - e$, $\beta \neq 0$, *where* $\eta_j = 1$ *if* $1 \leq \beta_j \leq \gamma_j - 1$, *and* $\beta_j = 0$ *otherwise.*

Proof. We adapt the argument used in [CuVa2, Lemma 3.8]. We only sketch the main steps.

It is easily seen that for $\delta' \leq \delta''$ one has $R_T^{\delta'}(1) \geq R_T^{\delta''}(1)$. Therefore, for a fixed $\epsilon \leq e$, $(R_T^{\epsilon \circ \delta})_\delta$ is a non-increasing sequence of positive operators in $\mathcal{L}(\mathcal{H})$, and so the limit (3.8) exists.

To obtain (3.9), assuming for the moment $\beta \geq e$, we have:

$$\sum_{\delta \in \mathbb{Z}_+^m} \rho_\beta(\delta) \left\langle R_T^\delta(\Delta_T^{p, \beta}) h, h \right\rangle \leq \|h\|^2 \quad (h \in \mathcal{H}) \ , \tag{3.10}$$

as a consequence of (3.7). If we fix an $h \in \mathcal{H}$, since $\delta^{\beta - e} = 0(\rho_\beta(\delta))$, we also have

$$\sum_{\delta \in \mathbb{Z}_+^m} \delta^{\beta - e} \left\langle R_T^\delta(\Delta_T^{p, \beta}) h, h \right\rangle < \infty \ ,$$

via (3.10). Notice that

$$\left\langle R_T^{\delta'}(\Delta_T^{p, \beta}) h, h \right\rangle - \left\langle R_T^{\delta''}(\Delta_T^{p, \beta}) h, h \right\rangle \geq 0$$

whenever $\delta' \leq \delta''$, since $\beta \leq \gamma - e$. In virtue of Lemma 3.4, we obtain

$$\lim_{\delta} \delta^{\beta} \langle R_T^{\delta}(\Delta_T^{p,\beta})h, h \rangle = 0 ,$$

i.e., (3.9) for $\beta \geq e$.

If β is not $\geq e$, by diminishing m if necessary, we apply the previous argument to those variables corresponding to $1 \leq \beta_j \leq \gamma_j - 1 (j = 1, \ldots, m)$. This finishes the proof of the lemma.

Let $T \in \mathcal{L}(\mathcal{H})^n$ be a c.m. satisfying the p.c. (p, γ) for some $\gamma \geq e$. If $\mathcal{K} = \ell^2(\mathbb{Z}_+^n, \mathcal{H})$, we define an operator $V_0 : \mathcal{H} \to \mathcal{K}$ by the relation

$$(V_0 h)(\alpha) := [b_{p,\gamma}(\alpha)\Delta_T^{p,\gamma}]^{\frac{1}{2}} T^{\alpha} h (h \in \mathcal{H}, \alpha \in \mathbb{Z}_+^n) . \tag{3.11}$$

Since

$$\|V_0 h\|^2 = \sum_{\alpha \in \mathbb{Z}_+^n} b_{p,\gamma}(\alpha) \|(\Delta_T^{p,\gamma})^{\frac{1}{2}} T^{\alpha} h\|^2 \quad (h \in \mathcal{H}) , \tag{3.12}$$

it follows from (3.6) that V_0 is well-defined and contractive.

We also have

$$\|V_0 h\|^2 = \sum_{\beta \in \mathbb{Z}_+^m} \rho_{\gamma}(\beta) \langle R_T^{\beta}(\Delta_T^{p,\gamma})h, h \rangle \quad (h \in \mathcal{H}) , \tag{3.13}$$

which follows from (3.7).

Notice that

$$(V_0 T_k h)(\alpha) = [b_{p,\gamma}(\alpha)\Delta_T^{p,\gamma}]^{\frac{1}{2}} T^{\alpha + e_k} h$$

$$= \left[\frac{b_{p,\gamma}(\alpha)}{b_{p,\gamma}(\alpha + e_k)} \right]^{\frac{1}{2}} [b_{p,\gamma}(\alpha + e_k)\Delta_T^{p,\gamma}]^{\frac{1}{2}} T^{\alpha + e_k} h = S_k V_0 h ,$$

for all $h \in \mathcal{H}$, $\alpha \in \mathbb{Z}_+^n$ and $k = 1, \ldots, n$, where S_k is given by Theorem 2.2 (see also (2.5)). Therefore

$$V_0 T_k = S_k^{(p,\gamma)} V_0 \quad (k = 1, \ldots, n) . \tag{3.14}$$

Lemma 3.7. *Let T and V_0 be as above. Then*

$$\|V_0 h\|^2 = \lim_{\delta} \left\langle \prod_{j=1}^{m} [I - (R_{T,j})^{\delta_j}](1)h, h \right\rangle \quad (h \in \mathcal{H}) . \tag{3.15}$$

Proof. The proof is similar to that of [CuVa2, Lemma 3.9].

Note first the identity

$$\sum_{k=0}^{q} \binom{s + k - 1}{s - 1} a^k (1 - a)^s = 1 - a^{q+1} \sum_{k=0}^{s-1} \binom{q + k}{k} (1 - a)^k$$

valid for all integers $q \geq 0$, $s \geq 1$, and an element a of a commutative algebra with unit. Using this identity and (3.13) we have:

$$\|V_0 h\|^2 = \lim_\delta \sum_{\beta \leq \delta} \rho_\gamma(\beta) \langle R_T^\beta(\Delta_T^{p,\gamma})h, h \rangle$$

$$= \lim_\delta \left\langle \prod_{j=1}^m [I - \sum_{\beta_j=0}^{\gamma_j-1} \binom{\delta_j + \beta_j}{\beta_j} R_{T,j}^{\delta_j+1}(I - R_{T,j})^{\beta_j}](1)h, h \right\rangle . \tag{3.16}$$

¿From (3.16) we obtain (3.15), via Lemma 3.6.

Lemma 3.8. *Let T and V_0 be as in Lemma 3.7. Then we have*

$$p^\delta(M_T)(1 - V_0^* V_0) \geq 0 \quad if \ \delta \leq \gamma ,$$
$$= 0 \quad if \ \delta \geq e^{(m)} .$$

Proof. We use some idea from [CuVa2, Lemmas 3.10, 3.13].

Assume first $\delta \leq \gamma$. Since the case $\delta = 0$ is trivial, we also suppose $\delta \neq 0$. Note that

$$p^\delta(M_T)(1 - V_0^* V_0) = p^\delta(M_T)(1) - p^\delta(M_T) \lim_{r \to 1^-} [p^\gamma(rM_T)]^{-1}(\Delta_T^{p,\gamma}) \tag{3.17}$$

by (3.13) (see also the proof of Corollary 3.4).

We shall use the equality

$$\lim_{r \to 1^-} (I - rC)^{-k}(I - C)^{k+1} = I - C ,$$

valid for every Banach space contraction C and each integer $k \geq 1$, whose proof is elementary. Since $R_{T,j}$ is a contraction for every j by (1.11), we have:

$$p^\delta(M_T) \lim_{r \to 1^-} [p^\gamma(rM_T)]^{-1}(\Delta_T^{p,\gamma}) = p^\delta(M_T) \lim_{r \to 1^-} [p^\epsilon(rM_T)]^{-1}(\Delta_T^{p,\epsilon}) , \tag{3.18}$$

where $\epsilon_j = \gamma_j$ if $\delta_j = 0$, and $\epsilon_j = 0$ if $\delta_j \neq 0 (j = 1, \ldots, m)$.

Note also that

$$p^\delta(M_T)[p^\epsilon(rM_T)]^{-1}(\Delta_T^{p,\epsilon}) = [p^\epsilon(rM_T)]^{-1}p^\epsilon(M_T)p^\delta(M_T)(1) \leq p^\delta(M_T)(1) ,$$

in virtue of Lemma 3.3. This remark, combined with (3.17), (3.18), proves our assertion for $\delta \leq \gamma$.

For the remaining part it suffices to assume $\delta = e$. Note that

$$p^e(M_T) \prod_{k=1}^m [I - (R_{T,k})^{\beta_k}] = \prod_{j=1}^m [I - R_{T,j} - R_{T,j}^{\beta_j}(I - R_{T,j})] \quad (\beta \in \mathbb{Z}_+^m) . \tag{3.19}$$

Since

$$s - \lim_{\beta_j \to \infty} [R_{T,j})^{\beta_j} - (R_{T,j})^{\beta_j+1}](1) = 0 \quad (j = 1, \dots, m) \ ,$$

as a consequence of (3.8), we can write

$$p^e(M_T)(V_0^* V_0) = p^e(M_T)\left\{ s - \lim_{\beta} \prod_{j=1}^{m} [I - (R_{T,j})^{\beta_j}](1) \right\}$$

$$= \prod_{j=1}^{m}(I - R_{T,j})(1) = p^e(M_T)(1) \ ,$$

via (3.15) and (3.19), which completes the proof of the lemma.

Definition 3.9. Let $T \in \mathcal{L}(\mathcal{H})^n$ be a p-c.c.m. We say that T is a *p-isometry* if $\Delta_T^{p,e} = 0$. If $\Delta_T^{p,e_j} = 0$ for all $j = 1, \dots, m$, we say that T is a *strong p-isometry*.

Note that a spherical isometry [Ath2] is a strong p-isometry, with p as in Remark 1.6.1. Note also that if T is a polydisc isometry [CuVa2] and T is a p-c.c.m. with p as in Remark 1.6.2 then T is a p-isometry.

We recall that a c.m. T is said to be normal if each T_j is normal, and T is said to be subnormal if it is the restriction of a normal c.m. to an invariant subspace.

Proposition 3.10. *Let $T \in \mathcal{L}(\mathcal{H})^n$ be a strong p-isometry. Then T is subnormal, and the minimal normal extension of T is also a strong p-isometry.*

Proof. We firstly prove a special version of this theorem.

Let $p := (p_1, \dots, p_m)$ be given by

$$p_j(z) := 1 - \sum_{k \in A_j} z_k (A_j \subset \{1, \dots, n\} \ , \quad j = 1, \dots, m) \ , \tag{3.20}$$

where for every $k \in \{1, \dots, n\}$ one can find $j \in \{1, \dots, m\}$ such that $k \in A_j$ (in other words, p satisfies the usual requirements from (1.1)).

With p as in (3.20), let $T \in \mathcal{L}(\mathcal{H})^n$ be a strong p-isometry. In this case we shall show that T is subnormal, and the minimal normal extension of T is also a strong p-isometry. We apply some results from [AtPe].

We show first that T is subnormal. Notice that

$$p_j(M_T)(1) = 0 \quad (j = 1, \dots, m) \ , \tag{3.21}$$

which follows from the hypothesis.

Now, let

$$p_{0,k}(z) := 1 - z_k \quad (k = 1, \dots, n) \ , \tag{3.22}$$

and set $p_0 := (p_{0,1}, \dots, p_{0,n})$. To prove that T is subnormal, it suffices to show (with our notation) that $p_0^\alpha(M_T)(1) \geq 0$ for all $\alpha \in \mathbb{Z}_+^n$. The subnormality of T then follows by [Ath1], Theorem 4.1 (or [AtPe, Prop. 0]). Indeed, we have

$$p_{0,k}(z) = p_{j_k}(z) + \sum_{\substack{q \in A_{j_k} \\ q \neq k}} z_q \ , \tag{3.23}$$

with $A_{j_k} \ni k$. Then, via (3.23),

$$p_0^\alpha(M_T)(1) = \sum_{k=1}^n [p_{j_k}(M_T) + \sum_{\substack{q \in A_{j_k} \\ q \neq k}} M_{T_q}]^{\alpha_k}(1) \geq 0 ,$$

in virtue of (3.21) and the fact that the nomomials $M_T^\alpha (\alpha \in \mathbb{Z}_+^n)$ are positive on $\mathcal{L}(\mathcal{H})$. Hence T is subnormal.

In virtue of [AtPe, Prop. 8], the minimal normal extension N of T also satisfies the equations $\Delta_N^{p,e_j} = 0$ $(j = 1, \dots, m)$, and so N is a strong p-isometry.

We can approach now the general case, i.e., the family p satisfies (1.1) rather than (3.20). With T a strong p-isometry, we define

$$V_{jk} := (c_{jk})^{1/2} T_k \quad (j = 1, \dots, m; k = 1, \dots, n) .$$

Then $V := (V_{jk})_{j,k} \in \mathcal{L}(\mathcal{H})^{mn}$ is a c.m. that satisfies the conditions required by the first part of the proof. Let $N_0 = (N_{jk})_{j,k}$ be the minimal normal extension of V, given by the previous argument. If we set $N_k := (c_{j_k k})^{-1/2} N_{j_k k}$, with $c_{j_k k} \neq 0$, then $N := (N_1, \dots, N_n)$ is the minimal normal extension of T, which is also a strong p-isometry.

Lemma 3.11. *Let T, V_0 be as in Lemma 3.7. Then there exists a Hilbert space $\mathcal{M}$, a p-isometry W on $\mathcal{M}$ and an operator $V_1 : \mathcal{H} \to \mathcal{M}$ such that $V_1 T_k = W_k V_1$ $(k = 1, \dots, n)$. In addition, W satisfies the p.c. (p, γ), and*

$$\|V_1 h\|^2 = \langle (1 - V_0^* V_0)h, h \rangle \quad (h \in \mathcal{H}) . \tag{3.24}$$

Proof. Similar results are [MüVa, Lemma 10] and [CuVa2, Lemma 3.14].

We define on $\mathcal{H}$ the seminorm

$$\|h\|_0^2 := \langle (1 - V_0^* V_0)h, h \rangle \quad (h \in \mathcal{H}) . \tag{3.25}$$

Let $\mathcal{N} := \{h \in \mathcal{H}; \|h\|_0 = 0\}$, and let $\mathcal{M}$ be the Hilbert space completion of $\mathcal{H}/\mathcal{N}$. Let also $V_1 : \mathcal{H} \to \mathcal{M}$ be given by $V_1 h := h + \mathcal{N}$. Then $\|V_1 h\| = \|h\|_0$, i.e., (3.24) holds.

We define now $W_k : \mathcal{H}/\mathcal{N} \to \mathcal{M}$ by the formula

$$W_k(h + \mathcal{N}) := T_k h + \mathcal{N} \quad (h \in \mathcal{H} , \ k = 1, \dots, n) . \tag{3.26}$$

Let us check the correctness of (3.26). For a fixed k we have

$$\|T_k h + \mathcal{N}\|^2 = \langle M_{T_k}(1 - V_0^* V_0)h, h \rangle$$
$$\leq (c_{j_k k})^{-1} \langle (1 - V_0^* V_0)h, h \rangle = (c_{j_k k})^{-1} \|h\|_0^2 ,$$

where $c_{j_k k} \neq 0$ (see also the proof of (2.14)), which follows from Lemma 3.8. Hence each W_k is correctly defined by (3.26), and it can be continuously extended to the space $\mathcal{M}$, by density. This extension will also be denoted by W_k. Set $W :=$ $(W_1, \ldots, W_n)$, which is easily seen to be a c.m. on $\mathcal{M}$. Notice that $W_k V_1 = V_1 T_k$ for all k, which follows from (3.26).

If $\beta \in \mathbb{Z}_+^m$ is fixed, from (3.26) we deduce that

$$\langle p^\beta(M_W)(1)(h + \mathcal{N}), h + \mathcal{N} \rangle = \langle p^\beta(M_T)(1)(1 - V_0^* V_0)h, h \rangle .$$

In particular, $p^\beta(M_W)(1) \geq 0$ for all $\beta \leq \gamma$, and $p^e(M_W)(1) = 0$, via Lemma 3.8. This shows that W is a p-isometry, and finishes the proof of the lemma.

We are now in the position to present a first description of the structure of a c.m. satisfying a given positivity condition.

Theorem 3.12.   *Let $T \in \mathcal{L}(\mathcal{H})^n$ be a c.m., let p be an m-tuple of complex polynomials satisfying (1.1), and let $\gamma \geq e^{(m)}$. The following two conditions are equivalent:*

(a) *T satisfies the p.c. (p, γ).*
(b) *T is unitarily equivalent to the restriction of $S^{(p,\gamma)} \oplus W$ to an invariant subspace, where W is a p-isometry that satisfies the p.c. (p, γ).*

If, moreover, W is a strong p-isometry, then T is unitarily equivalent to the restriction of $S^{(p,\gamma)} \oplus N$ to an invariant subspace, where N is a strong p- isometry consisting of normal operators.

Proof.   (a)$\Rightarrow$(b). Let V_0 be defined by (3.11), and let $\mathcal{M}$, W, and V_1 be as in Lemma 3.11. Then the operator $V : \mathcal{H} \to \mathcal{K} \oplus \mathcal{M}$ given by

$$Vh := V_0 h \oplus V_1 h \quad (h \in \mathcal{H}) , \tag{3.27}$$

is an isometry, via (3.24). Moreover,

$$VT_j h = (S_j^{(p,\gamma)} \oplus W_j)Vh \quad (h \in \mathcal{H}, \ j = 1, \ldots, n) ,$$

by (3.14) and Lemma 3.11. Therefore $V\mathcal{H}$ is invariant under $S^{(p,\gamma)} \oplus W$, and T is unitarily equivalent to $S^{(p,\gamma)} \oplus W|V\mathcal{H}$.

(b)$\Rightarrow$(a). Notice that

$$p^\beta(M_{S \oplus W}) = p^\beta(M_S) \oplus p^\beta(M_W) \quad (\beta \in \mathbb{Z}_+^m) , \tag{3.28}$$

where $S = S^{(p,\gamma)}$. Since both S and W satisfy the p.c. (p, γ) (in virtue of Theorem 2.2 and by the hypothesis, respectively) it follows from (3.28) that $S \oplus W$ also satisfies the p.c. (p, γ). Using Remark 1.3, we infer that T must satisfy the p.c. (p, γ).

The last assertion follows via Proposition 3.10.

Remark 3.13.   Theorem 3.12 extends and unifies the following results: [Vas, Theorem 4.9], [MüVa, Theorem 11], and [CuVa2, Theorem 3.16]. The starting point of all these results is the discussion from [SzFo, Section I.10.1]. For a different point of view, see also [Ag1], [Ath1], [Ath3].

The next result is a version of Theorem 3.12 that establishes conditions under which the additional term representing a p-isometry disappears. For similar results see also [Vas, Corollary 4.12], [MüVa, Theorem 9], and [CuVa2, Theorem 3.16].

Theorem 3.14.   *Let $T \in \mathcal{L}(\mathcal{H})^n$ be a c.m., let p be an m-tuple of complex polynomials satisfying (1.1), and let $\gamma \geq e^{(m)}$. The following conditions are equivalent:*

    (a) *T satisfies the p.c. (p, γ) and $s-\lim_{k\to\infty}(R_{T,j})^k(1) = 0$ for all $j = 1, \ldots, m$, where $R_{T,j}$ is given by (1.4).*

    (b) *T is unitarily equivalent to the restriction of $S^{(p,\gamma)}$ to an invariant subspace.*

Proof.   (a)$\Rightarrow$(b) It suffices to prove that the operator V_0, given by (3.11), is an isometry. Indeed, since $R_{T,j}$ are contractions by (1.11), it follows from (3.15) and the hypothesis that V_0 is in this case an isometry.

(b)$\Rightarrow$(a) If $S = S^{(p,\gamma)}$, it follows from Theorem 2.2 that $s - \lim_{k\to\infty}(R_{S,j})^k = 0$ $(j = 1, \ldots, m)$. Since this property is stable under both restrictions to invariant subspaces and unitary equivalence, the desired conclusion is obtained via Theorem 3.12.

A first consequence of Theorem 3.14 is the following *von Neumann type inequality*:

Proposition 3.15.   *Let $T \in \mathcal{L}(\mathcal{H})^n$ be a c.m. satisfying the p.c. (p, γ). Then for every function F holomorphic in a neighborhood of $\bar{\Omega}_p$ we have*

$$\|F(T)\| \leq \|F(S^{(p,\gamma)})\| \ . \tag{3.29}$$

Proof.   Both $F(T)$, $F(S^{(p,\gamma)})$ are defined by the holomorphic functional calculus, via Lemma 3.1.

Let $r \in (0, 1)$. Then the c.m. rT also satisfies the p.c. (p, γ). Indeed, as we have

$$p^\beta(r^2 z) = \prod_{j=1}^{m} [p_j(z) + (1 - r^2) \sum_{k=1}^{n} c_{jk} z_k]^{\beta_j} \ ,$$

we derive easily that $p^\beta(r^2 M_T)(1) \geq 0$ for all $\beta \leq \gamma$.

Note also that $R_{rT,j} = r^2 R_{T,j}$, and so $s - \lim_{k\to\infty}(R_{rT,j})^k = 0$ $(j = 1, \ldots, m)$. In virtue of Theorem 3.14, rT is unitarily equivalent to the restriction of $S^{(p,\gamma)}$ to an invariant subspace. Hence we must have

$$\|F(rT)\| \leq \|F(S^{(p,\gamma)})\| \ ,$$

whence, letting $r \to 1$, we derive (3.29).

Formula (3.29) implies, in particular, the classical von Neumann inequality for contractions, as well as some extensions of it, valid for one or several variables (see [Dru], [Vas2] etc.).

4. FINAL COMMENTS

The particular cases of Theorems 3.12 and 3.14, related to the geometry of the unit ball in $\mathbb{C}^n$ (see [Vas2], [MüVa], [CuVa1]) or to that of the unit polydisc in $\mathbb{C}^n$ (see [CuVa2]) have some useful applications. We mention especially the existence of the boundary limits for some operator-valued Poisson transforms in both cases, as well as the existence of the H^∞-functional calculus (see the works quoted above). It would be of interest to find versions of those results in the present framework.

Unfortunately, unlike in the case of the unit ball [Rud2], or that of the unit polydisc [Rud1], general results concerning the existence of the boundary values for complex functions do not seem to be available (although some results do exist, see for instance [BuMa], but they serve only partially our framework).

The structure of p-isometries is another problem of some interest, whose solution would complete the statement of Theorem 3.12.

We hope to present some (partial) answers to these questions in a forthcoming paper.

REFERENCES

[Ag1]   J. Agler, *Hypercontractions and subnormality*, J. Operator Theory **13** (1985), 203–217.

[Ath1] A. Athavale, *Holomorphic kernels and commuting operators*, Trans. Amer. Math. Soc. **304** (1987), 101–110.

[Ath2] A. Athavale, *On the intertwining of joint isometries*, J. Operator Theory **23** (1990), 339–350.

[Ath3] A. Athavale, *Model theory on the unit ball in $\mathbb{C}^m$*, J. Operator Theory (to appear).

[AtPe] A. Athavale and S. Pedersen, *Moment problems and subnormality*, J. Math. Anal. Appl. **146** (1990), 434–441.

[BuMa] J. Burbea and P. R. Masani, *Banach and Hilbert spaces of vector-valued functions*, Research Notes in Mathematics 90, Pitman Advanced Publishing Program, 1984.

[CuFi] R. Curto and L. Fialkow, *The spactral picture of (L_A, R_B)*, J. Funct. Anal. **71** (1987), 371–392.

[CuVa1] R. Curto and F.-H. Vasilescu, *Automorphism invariance of the operator-valued Poisson transform*, Acta Sci. Math. **57** (1993), 65–78.

[CuVa2] R. Curto and F.-H. Vasilescu, *Standard operator models in the polydisc*, Indiana Univ. Math. J. **42** (1993), 791–810.

[Dru]   S. W. Drury, *A generalization of von Neumann's inequality to complex ball*, Proc. Amer. Math. Soc. **68** (1978), 300–304.

[Mül]   V. Müller, *Models for operators using weighted shifts*, J. Operator Theory **20** (1988), 3–20.

[MüVa] V. Müller and F.-H. Vasilescu, *Standard models for some commuting multioperators*, Proc. Amer. Math. Soc. **117** (1993), 979–989.

[Rud1] W. Rudin, *Function theory in polydiscs*, Benjamin, 1969.

[Rud2] W. Rudin, *Function theory in the unit ball of $\mathbb{C}^n$*, Springer-Verlag, 1980.

[SzFo] B. Sz.-Nagy and C. Foias, *Analyse harmonique des opérateurs de l'espace de Hilbert*, Masson et Cie and Akadémiai Kiadó, 1967.

[Vas1] F.-H. Vasilescu, *Analytic functional calculus and spectral decompositions*, Reidel and Editura Academiei, 1982.

[Vas2] F.-H. Vasilescu, *An operator-valued Poisson kernel*, J. Funct. Anal. **110** (1992), 47–72.

INSTITUTE OF MATHEMATICS OF THE ROMANIAN ACADEMY, P.O. BOX 1-764, RO-70700, BUCHAREST, ROMANIA

Contemporary Mathematics
Volume **185**, 1995

TRACE FORMULAS AND COMPLETELY UNITARY INVARIANTS FOR SOME k-TUPLES OF COMMUTING OPERATORS

DAOXING XIA

ABSTRACT. Let $\mathbb{A} = (A_1, \ldots, A_k)$ be a k-tuple of commuting operators. Let A_1 be a pure subnormal operator with minimal normal extension N_1. Assume that $\mathrm{sp}(A_1)\backslash\mathrm{sp}(N_1)$ is a simply-connected domain with Jordan curve boundary satisfing certain smooth condition. Assume also that $[A_i^*, A_j] \in \mathcal{L}^1$, $i, j = 1, \ldots, k$. Then $\mathbb{A}$ is subnormal, and the set consisting of the $\mathrm{sp}(\mathbb{A})$ and the function $Q(\lambda), \lambda \in \sigma_p(A^*))^*$ is a complete unitary invariant for $\mathbb{A}$, where $Q(\lambda_1, \ldots, \lambda_k)$ is a parallel projection to $\ker(A_1^* - \overline{\lambda}_1 I) \cap \{f : (A_j^* - \overline{\lambda}_j I)^\ell f = 0, \text{ for some } \ell > 0, \ j = 2, \ldots, k\}$.

1. Introduction

Let $\mathcal{H}$ be a Hilbert space. Let $\mathcal{L}(\mathcal{H})$ be the algebra of all linear operators on $\mathcal{H}$ and let $\mathcal{L}^1(\mathcal{H})$ be the trace ideal of $\mathcal{L}(\mathcal{H})$. Carey and Pincus [4], [5], [12] established a theory on the principal current and local index of certain class of k-tuples of commuting operators $\mathbb{A} = (A_1, \ldots, A_k)$ satisfying conditions that (i) $[A_i^*, A_j] \in \mathcal{L}^1$ and (ii) the joint essential spectrum is a system of curves in $\mathbb{C}^n$. The one current ℓ is defined by setting

$$\ell(f dh) = i\mathrm{tr}\,[f(\mathbb{A}), h(\mathbb{A})]$$

for functions f, h in certain class. The ℓ is a real MC cycle in $\mathbb{C}^n$ which is the boundary in the sense of currents in $\mathbb{C}^n$ of a rectifiable current Γ, the so called principal current. With this Γ, they also defined the principal index for $\mathbb{A}$ at some points in the Taylor $\mathrm{sp}(\mathbb{A})$. Therefore it is worth to determine the explicit form of the current ℓ for some class of operator-tuples.

1991 *Mathematics Subject Classification*. Primary 47B20.

This work supported in part by a NSF grant no. DMS-9400766.

This paper is in final form and no version of it will submitted for publication elsewhere

A k-tuple $\mathbb{S} = (S_1, \dots, S_k)$ of operators on $\mathcal{H}$ is said to be subnormal ([7], [8], [13], [14], [15], [17]) if there is a k-tuple $\mathbb{N} = (N_1, \dots, N_k)$ of commuting normal operators on a Hilbert space $\mathcal{K} \supset \mathcal{H}$ such that

$$N_j \mathcal{H} \subset \mathcal{H} \qquad \text{and} \qquad S_j = N_j|_{\mathcal{H}}, j = 1, 2, \dots, k.$$

This $\mathbb{N}$ is said to be a normal extension of $\mathbb{S}$. If there is no proper subspace $\mathcal{H}_1 \subset \mathcal{K}$ satisfying conditions that $\mathcal{H}_1 \perp \mathcal{H}$ and $\mathcal{H}_1$ reduces N_j, $j = 1, 2, \dots, k$. Then $\mathbb{N}$ is said to be a m.n.e. (minimal normal extension) of $\mathbb{S}$.

THEOREM A. *(Pincus and Xia [13]). Let $\mathbb{S} = (S_1, \dots, S_k)$ be a subnormal k-tuple of operators with minimal normal extension $\mathbb{N}$. If rank $[S_i^*, S_i] < +\infty$ for $i = 1, 2, \dots, k$, then*

$$itr\, [f(\mathbb{S}), h(\mathbb{S})] = \frac{1}{2\pi} \int_L mfdh$$

where L is the union of a finite collection of closed curves, and is also the union of a finite collection of algebraic arcs such that $\mathrm{sp}(\mathbb{N})$ is the union of L and a finite set. Furthermore, $m(u)$ is an integer valued multiplicity function which is constant on the irreducible piece of L, and $f, h \in \mathcal{A}(\mathrm{sp}(\mathbb{S}))$.

Here, the algebra $\mathcal{A}(\sigma)$ is the algebra of functions on σ generated by all the analytic functions f and their conjugates $\overline{f}$ defined on a neighborhood of σ.

THEOREM B. *(Pincus and Zheng [14]). Under the condition of Theorem A,*

$$itr\, [f(\mathbb{A}), g(\mathbb{A})] = \frac{1}{2\pi} \sum_i \int_{C_j} m_j fdh$$

where $\{C_j\}$ are the collection of cycles in $\mathrm{sp}_{ess}(\mathbb{S})$ and the weights m_j are the spectral multiplicities of the m.n.e. $\mathbb{N}$ at any regular point ζ of C_j.

In Lemma 2 of this paper, we also get the form $\ell(fdh)$ in Theorem B for the case of a k-tuple of $\mathbb{A} = (A_1, \dots, A_k)$ of commuting operators satisfying the conditions that (i) $[A_i^*, A_j] \in \mathcal{L}^1$, $i, j = 1, 2, \dots, k$, and (ii) A_1 is a pure subnormal operator of which $\mathrm{sp}(N_1) \backslash \mathrm{sp}(A_1)$ is a simply connected domain with Jorden curve boundary satisfying certain smooth condition, where N_1 is the m.n.e. of A_1.

The main part of this paper is studying a complete unitary invariant for certain operator tuples.

Let $\mathcal{F}$ be a family of k-tuples $\mathbb{A} = (A_1, \dots, A_k)$ of operators. Let $CU(\mathbb{A})$ be a set of objects determined by $\mathbb{A} \in \mathcal{F}$. The object $CU(A)$ is said to be a complete unitary invariant for the k-tuple A in $\mathcal{F}$, if for $\mathbb{A} = (A_1, \dots, A_k)$ and $\mathbb{B} = (B_1, \dots, B_k)$ in $\mathcal{F}$, $CU(\mathbb{A}) = CU(\mathbb{B})$ is a necessary and sufficient condition for the existence of a unitary operator U satisfying $UA_jU^{-1} = B_j$, $j = 1, 2, \dots, k$.

Even in the case of single operator ($k = 1$), to find a simpler useful complete unitary invariant is one of the basic problems in the operator theory. It is classical that the family of measures associated with the spectral resolution is a complete unitary invariant for the normal operator (for example [11]), as well as the k-tuple of commuting normal operators. There are several directions in the theory of complete unitary invariants in the single operator case or k-tuple of commuting operators case, such as [3], [9], [10].

We are interested in finding some complete unitary invariants associated with some trace formulas or cyclic cohomology. It is well-known that the Pincus principal function is a complete unitary invariant for T, if T is pure and rank $[T^*, T] = 1$. For the pure subnormal operator with m.n.e. N, if $\mathrm{sp}(S)\backslash\mathrm{sp}(N)$ is a simply connected domain with boundary $\mathrm{sp}(N)$ satisfying certain smooth condition and $[S^*, S] \in \mathcal{L}^1$ (in [19], [20] and [21], we assumed that $[S^*, S]^{1/2} \in \mathcal{L}^1$, but this restriction may be removed) then the principal function (index $(S^* - \bar{z}I)$, $z \in \rho_{\mathrm{ess}}(S)$) is a complete unitary invariant for S. A theorem of Abrahamse and Douglas [1] may be also interpreted in this way with different conditions.

If $\mathrm{sp}(S)\backslash\mathrm{sp}(N)$ is not simply connected, then the principal function is no longer a complete unitary invariant. In [20] for some special case and then in [21] for the more general case, it proves that if S is a pure subnormal operator with m.n.e. N and satisfies conditions that (i) $\mathrm{sp}(S)\backslash\mathrm{sp}(N)$ is a m-connected domain with boundary $\mathrm{sp}(N)$ consisting of m Jordan curves satisfying certain smooth condition or an external Jordan curve satisfying certain smooth condition and $m - 1$ points and (ii) $[S^*, S] \in \mathcal{L}^1$, then the function (which is related to cyclic cocycle)

$$\mathrm{tr}\,[(S^* - \bar{\lambda}_0 I)^{-1},\ (S - \mu_0 I)^{-1}][(S^* - \bar{\lambda}_1 I)^{-1},\ (S - \mu_1 I)^{-1}]$$

$\lambda_i, \mu_j \in \rho(S)$ is a complete unitary invariant for S. In the Remark of §3, we will show that $\mathrm{tr}\,(Q(\lambda, S)Q(\mu, S))$, $\lambda, \mu \in \mathrm{sp}(S)\backslash\mathrm{sp}(N)$ is a complete unitary invariant for S, where $Q(\lambda, S)$ is a certain parallel projection to the eigenspace $\ker(S^* - \bar{\lambda}I)$.

Assume that $\mathbb{S} = (S_1, \ldots, S_k)$ is a pure subnormal tuple of operators with m.n.e. $\mathbb{N} = (N_1, \ldots, N_k)$ satisfying $[S_j^*, S_j] \in \mathcal{L}^1, j = 1, 2, \ldots, k$. In [22] it is proved that if $\mathrm{sp}(S_j)\backslash\mathrm{sp}(N_j)$ is a simply connected domain with Jordan curve boundary $\mathrm{sp}(N_j)$ satisfying certain smooth condition for $j = 1, 2, \ldots, k$, then

$$(1) \qquad \mathrm{tr}\,[f_0(\mathbb{A}), g_0(\mathbb{A})] = \frac{1}{2\pi}\int_{\mathrm{sp}(\mathbb{N})} m(u) f_0(u)\, dg_0(u)$$

and

$$\mathrm{tr}\,[f_0(\mathbb{A}), g_0(\mathbb{A})]\ldots[f_n(\mathbb{A}), g_n(\mathbb{A})] =$$

$$(2)$$

$$\frac{1}{(2\pi)^{n+1}}\int_{\mathrm{sp}(\mathbb{N})^{n+1}} m(u^{(0)}, \ldots, u^{(n)})\prod_{m=0}^{n} f_m(u^{(m)})\frac{g_m(u^{(m)}) - g_m(u^{(m-1)})}{u_1^{(m)} - u_1^{(m-1)}}\, du_1^{(m)}$$

where $u^{(j)} = (u_1^{(j)}, \ldots, u_k^{(j)}), u^{(-1)} = u^{(n)}$ for $n \geq 1$ and $f_j, g_j \in \mathcal{A}(\mathrm{sp}(\mathbb{S}))$. In [22], it is also proved that the sequence of functions $m(u^{(0)}, \ldots, u^{(n)})$ (which are related to cyclic n-cocycle), $n = 0, 1, \ldots$ is a complete unitary invariant for $\mathbb{S}$.

   In the present paper, we continue the study in [22]. We consider the class of k-tuples of commuting operators $\mathbb{A} = (A_1, \ldots, A_k)$ on $\mathcal{H}$ satisfying the conditions that (i) $[A_j^*, A_j] \in \mathcal{L}^1, j = 1, 2, \ldots, k$ and (ii) A_1 is a pure subnormal operator with m.n.e N_1 such that $\mathrm{sp}(N_1)$ is a Jordan curve satisfying certain smooth condition. In Theorem 1, it proves that the operator tuples $\mathbb{A}$ satisfying these conditions are subnormal, and the set consisting of the Taylor spectrum $\mathrm{sp}(A)$ and the sequence of n-points function

$$\text{(3)} \qquad \mathrm{tr}\,(Q(\lambda^{(1)}; \mathbb{A}) \ldots Q(\lambda^{(n)}, \mathbb{A})), \lambda^{(n)} \in \sigma_p(A^*)^*$$

(which is related to cyclic n-1-cocycle) is a complete unitary invariant for $\mathbb{A}$ where $Q(\lambda; \mathbb{A})$ is the parallel projection to the subspace $\ker(A_1^* - \bar{\lambda}_1 I) \cap \{f : (A_j^* - \bar{\lambda}_j I)^{\ell_j} f = 0, \ell_j > 0, j = 2, \cdots, k\}$. Or, the set consisting of the $\mathrm{sp}(\mathbb{A})$ and the function $Q(\cdot; \mathbb{A})$ is a complete unitary invariant.

   In §4 (cf. Corollary 1), we determine the form of the irreducible k-tuple of commuting operators $\mathbb{H} = (H_1, H_2, \cdots, H_k)$ satisfying the conditions that (i) $\mathrm{rank}[H_1^*, H_1] = 1$, (ii) $[H_j^*, H_j]$ is compact, $j = 2, \ldots, k$ and (iii) the pairs $(H_1, H_j), j = 2, \ldots, k$ are hyponormal. Under these conditions, either (i) H_1 is a linear combination of identity and unilateral shift with multiplicity one and $\mathbb{H}$ is subnormal, or (ii) H_1 is a non-subnormal hyponormal operator and there are $\alpha_j, \beta_j \in \mathbb{C}$ such that $H_j = \alpha_j H_1 + \beta_j I, j = 2, \cdots k$. This work is closely related to the Theorem 1 of [18].

2. A Lemma

   Let $H_{\mathcal{D}}^p(\mathbb{T})$, $p \geq 1$ be the Hardy space of $\mathcal{D}$-valued analytic functions on the open unit disk with boundary function on $\mathbb{T}$, where $\mathcal{D}$ is an auxiliary Hilbert space. Let $H^p(\mathbb{T}) = H_{\mathbb{C}}^p(\mathbb{T})$.

   LEMMA 1. *Let* $T(\cdot) \in H_{\mathcal{L}(\mathcal{D})}^\infty(\mathbb{T})$. *Let* T *be the operator*

$$(Tf)(\zeta) = T(\zeta) f(\zeta), \ f \in H_{\mathcal{D}}^2(\mathbb{T}).$$

If $[T^*, T] \in \mathcal{L}^1$, *then* $T(\zeta)$ *is a normal operator on* $\mathcal{D}$ *for almost every* $\zeta \in \mathbb{T}$.

   PROOF. There is an orthonormal sequence $\{h_j\}$ in $H_{\mathcal{D}}^2(T)$ and a sequence of real numbers $\{\lambda_j\}$, satisfying $\sum |\lambda_j| < \infty$, such that

$$\text{(4)} \qquad ([T^*, T] f)(z) = \sum \lambda_j (f, h_j) h_j(z)$$

for $|z| < 1$. In (4), let $f(z) = \alpha(z - \lambda)^{-1}$, $\alpha \in \mathcal{D}$, $|\lambda| > 1$, then (4) implies that

$$(F(z) - F(\lambda))\alpha - T(z)(H(z) - H(\lambda))\alpha =$$
$$\text{(5)} \qquad\qquad (\lambda - z)\lambda^{-1} \sum \lambda_j (\alpha, h_j(\bar{\lambda}^{-1}))_{\mathcal{D}} h_j(z)$$

for $|z| < 1$ where

$$F(z) = \frac{1}{2\pi i} \int \frac{T^*(\zeta) T(\zeta) d\zeta}{\zeta - z}, \quad H(z) = \frac{1}{2\pi i} \int \frac{T(\zeta)^* d\zeta}{\zeta - z},$$

since $T^*f = P(T^*(\cdot)f(\cdot))$ where P is the projection from L^2 to H^2.

For almost all $u \in \mathbb{T}$, in (5) let $z \to u$ and $\lambda \to u$. Then

$$([T^*(u), T(u)]\alpha, \beta) = 0$$

for $\alpha, \beta \in \mathcal{D}$ by Plemelj formula, since

$$\lim_{\substack{z \to u \\ \lambda \to u}} \sum \lambda_j(\alpha, h_j(\overline{\lambda}^{-1}))(h_j(z), \beta) = \sum \lambda_j(\alpha, h_j(u))(h_j(u), \beta)$$

is finite for almost all $u \in \mathbb{T}$.

3. Complete Unitary Invariant

For a k-tuple of operators $\mathbb{A} = (A_1, \ldots, A_k)$, let $\mathbb{A}^* = (A_1^*, \ldots, A_k^*)$ and $\sigma_p(\mathbb{A}) = \{(\lambda_1, \ldots, \lambda_k) : \text{there is } f \neq 0 \text{ such that } A_j f = \lambda_j f\}$. For any set $\sigma \subset \mathbb{C}^k$, let $\sigma^* = \{(\overline{\lambda}_1, \ldots, \overline{\lambda}_k) : (\lambda_1, \ldots, \lambda_k) \in \sigma\}$. Suppose $\mathbb{A} = (A_1, \ldots, A_k)$ is a k-tuple of commuting operators on $\mathcal{H}$, $\lambda = (\lambda_1, \ldots, \lambda_k) \in \sigma_p(\mathbb{A}^*)^*$ and $\dim \ker(\overline{\lambda}_1 I - A_1^*) < +\infty$. Let $P_{\mathcal{D}(\lambda_1)}$ be the orthonormal projection from $\mathcal{H}$ to $\mathcal{D}(\lambda_1) = \ker(\overline{\lambda}_1 I - A_1^*)$. Let

$$(6) \qquad Q(\lambda; \mathbb{A}) = \prod_{j=2}^k \frac{1}{2\pi i} \int_{|\zeta - \overline{\lambda}_j| = \delta} (\zeta I - A_j^*|_{\mathcal{D}(\lambda_1)})^{-1} d\zeta P_{\mathcal{D}(\lambda_1)},$$

where δ is a positive number such that there is no eigenvalue of A_j^* in $0 < |\zeta - \overline{\lambda}_j| \leq \delta$. Then $Q(\lambda; \mathbb{A})$ is a parallel projection from $\mathcal{H}$ onto

$$\bigcap_{j=2}^k \{f \in \ker(\overline{\lambda}_1 - A_1^*) : (\overline{\lambda}_j - A_j^*)^{\ell_j} f = 0 \text{ for some } \ell_j > 0\}.$$

A Jordan curve γ is said to satisfy the condition (CBI) if the univalent analytic mapping function $\phi(\cdot)$ from the interior domain of γ onto the open unit disk satisfying the condition that $\phi'(\cdot)$ and $\phi'(\cdot)^{-1}$ are bounded.

THEOREM 1. *Let $\mathbb{A} = (A_1, \ldots, A_k)$ be a k-tuple of commuting operators on $\mathcal{H}$. Assume that A_1 is a pure subnormal operator with minimal normal extension N_1 on $\mathcal{K} \supset \mathcal{H}$. Suppose $\mathrm{sp}(N_1)$ is a Jordan curve satisfying condition (CBI) and is the boundary of $\mathrm{sp}(A_1)$. Assume also that $[A_i^*, A_i] \in \mathcal{L}^1$, $i = 1, \ldots, k$. Then $\mathbb{A}$ is subnormal, and the set consisting of the $\mathrm{sp}(\mathbb{A})$ and the sequence of n-points function*

$$(7) \qquad \mathrm{tr}\,(Q(\lambda^{(1)}; \mathbb{A}) \ldots Q(\lambda^{(n)}; \mathbb{A})), \qquad \lambda^{(n)} \in \sigma_p(A^*)^*$$

is a complete unitary invariant for $\mathbb{A}$.

Or, the set consisting of the $\mathrm{sp}(\mathbb{A})$ and the function $Q(\cdot; \mathbb{A})$ is a complete unitary invariant.

PROOF. By the method in the proof of Theorem 3 in [19], where the condition $[A_1^*, A_1]^{1/2} \in \mathcal{L}^1$ can be changed to $[A_1^*, A_1] \in \mathcal{L}^1$, or by the Theorem 1 in [1], we may prove that there are a Hilbert space $\mathcal{D}$ and a unitary operator W from $\mathcal{K}$ onto $L_{\mathcal{D}}^2(\mathbb{T})$ such that $W\mathcal{H} = H_{\mathcal{D}}^2(\mathbb{T})$ and

$$(WN_1W^{-1}f)(z) = A_1(z)f(z), \; f \in L_{\mathcal{D}}^2(\mathbb{T})$$

where the function $A_1(z)$ is the conformal mapping from the open unit D disk onto $\mathrm{sp}(A_1)\backslash\mathrm{sp}(N_1)$. For the simplicity of notation, we assume that $\mathcal{K} = L_D^2(\mathbb{T})$, $\mathcal{H} = H_D^2(\mathbb{T})$ and $W = I$.

Let U_+ be the unilateral shift, $(U_+f)(z) = zf(z)$, $f \in H_D^2(\mathbb{T})$. Then $A_1 = A_1(U_+)$. It is easy to see that

$$\mathrm{index}\,(A_1^* - \overline{A_1(z)}I) = \mathrm{index}\,(U_+^* - \overline{z}I), \text{ for } |z| < 1$$

Since the mosaic (cf [16]) of A_1 is compact, we have

$$\mathrm{index}\,(A_1^* - \overline{A_1(z)}I) < +\infty, \text{ for } |z| < 1.$$

Therefore $\dim \mathcal{D} = \mathrm{index}\,(U_+^* - \overline{z}I) < +\infty$. There are functions $A_j(\cdot) \in H_{\mathcal{L}(\mathcal{D})}^\infty(\mathbb{T})$ such that

$$(8) \qquad\qquad (A_jf)(z) = A_j(z)f(z), \qquad j = 2,\dots,k.$$

By Lemma 1, $[A_j(z)^*, A_j(z)] = 0$ for a.e. $z \in \mathbb{T}$. Define

$$(N_jf)(z) = A_j(z)f(z), \; j = 2,\dots,k, \; f \in L_D^2(\mathbb{T}).$$

Then $\mathbb{N} = (N_1,\dots,N_k)$ is a normal extension of $\mathbb{A}$ on $L_D^2(\mathbb{T})$. Thus $\mathbb{A}$ is subnormal.

By the method of proving the Theorem 1 and the lemma 10 of [22], we may prove the following.

LEMMA 2. *Assume* $\mathbb{A} = (A_1,\dots,A_k)$ *satisfies the condition of this theorem and* $f,g \in \mathcal{A}(\mathrm{sp}(\mathbb{A}))$. *Then there is an integer valued measurable function* $m(\cdot)$ *which is the spectral multiplicity funciton of* $\mathbb{N}$ *at the regular points of* $\mathrm{sp}(\mathbb{N})$ *such that*

$$(9) \qquad\qquad \mathrm{tr}\,[f(\mathbb{A}),g(\mathbb{A})] = \frac{1}{2\pi i}\int_{\mathrm{sp}(\mathbb{N})} m(u)f(u)dg(u).$$

For $n \geq 1$, *if* $f_j, g_j \in \mathcal{A}(\mathrm{sp}(\mathbb{A}))$ *then there exists bounded measurable function* $m(u^{(0)},\dots,u^{(n)})$ *on* $\mathrm{sp}(\mathbb{N})^{n+1}$ *satisfying*

$$\mathrm{tr}\,[f_0(\mathbb{A}),g_0(\mathbb{A})]\dots[f_n(\mathbb{A}),g_n(\mathbb{A})] =$$

$$(10)$$

$$\frac{1}{(2\pi i)^{n+1}}\int_{\mathrm{sp}(\mathbb{N})^{n+1}} m_n(u^{(0)},\dots,u^{(n)})\prod_{j=0}^n f_j(u^{(j)})\frac{g_j(u^{(j)})-g_j(u^{(j-1)})}{u_1^{(j)}-u_1^{(j-1)}}du_1^{(j)}$$

where $u^{(j)} = (u_1^{(j)},\dots,u_k^{(j)})$, $u^{(-1)} = u^{(n)}$.

Furthermore, $\mathrm{tr}\,f_0(\mathbb{A})g_0(\mathbb{A})[f_1(\mathbb{A}),g_1(\mathbb{A})]\dots[f_n(\mathbb{A}),g_n(\mathbb{A})]$ *is also an integral expressed by functions* $f_j, g_j, j = 0,\dots,n$ *and* $m_n(u^{(0)},\dots,u^{(n)})$.

The integral in (10) is a multiple singular integral of Cauchy's type. Let $A_j(\zeta) = \sum \lambda_{jn}(\zeta)P_{jn}(\zeta)$ be the spectral resolution of the normal operator $A_j(\zeta)$,

$\zeta \in \mathbb{T}$, where $\lambda_{j_m}(\zeta)$ and $P_{j_m}(\zeta)$ are eigenvalues and spectral projections of $A_j(\zeta)$ respectively. Then for almost all $(A_1(\zeta), \lambda_{2m_2}(\zeta), \ldots, \lambda_{km_k}(\zeta)) \in \mathrm{sp}(\mathbb{N})$, define

$$P(A_1(\zeta), \ldots, \lambda_{km_k}(\zeta)) = P_{2m_2}(\zeta) \ldots P_{km_k}(\zeta).$$

Then

(11) $$m_n(u^{(0)}, \ldots, u^{(n)}) = \mathrm{tr}\ (P(u^{(0)}) \ldots P(u^{(n)})).$$

By the method of proving Theorem 2 in [22], we may prove the following:

LEMMA 3. *Under the condition of the Theorem 1*, $\{\mathrm{sp}(\mathbb{A}), m_n, n = 0, 1, \cdots\}$ *is a complete unitary invariant for the k-tuple $\mathbb{A}$ of operators.*

Now, we only have to prove that the function (7) determines $\{m, m_n : n = 1, 2, \ldots\}$.

For $z_1 \in D$, let $\lambda_1 = A_1(z_1)$ and $\mathcal{D}(\lambda_1) = \ker(A_1^* - \overline{\lambda}_1 I)$. It is easy to see that

(12) $$\mathcal{D}(\lambda_1) = \ker(U_+^* - \overline{z}_1 I) = \left\{ \frac{\alpha}{1 - \overline{z}_1(\cdot)}; \alpha \in \mathcal{D} \right\}$$

is of finite dimension, since $A_1 = A_1(U_+)$.

On the other hand

$$A_j^* \frac{\alpha}{1 - \overline{z}_1(\cdot)} = \frac{1}{2\pi i} \int \frac{A_j(\zeta)^* d\zeta}{(1 - \overline{z}_1 \zeta)(\zeta - (\cdot))} \alpha = \frac{A_j(z_1)^*}{1 - \overline{z}_1(\cdot)} \alpha.$$

Let $W(z_1)$ be the operator from $\mathcal{D}(\lambda_1)$ to $\mathcal{D}$ defined by

$$W(z_1) \frac{(1 - |z_1|^2)^{1/2} \alpha}{1 - \overline{z}_1(\cdot)} = \alpha$$

Then $W(z_1)$ is a unitary operator from $\mathcal{D}(\lambda_1)$ onto $\mathcal{D}$ and

$$W(z_1) A_j^* W(z_1)^{-1} = A_j(z_1)^*.$$

For $\lambda = (\lambda_1, \ldots, \lambda_n)$, define an operator

(13) $$E(\lambda) = \prod_{j=2}^{k} \frac{1}{2\pi i} \int_{|\zeta - \overline{\lambda}_j| = \delta} (\zeta I - A_j(z_1)^*)^{-1} d\zeta,$$

where δ is a small positive number satisfying the condition that there is no eigenvalue of $A_j(z_1)^*$ in $\{\zeta \in \mathbb{C} : 0 < |\zeta - \overline{\lambda}_j| \le \delta\}$. Then $E(\lambda)$ is the parallel projection from $\mathcal{D}$ to the intersection of root spaces

$$\{x \in \mathcal{D} : (A_j(z_1)^* - \overline{\lambda}_j I)^{\ell_j} x = 0, \text{ for some } \ell_j > 0, j = 2, \ldots, k\}.$$

Thus

(14) $$Q(\lambda) = W(z_1)^{-1} E(\lambda) W(z_1) P_{\mathcal{D}(\lambda_1)}$$

where $Q(\lambda) = Q(\lambda; \mathbb{A})$ and $P_{\mathcal{D}(\lambda_1)}$ is the orthogonal projection from $\mathcal{H}_D^2(\mathbb{T})$ to $\mathcal{D}(\lambda_1)$.

LEMMA 4. *Let* $\lambda^{(j)} = (A_1(z^{(j)}), \lambda_2^{(j)}, \ldots, \lambda_k^{(j)})$, *then*

$$\text{tr}\,(Q(\lambda^{(m)})\ldots Q(\lambda^{(1)})) = \prod_{j=1}^{m} \frac{(1-|z^{(j)}|^2)}{1-\overline{z}^{(j)}z^{(j+1)}}\, tr_{\mathcal{D}}(E(\lambda^{(m)})\ldots E(\lambda^{(1)}))$$

where $z^{(m+1)} = z^{(1)}$ *and* $z^{(j)} \in D$.

PROOF. Let $\{e_j\}$ be an orthonormal basis for $\mathcal{D}$. It is easy to see that

$$P_{\mathcal{D}(\mu^{(2)})}W(z^{(1)})^{-1}\alpha = \frac{(1-|z^{(1)}|^2)^{1/2}(1-|z^{(2)}|^2)^{1/2}}{1-\overline{z}^{(1)}z^{(2)}}W(z^{(2)})^{-1}\alpha,\ \alpha \in \mathcal{D},$$

where $\mu^{(2)} = A_1(z^{(2)})$. Thus by (14) we have

$$\text{tr}\,(Q(\lambda^{(m)})\ldots Q(\lambda^{(1)}))$$
$$= \sum_j (Q(\lambda^{(1)})W(z^{(1)})^{-1}e_j, Q(\lambda^{(2)})^* \ldots Q(\lambda^{(m)})^* W(z^{(1)})^{-1}e_j)$$
$$= \sum_j (Q(\lambda^{(2)})W(z^{(1)})^{-1}E(\lambda^{(1)})e_j, Q(\lambda^{(3)})^* \ldots Q(\lambda^{(m)})^* W(z^{(1)})^{-1}e_j)$$
$$= \frac{(1-|z^{(1)}|^2)^{1/2}(1-|z^{(2)}|^2)^{1/2}}{1-\overline{z}^{(1)}z^{(2)}}$$
$$\sum_j (Q(\lambda^{(3)})W(z^{(2)})^{-1}E(\lambda^{(2)})E(\lambda^{(1)})e_j, Q(\lambda^{(4)})^* \ldots Q(\lambda^{(m)})^* W(z^{(1)})^{-1}e_j).$$

continuing this process, we may prove Lemma 4.

Let $\mathbb{B}_n = (B_{1n},\ldots,B_{mn}), n=1,2,\ldots$ be a sequence of m-tuples of commuting operators on a finite dimensional inner product space $\mathcal{D}$ satisfying $\lim_{n\to\infty} B_{jn} = B_j$, $j=1,2,\ldots,m$, where $B_j, j=1,2,\ldots,m$ are commuting normal operators with common eigenvalue 0. For every $\lambda = (\lambda_1,\ldots,\lambda_m)$, let

$$E_n(\lambda) = \prod_{j=1}^{m}\frac{1}{2\pi i}\int_{|\zeta-\lambda_j|=\delta}(\lambda I - B_{jn})^{-1}d\lambda$$

be the parallel projection to $\{x \in \mathcal{D} : (B_{jn} - \lambda_j I)^{\ell_j}x = 0,\ \ell_j > 0,\ j=1,\ldots,m\}$, where δ is a small positive number such that there is no eigenvalue of B_{jn} in $\{\zeta : 0 < |\zeta - \lambda_j| \le \delta\}$. Let E be the orthogonal projection from $\mathcal{D}$ to $\bigcap_{j=1}^{m} \ker B_j$.

LEMMA 5. *There is a* $\epsilon > 0$ *such that*

$$(15) \qquad \lim_{n\to\infty} \sum_{|\lambda_j|<\epsilon} E_n(\lambda_1,\ldots,\lambda_m) = E.$$

PROOF. Let ϵ be a positive number satisfying the conditions that (i) $\{\zeta \in \mathbb{C} : 0 < |\zeta| \le \epsilon\} \cap \sigma_p(B_j) = \emptyset$, $j=1,\ldots,m$ and (ii)

$$\{\zeta \in \mathbb{C} : |\zeta| = \epsilon\} \cap \sigma_p(B_{jn}) = \emptyset,\ j=1,2,\ldots,m; n=1,2,\ldots.$$

It is obvious that this ϵ exists. It is easy to calculate that

$$(16) \qquad \sum_{|\lambda_j|<\epsilon} E_n(\lambda_1,\ldots,\lambda_m) = \prod_{j=1}^{m}\frac{1}{2\pi i}\int_{|\zeta|=\epsilon}(\zeta I - B_{jn})^{-1}d\zeta.$$

Letting $n \longrightarrow \infty$ in (16), we have

$$\lim_{n \to \infty} \sum_{|\lambda_j| < \epsilon} E_n(\lambda_1, \dots, \lambda_m) = \prod_{j=1}^{m} \frac{1}{2\pi i} \int_{|\zeta| = \epsilon} (\zeta I - B_j)^{-1} d\zeta$$

which proves (18).

From Lemma 4 and 5, we obtain the following.

LEMMA 6. *Under the conditions of the Theorem 1, the function*

$$m(u^{(0)}, u^{(1)}, \dots, u^{(p)}) =$$

$$\lim_{\epsilon \to 0} \overline{\lim_{r \to 1^-}} \frac{\prod_{j=0}^{p} (1 - r^2 \overline{z^{(j)}}) z^{(j+1)}}{(1 - r^2)^{p+1}} \sum_{\eta^{(\ell)} \in V(\epsilon, r, u^{(\ell)})} \mathrm{tr}\,(Q(\eta^{(0)}) \dots Q(\eta^{(p)})).$$

where $z^{(j)} = A_1^{-1}(u_1^{(j)})$, $u^{(p+1)} = u^{(0)}$ *and*

$$V(\epsilon, r, u) = \{A_1(rA_1^{-1}(u_1)), \eta_2, \dots, \eta_k) \in \sigma_p(\mathbb{A}^*)^* : |\eta_j - u_j| < \epsilon, j = 2, \dots, k\}.$$

From Lemma 3 and 6, it proves Theorem 1.

REMARK. In [20] and [21], it studies the complete unitary invariant for the pure subnormal operator S on $\mathcal{H}$ with the m.n.e. N on $\mathcal{K} \supset \mathcal{H}$ satisfying the conditions that (i) $\mathrm{sp}(S) \backslash \mathrm{sp}(N)$ is a k-connected domain with boundary $\mathrm{sp}(N)$ consisting of k Jordan curves satisfying smooth condition (CBI) or an external Jordan curve satisfying smooth condition (CBI) and $k - 1$ points and (ii) $[S^*, S]^{1/2} \in \mathcal{L}^1$. We may make small changes in the proofs of the theorems in [20] and [21], such that the most argument in [20] and [21] remain true if we release the condition $[S^*, S]^{1/2} \in \mathcal{L}^1$ to $[S^*, S] \in \mathcal{L}$. Besides, it is easy to see that the function

$$\mathrm{tr}\,(\mu(z)^* \mu(w)^*), \quad z, w \in \mathrm{sp}(S) \backslash \mathrm{sp}(N)$$

is a complete unitary invariant for this S, where

$$\mu(z)^* = P_M(N^* - \overline{z}I)^{-1}(N^* - S^*) \mid_M$$

$M = $ closure of $[S^*, S]\mathcal{H}$ and P_M is the projection from $\mathcal{K}$ to M. By means of the analytic model of subnormal operator, we may prove that

$$\mathrm{tr}\,(\mu(z)^* \mu(w)^*) = \mathrm{tr}\,(Q(z; S)Q(w; S))$$

where $Q(z, S)$ is the parallel projection from $\mathcal{H}$ to the eigenspace $\ker(S^* - \overline{z}I)$ satisfying the condition that

$$\ker Q(z, S) = \ker P_M(N^* - \overline{z}I)^{-1}(N^* - S^*)P_M.$$

As a matter of fact,

$$Q(z, S) = [(N^* - \overline{z}I)^{-1}(N^* - S^*)P_M]^2.$$

Therefore for this family of pure subnormal operators S satisfying the conditions (i) and $[S^*, S] \in \mathcal{L}^1$, the function

$$\operatorname{tr}\,(Q(\lambda; S)Q(\mu; S)), \ \lambda, \mu \in \operatorname{sp}(S) \backslash \operatorname{sp}(N)$$

is a complete unitary invariant for S.

4. Form of Some Hyponormal Tuples

A k-tuple of operators $\mathbb{A} = (A_1, \ldots, A_k)$ on $\mathcal{H}$ is said to be irreducible, if the only subspaces of $\mathcal{H}$ reducing $A_j, j = 1, 2, \ldots, k$ are trivial.

THEOREM 2. *Let* $\mathbb{H} = (H_1, \ldots, H_k)$ *be an irreducible k-tuple of commuting operators on* $\mathcal{H}$. *(a) Suppose that H_1 is hyponormal,* $\operatorname{rank}[H_1^*, H_1] = 1$, $\operatorname{ran}[H_j^*, H_1] \subset \operatorname{ran}[H_1^*, H_1]$ *and $[H_j^*, H_j]$ is compact, for $j = 2, \ldots, k$. Then either* (i) *H_1 is a linear combination of identity and unilateral shift with multiplicity one and $\mathbb{H}$ is subnormal, or* (ii) *H_1 is a non-subnormal and there are $\alpha_j, \beta_j \in \mathbb{C}$ such that*

$$(18) \qquad\qquad H_j = \alpha_j H_1 + \beta_j I, \quad j = 2, \cdots, k.$$

(b) suppose that H_1 is cohyponormal, $\operatorname{rank}[H_1^*, H_1] = 1$ *and* $\operatorname{ran}[H_j^*, H_1] \subset \operatorname{ran}[H_1^*, H_1]$, $j = 1, \cdots, k$. *Then there are $\alpha_j, \beta_j \in \mathbb{C}$ such that* (18) *holds.*

This theorem is an improvement of Theorem 1 in [18]. The proof of this theorem is closely related to the proof of Theorem 1 in [18]. In order to make the proof of this theorem readable, we have to copy some parts of the proof of Theorem 1 in [18].

PROOF. Case (a): (1) First let us study the pair (H_1, H_2). There is a vector $e_1 \in \mathcal{H}$, $e_1 \neq 0$ such that

$$(19) \qquad\qquad [H_1^*, H_1]x = (x, e_1)e_1, \quad \text{for } x \in \mathcal{H}.$$

Therefore there exists a vector e_2 such that

$$(20) \qquad\qquad [H_2^*, H_1]x = (x, e_2)e_1,$$

since $\operatorname{ran}[H_2^*, H_1] \subset \operatorname{ran}[H_1, H_1]$. Let $\mathcal{H}_1$ be the smallest subspace containing e_1 and reducing H_1. From (19) and (20) it is obvious that

$$(21) \qquad \|e_1\|^2 H_2^* e_1 = (e_1, e_2) H_1^* e_1 + (H_2^* e_1, e_1)e_1 - (H_1^* e_1, e_2)e_1 \in \mathcal{H}_1.$$

Also we have that

$$H_2^* H_1^{*m} H_1^n e_1 = \sum_{j=0}^{n-1} H_1^{*m} H_1^{n-j-1}[H_2^*, H_1] H_1^j e_1 + H_1^{*m} H_1^n H_2^* e_1.$$

Therefore $H_2^* H_1^{*m} H_1^n e_1 \in \mathcal{H}_1$ for $m, n = 0, 1, 2, \ldots$. Thus $\mathcal{H}_1$ is invariant with respect to H_2^*.

(ii) According to the decomposition $\mathcal{H} = \mathcal{H}_1 \oplus \mathcal{H}_1^\perp, H_j$ may be written as

$$H_j = \begin{pmatrix} A_j & 0 \\ B_j & C_j \end{pmatrix}, \; j = 1, 2,$$

where $A_j^* = H_j^* \mid_{\mathcal{H}_1}$, $B_1 = 0$ and C_1 is normal. Let e_2^1 be the projection of e_2 on $\mathcal{H}_1$. Let $e_2^2 = e_2 - e_2^1$. We have to prove that $e_2^2 = 0$. Suppose on contrary that $e_2^2 \neq 0$.

LEMMA 7. *If $e_2^2 \neq 0$, then $\mathcal{H}_1$ is the closure of the*

$$span\{(zI - H_1)^{-1}e_1, \; z \in \rho(H_1)\}$$

and

(22) $\quad ((zI - H_1)^{-1}e_1, (wI - H_1)^{-1}e_1) = \|e_1\|^2((\overline{w} - \overline{\beta})(z - \beta)I - \|e_1\|^2)^{-1}.$

where $\beta \in \mathbb{C}$.

The proof of this lemma is in [18]. See the part of [18] from lemma 1 of p. 427 to line 13 of p. 429. Especially notice the line 5 of p. 429.

Thus up to a unitary equivalence, we may assume that $\mathcal{H}_1 = H^2(\mathbb{T})$,

$$(A_1 f)(z) = (\alpha z + \beta) f(z) \quad \text{for } f \in H^2(\mathbb{T}),$$

and

$$(A_2 f)(z) = M(z) f(z), \quad \text{for } f \in H^2(\mathbb{T})$$

where $M(\cdot) \in H^2(\mathbb{T})$, since $[A_1, A_2] = 0$.

Let $\mathcal{H}_2$ be closure of the set $\{B_2 p(\cdot) + q(C_1^*)e_2^2 : p \text{ and } q \text{ are polynomials }\}$. Follow the steps from (20) through (24) in [18], we have the following.

LEMMA 8. *If $e_2^2 \neq 0$, then $\mathcal{H}_2$ reduces C_1, and up to a unitary equivalence, we may assume that $\mathcal{H}_2$ is the Hilbert space of all Borel measurable function $f(\cdot)$ on $\mathbb{T}$ satisfying*

$$(f, f)_{\mathcal{H}_2} = \frac{1}{2\pi} \int |f(e^{i\theta})|^2 F(e^{i\theta}) d\theta < +\infty$$

where $F(\cdot)$ is a bounded measurable function,

$$(B_2 f)(e^{i\theta}) = f(e^{i\theta}), \qquad f \in \mathcal{H}_1 = H^2(\mathbb{T}).$$

Thus, by (27) and (28) in [18], we have

$$([H_2^*, H_2]f, f) = ((A_2^* A_2 - A_2 A_2^* + B_2^* B_2)f, f) = \|(I - P)\overline{M}f\|^2 + \|f\|_{\mathcal{H}_2}^2,$$

for $f \in \mathcal{H}_1$. If F is not zero almost everywhere, then $[H_2^*, H_2]$ cannot be compact, which contradicts the assumption of the theorem. Thus $e_2^2 = 0$. Then follow the step (vi) of the proof of Theorem 1 in [18] we conclude that $\mathcal{H}_1 = \mathcal{H}$ and H_1 is pure.

(iii) As shown in (35) and (37) of [18], we have

(23) $$(H_2 - k_2 I)e_1 = (H_1 - k_1 I)e_2$$

(24) $$((H_2 - k_2 I)x, e_1)e_1 = ((H_1 - k_1, I)x, e_1)e_2$$

where $k_j = (H_j e_1, e_1)\|e_1\|^{-1}, j = 1, 2$.

Case 1. If $(H_1^* - \overline{k}_1 I)e_1 = 0$, then define

$$T(z, w) = (\overline{w}I - H_1^*)^{-1}e_1, (\overline{z}I - H_1^*)^{-1}e_1)$$
$$= (\overline{w} - \overline{k}_1)^{-1}(z - k_1)^{-1}\|e_1\|^2$$

By the commutator property

$$((zI - H_1)^{-1}e_1, (wI - H_1)^{-1}e_1) = T(z, w)(1 - T(z, w))^{-1}$$
$$= \|e_1\|^2((\overline{w} - \overline{k}_1)(z - k_1) - \|e_1\|^2)^{-1}.$$

Thus H_1 is a linear combination of the identical operator and the unilateral shift with multiplicity one. In that case, it is easy to see that $\mathbb{H}$ is subnormal, since $\mathbb{H}$ is a k-tuple of commuting operators.

Case 2. If $(H_1^* - \overline{k}_1 I)e_1 \neq 0$, then from (24), there is a $\alpha \in \mathbb{C}$ such that $\alpha_2 = \alpha e_1$. From (23), there is a $\beta \in \mathbb{C}$ such that

$$H_2 e_1 = (\alpha H_1 + \beta I)e_1.$$

Follow the step (viii) of [18], we may prove that $H_2 = \alpha H_1 + \beta I$. Case (a) is proved.

Case (b). In this case (19) and (20) become

(25) $$[H_1^*, H_1]x = -(x, e_1)e_1$$

and

$$\|e_1\|^2 H_2^* e_1 = -(e_1, e_2)H_1^* e_1 + (H_2^* e_1, e_1)e_1 + (H_1^* e_1, e_1)e_1 \in \mathcal{H}.$$

The left-hand sides of (14), (15), (16) of p. 428 in [18] have to be multiplied by (-1). But Corollary 1 in [18] is still true. The formula (21) becomes

$$((zI - H_1)^{-1}e_1, (wI - H_1)^{-1}e_1) = \|e_1\|^2((\overline{w} - \overline{\beta})(z - \beta) + \|e_1\|^2)^{-1}.$$

It leads to a contradiction that $\|(H_1 - \beta I)e_1\|^2 = -\|e_1\|^4$. Therefore $e_2^2 = 0$. By (vi) in the proof of Theorem 1 in [18]. We may prove that $\mathcal{H}_1 = \mathcal{H}$. The left-hand sides of (35) and (37) in [18] must be multiplied by -1 in this case. If $H_1^* e_1 = \overline{k}_1 e_j$, then we have

(26) $$((zI - H_1)^{-1}e_1, (wI - H_1)^{-1}e_1) = (z\overline{w} - 1)^{-1},$$

if we change H_j to $\alpha_j H_j + \beta_j$, so that $k_1 = k_2 = 0$ and $\|e_1\| = 1$. However (26) contradicts to (25). Thus $(H_1^* - \overline{k}_1 I)e_1 \neq 0$ and there is a $\alpha \in \mathbb{C}$ such that $e_2 = \alpha e_1$. By (viii) of [18], we prove that $H_2 = \alpha H_1 + \beta$. Theorem 3 is proved.

A k-tuple of operators $\mathbb{H} = (H_1, \ldots, H_k)$, $H_j \in \mathcal{L}(\mathcal{H})$ is said to be hyponormal (or jointly hyponormal [2]), if

$$(27) \qquad \sum_{i,j} ([H_i^*, H_j]x_i, x_j) \geq 0, \qquad x_i \in \mathcal{H}.$$

COROLLARY 1. *Let $\mathbb{H} = (H_1, \ldots, H_k)$ be an irreducible k-tuple of commuting operators. Suppose that $\mathrm{rank}[H_1^*, H_1] = 1$, (H_1, H_j) is hyponormal and $[H_j^*, H_j]$ is compact for $j = 2, \ldots, k$. Then the conclusion (a) of* Theorem 2 *holds.*

PROOF. From the hyponormalily (27) of every pair (H_1, H_j) we have

$$([H_2^*, H_1]x_2, x_1)^2 \leq ([H_1^*, H_1]x_1, x_1)([H_2^*, H_2]x_2, x_2).$$

for $x_1, x_2 \in \mathcal{H}$. This implies that ran $[H_2^*, H_1] \subset$ ran $[H_1^*, H_1]$ which returns to the case (a) of Theorem 2.

Part of this paper has been lectured at the Sum. Res. Conf. of Multivariable Operator Theory, Seattle, 1993. The author would like to thank the organizers for their invitation.

REFERENCES

1. M. B. Abrahamse and R. G. Douglas, *A class of subnormal operators related to multiply-connected domains*, Advances in Math. **19** (1976), 106–148.
2. A. Athavale, *On joint hyponormality of operators*, Proc. Math. Soc. **103** (1988), 417–423.
3. R. W. Carey, *A unitary invariant for pairs of self-adjoint operators*, J. Reine Angew. Math. **283/284** (1974), 294–312.
4. R. W. Carey and J. D. Pincus, *Principal currents*, Integr. Equat. Oper. Th. **8** (1985), 614–640.
5. ———, *Reciprocity for Fredholm operators*, Integr. Equat. Oper. Th. **9** (1986), 469–501.
6. A. Connes, *Non-commutative differential geometry. Ch. I, II*, Publ. Math. Inst. Hautes, Edu. dec. Sci. Publ. Math. **62** (1985), 257–260.
7. J. B. Conway, *Theory of Subnormal Operators*, Math. Surv. Mon., vol. 36, Amer. Math. Soc., Providence, RI., 1991.
8. ———, *Towards a functional calculus for subnormal tuples: The minimal normal extension*, Trans. Amer. Math. Soc. **326** (1991), 543–567.
9. M. J. Cowen and R. G. Douglas, *Complex geometry and operator theory*, Acta Math. **141** (1978), 187–261.
10. R. E. Curto, and M. Salinas, *Generalized Bergman kernels and the Cowen-Douglas theory*, Amer. J. Math. **106** (1984), 447–488.
11. P. R. Halmos, *A Hilbert Space Problem Book*, Von Nostrand, Princeton, 1967.
12. J. D. Pincus, *The principal index*, Proc. Symp. Pure Math. **51** (1990), 373–393.
13. J. D. Pincus and D. Xia, *A trace formula for subnormal operator tuples*, Integr. Equat. Oper. Th. **14** (1991), 390–398.
14. J. D. Pincus and D. Zheng, *A remark on the spectral multiplicity of normal extensions of commuting subnormal operator tuples*, Integr. Equat. Oper. Th. **16** (1993), 145–153.
15. M. Putinar, *Spectral inclusion for subnormal n-tuples*, Proc. Amer. Math. Soc. **90** (1984), 405–406.
16. D. Xia, *The analytic model of a subnormal operator*, Integr. Equat. Oper. Th. **10** (1987), 258–289.
17. ———, *Analytic theory of a subnormal n-tuple of operators*, Proc. Symp. Pure Math. **51** (1990), 617–640.
18. ———, *On some class of hyponormal tuples of commuting operators*, Operator Theory: Advances and Applications **48** (1990), 423–448.

19. _____ , *Trace formulas for almost commuting operators, cyclic cohomology and subnormal operators*, Integr. Eq. Oper. Th. **14** (1991), 276–278.

20. _____ , *Complete unitary invariant for some subnormal operators*, Integr. Eq. Oper. Th. **15** (1992), 154–166.

21. _____ , *A complete unitary invariant for subnormal operator with multipy connected spectrum*, (Preprint).

22. _____ , *Trace formulas for a class of subnormal tuples of operators*, Integr. Eq. Oper. Th. **17** (1993), 417–439.

DEPARTMENT OF MATHEMATICS, VANDERBILT UNIVERSITY, NASHVILLE, TN 37240

E-mail address: Xiad@ctrvax.Vanderbilt.Edu

Recent Titles in This Series

(Continued from the front of this publication)

(See the AMS catalog for earlier titles)